Advances in Steel–Concrete Composite Structures

Advances in Steel–Concrete Composite Structures

Guest Editors

Jiang Liu
Mingjin Zhang
Yinping Ma
Lipeng Sun

MDPI

Basel • Beijing • Wuhan • Barcelona • Belgrade • Novi Sad • Cluj • Manchester

Guest Editors
Jiang Liu
School of Highway
Chang'an University
Xi'an
China

Mingjin Zhang
Department of Bridge
Engineering
Southwest Jiaotong
University
Chengdu
China

Yinping Ma
School of Civil Engineering
Chongqing University
Chongqing
China

Lipeng Sun
Faculty of Architecture
Civil and Transportation
Engineering
Beijing University of
Technology
Beijing
China

Editorial Office
MDPI AG
Grosspeteranlage 5
4052 Basel, Switzerland

This is a reprint of the Special Issue, published open access by the journal *Buildings* (ISSN 2075-5309), freely accessible at: https://www.mdpi.com/journal/buildings/special_issues/106MC7997W.

For citation purposes, cite each article independently as indicated on the article page online and as indicated below:

Lastname, A.A.; Lastname, B.B. Article Title. *Journal Name* **Year**, *Volume Number*, Page Range.

ISBN 978-3-7258-4821-8 (Hbk)
ISBN 978-3-7258-4822-5 (PDF)
https://doi.org/10.3390/books978-3-7258-4822-5

Contents

About the Editors

Jiang Liu

Jiang Liu is an Associate Professor and Master's Supervisor holding a Doctor of Engineering degree. His research focuses on steel and composite structure bridges, as well as the theoretical study of bridge temperature calculation. He has led over 10 national, provincial, and enterprise-level projects, authored/co-authored more than 60 academic papers (including 31 SCI/EI publications as the first or corresponding author), co-compiled 2 technical codes and 3 textbooks/monographs, and received 4 provincial and academic awards, including the First Prize of Shaanxi Province, Special Prize, and First Prize from professional societies. In 2023, he was conferred the title of "Outstanding Doctoral Dissertation Recipient of Shaanxi Province". Currently, he serves as a Council Member of the Fatigue and Stability Branch of the China Steel Structure Society and a Young Editorial Board Member for the *Journal of Traffic and Transportation Engineering*, *Journal of Chang'an University* (Natural Science Edition), and *Journal of Architecture and Civil Engineering*.

Mingjin Zhang

Mingjin Zhang, Professor, Doctoral Supervisor, recipient of the National Ten Thousand Talents Program for outstanding young people, and Distinguished Youth of Sichuan Province. Mainly engaged in wind-resistant research of bridges. Hosted over four national scientific research projects and over 10 wind-resistant research projects for long-span bridges in mountainous areas. Published 50 SCI journal papers and received three first-prizes for scientific and technological progress at the provincial level. Served as the young editorial board of journals of *Southwest Jiaotong University* and *Acta Aerodynamica Sinica*, and served as the reviewer for more than 20 academic journals both domestically and internationally.

Yinping Ma

Yinping Ma, Ph.D. in Eng, research assistant at Chongqing University, specializing in bridge engineering research. Principal investigator for projects including the National Natural Science Foundation for Young Scholars, with three national invention patents, five utility model patents, and four software copyrights granted. Over 30 high-level academic papers have been published, including 1 paper included in the Ministry of Transport's Outstanding Scientific and Technological Achievements Repository and 1 Frontrunner 5000 (F5000) paper. Research achievements include the Special Prize for Science and Technology from the China Steel Construction Society (14/20) and the First Prize from the China Highway and Transportation Society (6/15).

Lipeng Sun

Lipeng Sun, Doctor of Engineering, Associate Professor, Doctoral Supervisor. He is a high-level talent (Outstanding Young Talent) of Beijing University of Technology and Secretary of the State Key Laboratory of Bridge Safety and Resilience. His main research focuses on the calculation theory, design methods and intelligent construction of high-performance steel-concrete composite structure bridges. He has presided over various scientific research projects, including the National Natural Science Foundation of China (NSFC), the General Project of China Postdoctoral Science Foundation, and the Science and Technology Plan Project of Beijing Municipal Education Commission. He was selected for the National Funded Postdoctoral Researchers Program. As first/corresponding author, he has published more than 20 papers in top SCI and EI journals. He has participated in the compilation of three group/local standards and two monographs. His research results have been applied to major engineering projects such as Guanyinsi Yangtze River Bridge and Taizhou Bay Cross-Sea Bridge.

buildings

MDPI

Article

Refined Analysis of Spatial Three-Curved Steel Box Girder Bridge and Temperature Stress Prediction Based on WOA-BPNN

Wei Hu [1], Zhongyong Zhang [1], Junwei Shi [1], Yulun Chen [1], Yixuan Li [2] and Qian Feng [2,*]

1 Hangzhou Grand Canal Group Construction Management Co., Ltd., Hangzhou 310011, China
2 College of Civil Engineering and Architecture, Zhejiang University, Hangzhou 310058, China
* Correspondence: fengqian@zju.edu.cn

Abstract: Bridges often improve the visual appeal of urban landscapes by incorporating curve elements to create iconic forms. However, it is noteworthy that curved bridges have unique mechanical properties under loads compared to straight bridges. This study analyzes a spatial three-curved steel box girder bridge based on an actual engineering case with a complex configuration. Initially, the finite element software Midas/Civil 2021 is utilized to establish a beam element model and a plate element model to examine the structural responses under dead loads in detail. Then, two different temperature gradient distribution models are employed for the temperature effect analysis. The back-propagation neural network (BPNN) optimized by the WOA algorithm is trained as a surrogate model for finite element models based on the results of temperature stress simulation. The results reveal that the bending–torsion coupling effect in the second span of the spatial three-curved steel box girder bridge is pronounced, with the maximum torque reaching 40% of the bending moment. The uneven distribution of cross-section stress is particularly significant at the vertices, where the shear lag coefficient exceeds 3. Under the action of temperature gradients, the bridge displays a warped stress state; the stress results obtained from the exponential model exhibit a 21% increase compared to BS-5400. Optimization of the weights by the WOA algorithm results in a significant improvement in prediction accuracy, and the convergence speed is improved by 30%. The coefficient of determination (R^2) for predicting temperature stress can reach as high as 0.99.

Keywords: refined analysis; spatial curve; steel box girder bridge; temperature stress; whale optimization algorithm; neural network

1. Introduction

The design of urban bridges requires not only the basic traffic function but also a certain degree of aesthetics in order to improve the appearance of the city. Incorporating elegant curve elements in design has become more prevalent due to the increasing demand for traffic flow and limited land availability. However, the curvature of curved bridges can cause a bending–torsion coupling effect due to the deviation of the center of gravity from the supporting line during service. Compared to straight bridges, curved counterparts exhibit complex stress characteristics and are more susceptible to damage, particularly under variable actions such as those induced by temperature [1–3].

Temperature loads are considered one of the most critical variable load types. In particular, solar thermal actions generate a three-dimensional temperature field with uneven distribution within the structure, leading to temperature-induced secondary stresses [4]. Scholars worldwide simplify the temperature field of the bridge into a two-dimensional distribution pattern along the longitudinal direction, comprising vertical and lateral temperature gradients. For small and medium-span bridges, there is little significant difference in lateral temperature distribution due to the relatively narrow width of the beam. Therefore, research on vertical temperature gradients is more widespread [5]. Both domestic and international standards have specific regulations for concrete structure bridges regarding temperature effect, whereas there are distinct differences in distribution patterns for steel structure bridges, such as specifications in China lacking explicit definitions of the vertical temperature gradient distribution [6]. Meanwhile, BS-5400 in the UK adopts a four-segment line to simulate this distribution [7]. Extensive research has been carried out at home and abroad on the temperature gradient of steel box girders based on measured data. Zhu et al. [8] and Teng et al. [9] compared measured data from an actual project to multinational specifications and found that the four-segment nonlinear mode provided the best fit for the temperature gradient. Wang et al. [10] used measured data from the steel box girder of a viaduct in Wuhan and discovered that the actual temperature distribution differed significantly from the specified linear distribution. The authors applied extreme value theory to derive the distribution pattern of an appropriate exponential function for the neighboring regions. Ding et al. [11] utilized extreme value theory to analyze ten years of temperature monitoring data for the Runyang Bridge, and found that the estimated extreme values of vertical temperature differences were in excellent agreement with the UK code. Guo et al. [12] analyzed in situ measurements of unpaved steel box girders and investigated the status of temperature gradient partitioning research in China [13]. It is evident that the four-segment model in the UK BS-5400 code provides a superior description of the vertical nonlinear temperature distribution pattern in steel box girders compared to other code. The aforementioned research primarily concentrated on straight bridges, emphasizing the necessity for further in-depth research on the coupled stress characteristics of curved steel bridges under temperature effects.

In recent years, computer science has experienced booming development, leading to increased interest in deep learning among scholars worldwide. The backpropagation neural network (BPNN) is a fundamental algorithm of deep learning that has gained attention in various research fields due to its exceptional applicability, nonlinearity, and robustness. In civil engineering, neural networks are primarily used for predicting structural responses and identifying damages [14]. Tian et al. [15] investigated the effect of temperature on bridges by training a neural network with temperature data from a large-span concrete bridge. They compared the predictions of the neural network with finite element results and demonstrated its high accuracy simultaneously with a significant reduction in computational resource consumption. Ying et al. [16] employed a BPNN to analyze and predict the temperature field distribution of a sea-crossing bridge, providing a reliable foundation for subsequent computations. Wang et al. [17] performed a finite element analysis to simulate temperature gradients at various locations of concrete box girders. They formed a comprehensive dataset and established a neural network to predict the vertical and horizontal temperature gradients of the top and bottom plates of the box girder with R^2 exceeding 0.9. Recent research on temperature effects has primarily focused on predicting temperature fields using traditional BPNNs, while comparatively less attention has been paid to predicting structural responses. Furthermore, the training of the BPNNs heavily relies on the selected initial weight values. Inappropriate values can cause the network to become trapped in a local optimal solution and fail to converge. To address this issue, an optimization algorithm is often introduced to determine the initial weight values.

This paper introduces the concept of a "three-curved spatial bridge" composed of cross-section, plane, and profile curves based on an actual engineering case. The alignment variations influence the mechanical characteristics of the bridge in three directions. Therefore, finite element models with different scale elements are established for a detailed analysis. A comparative analysis is conducted on the temperature gradient effect of spatial three-curved steel box girder bridges using two typical temperature gradient models: multi-segment piecewise linear and exponential. Based on simulated results, a WOA algorithm-optimized BPNN serving as a surrogate model is trained to predict stresses under temperature gradients.

2. Bridge Description

The Hanggang River Bridge is situated on Lishui Road, Hangzhou City, Zhejiang Province, China, just 300 m from the Beijing–Hangzhou Grand Canal. Upon completion, the bridge will serve as a main route to the Beijing–Hangzhou Grand Canal Museum. The bridge's design incorporates the flowing shape of the canal, resulting in a unique appearance with curve elements from every perspective. Additionally, the pedestrian and main bridges are designed separately, and the pedestrian bridges include resting platforms for pedestrians to appreciate the canal scenery. The rendering of the bridge is shown in Figure 1.

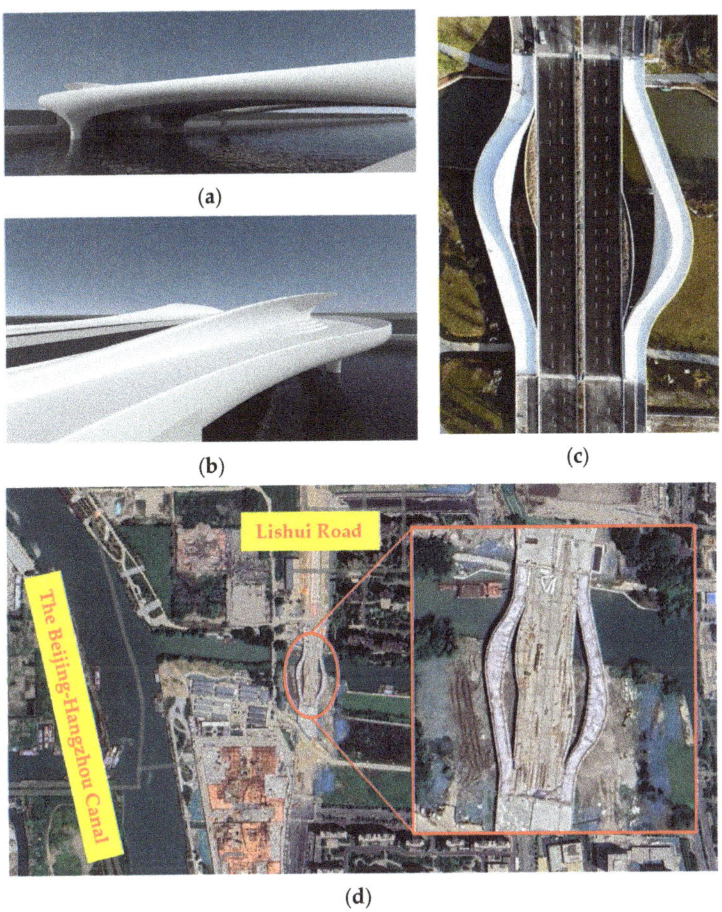

Figure 1. Depiction of the Hanggang River Bridge: (**a**) bottom configuration of the pedestrian bridge; (**b**) upper configuration of the pedestrian bridge; (**c**) as-built top view of the bridge; (**d**) site selection.

This paper discusses the West Pedestrian Bridge, which utilizes a variable cross-section steel box girder for the upper structure. The bottom plate comprises two segments of 1/4 elliptical curves with different widths and identical heights, forming a closed section. The cross-section of the bridge supports has a width of 6 m and a height of 2 m, while the standard section has a width of 7 m and a height of 2 m. The section has a maximum width variation of 10.5 m and a height of 2.2 m. The standard cross-section is depicted in Figure 2.

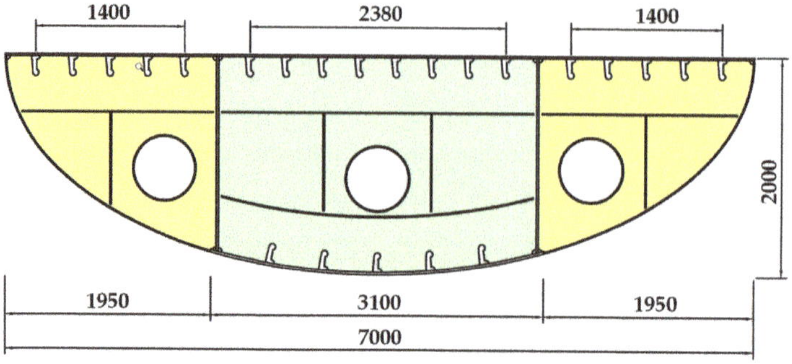

Figure 2. Standard cross-section of the pedestrian bridge (unit: mm).

The top surface of the steel box aligns with the road profile design, whereas the variation in the bottom plate employs curves for a smooth transition, as illustrated in Figure 3a. The top plate has a thickness of 20 mm, the bottom plate is 25 mm, the web measures 20 mm, and the longitudinal stiffeners have a thickness of 16 mm with a height of 150 mm. The practical standard spacing of the transverse diaphragm is 2.5 m, with a standard thickness of 12 mm. The plate thickness is 22 mm at the abutment supports and 25 mm at the pier supports.

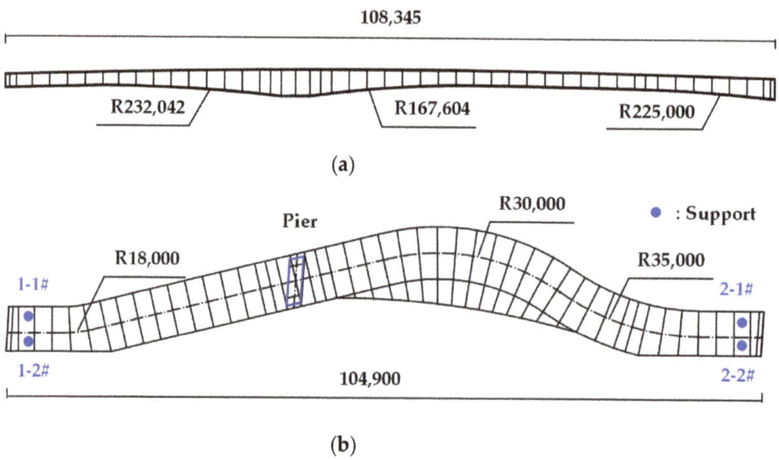

Figure 3. Diagram of the pedestrian bridge (unit: mm): (**a**) profile diagram and (**b**) general layout.

The bridge spans 104.9 m and has a layout alignment containing circular curves, straight lines, and two segments of tangent reverse circular curves. The general layout is shown in Figure 3b. Additionally, a total of 50 transverse diaphragms are evenly distributed along the bridge. The abutments adopt four spherical steel bearings, as shown by the blue dots in Figure 3b, and the pier employs an integral pier-beam connection form to ensure overall stability and prevent tilting.

The pedestrian bridge exhibits curved variations in the cross-section, plan alignment, and vertical alignment, creating a structural configuration known as a "spatial three-curved steel box girder." Compared to conventional steel box girder bridges, the design displays intricate spatial stress characteristics. Therefore, conducting a refined analysis and temperature effect analysis for a three-curved steel box girder bridge based on an actual engineering project is particularly crucial.

3. Refined Finite Element Analysis

3.1. Modeling

Although the three-curved bridge involves intricate geometries, its deformation under loads is minor and limited compared to the cable in cable system bridges, indicating that linear analysis is sufficient. An overall internal force analysis of the spatial three-curved bridge was conducted on a beam element model in Midas/Civil. Elements were assigned at the locations of the diaphragms and variations in the longitudinal stiffeners. Node creation was accomplished by importing coordinates. The cross-section exhibits spatial variations; thus, individual element sections must be drawn. To complete the establishment of variable cross-section elements, the Section Property Calculator (SPC) function was used to import external sections.

Elastic connections were used at the support locations to achieve constraints through different translational and rotational stiffness. The upper structure of the bridge was Q355 low-alloy high-tensile structural steel, while the bridge pier was C40 concrete. The cement was high-quality Portland cement, the coarse aggregate was continuously graded, and medium coarse sand was used as the fine aggregate. The different material properties are outlined in Table 1.

Table 1. Material properties of the bridge.

Material	E (MPa)	γ (N/m^3)	α (C^{-1})	μ
Q355	2.06×10^5	7.85×10^4	1.2×10^{-5}	0.31
C40	3.25×10^4	2.5×10^4	1×10^{-5}	0.2

As the steel box girder is thin-walled, detail component influences were not considered in the beam element model. A comprehensive plate element model was established to ensure a precise analysis of the stress distribution in the three-curved steel box girder bridge, applying the material characteristics mentioned earlier. The model comprised 7769 nodes and 11,640 elements, enabling accurate consideration of diaphragm and stiffener effects, as shown in Figure 4.

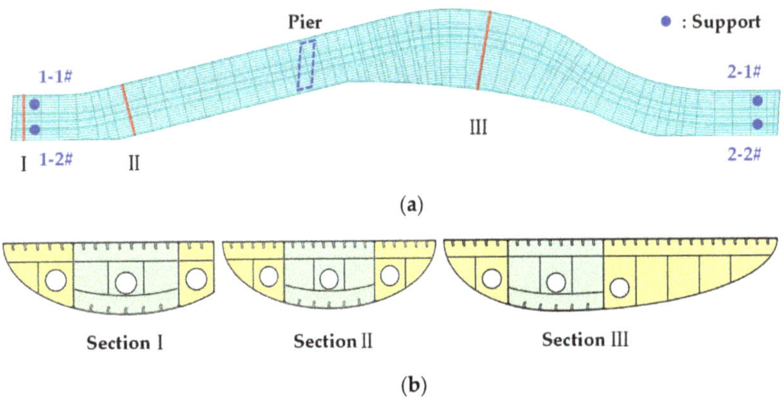

Figure 4. Plate element model of the pedestrian bridge: (**a**) top view and (**b**) three typical sections.

3.2. Force Analysis

The mechanical characteristics of the spatial three-curved steel box girder bridge are complex, and the bending-torsion coupling effect is significant. Therefore, it is crucial to first conduct an analysis under dead load conditions. The secondary dead loads of the bridge consist of the bridge deck paving, railings, and other accessories, and are considered an equivalent uniformly distributed load. The distribution of the bending and torsional moments across the entire bridge under dead loads is illustrated in Figure 5.

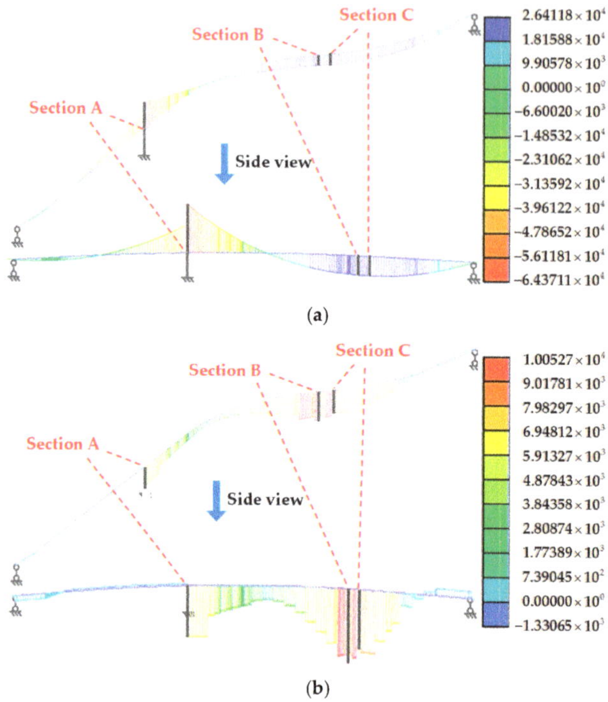

Figure 5. Distribution of moments under dead loads (unit: kN·m): (**a**) bending moment and (**b**) torsional moment.

The bridge forms a continuous beam system. The pier location (Section A) experiences the negative peak bending moment of −64,371.1 kN·m, while the reverse tangential point of the second span (Section B) experiences the positive peak bending moment of 26,411.8 kN·m. The overall torsional moment is relatively low due to the low non-uniformity and short length of the first span. However, the second-span section undergoes complex variations and has a longer span. The maximum positive torsional moment (Section C) is located near the peak position of the bending moment rather than at the widest section of the deck. It has a value of 10,052.7 kN·m, representing almost 40% of the maximum positive bending moment.

The deck width varies towards the inside of the curve, which balances the outward torsional moment generated by the curvature. This configuration reduces the negative torsional moment imposed on the structure. Compared to the first span, the cross-section and alignment of the second span are more complex. Therefore, the bending–torsion coupling effect is more pronounced. The subsequent detailed stress analysis of the bridge concentrates on control sections A, B, and C.

In contrast to traditional curved bridges, which experience positive torsion at the supports, the three-curved bridge generates negative torsion at both the starting and ending support locations. Hence, relative deviation (RD) is introduced to estimate the bending–torsion effect on the supports, as defined in Formula (1):

$$RD = \frac{|F_1 - F_2|}{F_1 + F_2} \times 100\% \tag{1}$$

where F_1 and F_2 are the reactions of the two supports, respectively.

The comparison of the support reactions at both ends is shown in Table 2. The layout and number of the supports can be found in Figure 3. The starting support has an RD of 29.6%; however, due to alignment and cross-section shape changes, the RD at the ending support is reduced to 19.8%. To mitigate significant differences in support reactions and prevent occurrences of overturning phenomena, introducing a reverse curve and changing the cross-section towards the inside of the curve can be effective.

Table 2. Comparison of support reaction.

Support	Force (kN)	RD (%)
1-1#	387.4	29.6
1-2#	712.8	
2-1#	973.4	19.8
2-2#	1452.8	

3.3. Stress Analysis

As a thin-walled component, the steel box girder exhibits a shear lag effect due to uneven cross-sectional stress distribution. The shear lag effect is exacerbated to a great extent in the case of a three-curved steel box girder bridge, where significant changes in cross-section size and the notable bending–torsion coupling effect make it more pronounced. Hence, a meticulous analysis of the shear lag effect based on a sophisticated model is imperative. Herein, the shear lag coefficient calculation is defined in Formula (2). The shear lag coefficient distribution for control sections A, B, and C under dead loads is calculated using the plate element model mentioned above, as depicted in Figure 6.

$$\lambda = \frac{\sigma_{xx}}{\int_0^b \sigma_{xx}(x,y)dy/b} \tag{2}$$

Here, σ_{xx} and b are the normal stress and width of the section, respectively.

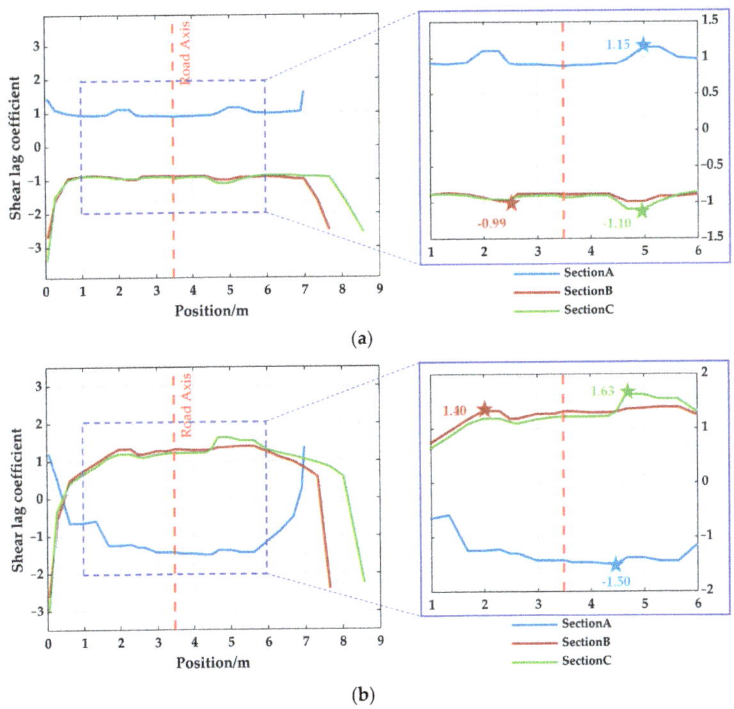

Figure 6. Shear lag coefficient of control sections: (**a**) top plate and (**b**) bottom plate. (star symbol represents the extreme value).

In order to depict the shear lag effect along with the width of the section, the abscissa represents the distance of each point from the left vertex of the section. The range of x corresponds to the section width, where $x = 3.5$ m is the road axis of the bridge. The results indicate that the shear lag effect is reflected in the top and bottom plates of the cross-section at the web position, while the stress is uniformly distributed between the webs. It is noteworthy that the bottom plates experience more pronounced stress changes along the section due to variations in section height compared to the top plates. Owing to the considerable bending–torsion coupling effect, the nonuniformity of the stress distribution is more severe in sections B and C than in section A. A comparison between sections B and C shows that the shear lag effect intensifies as the section width increases. Section C exhibits a more prominent shear lag effect in tensile and compressive stress, with the maximum absolute value exceeding 3.

Moreover, the shear lag coefficient exhibits significant variation near the vertices on both sides of the cross-section. The vertex position represents a higher critical stress concentration than at other locations. Additionally, the residual stress generated during the section welding process intensifies the stress at this location, leading to premature damage and destruction of the bridge deck. As a result, it is crucial to implement reasonable control measures during the construction and operation phases.

3.4. Curve Characterization

The mechanical properties of the three-curved steel box girder bridge are highly related to its spatial curve. Hence, the mechanical characteristics of the bridge have to be analyzed in relation to the impact of curve elements on the dimensions of the cross-section, the plane, and the profile. Finite element models were established for one- and two-curved steel box girder bridges with the same span as the three-curved model. The one-curved steel box girder bridge, with one curve element in the cross-section dimension, is a straight bridge composed of a semi-elliptical section. The two-curved bridge has curved forms in both the plane and cross-section aspects, of which the plane layout is consistent with the three-curved bridge and the cross-section is identical to the one-curved bridge.

The moment comparison of the three models under dead loads is depicted in Figure 7. Bridges with different curve characteristics exhibit the same variation trend in bending moment, although the value is slightly different. As the number of curve elements increases, the maximum positive bending moment gradually decreases while the absolute value of the maximum negative bending moment increases steadily. The maximum and minimum moments of the one-curved bridge are 31,081.1 kN·m and −50,103.6 kN·m, respectively. For the three-curved bridge, these values are 26,411.8 kN·m and −64,371.1 kN·m. The different curve characteristics also place higher demands on the mechanical properties of the pier. The bending moment of the pier on the one-curved bridge is −6138.3 kN·m, whereas the bending moment on the three-curved bridge is −7577.7 kN·m.

Because the one-curved bridge is straight, only the torsion of the two and three-curved bridges were compared, with the results shown in Figure 8. In the first span, the torsion value of the two bridges is relatively low; in the second span, however, the torsion increases, and the distribution is significantly different. In the second span of the three-curved bridge, the width of the cross-sections varies toward the inside of the curve. Increasing the torsion distribution range reduces the extreme torsion in the reverse curve section.

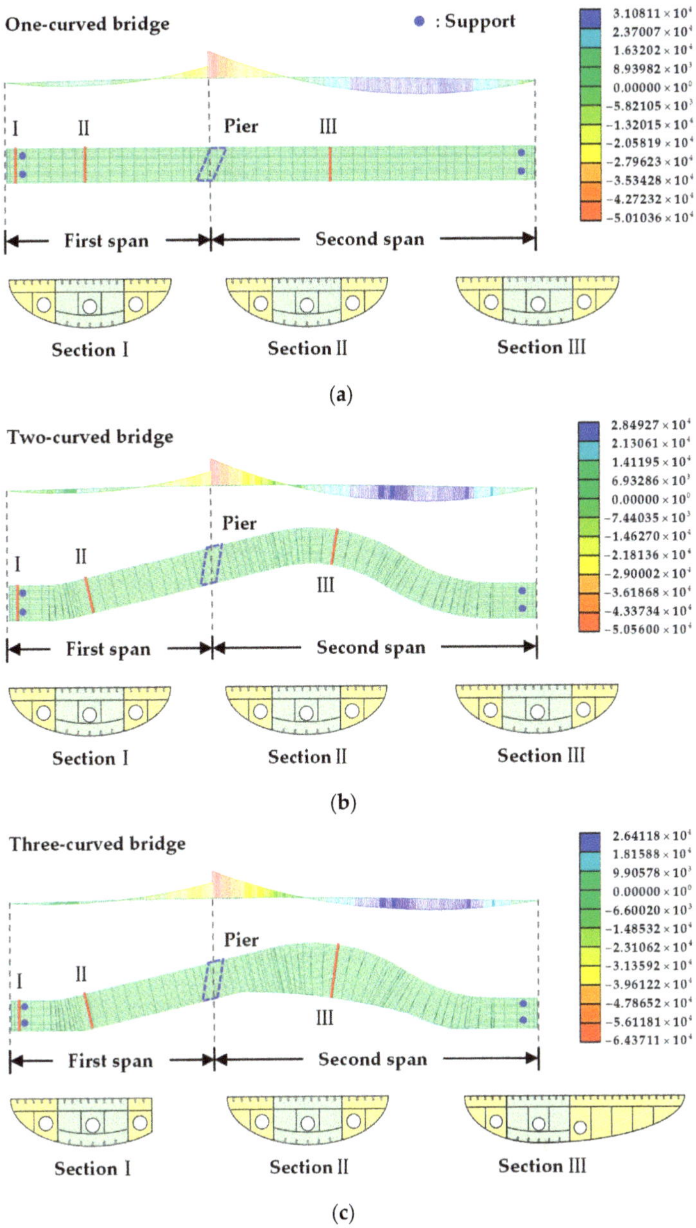

Figure 7. Moment comparison of bridges with different curve characteristics: (**a**) one-curved bridge, (**b**) two-curved bridge, and (**c**) three-curved bridge.

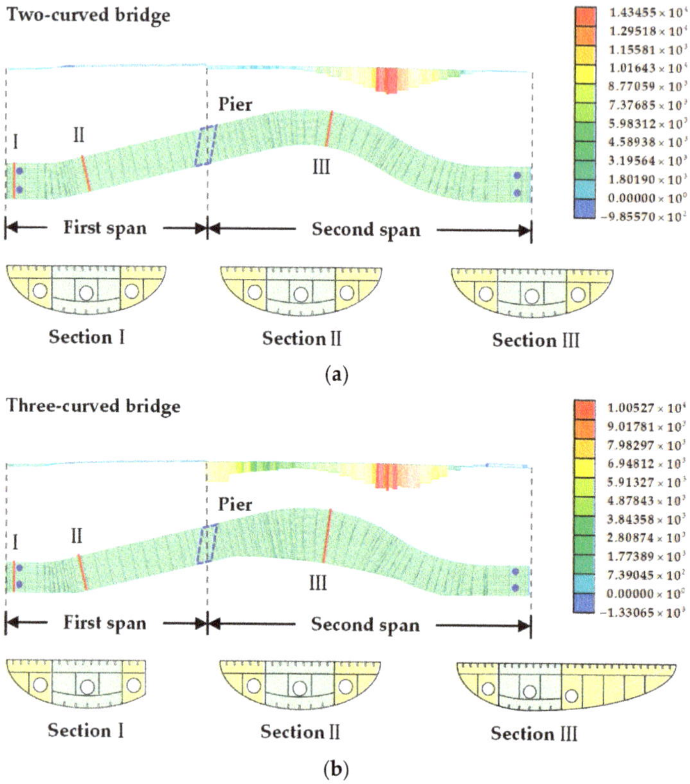

Figure 8. Torsion comparison of bridges with different curve characteristics: (**a**) two-curved bridge and (**b**) three-curved bridge.

4. Temperature Gradient Effect Analysis

4.1. Temperature Gradient Distribution

Currently, the specifications adopted in China provide explicit temperature gradient modes for concrete bridges and steel–concrete composite structure bridges. However, there is no specific code regarding the temperature gradient model of steel box girder bridges. Therefore, the analysis of the temperature effect on steel box girder bridges often references the BS-5400 [11]. Additionally, engineers develop temperature gradient models applicable to specific regions based on measured data from real engineering projects.

The BS-5400 code specifies the temperature distribution in steel and concrete structure bridges under various bridge deck pavements. For steel box girder bridges, the temperature distribution pattern is segmented into four linear segments extending from the top to the bottom, as illustrated in Figure 9a. In particular, no finite temperature adjustment values are required for steel box girder bridges beneath a 200 mm pavement layer.

Wang et al. [18,19] proposed a temperature gradient distribution model for curved steel box girder bridges with latitudes similar to those in Hangzhou and its surrounding regions. The model, represented by Equation (3), was validated for application in pedestrian bridges. As depicted in Figure 9b, the introduced temperature gradient follows an exponential distribution in the vertical direction. The constants within the model were determined by fitting measured data, resulting in a temperature difference of 27.6 °C between the top and bottom plates.

$$T = 27.6e^{-\frac{x}{0.18}} \quad (3)$$

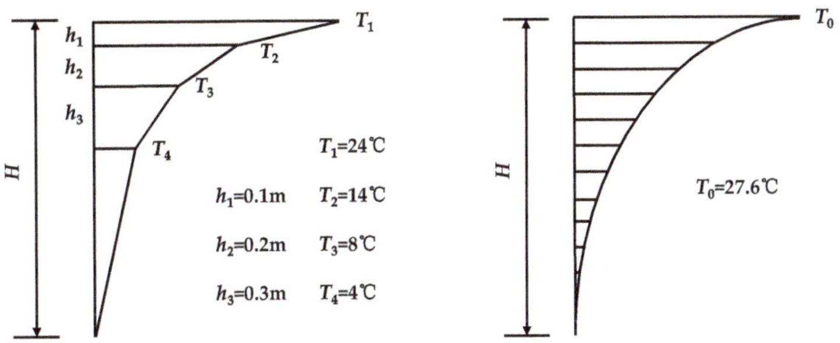

(**a**) (**b**)

Figure 9. Temperature distribution models: (**a**) BS-5400; (**b**) Wang et al. [18,19].

4.2. Force Analysis

To account for the varying width and height of the sections, it is imperative to implement the temperature conditions of the beam cross-sections by calculating the temperature corresponding to the change points of the section width and the temperature gradient segment points. For the boundary conditions, elastic connections between the corresponding element nodes and the fixed nodes were employed to simulate the spherical bearings. The torsional moment distribution influenced by the temperature gradient is shown in Figure 10. The bending and torsional moment variation trend under different temperature gradients exhibits general consistency, with relatively lower torsion levels in the first span. Section A experiences the negative peak torsional moment, while section B displays the positive peak value. In terms of the magnitudes for bending and torsional moment, as presented in Table 3, the values prescribed by Wang's model are more than 10% higher than those stipulated by the British code. The extreme values are increased by more than 24% due to the superposition of internal forces under temperature gradients and dead loads, further exacerbating the bending–torsion coupling effect of the structure.

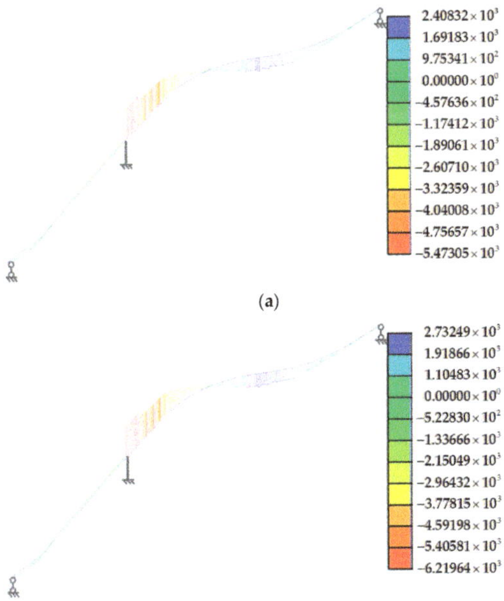

(**b**)

Figure 10. *Cont.*

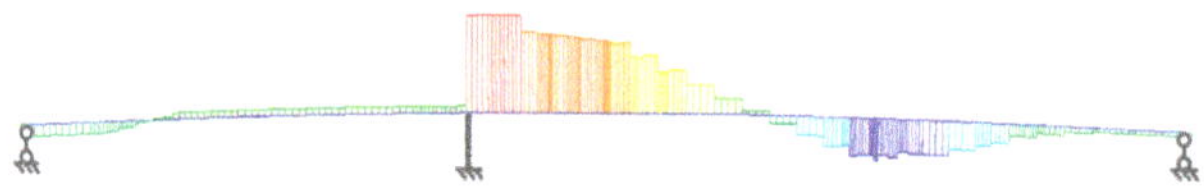

(c)

Figure 10. Distribution of the torsional moment under temperature gradients (unit: kN·m): (**a**) BS-5400; (**b**) Wang et al. [18,19]; (**c**) side view.

Table 3. Force comparison of different temperature distributions.

Moment (kN·m)	BS-5400	Wang et al. [18,19]
M_{max}	16,612.1	18,822.5
T_{max}	2408.3	2732.5
T_{min}	5473.1	6219.6

Under the influence of two temperature distribution patterns, torsional moments are generated in opposite directions at both ends, resulting in an overall warped stress state in the bridge. The location at which the cross-section undergoes the maximum variation experiences zero torsion. The support reactions exhibit a greater difference when subjected to a temperature gradient, as shown in Table 4. The RD value exceeds 70%, which is more than double the difference observed under the dead loads condition.

Table 4. Support reaction of different temperature distributions.

Method	Support	Force (kN)	RD
BS-5400	1-1#	28.6	86.0%
	1-2#	379.5	
	2-1#	232.3	75.2%
	2-2#	33	
Wang et al. [18,19]	1-1#	31.6	86.3%
	1-2#	430.8	
	2-1#	262.8	74.4%
	2-2#	38.6	

4.3. Stress Analysis

The section with the most significant lateral variation in the second span shows the maximum compressive stress in the top plate. According to BS-5400, the compressive stress is −40.4 MPa, while the value calculated by Wang's model is −49.1 MPa. According to BS-5400, the maximum tensile stress in the bottom plate is located in section A with a value of 26.5 MPa, whereas the other method shows 29.6 MPa in the section with the most significant lateral variation in the second span.

Furthermore, a comparative analysis of tensile and compressive stresses in the control sections under different temperature gradient distribution patterns is presented in Table 5. Per the table, BS-5400 specifies values of 26.5 MPa for the maximum tensile stress and 39.2 MPa for the maximum compressive stress, while Wang's model shows 27.8 MPa and 48.0 MPa. The temperature distribution proposed by Wang et al. expresses an average stress increase of approximately 21% compared to BS-5400. Compared to BS-5400, Wang's model exhibits a larger temperature difference between the top and bottom plates. The temperature near the bottom plate in both models remains relatively small, with no significant changes. Therefore, the temperature change near the top plate is the crucial factor affecting the temperature stress.

Table 5. Temperature effect on the stresses of the control sections.

Section	BS-5400		Wang et al. [18,19]	
	Tensile Stress (MPa)	Compressive Stress (MPa)	Tensile Stress (MPa)	Compressive Stress (MPa)
A	26.5	39.2	27.8	48.0
B	15.3	26.6	17.0	33.0
C	19.0	27.2	26.8	33.5

5. Neural Network Prediction

5.1. Methods

5.1.1. BPNN

A BPNN utilizes input features and implements nonlinear mapping through weights and activation functions, as illustrated in Figure 11. The weights between neurons are rapidly updated based on the loss backpropagation, typically the Mean Square Error (MSE). It is worth noting that the training process of neural networks is closely related to the initialization of the weight values. Inappropriate weights can lead to local optimum solutions and unsatisfactory astringency and accuracy. To this end, researchers have introduced GA [20], SSA [21], and PSO [22] optimization methods to enhance the searchability of the entire region and effectively increase the speed of convergence.

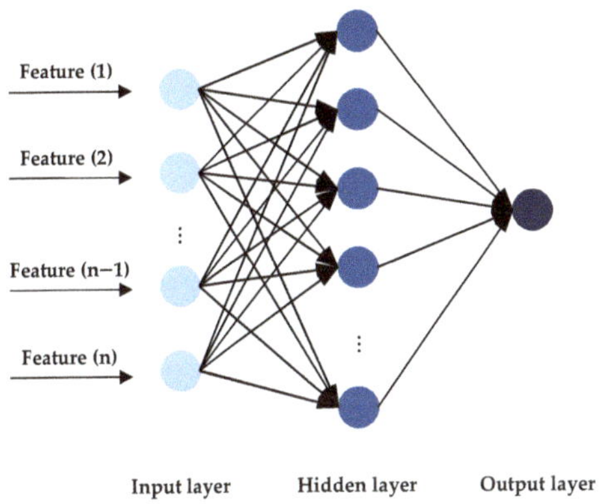

Figure 11. Architecture of BPNN.

5.1.2. WOA

The use of WOA as a meta-heuristic optimization algorithm was first proposed by Mirjalili [23] in 2016, inspired by the hunting process of the whale populations. The hunting process involves three different behaviors: encircling, attacking, and searching for prey. Encircling and searching ensure global search ability, while the bubble-net attack ensures local search ability. The distance between the prey and the whale is evaluated by a fitness function based on their high-dimensional coordinates. The formula for updating the whale's position under each behavior is as follows.

Encircling the prey. The target direction at time t is the whale X^* that is closest to the prey, and the remaining individuals move towards that position:

$$D = |C \cdot X^*(t) - X(t)| \quad (4)$$

$$X(t+1) = X^*(t) - A \cdot D \quad (5)$$

$$A = 2ar_1 - a \quad (6)$$

$$a = 2 - \frac{2t}{t_{\max}} \tag{7}$$

$$C = 2r_2 \tag{8}$$

where r_1 and r_2 are random numbers in [0, 1] and $t_{\max}$ is the maximum iteration number.

Searching for prey. The global search ability is demonstrated by randomly selecting another individual X_{rand} and swimming towards it.

$$X(t+1) = X_{rand} - A \cdot |C \cdot X_{rand} - X(t)| \tag{9}$$

Bubble-net attack. Whales swim in a spiral path and use bubbles to shrink the range of prey, allowing them to hunt:

$$D' = |X^*(t) - X(t)| \tag{10}$$

$$X(t+1) = D' \cdot e^l \cdot \cos(2\pi l) + X^*(t) \tag{11}$$

where l is randomly $[-1, 1]$.

5.1.3. WOA-BPNN

The optimal ability of the WOA algorithm makes it suitable for initializing the weight values of the neural network; the flow chart of the WOA-BPNN is depicted in Figure 12. The first step is to preprocess the data and determine the network architecture. Based on the outcomes of the finite element calculations mentioned above, 250 data items from stress points on various cross-sections were extracted under two temperature gradient patterns. The dataset incorporated stress as output, with input features including the distance L of the section from the starting point, temperature T at the stress point, section width D, and height from the bottom plate H. The architecture utilized a single hidden layer of eight neurons, and the learning rate was 0.01. Then, the whale populations were initialized, each representing a set of parameters for the neural network. The WOA algorithm was iterated until the optimal parameters were obtained and the neural network had completed the follow-up process of updating the parameters. This process was repeated until the maximum number of iterations was reached, at which point the network training was complete.

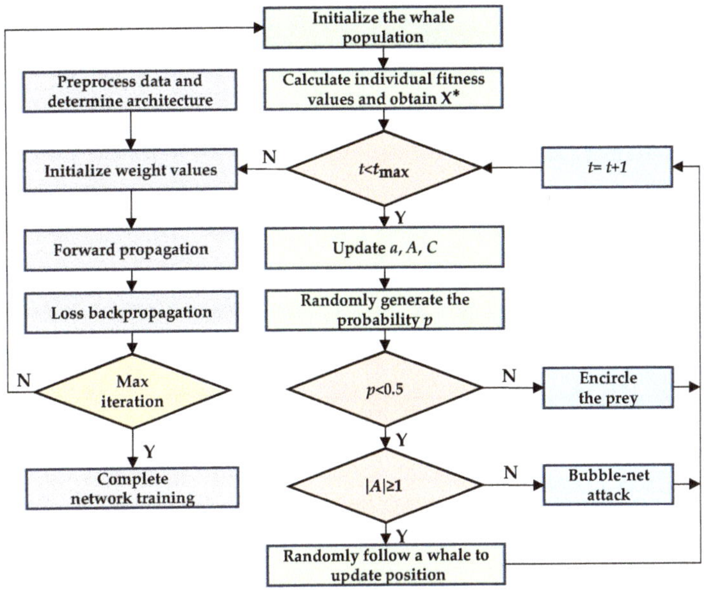

Figure 12. Flow chart of the WOA-BPNN algorithm.

5.2. Evaluation Metric

For comprehensive evaluation of the fitting performance of the BPNN model, metrics including a10, a20, coefficient of determination (R^2), mean absolute error (MAE), and mean absolute percentage error (MAPE) were employed. The calculation methods for each evaluation metric are outlined below. A higher proximity of a10, a20, and R^2 to 1, along with more petite MAE and MAPE values, indicates better reliability. A minor fitting error of the neural network, with the predicted values of $\hat{y}_i$ closely aligning with the actual values of y_i, reflects a superior fitting performance.

$$a10 = \frac{N_{10}}{N} \times 100\%,\ N_{10} \text{ is the numbers of data with } 0.9 \leq \frac{\hat{y}_i}{y_i} \leq 1.1 \quad (12)$$

$$a20 = \frac{N_{20}}{N} \times 100\%,\ N_{20} \text{ is the numbers of data with } 0.8 \leq \frac{\hat{y}_i}{y_i} \leq 1.2 \quad (13)$$

$$R^2 = 1 - \frac{\sum(y_i - \hat{y}_i)^2}{\sum(y_i - \overline{y}_i)^2} \quad (14)$$

$$MAE = \frac{1}{N}\sum|\hat{y}_i - y_i| \quad (15)$$

$$MAPE = \frac{1}{N}\sum\left|\frac{\hat{y}_i - y_i}{y_i}\right| \times 100\% \quad (16)$$

5.3. Prediction Results

The predictive performance of the neural network on the training dataset indicates the ability to fit the given data. The test dataset, which comprises data not seen during training, is used to evaluate the capacity for generalization. To better compare the optimization ability of WOA algorithm, the architecture of the BPNN was consistent with that of the WOA-BPNN. The learning rate hyperparameter was optimized to 0.002 through parameter search and comparison in order to improve performance. The performance comparison between WOA-BPNN and BPNN in predicting stress under the temperature gradient in the three-curved steel box girder bridge is shown in Table 6. Herein, the model was trained on 70% of the available dataset, while the remaining 30% was used to evaluate its generalizability. WOA-BPNN outperforms BPNN in terms of a10, a20, R^2, and MAE on each dataset. Although the MAPE value is slightly higher than that of BPNN, it remains low. The a10 index of BPNN achieves only 75.00% accuracy on the entire dataset, whereas WOA-BPNN obtains 91.05%. The R^2 for WOA-BPNN is 0.99, indicating a substantial correlation between the predicted results and the ground truth.

Table 6. Performance evaluation of WOA-BPNN and BPNN.

Dataset		a10	a20	R^2	MAE (MPa)	MAPE
BPNN	Training	75.75%	84.60%	0.97	3.33	4.27
	Test	72.60%	86.60%	0.97	3.28	4.44
	All	75.00%	85.24%	0.97	3.30	4.29
WOA-BPNN	Training	90.55%	92.45%	0.99	2.13	4.87
	Test	92.36%	93.64%	0.99	2.02	5.78
	All	91.05%	92.98%	0.99	2.08	4.94

Furthermore, the comparison of prediction accuracy between two networks with different allowable errors is illustrated in Figure 13. WOA-BPNN covers a larger area of the radar map in each dataset, providing more precise temperature stress predictions. In addition, the prediction accuracy of WOA-BPNN exceeds 60% for the metric with a 5% allowable error, which suggests that the network has higher global prediction accuracy and performs better on small-scale data.

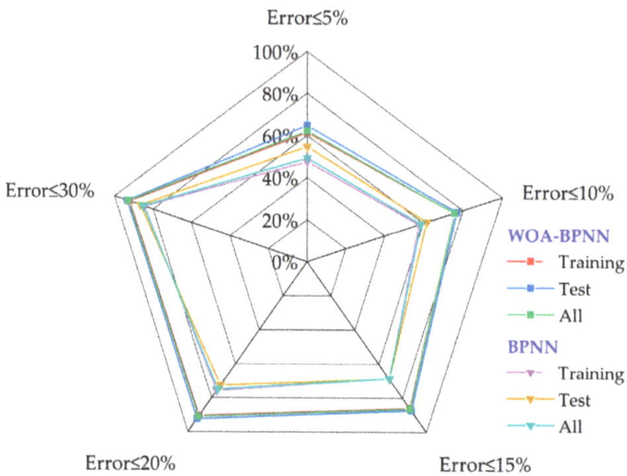

Figure 13. Comparison of network prediction accuracy with different allowable errors.

The WOA algorithm significantly enhances both the prediction accuracy and convergence speed of the network. BPNN and WOA-BPNN were each run for a maximum of 100 epochs, and training was repeated ten times. The comparison diagram illustrating the convergence speed of the two methods on the test set is shown in Figure 14. From the start of the training, the WOA algorithm reduces the MSE of the neural network, and the MSE remains lower than that of the BPNN in the following training process. The MSE of WOA-BPNN at the time of the training stop is 76.47, while that of BPNN is 140.55, representing a reduction of nearly 50%. Additionally, the fluctuation of the repeated training is minor, manifested as a narrow range of the distribution of the error bars in the figure. It is imperative to note that the convergence trend of the two networks remains consistent.

Therefore, the WOA algorithm optimizes weight initialization to reduce loss value, which is crucial for speeding up the convergence of the network. Compared to BPNN, which has an epoch number of over 100 at the end of the training, WOA-BPNN has an epoch number of only 70, demonstrating an improvement in convergence speed of over 30% through implementing the WOA algorithm.

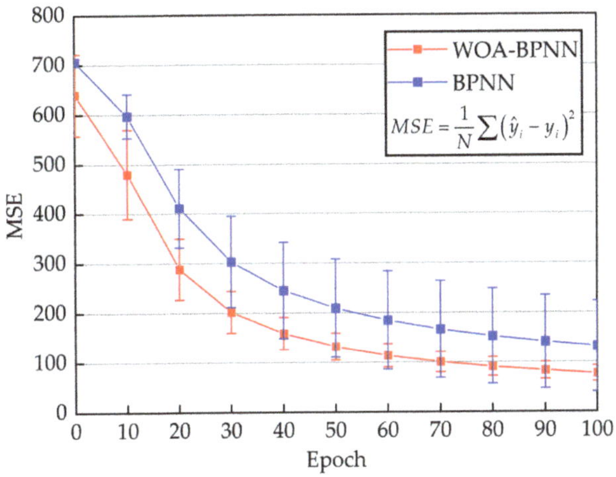

Figure 14. Comparison of convergence between WOA-BPNN and BPNN on the test dataset.

6. Conclusions

Based on a practical engineering project, this paper first establishes finite element models of different scales for a spatial three-curved steel box girder bridge in order to perform refined analysis. Then, a study of the temperature gradient effect under different temperature distribution patterns is conducted. With the obtained temperature stress dataset, a BPNN and WOA-BPNN are trained and their predictions are compared. The conclusions are as follows:

1. A beam element model is established to analyze the general internal force of the bridge. The spatial three-curved steel box girder bridge exhibits spatial stress characteristics and a notable bending–torsion coupling effect. The bending and torsional moments at the cross-sections of the reverse curve section are relatively high.
2. A detailed stress analysis is performed on a plate element model. The shear lag effect is significant on the top and bottom plates of the steel box girder, and the widening of the cross-section aggravates the uneven stress distribution. In addition, severe stress concentration occurs at the vertices at both ends of the cross-section, with the maximum shear lag coefficient exceeding 3.
3. Under the temperature effect, the three-curved steel box girder bridge displays a warped stress state. The torsion asymmetry phenomenon in the second span of the bridge intensifies the coupling effect of the bending and torsional moment.
4. Under different temperature gradient modes, the temperature of the steel box girder's bottom plate remains low and exhibits minimal changes. Compared to the BS-5400, the exponential temperature gradient pattern emphasizes a higher temperature gradient near the top plate of the cross-section, resulting in a 21% increase in average stress.
5. Compared to BPNN, the use of the WOA algorithm to optimize the initial weight value can greatly improve the convergence speed of neural network training and the prediction accuracy of temperature gradient stress. The prediction accuracy of WOA-BPNN can reach over 90% with a 10% error tolerance and R^2 of 0.99.

Although a refined analysis for the spatial three-curved steel box girder bridge has been carried out, a number of shortcomings are worthy of note. The analysis simplifies the substructure of the bridge by considering the pier and the beam as a fixed form. However, in actual engineering there is a certain range of contact between the two components. Therefore, a more refined model is required to fully simulate their interaction. Moreover, only the simple temperature gradient model has been considered in this work. Analysis of mechanical properties under complex temperature fields will be included in future work.

Author Contributions: Conceptualization, W.H., Z.Z. and Q.F.; methodology, J.S., Y.L. and Q.F.; software, Y.L.; validation, W.H., Q.F., Z.Z. and Y.C.; investigation, J.S.; resources, W.H. and Z.Z.; writing—original draft preparation, Y.C., Y.L. and Q.F.; writing—review and editing, W.H., J.S., Y.L. and Q.F.; project administration, W.H., Z.Z. and Q.F. All authors have read and agreed to the published version of the manuscript.

Funding: This work was supported by the National Natural Science Foundation of China (grant No. U23A20659).

Data Availability Statement: Data are available on request from the corresponding author due to privacy.

Conflicts of Interest: Authors Wei Hu, Zhongyong Zhang, Junwei Shi, Yulun Chen were employed by the company Hangzhou Grand Canal Group Construction Management Co., Ltd. The remaining authors declare that the research was conducted in the absence of any commercial or financial relationships that could be construed as a potential conflict of interest.

References

1. Li, G.H.; Yu, Z.W.; Wang, Y.Z. Diseases of Curved Bridge and Temperature Effect. *J. Highw. Trans. Res. Dev.* **2008**, *25*, 58–63.
2. Tong, Z.Y. Influence of Temperature Effect on Stress of Continuous Curved Box Girde. *Railw. Eng.* **2008**, *11*, 4–6.
3. Zhang, J.D.; Zhang, Y.W.; Xing, S.L.; Zhu, J.P. Study of Temperature Effect of Concrete Curved Beam Bridge. *World Bridge* **2014**, *42*, 47–52.
4. Liu, Y.J.; Liu, J.; Zhang, N. Review on Solar Thermal Actions of Bridge Structures. *Civ. Eng. J.* **2019**, *52*, 59–78.
5. Liu, C. The Temperature Field and Thermal Effect of Steel-Concrete Composite Bridge. Ph.D. Thesis, Tsinghua University, Beijing, China, 2018.
6. *JTG D60-2015*; General Specifications for Design of Highway Bridge and Culverts. China Communications Press: Beijing, China, 2015.
7. *BS 5400-2*; Steel, Concrete and Composite Bridges (Part 2: Specification for Loads). Board of BSI: London, UK, 1978.
8. Zhu, Z.W.; Wang, Y.J.; Li, J.P.; Teng, H.J. Comparison of Different Specification and Field Measurement on Vertical Temperature Gradient of Steel Box Girder. *Stl. Constr.* **2022**, *37*, 40–62.
9. Teng, H.J.; Zhu, Z.W.; Li, J.P. Investigation on Vertical Temperature Gradient of Steel Box Girder with Steel Deck under Solar Radiation. *J. Railw. Sci. Eng.* **2021**, *18*, 3267–3277.
10. Wang, Y.F.; Yang, H.Y.; Wang, Y.; Hao, J.; Xu, F.; Bai, C. Vertical Temperature Gradient Model of Steel Box Girder: Taking Steel Box Girder of Viaduct in Wuhan as Study Case. *J. Wuhan Inst. Technol.* **2021**, *43*, 202–220.
11. Ding, Y.; Zhou, G.; Li, A.; Wang, G. Thermal Field Characteristic Analysis of Steel Box Girder Based on Long-Term Measurement Data. *Int. J. Steel Struct.* **2012**, *12*, 219–223. [CrossRef]
12. Guo, F.; Zhang, S.; Duan, S. Analysis of Measured Temperature Field of Unpaved Steel Box Girder. *Appl. Sci.* **2022**, *12*, 8417. [CrossRef]
13. Guo, F.; Zhang, S.; Duan, S.; Shen, Z.; Yu, Z.; Jiang, L.; He, C. Temperature Gradient Zoning of Steel Beams without Paving Layers in China. *Case Stud. Constr. Mater.* **2023**, *18*, e02054. [CrossRef]
14. Editorial Department of China Journal of Highway and Transport. Review on China's Bridge Engineering Research: 2021. *J. Highw. Trans.* **2021**, *34*, 1–97.
15. Tian, G.; Xu, R.N.; Liao, C.Q.; Sun, C.Z. Temperature Estimation of Mass Concrete Structure Based on BP Neural Network. *J. Water Resour. Archit. Eng.* **2010**, *8*, 107–109, 123.
16. Ying, M.; Zheng, Y.F.; Wang, D.J. Research of Arch Bridge Ring Temperature Field Based on BP Method. *J. Harbin Inst. Technol.* **2011**, *43*, 134–137.
17. Wang, K.; Zhang, Y.; Liu, J.L.; He, X.H.; Cai, C.Z.; Huang, S.J. Prediction of Concrete Box-girder Maximum Temperature Gradientbased on BP Neural Network. *J. Railway Sci. Eng.* **2023**, 1–11. [CrossRef]
18. Wang, J.F.; Xu, Z.Y.; Fan, X.L.; Lin, J.P. Thermal Effects on Curved Steel Box Girder Bridges and Their Countermeasures. *J. Perform. Constr. Facil.* **2017**, *31*, 04016091. [CrossRef]
19. Xu, Z.Y. The Temperature Fields of Curved Bridges with Steel Box Girder and their Effect. Master's Thesis, Zhejiang University, Hangzhou, China, 2017.
20. Yuan, J.; Lv, S.; Peng, X.; You, L.; Borges Cabrera, M. Investigation of Strength and Fatigue Life of Rubber Asphalt Mixture. *Materials* **2020**, *13*, 3325. [CrossRef] [PubMed]

21. Mei, X.; Li, C.; Sheng, Q.; Cui, Z.; Zhou, J.; Dias, D. Development of a Hybrid Artificial Intelligence Model to Predict the Uniaxial Compressive Strength of a New Aseismic Layer Made of Rubber-Sand Concrete. *Mech. Adv. Mater. Struct.* **2023**, *30*, 2185–2202. [CrossRef]
22. Liu, X.; Liu, Z.; Liang, Z.; Zhu, S.-P.; Correia, J.A.F.O.; De Jesus, A.M.P. PSO-BP Neural Network-Based Strain Prediction of Wind Turbine Blades. *Materials* **2019**, *12*, 1889. [CrossRef] [PubMed]
23. Mirjalili, S.; Lewis, A. The Whale Optimization Algorithm. *Adv. Eng. Softw.* **2016**, *95*, 51–67. [CrossRef]

buildings

MDPI

Article

Finite Element Analysis of Axial Compression Behavior of L-Shaped Concrete-Filled Steel Tubular Columns with Different Combinations

Hexiao Li [1,2,3], Zhong Tao [1,3,*] and Dongji Han [1,4]

[1] Faculty of Civil Engineering and Mechanics, Kunming University of Science and Technology, Kunming 650500, China; lihexiao@ynou.edu.cn (H.L.); handongji@cscec.com (D.H.)
[2] College of Urban Construction, Yunnan Open University, Kunming 650223, China
[3] Yunnan Earthquake Engineering Research Institute, Kunming 650000, China
[4] 3rd Construction Co., Ltd. of China Construction 5th Engineering Bureau, Changsha 410000, China
* Correspondence: taozhong@kust.edu.cn

Abstract: L-shaped concrete-filled steel tubular (CFST) columns, a kind of structural member appropriate for high-rise buildings, not only avoid the defect of conventional square columns protruding from the wall but also have the green and low-carbon properties of steel structures appropriate for fabricated construction. To learn more about their axial compression behavior, refined 3D finite element models were established using the general finite element software ABAQUS. The reliability of the models was subsequently verified based on failure tests and load–displacement relation tests on eight L-shaped specimens. The axial compression mechanism of L-shaped CFST columns was investigated using the verified finite element models. Further systematic parameter analysis was carried out to investigate the influence of parameters such as steel strength, concrete strength, length ratio of long limb to short limb, the angle between the two limbs, and combination methods on the axial compression behavior of L-shaped CFST columns. The results demonstrate that the angle between the two limbs has a significant impact on the stress distribution of concrete and steel pipes. The corner effect increases as the angle between the two limbs decreases. The combination of F-type specimens can better exert the constraint effect of steel pipes on concrete, while the triangular cavity of unequal-limb specimens and specimens with an included angle of 60° cannot effectively trigger the interaction between steel pipes and concrete. The initial stiffness of L-shaped CFST columns increases with an increase in concrete strength and a decrease in limb length ratio, which is not sensitive to changes in steel strength and the included angle. The peak bearing capacity of the specimens increases with increases in steel strength and concrete strength and a decrease in the limb length ratio. Compared to C-type and Z-type specimens, the initial stiffness of F-type specimens is slightly higher, and the peak bearing capacity is significantly increased.

Keywords: L-shaped concrete-filled steel tubular column; axial compression behavior; finite element analysis; combination method

1. Introduction

The annual carbon emissions of the construction industry account for more than 40% of the total [1] when upstream and downstream businesses like transportation and the production of construction materials are taken into account. To achieve the objective of "carbon peaking and neutrality", it is crucial to promote the green transformation of the construction industry. When compared to concrete structures, steel structures are more ecologically friendly and have lower whole-life-cycle carbon emissions. W. Hawkins [2] compared the whole-life-cycle carbon emissions of concrete, steel, and timber structures and found that steel structures reduced carbon emissions by about 22% compared to concrete structures at all stages. Su [3] concluded that steel structures were more environmentally friendly with less wet work than concrete structures, and a case study showed that steel structures emit 48.1% less carbon than concrete structures. After taking steel recycling into account, A. Zeitz's [4] study on four different structural forms of parking lots revealed that the carbon emissions of steel structures were only 38.3% of their original carbon content, a significantly lower percentage than for concrete structures. It is evident from the above study that steel structures fit the development needs of the green transformation in the construction sector since they are labor-saving, resource-conserving, and have a lower environmental impact during construction than traditional reinforced concrete structures cast in situ. Steel structures serve as crucial carriers for the advancement of a low-carbon economy and green environmental protection [5,6].

Due to the use of standard sections such as rectangular and H-shaped columns in traditional residential buildings with steel structures, the column ribs frequently protrude from the walls, which not only takes up limited indoor space but also impedes furniture deployment. Indoor appearance and functioning are negatively affected, posing a significant barrier to the development and application of prefabricated steel constructions in residential buildings. The advantages of high steel structure assembly level and great fire resistance of concrete structure are combined in the concrete-filled steel tubular structure, which has good economic benefits. For example, the interior convex angle achieved by utilizing square CFST columns can be handled by concealing the structural columns inside the wall with L-shaped or T-shaped CFST columns. CFST columns have been used in practical applications [7], as shown in Figure 1.

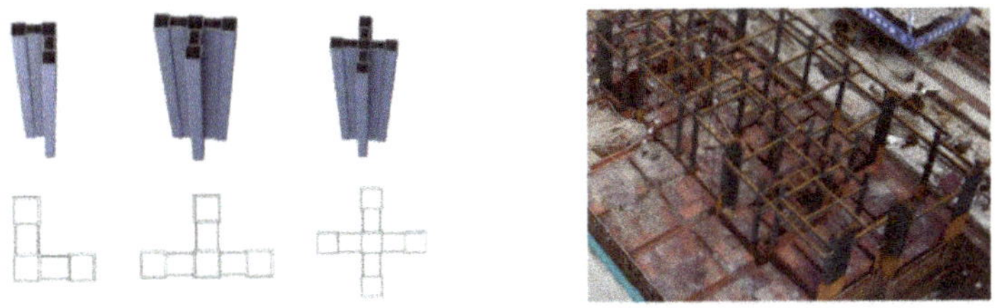

(**a**) CFST columns (**b**) Practical project

Figure 1. Engineering applications of CFST columns.

Numerous researchers have conducted research on the axial compression, eccentric compression, and seismic performance of L-shaped CFST columns. Zuo [8] conducted experiments on special-shaped columns with binding bars, and the results showed that the binding bars can effectively enhance the binding force of steel pipes on concrete. Zhou [9] investigated the effects of axial compression ratio and aspect ratio on the seismic performance of L-shaped CFST columns through experiments, studied the asymmetry character of the column, and carried out finite element analysis to simulate the behavior of SCFST columns. Shen [10] tested six specimens under constant axial and cyclically varying lateral loads to investigate the effects of width to depth ratio of section, depth to thickness ratio of steel tube, and axial load level on the strength, as well as stiffness, ductility, and energy dissipation of L-shaped CFST columns. Zheng [11] presented four tests carried out on multi-cell L-shaped CFST columns under combined constant compression and lateral cyclic loads. The main variables were loading angles and axial load levels. A finite element model was developed to simulate the multi-cell L-shaped CFST columns at different loading angles, and parametric analysis was conducted to evaluate the impact of various parameters on the lateral load–displacement curves of the columns under diagonal loading. Zhang [12] studied the seismic performance of nine L-shaped concrete-filled steel tubular columns and proposed the calculation formulas on the model curve based on the parameters, load, displacement, and stiffness degradation. Chen [13] proposed a prefabricated special-shaped concrete-filled steel tube with a rectangular multi-cell (S-CFST-R) column for residential systems, performed seismic testing of S-CFST-R columns, and verified the strong constraint effect of multi-cavity special-shaped CFST columns on concrete. Zhang [14] examined the performance of a novel type of composite column: L-shaped columns comprising concrete-filled steel tubes joined by double-vertical steel plates (LCFST-D). Seven LCFST-D columns were tested under axial compression. Parametric experiments using the finite element model were conducted to investigate the effects of the thickness and width of the vertical steel plates. Wang [15] analyzed the axial behavior of an innovative L-shaped composite column. Seven specimens were evaluated to determine the failure progress, load capacity, and failure modes. The tested specimens exhibited shear failure with local buckling. Wang [16] tested eight SS-CFST columns to determine their axial compressive properties. Zhang [17] investigated the effects of eccentricity, column height, and sectional dimension on the stability performance of MCL-CFSTs columns.

Furthermore, some researchers have explored the constitutive relation and calculation formula of special-shaped CFST columns. Zhou [18] conducted experiments to investigate the biaxial loading behavior of L-shaped SCFT columns. Based on the findings, a reasonably simplified formula was proposed and proved effectively by comparing test results. Zheng [19] developed finite element models to simulate the axial behavior of unstiffened, stiffened, and multi-cell L-shaped and T-shaped CFST stub columns. A parametric study of 804 cases covering a wide range of parameters was conducted based on the FE models. New design models of axial compressive strength and stiffness were proposed. Lei et al. [20–22] proposed a uniaxial constitutive relation for T-shaped CFST columns based on fiber model analysis. Liu [23] experimented with six L-shaped and twelve T-shaped CFST stub columns under axial compression. Further parameter analysis with ABAQUS was carried out to study the influences of the steel ratio α, steel yield strength f_y, concrete strength f_{ck}, and slenderness ratio λ. Based on the experimental results and FE analysis, design formulas for calculating sectional bearing capacity and stability bearing capacity of special-shaped CFST columns were proposed. Liu [24] calculated the N–M correlation curves of the L-shaped CFST columns by using a full-section plasticity method and proposed formulas for calculating the eccentric compression ultimate bearing capacity of L-shaped CFST columns. Xiao [25] carried out experiments on 32 specimens, including 16 short columns and 16 medium–long columns, and provided formulas for estimating the ultimate bearing capacity of short and medium–long composite columns. Based on limit state theory, Zhang [26] established a calculation method for the sectional flexural capacity of the W-LCFST column. Additionally, they analyzed the accuracy of predicting the compression and bending capacity of the W-LCFST column under positive and negative loads using Standard CECS159–2004, with the recommended correction value provided. Chen [27] designed formulas for the stable bearing capacity of SCFST columns. Tu et al. [28,29] analyzed the influence of parameters such as concrete and steel strength on L-shaped CFST columns through FEA and proposed a formula for calculating the axial compressive bearing capacity. Huang [30] proposed a design formula for the bearing capacity of ribbed L-shaped CFST columns that took into account improvements in concrete performance. Chen [31] proposed a simplified strength model of ECB for L-shaped CFST based on the composite yielding method. Zheng [32] proposed a practical design method for predicting the bearing capacity of L-shaped CFST columns.

In the current research, researchers typically use methods such as tied reinforcement, tension bars, welded stiffening ribs, and so on to postpone the local buckling of the steel pipe of the L-shaped CFST column and enhance the restraining effect of the steel pipe on the concrete. However, these methods are complicated in construction, inconvenient in construction, and have an impact on the appearance of the buildings. In addition, when carrying out building design, building designers usually need to consider lighting, ventilation, and other factors, resulting in the corners of the building plan not necessarily having right angles. However, most of the L-shaped CFST columns studied so far are equal-limb cross-sections with right angles, and research on L-shaped CFST columns with special angles (non-right angles) existing in engineering practice has not been reported.

Based on this, we proposed three combination forms of L-shaped CFST columns with cross-section combinations. These forms are simple to assemble and do not alter the building's appearance. The cross-sectional form is shown in Figure 2, where the F-type specimen is composed of square steel pipes, the C-type specimen is composed of square steel pipes and C-shaped cold-formed components, and the Z-type specimen is composed of C-shaped and Z-shaped cold-formed components. Experiments to investigate the mechanical behavior of the three combination forms of L-shaped CFST columns were conducted by our research team.

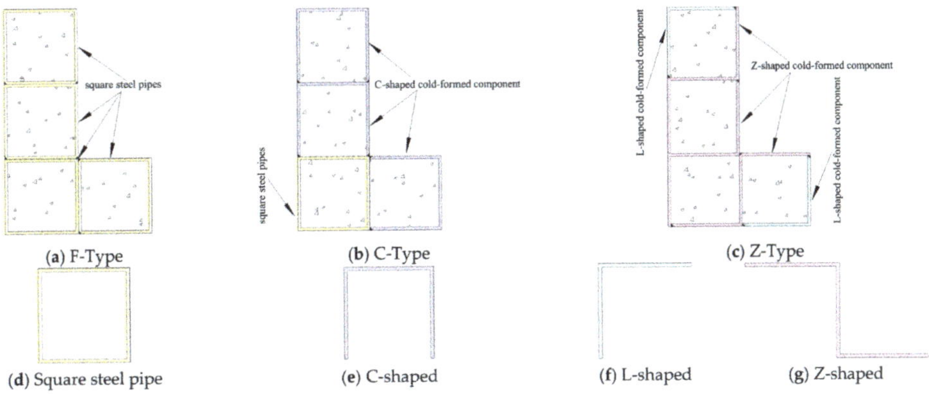

Figure 2. Types and components of L-shaped CFST columns.

This study proposed a refined numerical model of the specimens using ABAQUS to further investigate the axial compression mechanism of L-shaped CFST columns. The reliability of the model was verified using the obtained test results, and the impact patterns of different parameters were studied through extensive parameter analysis to clarify the interaction mechanism between steel pipes and concrete in L-shaped CFST.

2. Test Profile

Eight specimens were designed and manufactured to investigate the effects of combination methods, the length ratio of long limb to short limb, and the angle between the two limbs on the axial compression behavior of L-shaped CFST columns. The major parameters and test results of the specimens are listed in Table 1. To investigate the axial bearing capacity of L-shaped CFST columns, a 1000 kN compression testing machine was employed, as shown in Figure 3. The axial displacement of the specimen was measured by the displacement capture device at the loading end. Vertical and transverse strain gauges were attached at the mid-section of each specimen.

Table 1. Major parameters and results of test specimens.

Specimen No.	t/mm	H/mm	Concrete Grade	Steel Grade	d_u/mm	N_u/KN
AF-1.5-90°	3.75	600	C35	Q235	11.72	2685.8
AFD-1-90°	3.75	600	C35	Q235	10.12	3224.8
AC-1.5-90°	3.75	600	C35	Q235	7.46	2311.9
ACD-1-90°	3.75	600	C35	Q235	10.55	2975.0
AZ-1.5-90°	3.75	600	C35	Q235	7.97	2229.5
AZD-1-90°	3.75	600	C35	Q235	8.54	2945.8
ACD-1-60°	3.75	1200	C35	Q235	11.15	2315.2
ACD-1-135°	3.75	1200	C35	Q235	16.61	3352.9

Where A represents the axially loaded specimen; F, C, and Z refer to the type of specimen; D indicates an equal length of the two limbs; 1.5 and 1 represent the length ratio of the long limb to short limb; 90°, 60°, and 135° represent the angle between the two limbs; t is the thickness of the steel; H is the height of the column; d_u represents the peak displacement of the specimen; N_u represents the peak bearing capacity of the specimen.

Figure 3. Typical loading diagram.

Using a graded loading approach, the test was run for two minutes, with each stress level representing about one-tenth of the anticipated axial compressive load capacity. Once the specimen reached its yield strength, the load range was whittled down to one-fifteenth of its axial compressive load capacity. Loading proceeded gradually as the anticipated ultimate load approached, and it ceased when the load capacity decreased to 60% of the ultimate load capacity or there was noticeable deformation.

3. Finite Element Model (FEM)

Finite element analysis (FEA) has become an invaluable part of most structural studies since it can be used as an efficient tool to investigate structural behaviors, especially the composite actions between components of CFST specimens. Numerous researchers have investigated the mechanical performance of L-shaped CFST columns through ABAQUS. Zheng [11] investigated the effects of various parameters on lateral load–displacement curves of L-shaped CFST columns under diagonal loading. Chen [13] analyzed the influence of different loading directions on the ultimate bearing capacity of L-shaped CFST columns. Wang [15] conducted parametric analysis to investigate the influence on the axial behavior of L-shaped CFST columns. Liu [23] analyzed the axial compression behavior of L-shaped CFST columns through ABAQUS.

According to the aforementioned research, ABAQUS was able to accurately model the mechanical behavior of L-shaped CFST columns. In this paper, refined numerical models of the specimens were proposed using ABAQUS to further investigate the axial compression mechanism of L-shaped CFST columns.

3.1. Model Unit Selection and Meshing

The same model was employed for C-type and Z-type specimens due to their similar ultimate bearing capacities and failure modes. ABAQUS models were established for all specimens. Each component of the FEM adopts a 3D solid (C3D8R) in a reduced integral format and is meshed based on the following principles: (1) the number of grid layers in the thickness direction of steel pipe components shall not be less than 2; (2) to achieve reliable modeling with minimum increase in computational consumption, mesh convergence studies were conducted to determine the appropriate mesh size; (3) the steel pipe serves as the master surface, with a mesh size of 30 mm; the concrete serves as the slave surface, with a mesh size of 28 mm. The components and meshing are displayed in Figure 4.

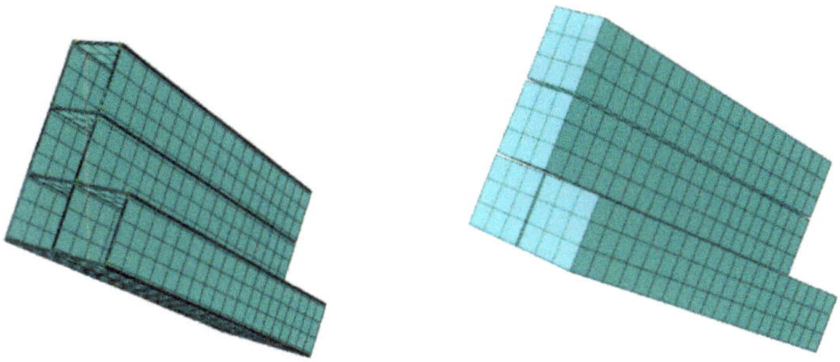

(**a**) Steel pipe (**b**) Concrete

Figure 4. Components and meshing.

3.2. Material Constitutive Relations

3.2.1. Steel Constitutive Relation

Based on the constitutive relation of steel proposed in the literature [33], this study simulated the elastic–plastic properties of steel used in axial compression tests. Figure 5 displays the uniaxial stress–strain relation curve of steel, with f_p, f_y, and f_u representing the proportional limit, yield strength, and tensile strength of the steel, respectively; ε_e is the strain corresponding to f_p; ε_{y1} and ε_{y2} represent the strain corresponding to the starting and ending points of the yield plateau; ε_u is the strain corresponding to f_u.

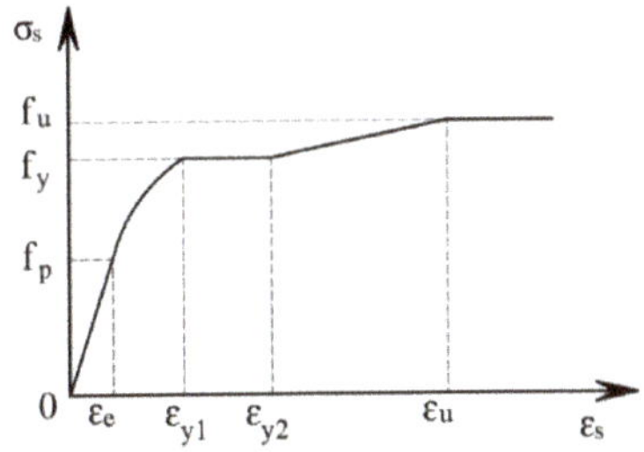

Figure 5. Stress–strain curve of Q235 steel.

The stress–strain relationship of steel is described as follows:

$$\sigma_s = \begin{cases} E_s\varepsilon_s & (\varepsilon_s < \varepsilon_e) \\ -A{\varepsilon_s}^2 + B\varepsilon_s + C & (\varepsilon_e < \varepsilon_s \le \varepsilon_{y1}) \\ f_y[1 + 0.6\frac{\varepsilon_s - \varepsilon_{y2}}{\varepsilon_{y3} - \varepsilon_u}] & (\varepsilon_{y2} < \varepsilon_s \le \varepsilon_u) \\ 1.6f_y & (\varepsilon_s \ge \varepsilon_u) \end{cases} \quad (1)$$

where the elastic modulus of steel $E_s = 2.06 \times 10^5$ MPa; Poisson's ratio is 0.3; $\varepsilon_e = 0.8 f_y/E_s$; $\varepsilon_{y1} = 1.5\ \varepsilon_e$; $\varepsilon_{y2} = 10\ \varepsilon_{y1}$; $A = 0.2 f_y/(\varepsilon_{y1} - \varepsilon_e)^2$; $B = 2\ A\varepsilon_{y1}$; $C = 0.8 f_y + A{\varepsilon_e}^2 - B\varepsilon_e$.

3.2.2. Concrete Constitutive Relation

This study adopts the constitutive relation of concrete proposed in the literature [33,34] based on the research of CFST columns, as shown in Figure 6.

The compressive stress–strain relationship of concrete is described as follows:

$$y = \begin{cases} 2x - x^2 & (x \le 1) \\ \frac{x}{\beta_0 (x-1)^\eta + x} & (x > 1) \end{cases} \quad (2)$$

where $x = \varepsilon/\varepsilon_0$; $y = \sigma/f_{c0}$; $f_{c0} = f_c\ [1 + (-0.0291\ \xi^2 + 0.1189\ \xi) \times (22.8/f_c)^{0.3048}]$; $\varepsilon_0 = [(1300 + 12.5 f_{c0}) + 800\ \xi^{0.2}] \times 10^{-6}$; $\eta = 1.6 + 1.5/x$; $\beta_0 = (f_c)^{0.1}/1.2\ \sqrt{1+\xi}$. $\xi = A_s f_y / A_c f_{ck}$. ε is the compressive strain of concrete; ε_0 is the peak compressive strain of concrete considering the constraint effect; σ is the compressive stress of concrete; f_{c0} is the peak compressive stress of concrete after considering the constraint effect; f_c is the compressive strength of the concrete prism; f_{ck} is the standard value of axial compressive strength of concrete; A_s and A_c represent the areas of steel and concrete, respectively; ξ is the equivalent coefficient of constraint effect; and η and β_0 are coefficients.

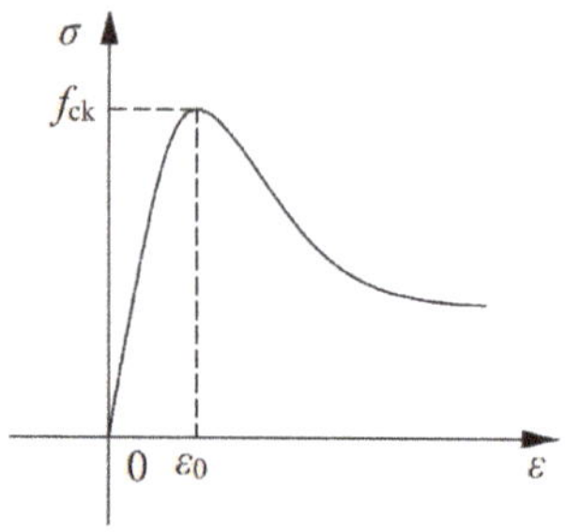

Figure 6. Stress–strain curve of concrete.

The concrete damage plasticity (CDP) module available in the material library of ABAQUS was used to describe the plastic behavior of the concrete infill in the FE model. For the inputs of the CDP module, constant values of 30, 0.1, 1.16, 2/3, and 0.0 were used for dilation angle (ψ), flow potential eccentricity (e), ratio of the compressive cylinder strength under biaxial loading to uniaxial compressive strength ($f_{b0}/f_{c'}$), and viscosity, respectively.

3.3. Model Interactions and Boundary Conditions

Surface-to-surface contact was adopted between the outer steel pipe and the in-fill concrete. Steel pipes and concrete are defined as the master surface and the slave surface, respectively, allowing the master surface to invade the slave surface but not vice versa. When the concrete contacts the steel tube due to its transverse deformation under axial compression, the concrete will be partially confined by the steel tube. To ensure that the two surfaces can separate from each other after contact, the normal contact attribute adopts hard contact, and the tangential contact attribute adopts a "penalty" function, with the friction coefficient being 0.6. The boundary condition and loading criterion are modeled completely according to those in the experiment. The upper and lower end plates were omitted in this model, making the same side of the concrete and steel pipe coupled with the coupling point. All the degrees of freedom (DOF) at the bottom of the model were confined throughout the calculation, while the top was totally constrained with the exception of releasing the Z-direction DOF. Axial displacement was used at the free end. The boundary conditions of the finite element model are displayed in Figure 7.

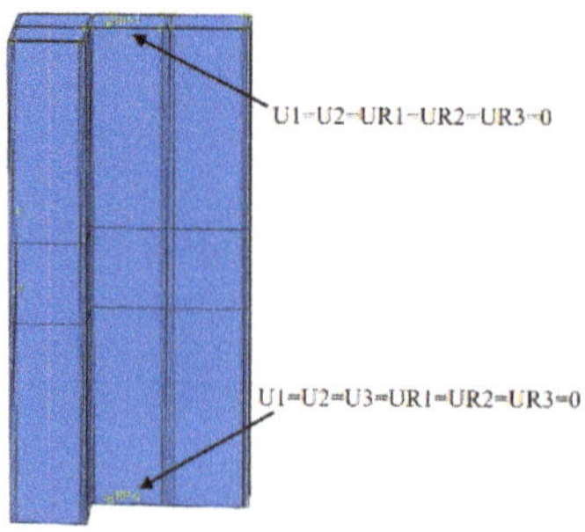

Figure 7. Boundary conditions of the finite element model.

4. Model Verification

4.1. Load–Displacement Curve Comparison

To verify the reliability of the FEM, the FEA results were compared with test results, as shown in Figure 8 and Table 2. It can be seen that the variation trend of the finite element calculation curve is basically consistent with the curve of the test results. In the elastic stage, the measured stiffness is slightly smaller than that obtained through the FEM. This is mainly due to a certain gap between the steel pipe and the concrete. Therefore, there is no cooperative bearing between the steel pipe and concrete in the early stage of loading. In the elastoplastic stage, there are slight differences between the test and FEM calculation curves, with a small difference in ultimate bearing capacity between the two. According to Table 2, the maximum deviation is 8.7%, and the standard deviation is 0.12.

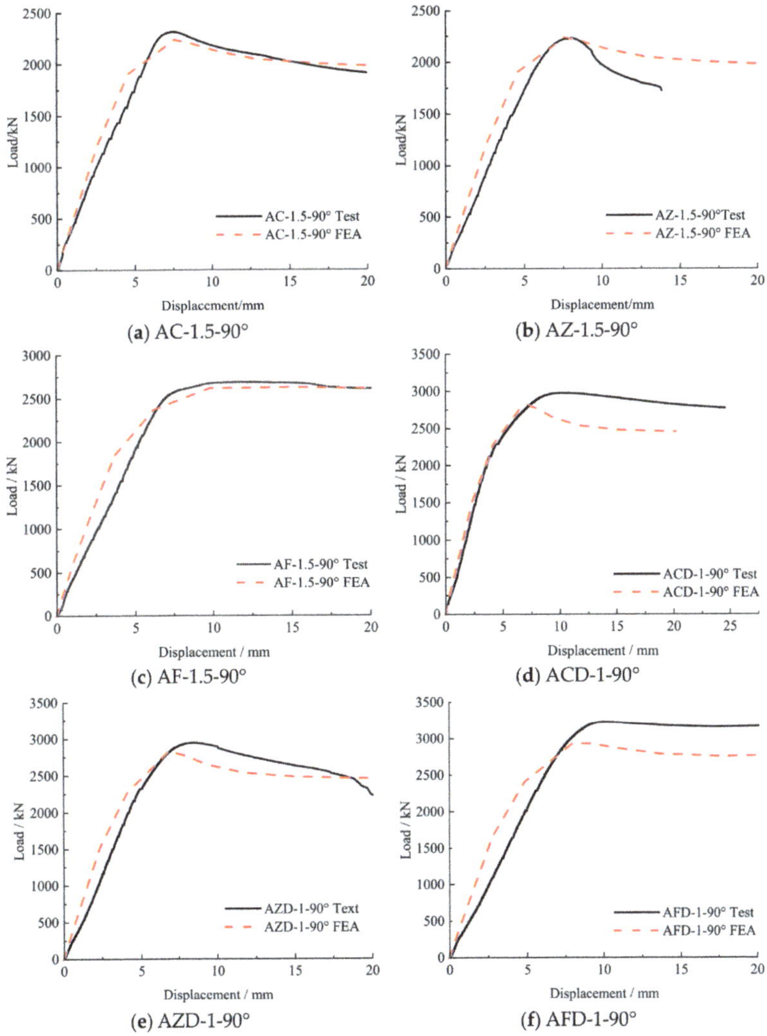

(**a**) AC-1.5-90° (**b**) AZ-1.5-90° (**c**) AF-1.5-90° (**d**) ACD-1-90° (**e**) AZD-1-90° (**f**) AFD-1-90°

Figure 8. *Cont.*

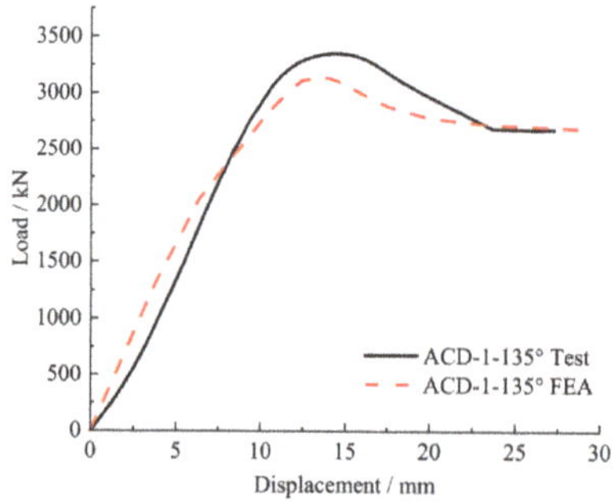

(**g**) ACD-1-135°

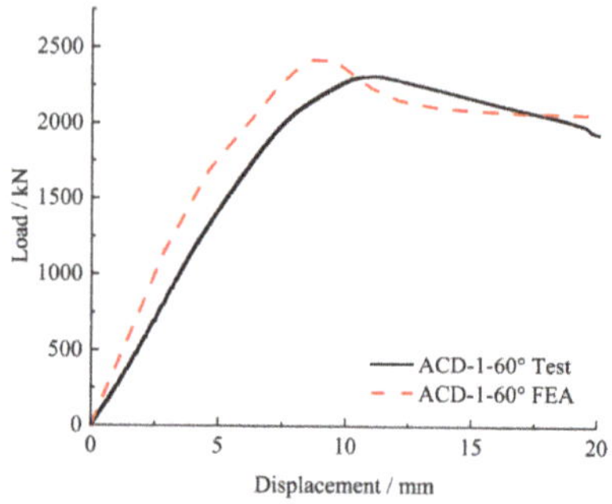

(**h**) ACD-1-60°

Figure 8. Load–displacement curve comparison.

Table 2. Bearing capacity comparison.

Specimen No.	Specimen Size	K_{FE}	K_A	N_{FE}	N_A	N_{FE}/N_A	K_{FE}/K_A
AF-1.5-90°	300 × 200 × 100 × 600	490	374	2619.70	2679.30	1.023	1.309
AFD-1-90°	300 × 300 × 100 × 600	589	407	2938.96	3217.85	1.095	1.448
AC-1.5-90°	300 × 200 × 100 × 600	412	349	2238.21	2310.85	1.032	1.181
ACD-1-90°	300 × 300 × 100 × 600	632	582	2837.55	2969.25	1.046	1.086
AZ-1.5-90°	300 × 200 × 100 × 600	412	353	2238.21	2229.53	0.996	1.167
AZD-1-90°	300 × 300 × 100 × 600	632	459	2837.55	2940.52	1.036	1.377
ACD-1-60°	300 × 300 × 100 × 1200	367	290	2418.16	2314.62	0.957	1.262
ACD-1-135°	300 × 300 × 100 × 1200	325	287	3136.39	3352.61	1.069	1.133

Where K_{FE} represents the initial stiffness of the specimen in the FEA; K_A represents the initial stiffness of the test specimen; N_A represents the peak load of the test specimen; N_{FE} represents the peak load of the specimen in the FEA.

4.2. Failure Mode Comparison

Figure 9 compares the usual failure modes of the test specimens and the FEM. It is clear that the specimen bulges mostly at the ends and in the middle, not at the internal corners, which is typical in regular L-shaped CFST columns. This is due to the constraint effect of cavity walls on the internal corners in specimens. Overall, the FEM and material constitutive relation established in this study are reliable. Therefore, the proposed FEM can be a suitable tool for investigating the axial compression mechanism of L-shaped CFST columns.

(**a**) Equal-limb specimen (**b**) Unequal-limb specimen

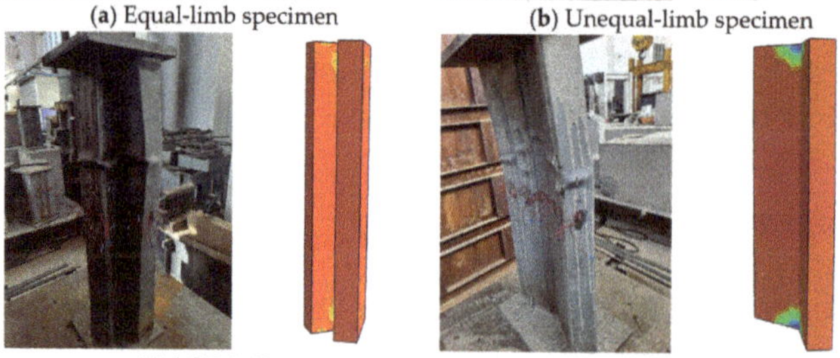

(**c**) ACD-1-60° (**d**) ACD-1-135°

Figure 9. Comparison of failure modes of test specimens and FEM.

5. Axial Compression Mechanism Analysis

5.1. Analysis of the Whole Stress Process of the Specimens

Based on the proposed FEM, axial compression analysis was conducted on L-shaped CFST columns. Figure 10 depicts the load–displacement curves of the specimens, including the load of concrete, outer steel pipes, and internal steel pipe walls. The load proportions of each part of the specimens are listed in Table 3. The loading procedure of all the specimens consists of four stages, with the four characteristic points marked on the curve. A depicts the elastic limit of the steel pipe. B indicates the point of buckling and local buckling of the steel pipe. C refers to the ultimate bearing capacity of the specimen. The displacement corresponding to point D is the failure displacement of the specimen.

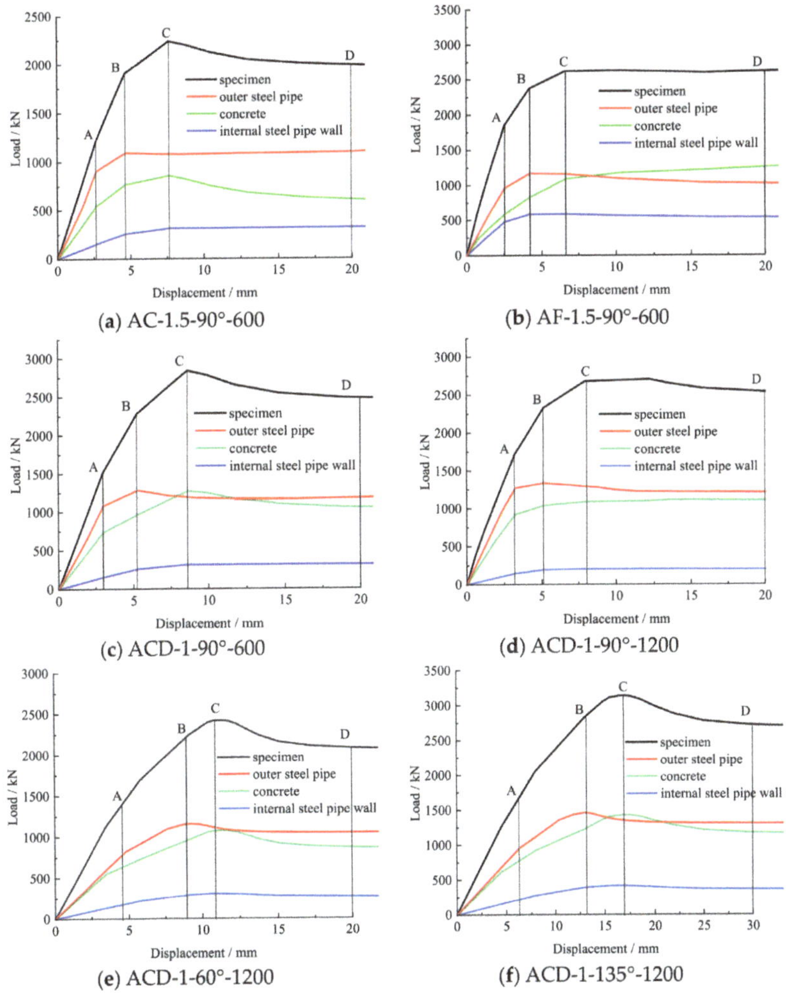

Figure 10. Load–displacement curves.

(1) O–A: The specimen is essentially in the elastic stage, where the load is increasing linearly and the steel pipes and concrete are each in a state of independent compression. The load–displacement curve of the steel pipe turns at point A, when the axial stress meets the yield stress of the steel pipe, and the slope of the specimen curve continues to decline.

(2) A–B: The specimen is in an elastic–plastic stage. Following point A, the load of the steel pipe still increases to a certain extent, and the load proportion of concrete continues to increase. The small cracks and lateral deformation inside the concrete gradually increase, and the stiffness of the specimen decreases to a certain extent.
(3) B–C: The specimen is in a plastic stage. Following point B, the overall bearing capacity of the specimen and the load of the concrete continue to increase. Moreover, due to the constraint effect of the steel pipe on the concrete, the bearing capacity of the concrete exceeds the compressive strength of the concrete block, while the load of the steel pipe begins to decrease to varying degrees. Compared to the other specimens, the 60° and 135° specimens exhibited more local buckling of the outer steel pipe, resulting in a more pronounced decrease in the load of the outer steel pipe. The outer steel pipes of the equal-limb specimen and the F-type specimen have a stronger constraint effect on the concrete than the unequal-limb specimen and the C-type specimen, resulting in a more significant rise in the load of the concrete of specimens ACD-1-90°-600 and AF-1.5-90°-600.
(4) C–D: The specimen is in the softening stage. Following point C, except Figure 10b,d, the overall bearing capacity of the rest of the specimens first shows a significant decrease and then tends to stabilize. The load of concrete gradually decreases due to intensified damage, while the load of the outer and inner steel pipe walls tends to stabilize.

According to Figure 10b, following point C, the load of concrete does not appear to be significantly reduced, and the F-type specimen can be preferred in the actual project due to the stronger restraining effect of the steel pipe on the concrete in the F-type specimen and the difficulty of occurrence of unfavorable situations like weld decay during loading. According to Figure 10d, the reason for the apparent lack of significant reduction in the load of concrete could be attributed to the specimen's slight bending during the compression process after its height was increased. This led to a certain enhancement of the steel pipe's restraining effect on the concrete. Subsequent studies will delve into this phenomenon in further detail.

Table 3. Load proportions of each specimen part.

Specimen No.	H/mm	Outer Steel Pipe	Concrete	Internal Steel Pipe Wall
AC-1.5-90°-600	600	48.71%	38.20%	13.80%
AF-1.5-90°-600	600	44.29%	44.66%	22.32%
ACD-1-90°-600	600	45.08%	44.66%	10.89%
ACD-1-90°-1200	1200	49.86%	40.46%	7.54%
ACD-1-60°-1200	1200	47.27%	44.95%	12.41%
ACD-1-135°-1200	1200	46.52%	45.65%	13.23%

5.2. Stress Analysis of Concrete

The stress states of each concrete specimen at four characteristic points are shown in Figure 11. From the beginning of loading to point A, the vertical stress distribution of the concrete in all the other specimens is relatively uniform except for specimen ACD-1-60°-1200, with a relatively large stress at its corner cavity, indicating that reducing the angle between the two limbs may easily lead to the corner stress concentration of the specimen in the initial loading stage. From A to B, the maximum vertical stress of the concrete gradually diffuses towards the outer edges and corners. Due to the corner effect, the stress in the specimen's corner cavity considerably rises when the included angle varies. When loaded to point B, the steel pipe reaches its maximum bearing capacity, while the concrete only reaches its maximum bearing capacity after point B. This is primarily due to the steel pipe's constraint effect on the concrete.

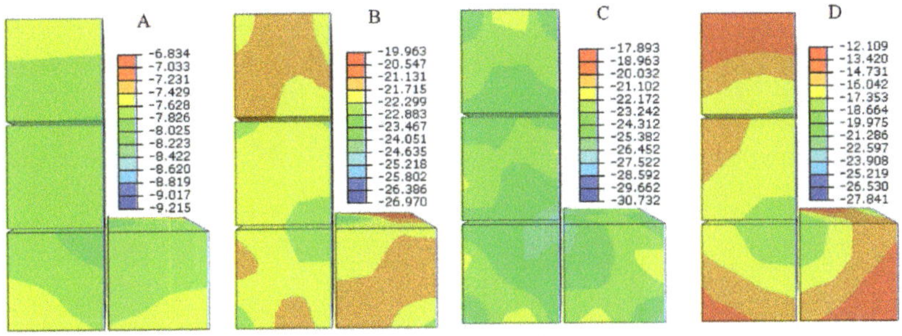

(**a**) AC-1.5-90°-600

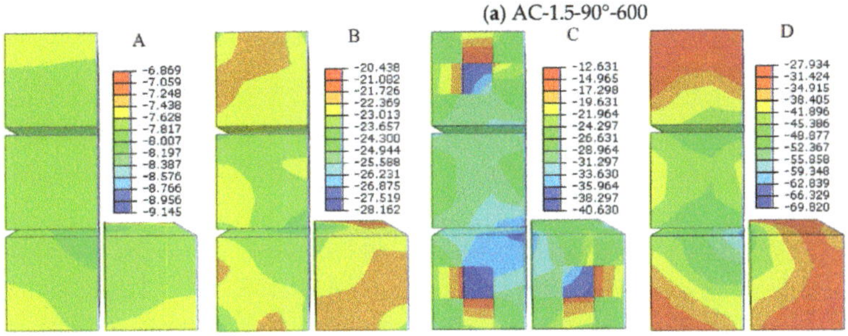

(**b**) AF-1.5-90°-600

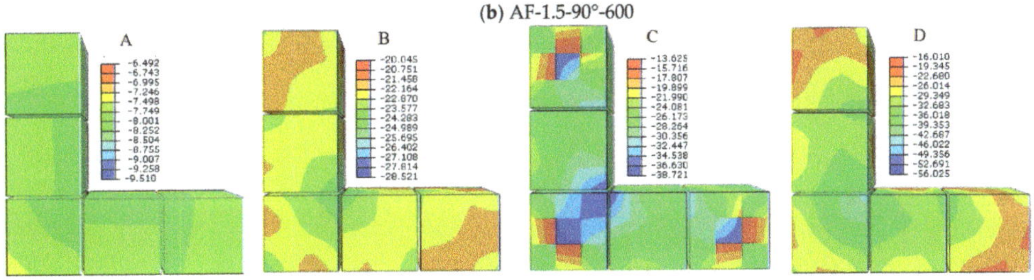

(**c**) ACD-1-90°-600

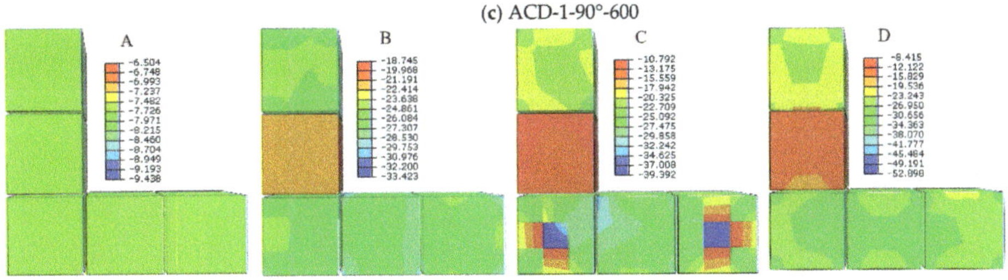

(**d**) ACD-1-90°-1200

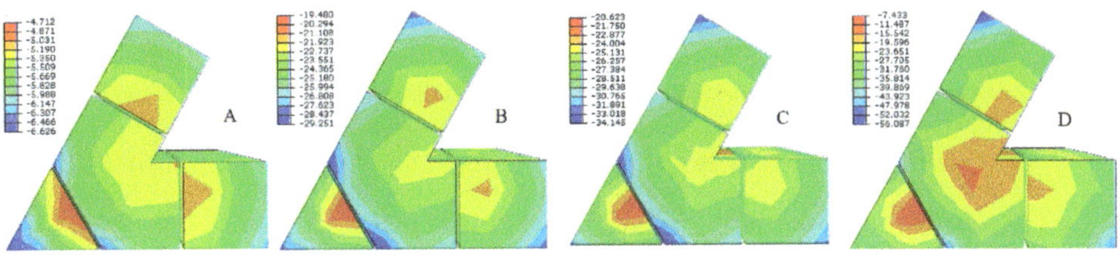

(**e**) ACD-1-60°-1200

Figure 11. *Cont.*

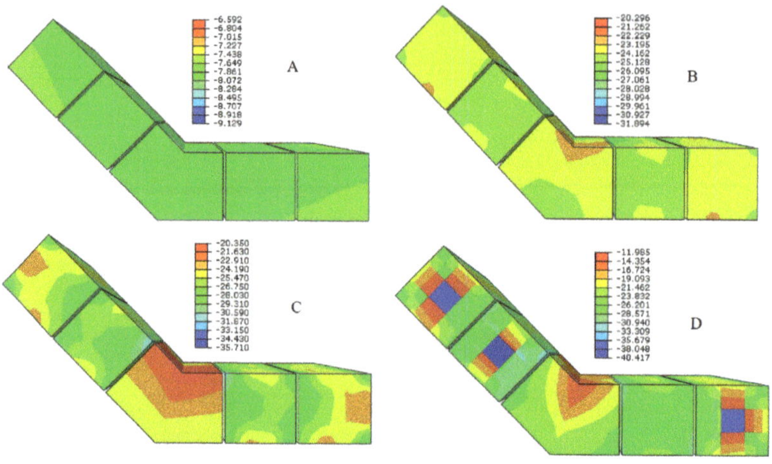

(f) ACD-1-135°-1200

Figure 11. Stress nephogram of concrete.

From B to C, the stress at the internal corner of the specimen with a 90° included angle rises from 24 MPa to 26 MPa, slightly higher than the stress at the external corner. As for the specimen with a 60° included angle, the stress at the internal corner increases from 21.92 MPa to 24.00 MPa, whereas the stress at the external corner increases from 29.25 MPa to 34.14 MPa. The greatest stress area at the external corner steadily diminishes, showing that decreasing the angle between the two limbs increases the corner effect of specimens. The stress at the internal corner of the 135° specimen drops from 22.22 MPa to 20.53 MPa, while the stress at the external corner remains unchanged. As can be observed, the angle between the two limbs has a major impact on the specimen's corner effect and stress distribution. From C to D, each specimen's high-stress zone appears at the 90° angle of the square cavity, with considerable fluctuations in stress at the internal and external corners. The stress in the corner cavity of some concrete specimens with the included angles of 60° and 135° diminishes because the stress area of the 135° and 60° corner cavities is smaller than that of the two flanges, with the other square cavities bearing the majority of the load. Therefore, in the actual engineering design for the special-angled L-shaped CFST columns, the angle between the two limbs should be considered in terms of the specimen's corner effect and stress distribution, and certain strengthening measures should be taken.

5.3. Stress Analysis of Steel Pipe

Figure 12 depicts the longitudinal stress nephogram of the steel pipe under the ultimate load for each specimen. Since there is little difference in the stress nephogram of the steel pipe between specimens AF-1.5-90° and AC-1.5-90°, specimen AC-1.5-90° was used as an example in this study. The figure shows that the steel pipe's stress distribution is symmetrical along the specimen's length from top to bottom. However, there are high-stress areas at the connection between the specimen and the upper and lower end plates, primarily because of the end plate's confinement effect on the steel pipe. Tensile stress appears in the middle of the specimen, mainly caused by local buckling of the steel pipe, and the maximum tensile stress zone is located in the middle of the specimen's cavity, indicating that the cavity can improve from single-wave buckling to multi-wave buckling.

As the limb length ratio increases, the long limbs of the unequal-limb specimen are characterized by multi-wave bulging, and the tensile stress in the middle of the cavity significantly increases. As the height increases, the tensile stress on the left flange of ACD-1-90°-1200 gradually concentrates in the middle of the two end cavities, while the tensile stress and distribution area in the middle cavity decrease, demonstrating that the two end cavities bear the majority of the load. As the included angle decreases, tensile stress gradually moves towards the two cavities at the end of the left flange of ACD-1-60°-1200, with an increase to a certain extent, indicating that the cavity at the end of the flange serves as the primary load-bearing component of the specimen. As the included angle increases, the outer edge of the steel pipe of ACD-1-135°-1200 remains in a relatively uniform state in terms of tensile stress distribution. The tensile stress progressively increases with increasing load, owing primarily to the bending of the specimen, which is consistent with the test results.

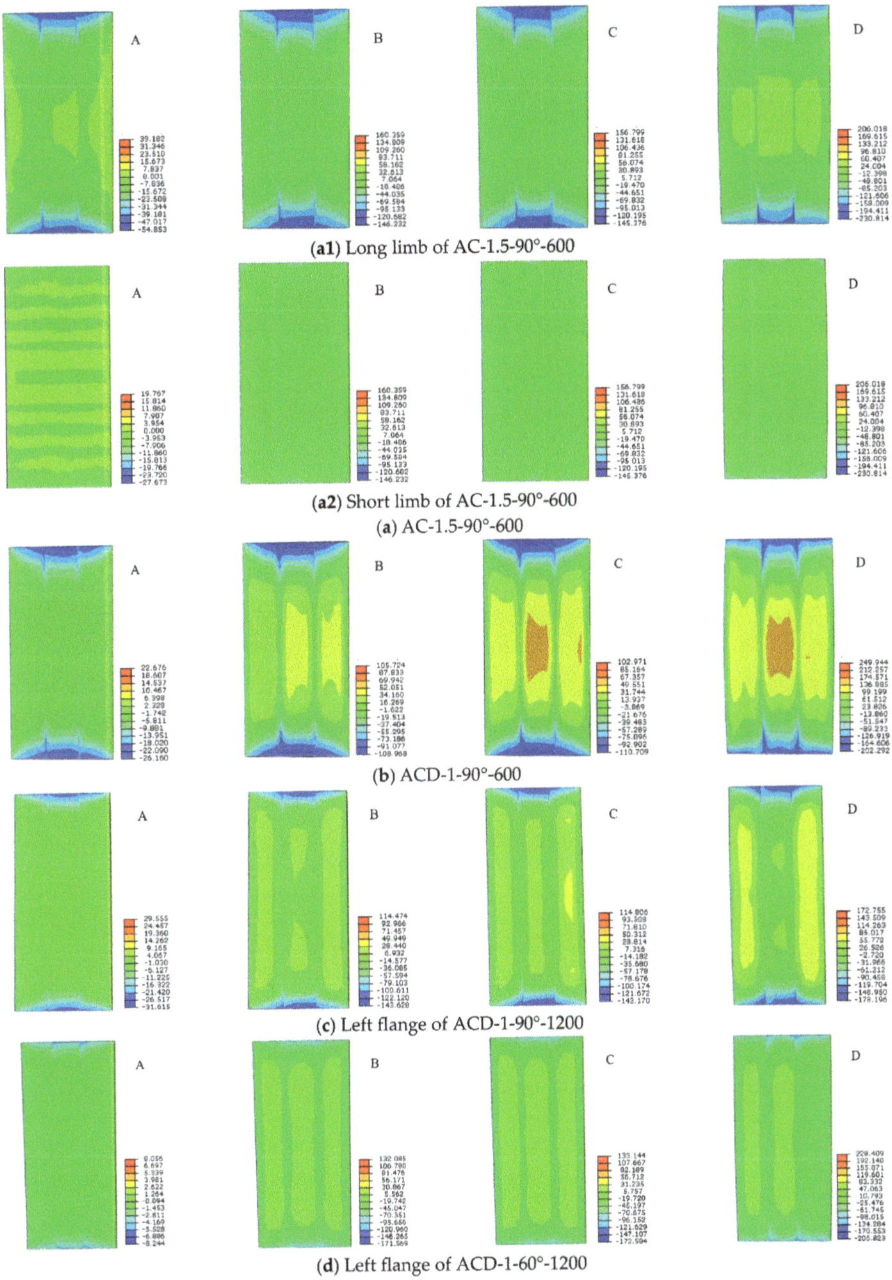

(**a1**) Long limb of AC-1.5-90°-600

(**a2**) Short limb of AC-1.5-90°-600

(**a**) AC-1.5-90°-600

(**b**) ACD-1-90°-600

(**c**) Left flange of ACD-1-90°-1200

(**d**) Left flange of ACD-1-60°-1200

Figure 12. *Cont.*

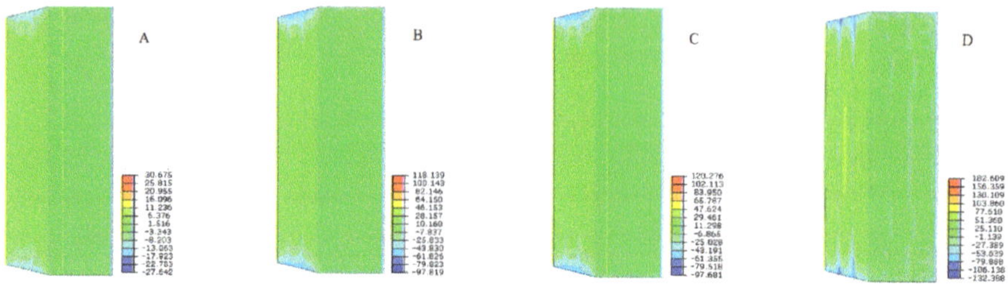

(e) ACD-1-135°-1200

Figure 12. Stress nephogram of the steel pipes of specimens.

5.4. Interaction between Steel Pipes and Concrete

To further analyze the effects of the length ratio of long limb to short limb, column height, and the angle between the two limbs on the interaction between steel pipes and concrete, a curve was drawn to describe the magnitude and distribution of constraint stress by taking P_i and Δ as the horizontal and vertical coordinates. P_i is the contact stress, which can directly reflect the constraint effect; Δ represents the displacement. To avoid the corner effect of the steel pipe, the value of Pi was obtained from the middle of the specimen. The location of the contact stress point is shown in Figure 13. Figure 14 illustrates the P_i–Δ curves of the specimens.

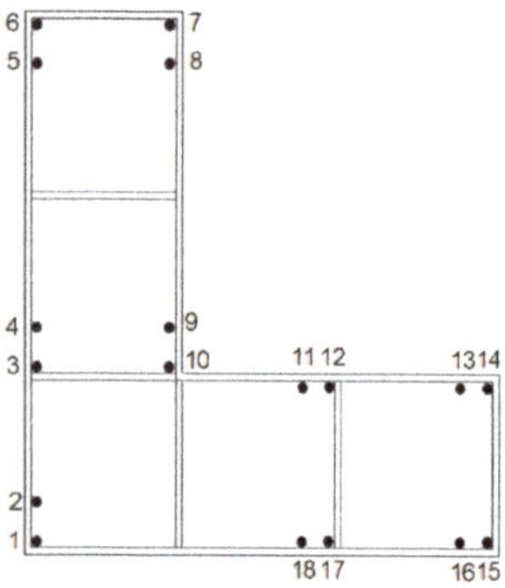

(a) Equal-limb specimens

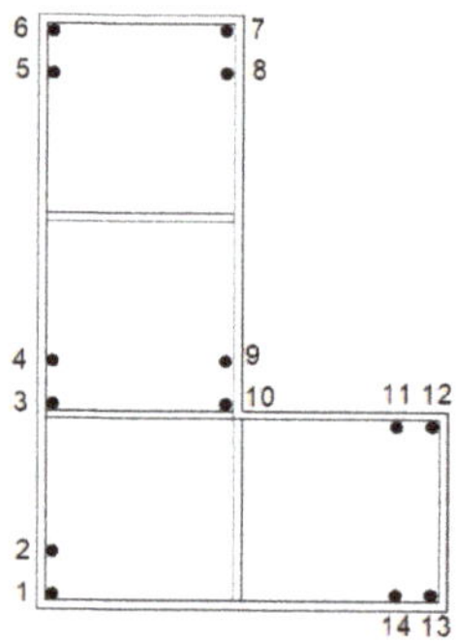

(b) Unequal-limb specimens

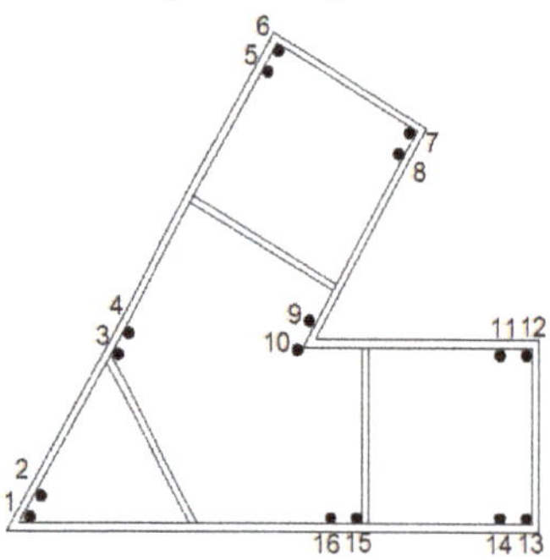

(c) ACD-1-60°-1200

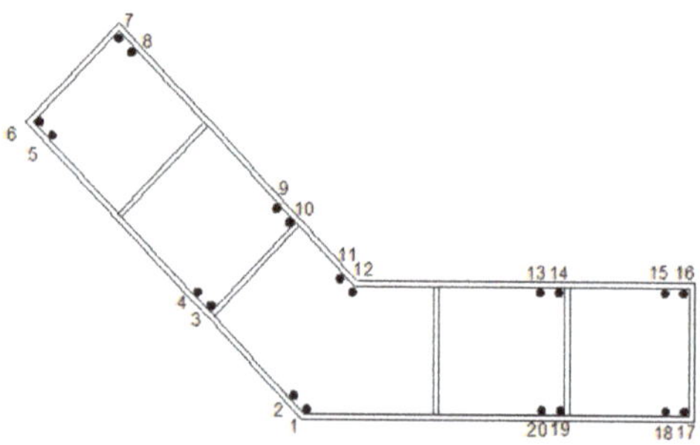

(d) ACD-1-135°-1200

Figure 13. Locations of the contact stress points.

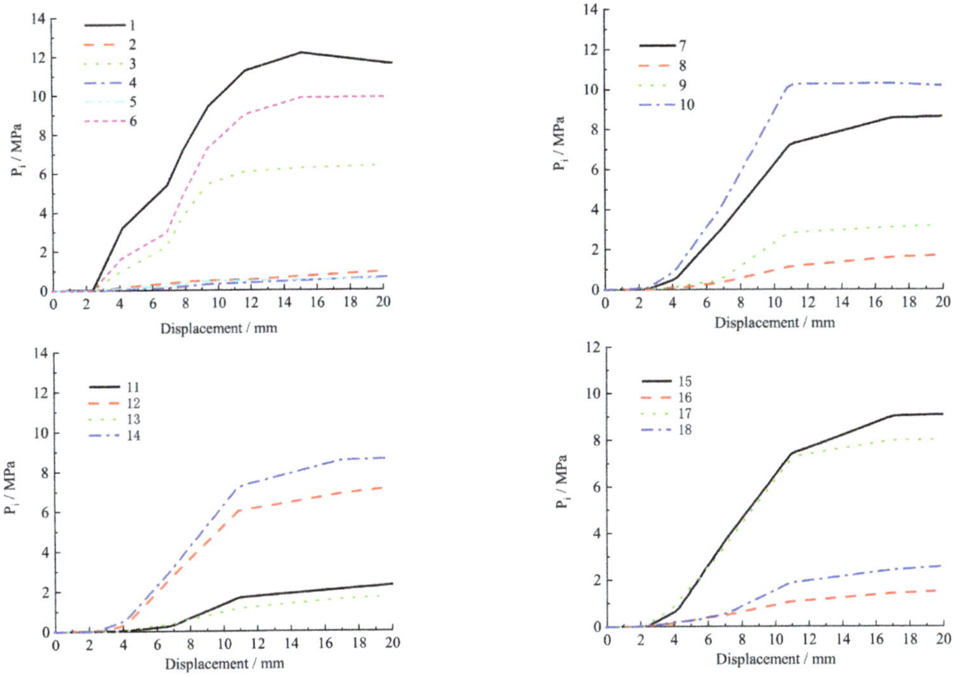

(a) ACD-1-90°-600

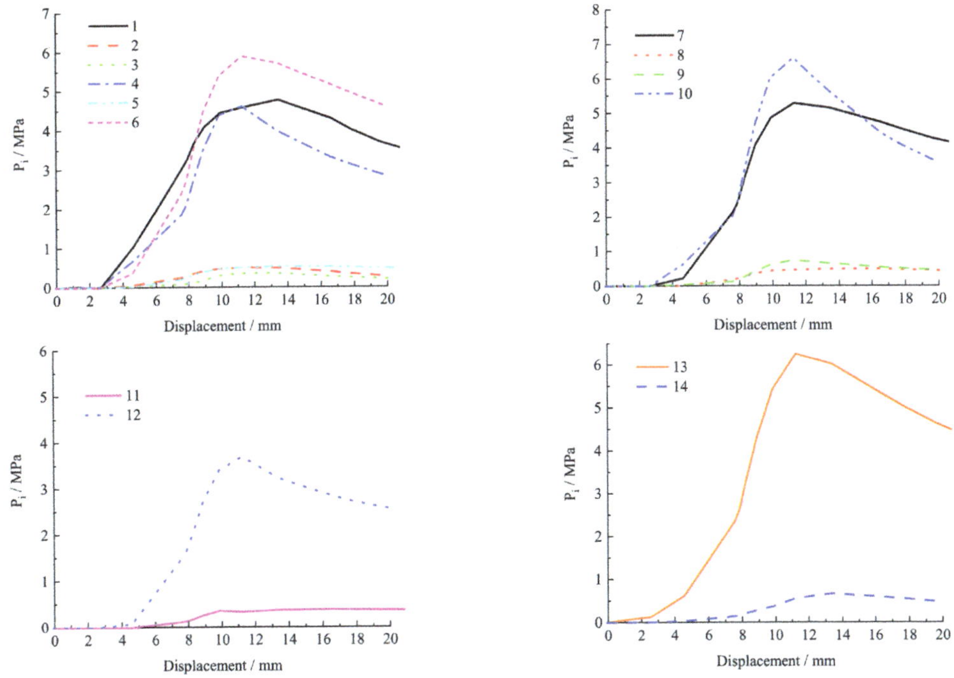

(b) AC-1.5-90°-600

Figure 14. *Cont.*

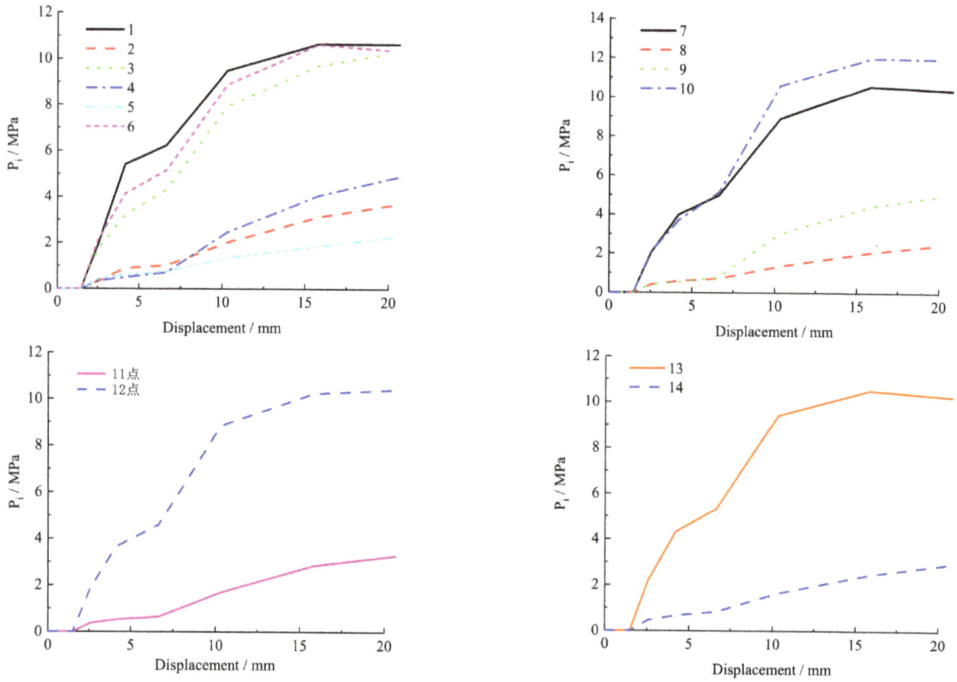

(**c**) AF-1.5-90°-600

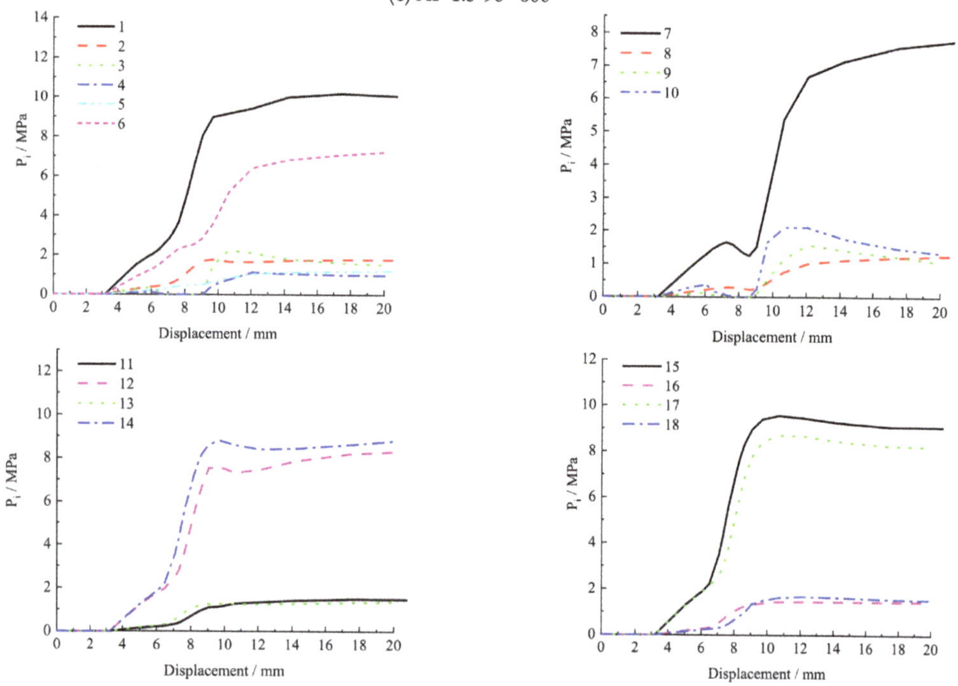

(**d**) ACD-1-90°-1200

Figure 14. *Cont.*

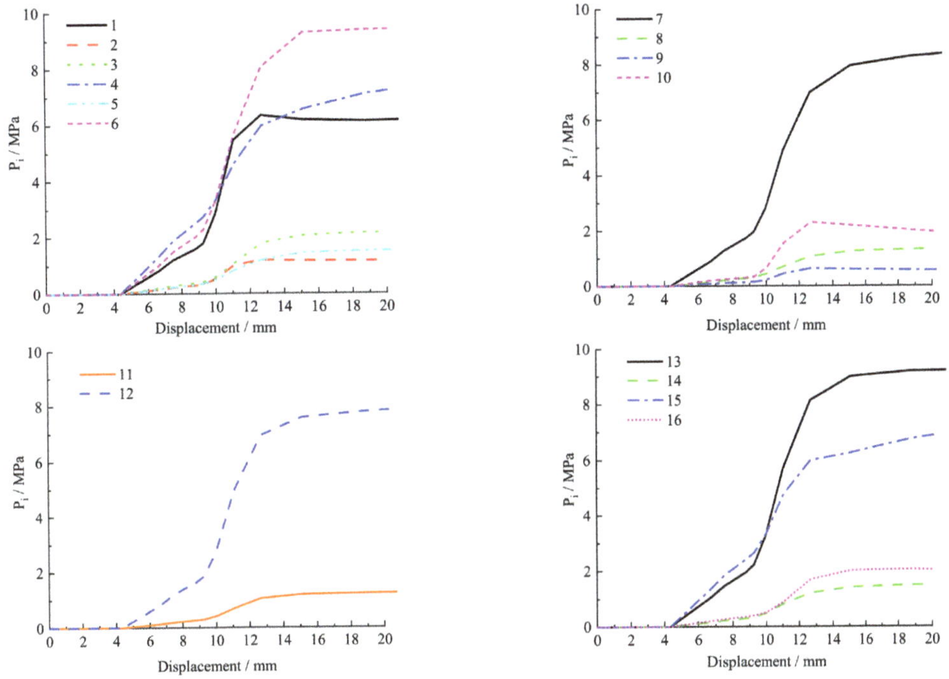

(e) ACD-1-60°-1200

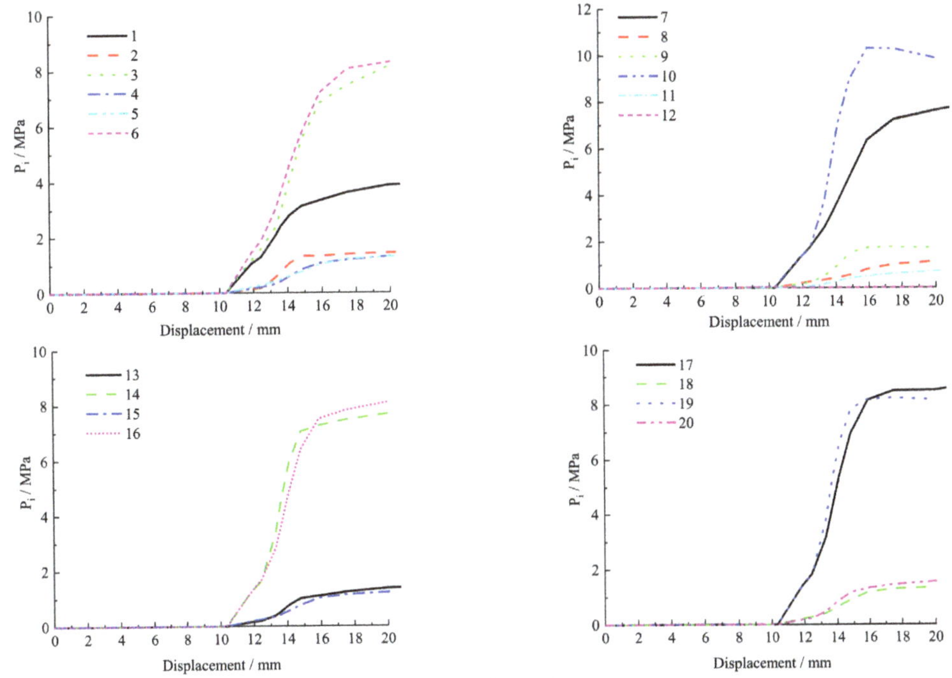

(f) ACD-1-135°-1200

Figure 14. P_i–Δ curves of the specimens.

According to Figure 14, there was no contact stress between the steel pipe and the concrete during the initial loading stage. The high Poisson ratio of steel pipes resulted in more lateral deformation when compared to concrete. Small fractures arose in the concrete when the axial displacement increased, and the lateral deformation of the concrete exceeded that of the steel pipe, at which point contact stress occurred.

When compared to specimen ACD-1-90°-600, the contact stress of AC-1.5-90°-600 significantly declines after reaching its peak, as does the peak contact stress of corner points (such as points 1, 6, 7, and so on). At the later loading stage, the stiffness and bearing capacity fall faster. Furthermore, unlike in the case of equal-limb specimens, contact stress near the corners did not increase further in the later stage of loading. The circumstance described above shows that unequal-limb specimens cannot successfully exert an interaction between steel pipes and concrete.

The contact stress at and near the corners of specimen AF-1.5-90°-600 is significantly higher than that of AC-1.5-90°-600, and even slightly higher than that of ACD-1-90°-600, indicating that the square steel tube combination can better exert the contact effect between steel pipes and concrete. The corner contact stress of ACD-1-90°-1200 shows a similar development trend to that of ACD-1-90°-600, with the exception of a little drop in the later stage of loading, demonstrating that column height has a minimal effect on the interaction between steel pipes and concrete.

When the angle between the two limbs decreases to 60°, the peak contact stress at the external corner point (point 1) decreases, while the contact stress at the internal corner point (point 10) is at a lower level, indicating that the triangular cavity has a weaker constraint effect on the concrete. When the angle is increased to 135°, the occurrence time of contact stress at each point is much later than that of other specimens, indicating that the concrete failure of the specimens is relatively slow. In addition, the contact stress of each cavity on the two flanges of the specimen is not significantly different, indicating that the internal forces of each cavity in the specimen are relatively uniform.

6. Parametric Analysis

It has been shown that steel strength and concrete strength have an effect on the mechanical properties of L-shaped CFST columns, including initial stiffness, load capacity, and ductility [8,11,14,23]. In addition, the results of Wang [15] showed that the limb length ratio has an effect on the stiffness and load capacity of L-shaped CFST columns under axial pressure. Wang [35] found that the mechanical properties of specially angled columns differ from those of right-angled columns.

This study conducted parameter analysis based on the FEM proposed above to examine the influence of different parameters on the axial compression behavior of L-shaped CFST columns. The parameters include steel strength, concrete strength, length ratio of long limb to short limb, the angle between the two limbs, and the combination method.

6.1. Steel Strength

Figure 15 depicts the load–displacement curves of L-shaped CFST columns composed of different steel strengths. There is no significant difference in the initial stiffness of the columns, and the peak bearing capacity increases as the steel strength increases, which is consistent with Wang's findings [15]. Therefore, increasing the strength of steel does not effectively control the elastic displacement of L-shaped CFST columns in practical engineering design.

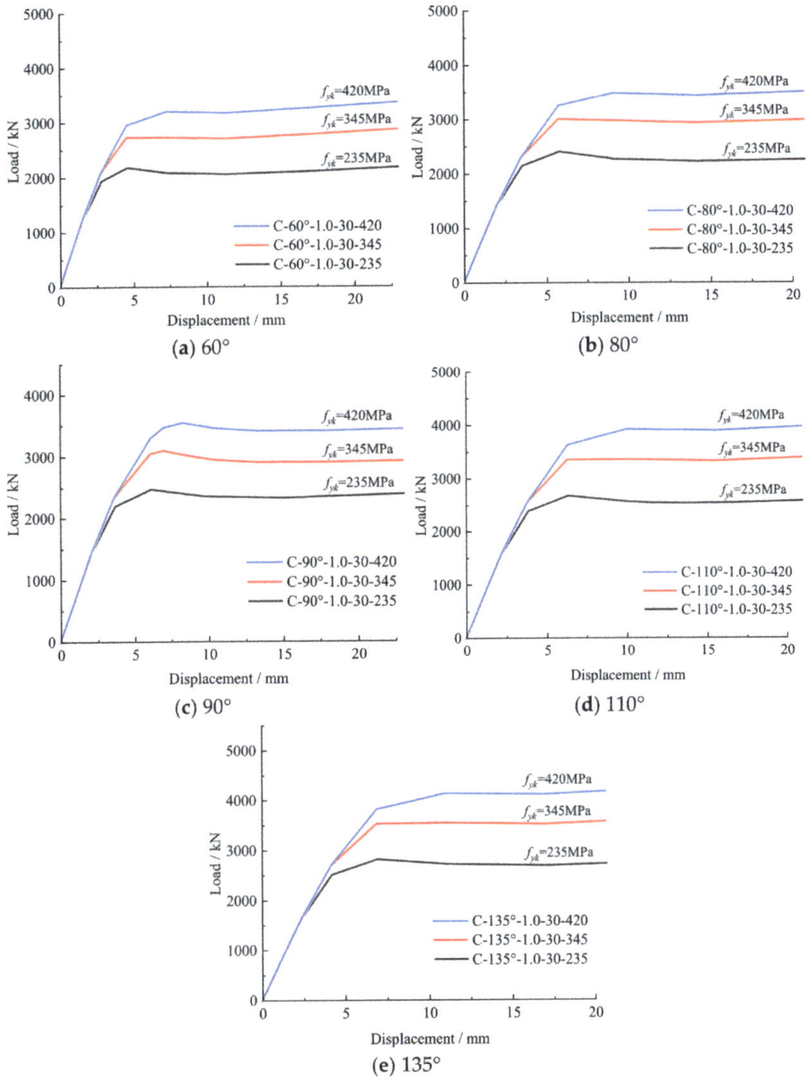

Figure 15. Load–displacement curves of specimens with different steel strength. Note: taking specimen C-60°-1.0-30-235 as an example, C represents the type of specimen; 60° represents the angle between the two limbs; 1.0 represents the length ratio of long limb to short limb; 30 represents the concrete strength; 235 represents the steel strength.

6.2. Concrete Strength

In the case of different angles between the two limbs, changing the concrete strength while keeping the other parameters unchanged, the load–displacement curves of each specimen can be obtained as shown in Figure 16. It can be seen that increasing concrete strength can significantly improve the initial stiffness of the specimen, and the peak bearing capacity of the specimen increases as the concrete strength increases, which is consistent with the conclusions in the literature [15]. In addition, the degree of reduction in the peak bearing capacity of the specimens increases with the increase in concrete strength. Therefore, this phenomenon should be noted if high-strength concrete is used in practical applications.

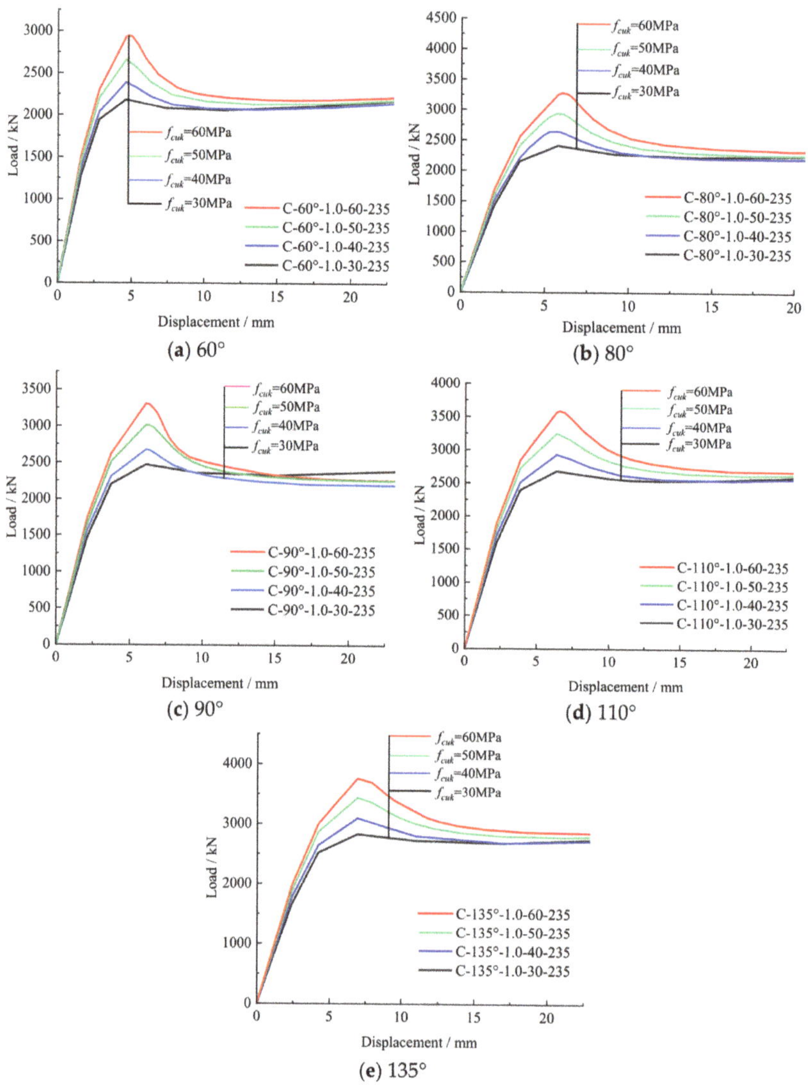

Figure 16. Load–displacement curves of specimens with different concrete strength.

6.3. Length Ratio of Long Limb to Short Limb

Figure 17 depicts the load–displacement curves of specimens with different angles between the two limbs and limb length ratios of 1.0, 1.2, 1.5, 1.7, and 2.0. It can be seen that, as the limb length ratio increases, the initial stiffness and the peak bearing capacity of the specimen both decrease. When the limb length ratio reaches 1.5, the loss in peak bearing capacity increases abruptly due to the reduction in one cavity in the short limb. Notably, there was no significant change in the peak displacement of specimens with different limb length ratios. Based on the aforementioned results, it can be concluded that an increase in limb length ratio has a significant impact on the initial stiffness and peak bearing capacity of L-shaped CFST columns, which is in line with the findings in the literature [15]. To guarantee superior mechanical qualities, L-shaped steel pipe concrete columns should have equal limb sections in real engineering design.

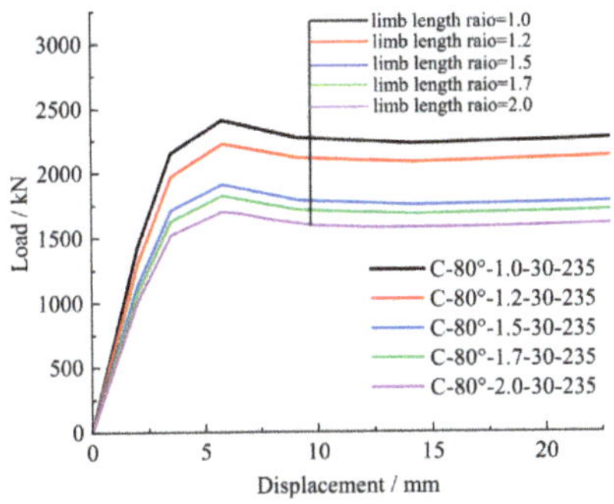

(**a**) 80°

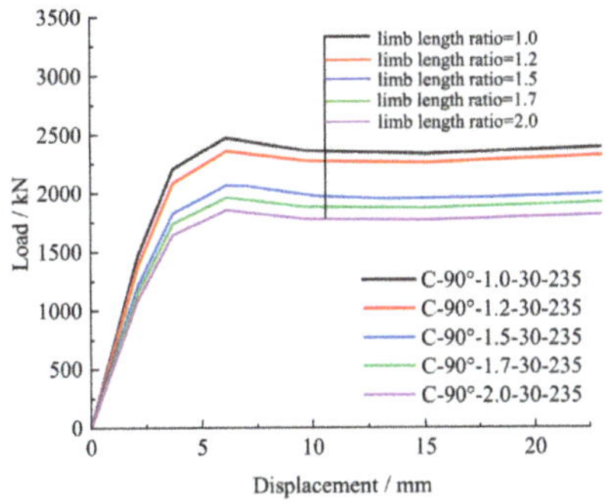

(**b**) 90°

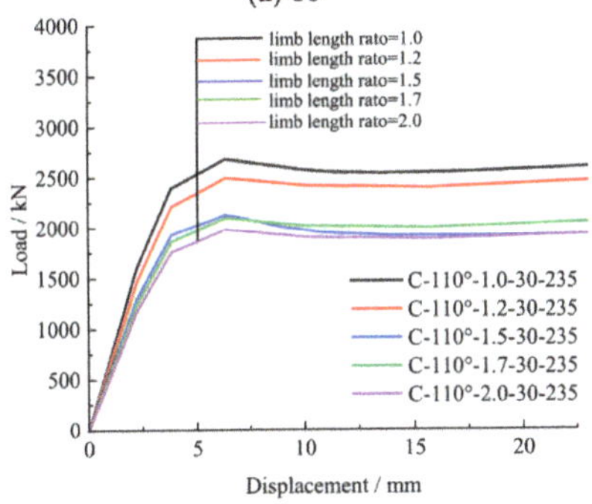

(**c**) 110°

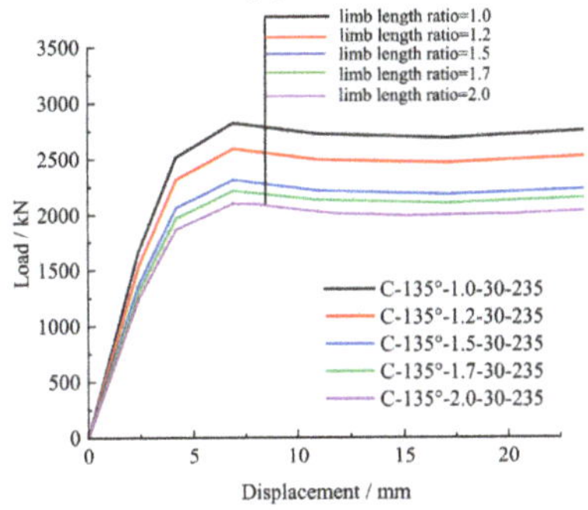

(**d**) 135°

Figure 17. Load–displacement curves of specimens with different length ratios of long limb to short limb.

6.4. Angle between Two Limbs

Figure 18 depicts the impact of the angle between the two limbs on the load–displacement curves of specimens with different limb length ratios. It can be inferred from the figure that specimens with different included angles have similar initial stiffness. The peak bearing capacity of the specimen increases as the included angle grows, as does the peak displacement. The angle between the two limbs has no significant effect on the loss of peak bearing capacity of the specimens.

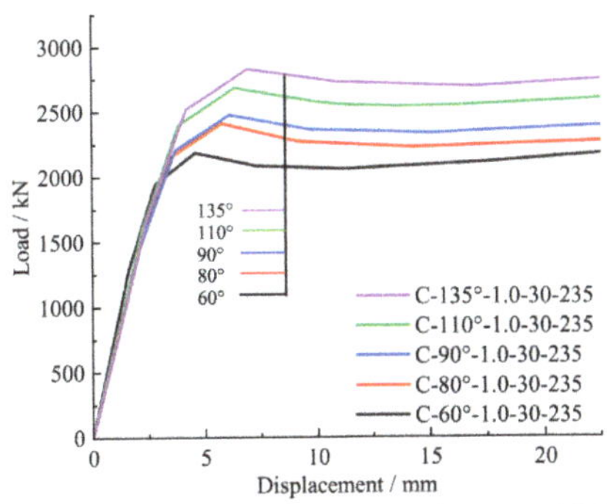

(**a**) Specimens with a limb length ratio of 1.0

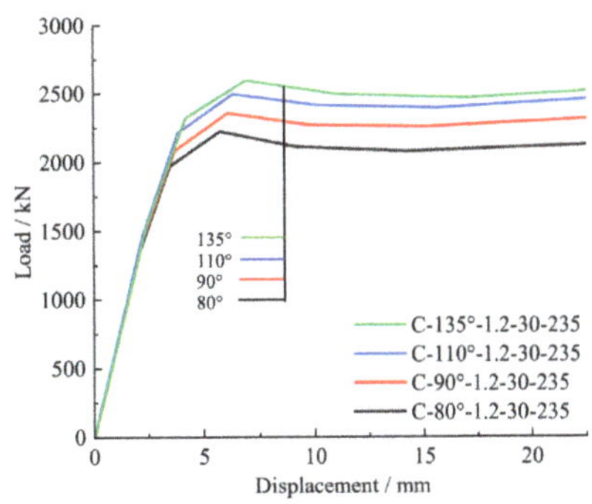

(**b**) Specimens with a limb length ratio of 1.2

Figure 18. *Cont.*

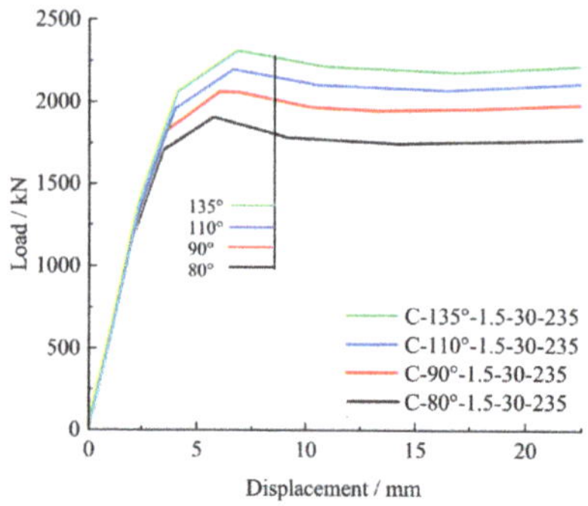

(**c**) Specimens with a limb length ratio of 1.5

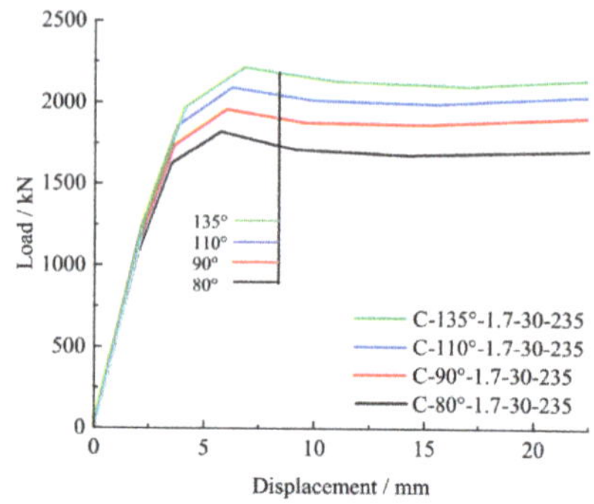

(**d**) Specimens with a limb length ratio of 1.7

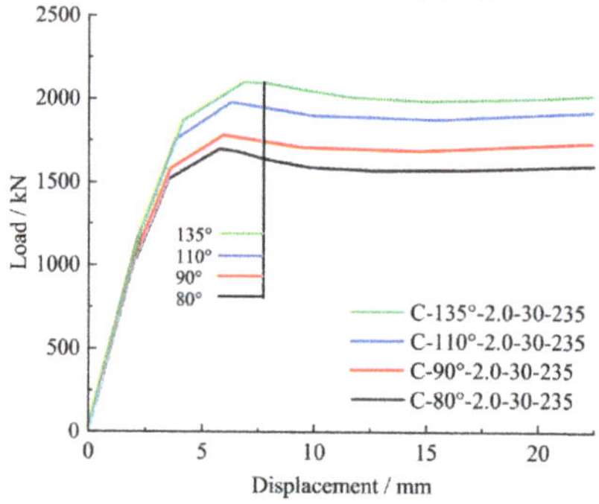

(**e**) Specimens with a limb length ratio of 2.0

Figure 18. Load–displacement curves of specimens with different angles between the two limbs.

6.5. Combination Method

Figure 19 displays the load–displacement curves for L-shaped CFST column specimens with different combinations. The figure shows that the F-type specimens have a much larger peak bearing capacity and slightly higher initial stiffness compared to the C-type specimens. This is because the steel pipe in F-type specimens has a stronger constraint effect on concrete, and the F-type specimens are less prone to adverse situations such as weld decay during loading. There is no discernible change in the peak displacement of specimens with different combinations and the rate of peak load reduction.

In order to investigate the sensitivity of the initial stiffness and peak load capacity of L-shaped CFST columns to parameter variations, the data from the above parametric analyses were analyzed in this paper as shown in Table 4. It is worth noting that the change rate of the initial stiffness is calculated as the ratio of the percentage change in the initial stiffness to the percentage change in the parameter. This relative ratio makes it easier to compare the impacts of various parameters quantitatively and consistently.

Table 4. Parameter impact analyses.

Parameter	Rate of Change in Initial Stiffness	Rate of Change in Peak Bearing Capacity
Steel strength	0.12%	64.41%
Concrete strength	17.92%	41.75%
Limb length ratio	−47.40%	−36.80%
Angle between two limbs	−2.63%	34.74%

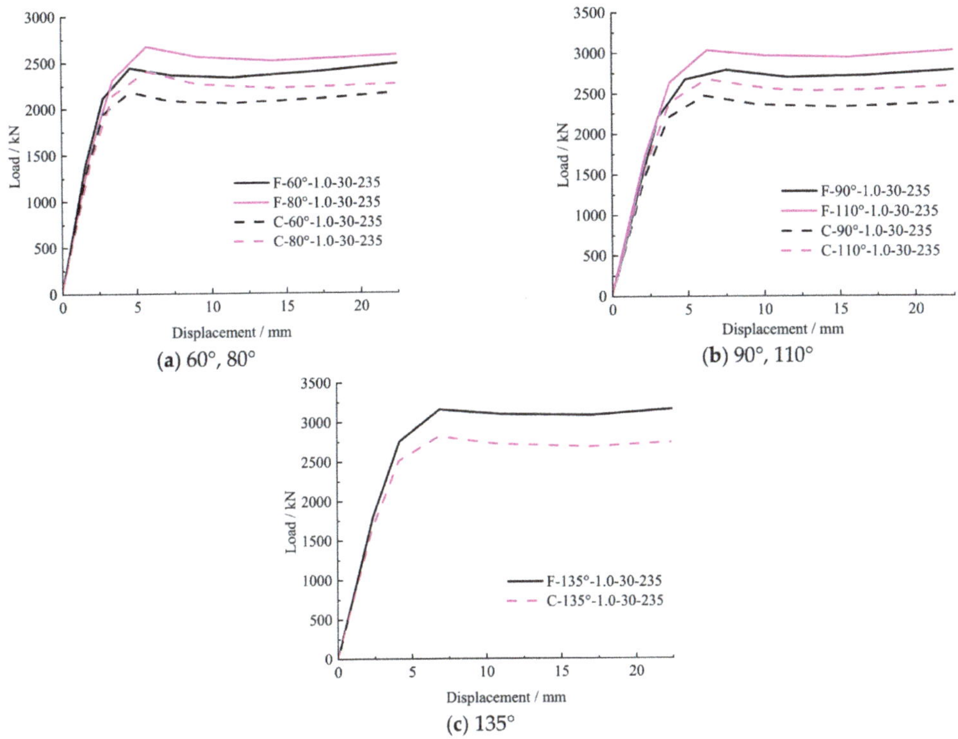

Figure 19. Load–displacement curves of specimens with different combinations.

Table 4 does not include an analysis of changes in combination methods since they are difficult to quantify numerically. Table 4 shows that limb length ratio and steel strength are the two key parameters affecting the initial stiffness and peak bearing capacity of the L-shaped CFST columns. Therefore, in practical engineering design, if the initial stiffness or peak bearing capacity of L-shaped CFST columns needs to be adjusted, priority should be given to changing the limb length ratio of the specimen or the strength of the steel.

7. Conclusions

This study employed ABAQUS to establish finite element models of L-shaped CFST columns with different combinations and compared the calculated results with the test results. The axial compression mechanism of L-shaped CFST columns was investigated in this study through the finite element models. Based on the comprehensive analysis conducted on parameters such as the length ratio of long limb to short limb, the angle between the two limbs, concrete strength, and steel strength, the following conclusions were obtained:

(1) The results obtained through the FEM are in good agreement with the test results in terms of axial compression load–displacement curves and failure modes. The maximum difference between the ultimate bearing capacity results of FEA and testing is 8.7%, with a standard deviation of 0.12, indicating that this modeling method can be used for analyzing the axial compression mechanism of L-shaped CFST columns.

(2) The angle between the two limbs has a significant impact on the stress distribution of concrete and steel pipes. The corner effect increases as the angle between the two limbs decreases. In the elastic–plastic stage, the concrete stress at the internal corner of the specimen with an included angle of 60° increases from 21.92 MPa to 24.00 MPa, and the stress at the external corner increases from 29.25 MPa to 34.14

MPa, and the maximum stress area at the external corner gradually decreases. As for the specimen with an angle of 135°, the stress at the internal corner decreases from 22.22 MPa to 20.53 MPa. The steel pipe stress of the specimen with an angle of 60° concentrates towards the end cavity, while the outer edge of the steel pipe is always in a condition of tensile stress due to the overall bending of the 135° specimen.

(3) In terms of the contact stress between steel pipes and concrete, the combination of F-type specimens can better impose the constraint effect of steel pipes on concrete. The triangular cavity of unequal-limb specimens and specimens with an included angle of 60° cannot effectively exert the interaction between steel pipes and concrete.

(4) The initial stiffness of the L-shaped CFST columns increases by 17.92% with increased concrete strength and decreases by 47.40% with increased limb length ratio. It is not sensitive to changes in steel strength or the angle between the two limbs. The peak bearing capacity increases by 64.41% and 41.75% with increased steel strength and concrete strength, respectively. Compared to C-type and Z-type specimens, the initial stiffness of F-type specimens is slightly higher, and its peak bearing capacity is significantly increased.

(5) L-shaped CFST columns exhibit favorable axial compressive behavior, and their research and engineering applications support the advancement of assembled steel structure residential buildings. In actual projects, F-type equal-limb specimens can be used in priority; for special-angled L-shaped CFST columns, the angle between the two limbs should be considered in terms of the specimen's corner effect and stress distribution, and certain strengthening measures should be taken.

Author Contributions: Conceptualization, Z.T.; Data curation, H.L. and D.H.; Formal analysis, H.L.; Funding acquisition, Z.T.; Investigation, D.H.; Methodology, H.L. and D.H.; Project administration, Z.T.; Resources, Z.T.; Supervision, Z.T.; Validation, H.L.; Visualization, H.L.; Writing—original draft, H.L. and D.H.; Writing—review and editing, Z.T. All authors have read and agreed to the published version of the manuscript.

Funding: This research was funded by the Key Research and Development Programs (Key R and D programs) department of the Science and Technology Department of Yunnan Province, the Faculty of Civil Engineering and Architecture, Kunming University of Science and Technology, number 202003AC100001.

Data Availability Statement: The data supporting this study's findings are available on request from the corresponding author.

Conflicts of Interest: Author Dongji Han was employed by the company 3rd Construction Co., Ltd. of China Construction 5th Engineering Bureau. The remaining authors declare that the research was conducted in the absence of any commercial or financial relationships that could be construed as a potential conflict of interest.

References

1. United Nations Environment Programme. *Emissions Gap Report 2020–Executive Summary: DEW/2310/NA*; UNEP: Nairobi, Kenya, 2020.
2. Hawkins, W.; Cooper, S.; Allen, S.; Roynon, J.; Ibell, T. Embodied carbon assessment using a dynamic climate model: Case-study comparison of a concrete, steel and timber building structure. *Structures* **2021**, *33*, 90–98. [CrossRef]
3. Su, X.; Zhang, X. A detailed analysis of the embodied energy and carbon emissions of steel-construction residential buildings in China. *Energy Build.* **2016**, *119*, 323–330. [CrossRef]
4. Zeitz, A.; Griffin, C.T.; Dusicka, P. Comparing the embodied carbon and energy of a mass timber structure system to typical steel and concrete alternatives for parking garages. *Energy Build.* **2019**, *199*, 126–133. [CrossRef]
5. Song, Z.; Luo, X.; Ge, J. Research on carbon emission in materialization stage of steel structure fabricated residence. *Archit. Technol.* **2023**, *54*, 158–162.
6. Li, Q.; Yue, Q.; Feng, P.; Xie, N.; Liu, Y. Development status and prospect of steel structure industry based on carbon peak and carbon neutrality target. *Prog. Steel Build. Struct.* **2022**, *24*, 1–6+23. [CrossRef]
7. Zong, D. Research on Design of Prefabricated Steel Structure Residence Based on DFMA. Ph.D. Thesis, Chongqing University, Chongqing, China, 2022.
8. Zuo, Z.-L.; Cai, J.; Yang, C.; Chen, Q.-J. Eccentric load behavior of L-shaped CFT stub columns with binding bars. *J. Constr. Steel Res.* **2012**, *72*, 105–118. [CrossRef]

9. Zhou, T.; Chen, Z.; Liu, H. Seismic behavior of special shaped column composed of concrete filled steel tubes. *J. Constr. Steel Res.* **2012**, *75*, 131–141. [CrossRef]
10. Shen, Z.-Y.; Lei, M.; Li, Y.-Q.; Lin, Z.-Y.; Luo, J.-H. Experimental Study on Seismic Behavior of Concrete-Filled L-Shaped Steel Tube Columns. *Adv. Struct. Eng.* **2013**, *16*, 1235–1247. [CrossRef]
11. Zheng, Y.; Yang, S.; Lai, P. Hysteretic behavior of multi-cell L-shaped concrete-filled steel tubular columns at different loading angles. *Eng. Struct.* **2020**, *202*, 109887. [CrossRef]
12. Du, G.; Zhang, J.; Li, Y.; Zhang, J.; Zeng, L. Experimental study on hysteretic model for L-shaped concrete-filled steel tubular column subjected to cyclic loading. *Thin-Walled Struct.* **2019**, *144*, 106278. [CrossRef]
13. Chen, H.; Wang, L.; Chen, H.; Cui, W. Experimental study on the seismic behavior of prefabricated L-shaped concrete-filled steel tube with rectangular multi-cell columns under different lateral loading directions. *J. Constr. Steel Res.* **2021**, *177*, 106480. [CrossRef]
14. Zhang, W.; Chen, Z.; Xiong, Q. Performance of L-shaped columns comprising concrete-filled steel tubes under axial compression. *J. Constr. Steel Res.* **2018**, *145*, 573–590. [CrossRef]
15. Wang, Y.; Guo, L.; Li, H. L-shaped steel-concrete composite columns under axial load: Experiment, simulations and design method. *J. Constr. Steel Res.* **2021**, *185*, 106871. [CrossRef]
16. Wang, Z.; Liu, Z.; Zhou, X.J. Experimental Investigation of Special-Shaped Concrete-Filled Square Steel Tube Composite Columns with Steel Hoops under Axial Loads. *Materials* **2022**, *15*, 4179. [CrossRef]
17. Zhang, L.; Zhang, Y.J.; Tong, J.Z.; Tong, G.S.; Fu, B.; Xu, Y.M. Experimental study on multi-cellular L-shaped CFST slender columns under biaxial eccentric compression. *Structures* **2022**, *45*, 2210–2225. [CrossRef]
18. Zhou, T.; Xu, M.Y.; Chen, Z.H.; Wang, X.D.; Wang, Y.W. Eccentric loading behavior of L-Shaped columns composed of concrete-filled steel tubes. *Adv. Steel Constr.* **2016**, *12*, 227–244. [CrossRef]
19. Zheng, Y.; Zeng, S. Design of L-shaped and T-shaped concrete-filled steel tubular stub columns under axial compression. *Eng. Struct.* **2020**, *207*, 110262. [CrossRef]
20. Lei, M.; Shen, Z.; Li, Y.; Luo, J. Behavior of Concrete-filled T-shaped Steel Tube Intermediate Long Columns Subjected to Axial Loading. *J. Tongji Univ. Nat. Sci.* **2016**, *44*, 520–527.
21. Lei, M.; Shen, Z.; Li, Y.; Luo, J. Sectional Strength of Concrete-filled T-shaped Steel Tube Column Subjected to Axial Compression and Bending Moment. *J. Tongji Univ. Nat. Sci.* **2016**, *44*, 348–354.
22. Lei, M.; Shen, Z.; Li, Y.; Luo, J. Analysis of Behavior of Concrete-filled T-shaped Steel Tube Tall Column Subjected to Uniaxial Eccentric Load. *J. Tongji Univ. Nat. Sci.* **2016**, *44*, 207–212.
23. Liu, X.; Xu, C.; Liu, J.; Yang, Y. Research on special-shaped concrete-filled steel tubular columns under axial compression. *J. Constr. Steel Res.* **2018**, *147*, 203–223. [CrossRef]
24. Liu, J.; Zhou, T.; Lei, Z.; Chen, X.; Chen, Z. Eccentric compression performance of slender L-shaped column composed of concrete-filled steel tubes connected by steel linking plates. *J. Constr. Steel Res.* **2019**, *162*, 105675. [CrossRef]
25. Xiao, Y.; Bie, X.-M.; Song, X.; Zhang, J.; Du, G. Performance of composite L-shaped CFST columns with inner I-shaped steel under axial compression. *J. Constr. Steel Res.* **2020**, *170*, 106138. [CrossRef]
26. Zhang, W.; Li, G.; Xiong, Q.; Gui, H. Seismic behavior of wide-limb special-shaped columns composed of concrete-filled steel tubes. *J. Constr. Steel Res.* **2023**, *205*, 107887. [CrossRef]
27. Chen, X.; Zhou, T.; Chen, Z.; Liu, J.; Jiang, B. Mechanical properties of special-shaped concrete-filled steel tube columns under eccentric compression. *J. Constr. Steel Res.* **2020**, *167*, 105779. [CrossRef]
28. Liu, L.; Tu, Y.; Ye, Y. Finite element analysis of L-shaped concrete filled steel tubular column based on ABAOUS. *J. Shenyang Univ. Technol.* **2011**, *33*, 349–354.
29. Tu, Y.; Wen, Q. Bearing capacity calculation of L-shaped concrete filled steel tubular columns subjected to axial compression. *J. Build. Struct.* **2013**, *34*, 314–320. [CrossRef]
30. Huang, H.; Wang, Y.; Zhang, L.Q.; Chen, M.C.; Zhu, Y.Q. Study on the mechanical behavior of L-shaped concrete-filled steel tubular columns under axial compression. *Struct. Concr.* **2023**, *24*, 2475–2489. [CrossRef]
31. Chen, Z.H.; Xu, J.H.; Zhou, T.; Su, J. Seismic research on column base joint of L-shaped CFST columns under cyclic loading. *Structures* **2022**, *45*, 1212–1224. [CrossRef]
32. Zheng, Y.Q.; Liang, W.G.; Ma, S.S.; Zeng, S.X. Behavior of stiffened and multi-cell L-shaped CFST columns under eccentric compression. *Thin-Walled Struct.* **2022**, *174*, 109156. [CrossRef]
33. Han, L.-H. *Concrete Filled Steel Tubular Structures from Theory to Practice*, 3rd ed.; Science Press: Beijing, China, 2016.
34. Wang, F.-C.; Han, L.-H. Analytical behavior of special-shaped CFST stub columns under axial compression. *Thin-Walled Struct.* **2018**, *129*, 404–417. [CrossRef]
35. Wang, J.; Tian, C. Seismic Behavior Study of 135 Degree Shaped Column. *World Sci-Tech R D* **2013**, *35*, 357–359.

buildings

MDPI

Article

Experimental Investigation and Numerical Analysis of the Axial Load Capacity of Circular Concrete-Filled Tubular Columns

Marija M. Lazović Radovanović * and Jelena Z. Nikolić

Faculty of Civil Engineering, The University of Belgrade, Bulevar Kralja Aleksandra 73, 11000 Belgrade, Serbia; jnikolic@grf.bg.ac.rs
* Correspondence: mlazovic@grf.bg.ac.rs

Abstract: This paper focuses on the experimental investigation of the axial load capacity of CFT (concrete-filled steel tube) columns under actual construction conditions during building reconstruction. A total of four samples were loaded up to failure. The varied parameters were the column length and absence/presence of the concrete infill within the steel tube. Further, the analysis is extended to developing a numerical model in the finite element-based software ABAQUS version 6.9. This numerical model includes material and geometrical nonlinearities and was validated with the experimental results. The contribution of the concrete core to the column capacity and the concrete core confinement effect are discussed. Finally, the column capacity was calculated according to several design codes: the Eurocode 4 with and without considering the confinement effect, American specifications, Australian standards, the American Institute of Steel Construction, and the Architectural Institute of Japan. The Eurocode 4 considering the confinement effect provides the closest results to those obtained in the tests.

Keywords: CFT columns; experimental tests; nonlinear analysis; numerical simulations

1. Introduction

Due to their constructive advantages, columns derived from steel hollow profiles filled with concrete (CFT columns) have found wide application in engineering practice. Compared to traditional concrete or steel columns, the advantages of these columns are numerous: higher load capacity, rigidity and ductility, and savings in space, material, and construction time.

Determining the ultimate bearing capacity of CFT columns is challenging due to many parameters which need to be investigated. First, the nonlinear behavior of the constitutive materials, steel and concrete, needs to be defined for application in numerical analysis. Further, the study should include the steel profile's second-order influences, geometric imperfections, and residual stresses. Calculating the bearing capacity of CFT columns by current design codes and provisions uses simplified methods and is based on the calculation according to limit states.

In recent years, many authors have published works on the experimental and numerical analysis of the bearing capacity of CFT columns. Numerical analyses are mostly conducted using available finite element method (FEM) software. The majority of studies employed ABAQUS, as in [1–4], while certain studies utilized ANSYS [5–7] or LS-DYNA [8–10]. Further, many researchers provided a comparative analysis of their results with current design codes. The accuracy of calculating the bearing capacity of CFT columns is significantly affected by the adopted constitutive models for concrete and steel. Whole series of proposals for these relationships that more or less accurately describe the behavior of concrete and steel can be found in the literature [3,11–17]. Schneider [11] analyzed the influence of the section shape of the hollow steel profile and the steel quality on the CFT column load capacity on fourteen samples.

Ehab Ellobody et al. [3] proposed constitutive models for confined concrete and steel for ABAQUS use. The numerically obtained results of the axial bearing capacity of CFT columns are compared with the results of experiments, as well as with the following design codes: EC4, ACI, and AISC. Giakoumelis and Lam [18] examined the effect of wall thickness, bond behavior of the hollow steel profile and concrete core, concrete shrinkage, and creep on the axial load capacity of circular CFT columns. In the paper [13], Hu proposed a constitutive model for the concrete core inside the CFT column, separately for circular and square sections. Liang and Fragomeni [19] presented constitutive models for the concrete and steel of normal- and high-strength materials. They conducted a parametric analysis to determine the influence of concrete confinement by the steel profile, the diameter-to-wall thickness ratio of the steel profile D/t, and the impact of concrete strength and steel quality on the bearing capacity of the CFT column of the circular section. Li et al. [20] analyzed the fundamental behavior of high-strength concrete-filled high-strength square steel tubular (HCFHSST) long columns under eccentric compression. Lin et al. [16] developed a new fiber beam element (FBE) model that applies to both small and large CFT and slender columns. Finally, attention should also be directed towards exploring new solutions and approaches, including innovative materials and sustainable solutions [21–25].

The CFT column's slenderness, defined as the column length-to-dimeter ratio L/D, significantly influences its bearing capacity [26–29]. In the case of short columns ($L/D \leq 3.00$) [26, 30] the loss of bearing capacity may occur due to failure in the concrete material or yield of the steel profile.

The concrete core confinement effect-related research can be found in [18,19,31–34]. During the axial load, the inner concrete core gets strengthened by an outer steel profile that affects the bearing capacity of a composite column. This confinement decreases with the increased diameter-to-profile thickness ratio D/t, or with concrete strength, column slenderness, and load eccentricity. On the contrary, it rises with higher steel quality.

Oliveira et al. [26] presented experimental results from 32 axially loaded CFT columns, where the load was applied solely on the concrete core to assess the passive confinement provided by the steel tube. Abed et al. [35] conducted experimental and numerical studies on 16 CFT specimens, identifying the most influential factor on compressive behavior as D/t. Tao et al. [1] proposed an expression for determining lateral pressure in circular columns based on numerical tests with varying parameters. Han et al. [36] introduced the confinement factor, which relies on the cross-sectional areas and strengths of steel and concrete. However, instructions in the literature regarding when and how to consider confinement remain ambiguous today. This is partly due to challenges in direct measurement in experiments, leading some authors to provide algorithms for estimating this strengthening effect.

When creating a numerical model of the CFT column in ABAQUS, the bond behavior of two surfaces in a composite element should be defined. This contact is formed at the connection of the inner steel profile surface and the outer concrete core surface. This connection can be modeled using gap elements representing contact behavior in the longitudinal and tangential directions. It can be defined by setting the appropriate friction coefficient in the range of 0.2–0.6 [3,13,19]. The bond between the concrete core and the hollow steel profile decreases with increasing concrete strength [26,27]. This phenomenon occurs since higher-strength concrete has more significant shrinkage [3].

This research aimed to perform experimental tests on CFT columns on an actual reconstruction site to validate previous design calculations. Generally, there is a need for studies dealing with an empirical investigation conducted in conditions that may significantly differ from those achieved in laboratory settings. Moreover, this data may be valuable for validating other laboratory-performed tests or numerically obtained results. Finally, this study aimed to define a numerical 3D model which would successfully predict the structural behavior of a circular section's axially loaded CFT column with reasonable accuracy. Nonlinear numerical analysis was performed in ABAQUS [37] using the proper concrete core and steel profile constitutive models, and aimed to include concrete confinement. The results of the study were validated with test results and compared with the relevant design codes: the EN-1994-1-1 (EC4) with and without considering the confinement effect [38], American specifications (ACI) [39], Australian standards (AS) [40,41], the American Institute of Steel Construction (AISC) [42], and the Architectural Institute of Japan (AIJ) [43]. Besides, design codes offer diverse instructions and criteria on how to account for the confinement effect. For instance, EC4 utilizes relative column slenderness, whereas AIJ employs the length-to-diameter (L/D) ratio.

The study introduces an experimental program designed to assess the axial load capacity of CFT columns under real construction conditions. The primary objective was to evaluate the effectiveness of concrete infill and steel tube confinement in strengthening the columns. Key parameters under investigation included column length and the impact of concrete infill on axial load capacity. The study tested the axial bearing capacity of four specimens: two hollow steel profiles with a D/t ratio of 101.6/2.7 mm and lengths of 0.5 m and 1.0 m, and two CFT columns consisting of identical steel profiles filled with C25/30 concrete. This approach allowed for measurement of the contributions of concrete infill and the calculation of steel tube confinement.

2. Experimental Research Program

2.1. Test Setup

The axial load capacity of steel and CFT columns was tested using a particular constructed set (Figure 1). It consists of a steel frame 2400 × 600 × 600 mm with two hydraulic 200 mm diameter presses. Hydraulic presses are driven by two electric motors (power 7 kW and 10 kW) with two hydraulic pumps (for lower and higher pressure). The total capacity of the press is 300 bar. The force was continuously measured using the C6A2MN dose device, which can measure a load of up to 2000 kN with an accuracy of ±0.1 kN. The column shortening was continuously measured with the W100 displacement gauge (inductive standard displacement transducer), which can measure the vertical displacement or shortening of the column with an accuracy of ±0.01 mm. Four gauges are placed on the top of the circular specimen in two perpendicular directions.

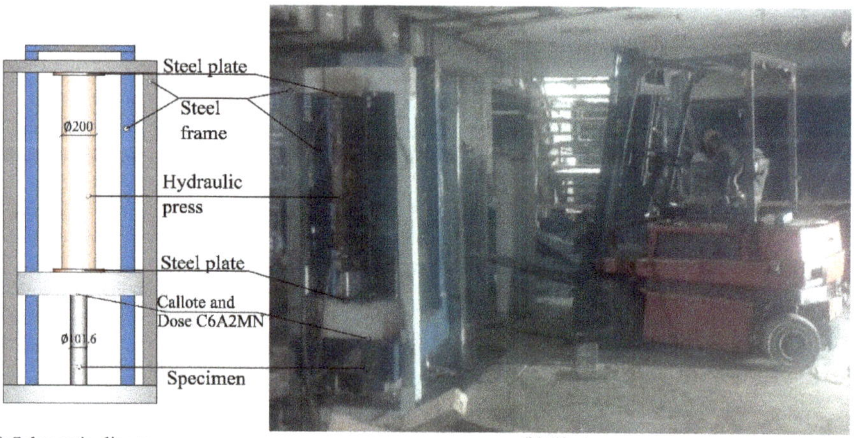

(a) Schematic diagram (b) Photo

Figure 1. Testing set for applying the axial load to CFT columns.

During the column's concreting, the concrete core was left 1–2 cm longer than the steel profile and finely aligned with a concrete grinder on the testing day. The load was applied in steps via a rigid plate welded to the bottom and top ends of the steel column and the CFT column. This way, the load was applied simultaneously to the steel and concrete part of the column cross-section. The load applied in the tests was a quasistatic monotonically increasing centric axial load. A calotte device was used to achieve the centric introduction of the load. The loading process can be considered short-term as each sample's failure took about 3 min.

The axial load capacity of four specimens was tested: two hollow steel profiles denoted as S_1 and S_2 and two CFT columns denoted as C_1 and C_2, as in Table 1. The ultimate load capacity of the steel columns and the corresponding CFT columns was analyzed, and the concrete infill's contribution to the load capacity was eventually determined.

Table 1. Specimen details.

Specimen	L	D	t	$\overline{\lambda}$	N_{TEST}	N_{FEM}	N_{FEM}/N_{TEST}
-	[m]	[mm]	[mm]	-	[kN]	[kN]	[%]
S_1	0.5	101.6	2.7	0.094	315.78	329.10	-
S_2	1.0	101.6	2.7	0.187	297.84	328.09	-
C_1	0.5	101.6	2.7	0.113	701.71	697.52	99.40
C_2	1.0	101.6	2.7	0.227	555.57	568.65	102.35

The specimen D/t ratio of all specimens was 37.63. The steel was S355 quality, and its characteristics were taken from the material specification. CFT column specimens C_1 and C_2 have the same length and steel characteristics as their empty reference columns S_1 and S_2. The concrete infill was C25/30 with a three-fraction granulometric composition made at the concrete plant. The specimen testing was performed on the construction site. The old "NAPRED" building in Belgrade was reconstructed using specifically CFT columns. The specimens were concreted in several layers and slumped with a metal rod. The concrete compressive strength was checked for three control cylinder samples 150 mm·300 mm, and the mean value f_c' was 30.5 MPa.

A rigid steel plate at the bottom end of the specimens was also welded to the testing set. This way, the displacements and rotations of the samples are prevented by simulating fixed-end conditions. A rigid steel plate at the upper end of the specimen was also welded to the inner frame of the testing set to avoid rotation but allow movement in the loading direction. According to EC4 [38], the relative slenderness of the column was calculated using the following expression:

$$\overline{\lambda} = \sqrt{\frac{N_{pl,Rk}}{N_{cr}}} \tag{1}$$

where $N_{pl,Rk}$ is the characteristic value of the plastic resistance of the composite section to the compressive normal force, and N_{cr} is the elastic critical normal force. The relative slenderness of the specimens is also shown in Table 1.

2.2. Failure Modes and Results

The axial load capacities of the four specimens obtained in tests N_{test} are reported in Table 1. Figure 2a shows the 0.5 m long hollow specimen S_1 and CFT column C_1 failure pattern. In these tests, the failure occurred due to the yielding of the steel profile and outward local buckling at both ends of the steel tube. Figure 2b shows the 1.0 m long hollow specimen S_2 and CFT column C_2 after testing. Here, the failure also occurred due to the outward local buckling of the steel profile at both ends of the column. The tests concluded once the pressure on the manometer began to decrease. At the same time, significant deformations in the steel tube were noted. Additionally, post-failure, the steel tubes of the CFT column specimens C_1 and C_2 were sectioned to inspect the inner concrete core. It was affirmed that the failure resulted from the yielding of the steel tube, as no evidence of concrete crushing was detected within the core. For longer specimens, besides the yielding of the steel tube, the signs of the overall column's buckling could also be observed.

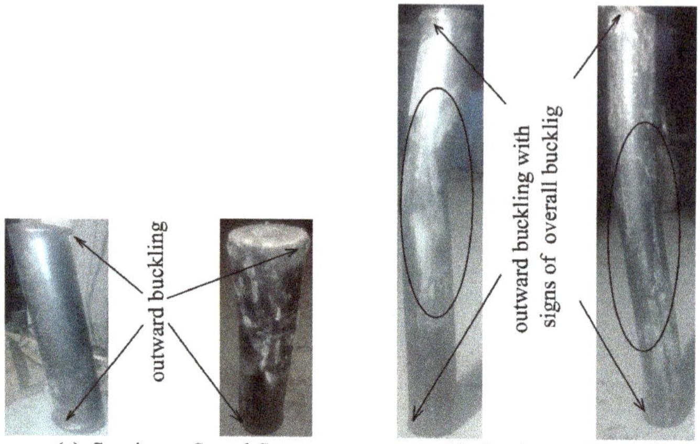

(a) Specimens S_1 and C_1 (b) Specimens S_2 and C_2

Figure 2. Specimens after failure.

3. Numerical Modeling of Axial Bearing Capacity of CFT Columns Using FEM

3.1. General

The axial load capacity of circular CFT columns was determined using the finite element method-based software ABAQUS version 6.9 [37] (Figure 3). Three-dimensional C3D8R finite elements (8-node linear brick, reduced integration with hourglass control) were used for the concrete core. S4R finite shell elements (4-node general-purpose shell, reduced integration with hourglass control, finite membrane strains) were used to model the steel profile. An element size of 10 mm was chosen for both concrete and steel components based on the mesh sensitivity analysis [2]. The material nonlinear stress–strain relations are included within the constitutive materials. Finally, the connection of the steel profile with the concrete core was modeled using "surface-to-surface contact" elements [37].

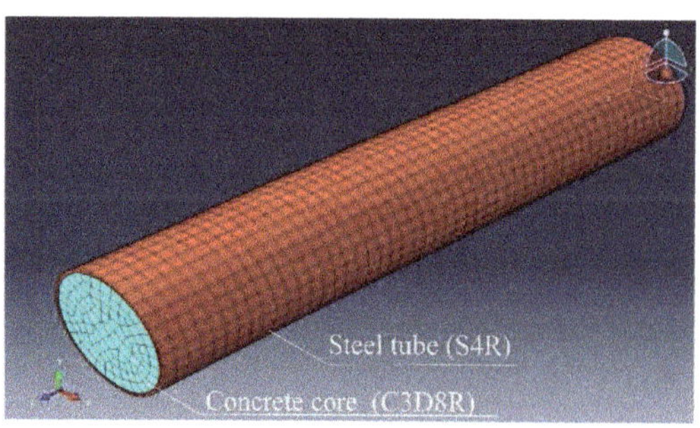

Figure 3. CFT column 3D model and meshing.

The nonlinear analysis uses Newton's direct method with a maximum of ten increments. The initial increment size is 0.1, the minimum increment size is 0.0001, and the maximum is 1.0. A linear extrapolation of the previous increment is adopted at the beginning of the next increment. The load was applied as a surface load in steps equivalent to the test's applied pressure. The boundary conditions match the conditions from the experiment. At the bottom end of the CFT column, translation and rotation in all three directions are restrained, representing a rigid fastening. To apply the axial load, only vertical translation of the column is allowed at the upper end of the CFT column.

3.2. Constitutive Model for Concrete

The constitutive model for concrete significantly influences the accuracy of the obtained results. The model applied in this study was recommended by [3] (Figure 4), and it presents stress–strain (σ–ε) diagrams for both unconfined and confined concrete. The confined concrete model incorporates an increase in peak compressive concrete strength as well as improved ductility due to the presence of the steel casing.

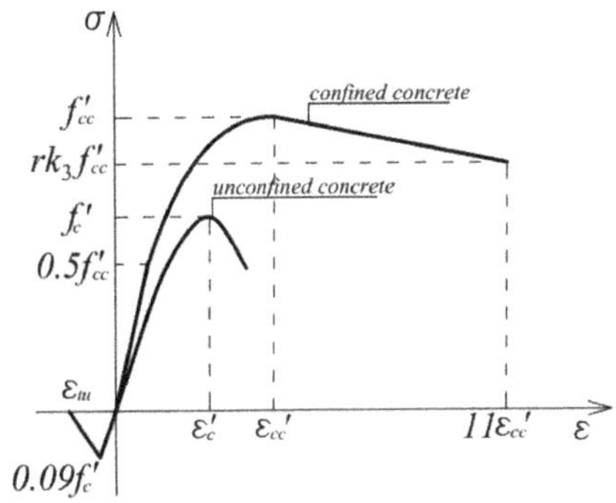

Figure 4. Stress–strain diagram for confined and unconfined concrete.

The uniaxial compressive strength of unconfined concrete $f_c{}'$ was obtained in control tests as 30.5 MPa. The corresponding strain $\varepsilon_c{}'$ is taken as 0.003. The stress–strain relation of unconfined concrete is taken as linear until the value of $0.5{\cdot}f_c{}'$. The modulus of elasticity of concrete E_c is determined according to EN 1992-1-1 [44] based on the following expression:

$$E_c = 22,000 \cdot \left[\frac{f_c{}' + 8}{10}\right]^{0.3} \text{[MPa]} \tag{2}$$

The nonlinear part of the σ–ε diagram for unconfined concrete beyond $0.5{\cdot}f_c{}'$ is defined by the following equation, initially proposed by the author [45]:

$$\sigma = \frac{E_c \cdot \varepsilon}{1 + (R + R_E - 2) \cdot \left(\frac{\varepsilon}{\varepsilon_c{}'}\right) - (2 \cdot R - 1) \cdot \left(\frac{\varepsilon}{\varepsilon_c{}'}\right)^2 + R \cdot \left(\frac{\varepsilon}{\varepsilon_c{}'}\right)^3} \tag{3}$$

where R_E and R are coefficients calculated as follows:

$$R_E = \frac{E_c \cdot \varepsilon_c{}'}{f_c{}'} \tag{4}$$

$$R = \frac{R_E \cdot (R_\sigma - 1)}{(R_\varepsilon - 1)^2} - \frac{1}{R_\varepsilon} \tag{5}$$

where R_σ and R_ε in (5) equal 4 [46].

The tensile strength of concrete is taken as approximately 9% of the compressive strength of concrete $f_c{}'$. After reaching the peak value, the tensile stress decreases to zero at the strain of 0.001 [37].

The confinement effect significantly influences the axial bearing capacity of short circular CFT columns [30,47]. At the initial loading level, the influence of the steel part of the section on the concrete part is small, considering that steel has a higher Poisson coefficient. At an axial strain of approximately 0.001, micro-cracks appear in concrete, with the transverse strains in the concrete growing rapidly [48]. With a further increase in the load, the concrete gets strengthened by the steel profile. Consequently, the steel part of the section is in a biaxial stress state, while the concrete is in a triaxial stress state. It is known that the compressive strength of concrete increases in this condition. When this strengthening effect is considered, the CFT column's load capacity is greater than the sum of the load capacity of the individual column elements.

To generate the σ–ε diagram of confined concrete, the previous E_c, f_c', and ε_c' defined in Equations (2)–(5) are replaced with a modulus of elasticity of confined concrete E_{cc}, compressive strength of confined concrete f_{cc}', and the corresponding strain ε_{cc}' [3,13,49], defined by the following expressions [50]:

$$f_{cc}' = f_c' + k_1 \cdot f_l \tag{6}$$

$$\varepsilon_{cc}' = \varepsilon_c' \cdot \left(1 + k_2 \cdot \frac{f_l}{f_c'}\right) \tag{7}$$

where k_1 and k_2 are coefficients 4.1 and 20.5, respectively [51], while f_l is the lateral confining pressure, which depends on the ratio D/t and on the yield strength of steel f_y. It can be calculated in the following way [13]:

$$f_l = \left(0.043646 - 0.000832 \cdot \frac{D}{t}\right) \cdot f_y \tag{8}$$

The decreasing branch of the confined concrete diagram is linear in the interval from ε_{cc}' to $11 \cdot \varepsilon_{cc}'$. The reduction coefficient k_3 depends on the ratio D/t and the f_y and can be calculated as follows [13]:

$$k_3 = 1.00 \text{ for } D/t \leq 40 \tag{9}$$

$$k_3 = 0.0000339 \cdot \left(\frac{D}{t}\right)^2 - 0.010085 \cdot \left(\frac{D}{t}\right) + 1.3491 \text{ for } 40 < D/t \leq 150 \tag{10}$$

while r is the reduction coefficient that depends on the class of concrete. The reduction coefficient for concrete type C25/30 is 1.00 [13].

This model was then utilized within the concrete damaged plasticity formulation implemented in ABAQUS, with the dilation angle set at 20° [45]. The adopted Poisson's ratio ν is 0.2 [1,3,13,52,53]. The material properties used in ABAQUS are shown in Table 2.

Table 2. Material properties used in ABAQUS.

Concrete					Steel		
Unconfined		Confined					
f_c'	E_c	f_{cc}'	E_{cc}	ν	f_y	E	ν
[MPa]	[MPa]	[MPa]	[MPa]	[-]	[MPa]	[MPa]	[-]
30.5	32,965	48.5	36,986	0.2	355	210,000	0.3

3.3. Constitutive Model for Steel

The von Mises model with isotropic hardening was applied in the paper. Figure 5 shows the adopted constitutive uniaxial model for structural steel [54].

Here, f_y is the yield strength 355 MPa for S355 steel quality, ε_y is the corresponding strain, f_u is the ultimate strength 490 MPa, and ε_u is the corresponding ultimate strain. Young's modulus of elasticity of steel is 210 GPa, while Poisson's ratio is 0.3.

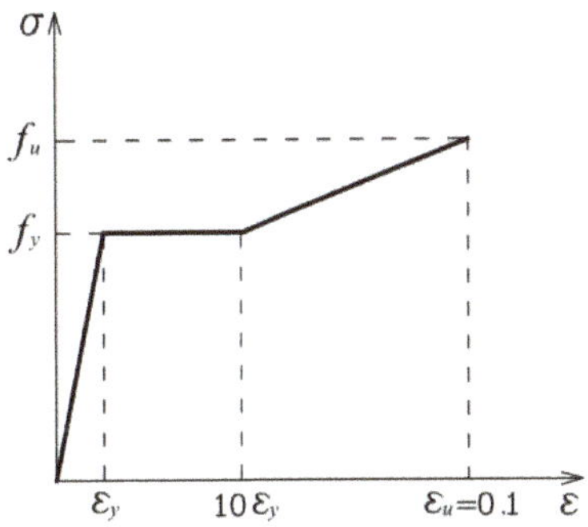

Figure 5. Stress–strain diagram for steel.

3.4. The Connection between the Concrete Core and the Steel Profile

The bond connection of the steel profile with the concrete core can be modeled using "surface-to-surface contact" elements [11]. The model uses contact elements that simulate the behavior of the connection in the longitudinal and tangential directions. The first is defined using the "Hard Contact" option available in ABAQUS. It allows separation of the contact surfaces in tension but prevents penetration of materials in compression. For the tangential direction, the "Columb friction model" was adopted, whereby the coefficient of friction was set at 0.47 [55]. According to numerous authors, this coefficient is usually in the range of 0.2–0.6 [3,13,19], although it has little impact on axial response [2].

3.5. Numerical Results

The failure of both samples C_1 and C_2 occurred due to the yielding of the steel profile, and the axial load capacities N_{FEM} are reported in Table 1. The ultimate axial strength obtained for the C_1 is 697.52 kN. The ultimate strength sustained by the steel part of the column can be obtained by integrating the stress along the steel part in the cross-section. As it equals 329.10 kN, it was evident that the contribution of the concrete filling to the load capacity of the column was 53%. The compressive strength of the confined concrete is $f_{cc}{}' = 50.69$ MPa. In this case, the lateral confining pressure is $f_l = 4.92$ MPa. The stress results for the specimen C_1 are shown in Figure 6a.

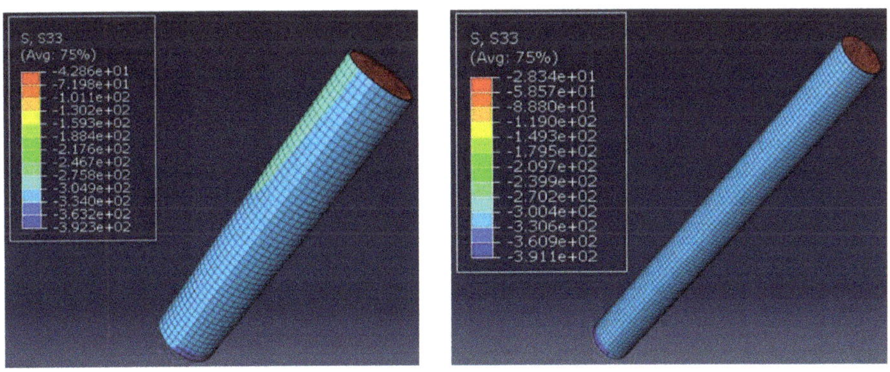

(**a**) (**b**)

Figure 6. Specimen FEM results: (**a**) specimen C_1; (**b**) specimen C_2.

The ultimate axial strength obtained for the C_2 specimen is 568.65 kN, while the strength sustained by the steel part of the column is 328.09 kN. The contribution of the concrete infill to the bearing capacity of the column is 42%, meaning that the strength of the confined concrete is $f_{cc}{}' = 33.10$ MPa. In this case, the lateral tension pressure is $f_l = 0.63$ MPa. The stress results for the specimen C_2 calculations are shown in Figure 6b.

4. Validation of Results

4.1. Comparison of N-ε Relations and Ultimate Strength Capacity Obtained in Tests and FEM

To validate the model presented in chapter 3 calculated using FEM, the results were compared with the experimental results. Figure 7 shows the ultimate force N-ε strain response for CFT specimens C_1 and C_2. A good agreement between the results can be observed. For CFT column C_1, the ultimate force N_{FEM} obtained by ABAQUS is 0.6% lower than the limit force obtained by experimental research. In Figure 7a, the axial force increases approximately linearly to about 400 kN and 0.0025 strain. With a further load increase, the specimen demonstrates plastic behavior with hardening and high strain levels achieved. For CFT column C_2, the ultimate force N_{FEM} is 2.3% higher than N_{test}. However, the inelastic behavior of the column occurs much earlier, at a force of about 200 kN and a strain of approximately 0.001, as shown in Figure 7b. This occurs because of the column's global instabilities.

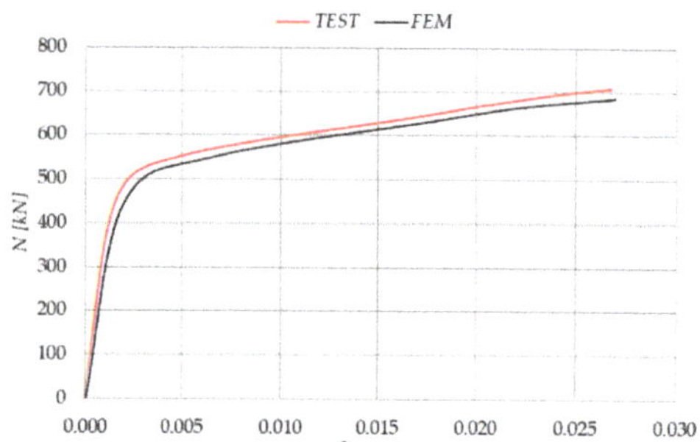

(**a**)

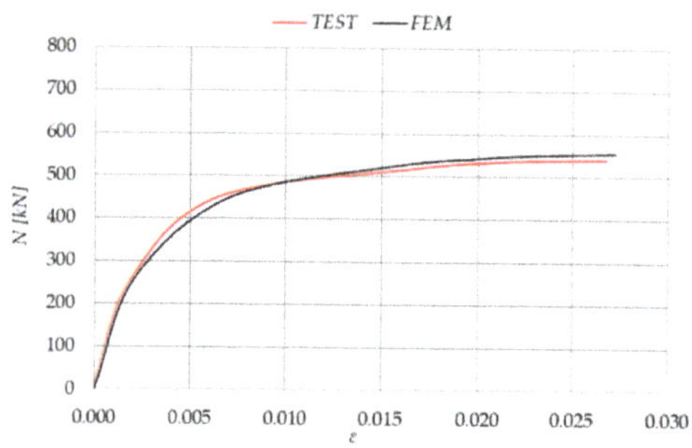

(**b**)

Figure 7. N-ε relation obtained in tests and FEM: (**a**) specimen C_1; (**b**) specimen C_2.

Moreover, it can be concluded that the column's length significantly influences the axial load capacity of the CFT column. With a double increase in the column length from 0.50 m to 1.00 m, the limit force of the CFT column decreases 1.26 times.

4.2. Comparison of Steel Profile Contribution and Confinement Effect

Based on the results of the experimental tests, the ultimate axial capacity of steel column S_1 and CFT column C_1 are 315.78 kN and 701.71 kN, respectively. Based on these results, it can be concluded that the contribution of the concrete core to the load capacity of the column is 55%. Thus, the compressive strength of the concrete, initially measured at 30.5 MPa, increased to 53.10 MPa due to confinement provided by the outer steel tube. For the accepted value of $k_1 = 4.1$, the lateral confining pressure f_l according to Equation (6) is 5.51 MPa. A good agreement of the results can be observed considering the following:

$$\frac{f'_{cc,test}}{f'_{cc,FEM}} = \frac{53.10}{50.69} = 1.047 \tag{11}$$

Similarly, based on the results of the experimental tests, the ultimate axial capacity of steel column S_2 is 297.84 kN, while the capacity of CFT column C_2 is 555.57 kN. The concrete infill contribution to the column's total load capacity is 46.4%. The strength of the confined concrete increased to 35.46 MPa, so the lateral confining pressure is 1.21 MPa. A good agreement between the results can also be noted, given the following:

$$\frac{f'_{cc,test}}{f'_{cc,FEM}} = \frac{35.46}{33.10} = 1.071 \tag{12}$$

Finally, the lateral confining pressure f_l registered for specimens C_1 and C_2 as 0.5 m and 1.0 m long, respectively, is compared in Equation (13). As also previously reported in the literature, it is confirmed that concrete confinement has a greater influence in shorter rather than in slender CFT columns, and so should be taken into account.

$$\frac{f_{l,test,\ C_1}}{f_{l,test,\ C_2}} = \frac{5.51}{1.21} = 4.55 \tag{13}$$

4.3. Comparison of the Obtained Results with Current Design Codes

Tables 3 and 4 show the comparison of experimental results N_{TEST}, results obtained in ABAQUS N_{FEM} for CFT specimens C_1 and C_2 with the N values calculated by the following design codes: EC4, EC4 considering the confinement, ACI/AS, AICS, and AIJ. Unit partial safety factors for the material are adopted.

Table 3. Ultimate values of axial forces of CFT column specimen C_1.

N_{TEST} [kN]	N_{TEST}/N_{FEM}	N_{TEST}/N_{EC4}	$N_{TEST}/N_{EC4,\ conf.}$	$N_{TEST}/N_{ACI/AS}$	N_{TEST}/N_{AISC}	N_{TEST}/N_{AIJ}
701.71	1.006	1.351	1.049	1.443	1.380	1.420

Table 4. Ultimate values of axial forces of CFT column specimen C_2.

N_{TEST} [kN]	N_{TEST}/N_{FEM}	N_{TEST}/N_{EC4}	$N_{TEST}/N_{EC4,\ conf.}$	$N_{TEST}/N_{ACI/AS}$	N_{TEST}/N_{AISC}	N_{TEST}/N_{AIJ}
555.57	0.977	1.069	0.946	1.143	1.093	1.124

Following the EC4 [38], the plastic resistance to compression $N_{pl,Rd}$ of a composite cross-section should be calculated by adding the plastic resistances of its components as follows:

$$N_{pl,Rd} = A_a \cdot f_{yd} + A_c \cdot f_{cd} + A_s \cdot f_{sd} \tag{14}$$

where A_a, A_c, and A_s are the cross-sectional area of the steel profile, concrete, and reinforcement, respectively, and $f_{yd} = f_y/\gamma_a$, $f_{sd} = f_{sk}/\gamma_s$, and $f_{cd} = f_{ck}/\gamma_c$ are the corresponding design values of the yield strength for steel and reinforcement and the compressive strength of concrete; f_y, f_{sk}, and f_{ck} are their characteristic values; γ_a, γ_s, and γ_c are the recommended values of the partial safety coefficients for the corresponding materials provided in EC2 [44] and EC3 [56], but here for validation all are taken as unity.

According to EC4, in the case of circular hollow steel profiles filled with concrete, the increase in concrete strength under pressure caused by the confining of the concrete by the steel profile can be taken into account if the following conditions are met [38]: relative slenderness $\bar{\lambda} \le 0.5$, $e/D < 0.1$, where e is the load eccentricity. In that case, the plastic resistance to compression $N_{pl,Rd}$ of a composite cross-section should be calculated as follows:

$$N_{pl,Rd} = \eta_a \cdot A_a \cdot f_{yd} + A_c \cdot f_{cd} \cdot [1 + \eta_c \cdot \frac{t}{D} \cdot \frac{f_y}{f_{ck}}] + A_s \cdot f_{sd} \tag{15}$$

Coefficients η_c and η_a include the concrete confinement effects introduced and depend on the relative slenderness and e/D ratio [16].

Applying ACI [39] and AS [40,41] the resistance to compression of the CFT column cross-section is also calculated without appropriate safety factors for the materials (unfactored axial bearing capacity) and is given by the following expression:

$$N_{ACI,AS} = A_a \cdot f_y + 0.85 \cdot A_c \cdot f_{ck} \tag{16}$$

As per AISC [42], the resistance of the cross-section under axial pressure is determined based on the following expression:

$$N_{AISC} = A_a \cdot f_y + C \cdot A_c \tag{17}$$

where C is the strength factor for concrete, and the rectangular CFT column amounts to $0.85 \cdot f_{ck}$, while for the circular CFT column, $0.95 \cdot f_{ck}$.

AIJ suggests the following expression for determining the ultimate bearing capacity of the cross-section of a CFT column under axial pressure [43]:

$$N_{AIJ} = 0.85 \cdot f_{cyl,100} \cdot A_c + (1 + \eta) \cdot f_y \cdot A_a \tag{18}$$

where $f_{cyl,100}$ is the compressive strength of concrete tested on a cylinder sample with dimensions of 100 mm·200 mm, and η is the coefficient that considers the confinement effect and is 0.27. The confining effect can be considered if the following condition is met: $L/D \leq 4$.

It can be observed that for a CFT column C_1, the EC4, taking into account the confinement effect, gives the closest results to the results obtained by experimental testing. On the other hand, the ACI/AS regulations give the most conservative results.

For a CFT column C_2, the confinement effect is significantly lower due to the increase in column slenderness. In this case, the ultimate value of the axial force of the CFT column obtained by experimental testing is lower than the calculated value proposed by EC4, considering the confinement effect. However, the same value $N_{EC4,\,conf}$ again gives the closest results to the results of the experimental tests, while the ACI/AS norms provide the most conservative results.

5. Conclusions

This paper presents the experimental investigation of the CFT column's axial capacity under construction conditions. The influence of the column's length, the concrete infill on the total bearing capacity of the CFT column, and the concrete confinement effect were analyzed experimentally and numerically.

Using the developed numerical models in the finite element-based software ABAQUS, it is possible to determine the bearing capacity of CFT columns with satisfactory accuracy, considering geometric and material nonlinearities. The differences between the ultimate limit force of the CFT column obtained in FEM models and tests are within 0.6–2.3%. Regarding an increase in the CFT column's length from 0.50 m to 1.00 m, the total limit force decreases by 26%. The ultimate axial capacity of a CFT column 0.50 m long compared to an empty steel column improves by 55%. For 1.0 m long specimens with greater relative slenderness, concrete infill improves the strength by 46.4%.

The obtained results of the experimental tests were also compared with relevant design codes: EC4, EC4 taking into account the confinement effect, ACI, AS, AICS, and AIJ. The EC4 considering the confinement effect gives the closest results to the results of the experimental tests, while the ACI/AS norms provide the most conservative results. Based on the experimental and numerical results, it is demonstrated that the confinement effect has a greater influence on CFT columns with the L/D ratio close to 4. As the column's length increases, the confinement effect decreases significantly. This was obtained for specimen C_2 with an L/D ratio of approximately 10, where the lateral confinement pressure was found to be insignificant. However, it is recommended to consider the confinement effect in this specimen as per EC4 guidelines, and this criterion should be further examined.

Future research should focus on a larger sample size of specimens with L/D ratios within the range of 4 to 10. This expanded scope of study would provide more accurate and informed design decisions.

Author Contributions: Conceptualization, M.M.L.R. and J.Z.N.; methodology, M.M.L.R.; software, M.M.L.R.; validation, M.M.L.R.; formal analysis, M.M.L.R. and J.Z.N.; investigation, M.M.L.R.; resources, M.M.L.R.; data curation, M.M.L.R. and J.Z.N.; writing—original draft preparation, M.M.L.R.; writing—review and editing, J.Z.N.; visualization, J.Z.N.; supervision, M.M.L.R.; project administration, M.M.L.R. and J.Z.N.; funding acquisition, M.M.L.R. All authors have read and agreed to the published version of the manuscript.

Funding: This research received no external funding.

Data Availability Statement: Dataset available on request from the authors. The raw data supporting the conclusions of this article will be made available by the authors on request.

Acknowledgments: The authors thank the Ministry of Science and Technology of the Republic of Serbia for the financial support through project number 2000092.

Conflicts of Interest: The authors declare no conflicts of interest.

References

1. Tao, Z.; Wang, Z.B.; Yu, Q. Finite element modelling of concrete-filled steel stub columns under axial compression. *J. Constr. Steel Res.* **2013**, *89*, 121–131. [CrossRef]
2. Nikolić, J.; Kostić, S.; Stošić, S. Numerical modelling of concrete-filled steel tubular short columns under axial compression. *Gradjevinski Mater. Konstr.* **2023**, *66*, 93–106. [CrossRef]
3. Ellobody, E.; Young, B.; Lam, D. Behaviour of normal and high strength concrete-filled compact steel tube circular stub columns. *J. Constr. Steel Res.* **2006**, *62*, 706–715. [CrossRef]
4. Liu, J.; Yu, W.; Fang, Y.; Pan, Z.; Cao, G. Finite Element Analysis on the Seismic Performance of Concrete-Filled Steel Tube Columns with a Multiple-Chamber Round-Ended Cross-Section. *Buildings* **2024**, *14*, 1154. [CrossRef]
5. Fellouh, A.; Bougara, A.; Piloto, P.A.G.; Benlakehal, N. Non-linear buckling analysis of composite columns made from high and normal strength concrete under fire. *Asian J. Civ. Eng.* **2020**, *21*, 17–27. [CrossRef]
6. Taufik, S.; Sugianto, A.; Utomo, G. Behavior of Filled and Encased Composite Column Using 3D Numerical Modelling ANSYS. *Int. J. Compos. Mater.* **2018**, *8*, 18–23. [CrossRef]
7. Hussain, A.; Sethi, H.; Shams, R.; Yadav, I.K. Analysis of the Composite Columns using Finite Element Modelling in Ansys Environment. *Int. J. Eng. Res. Technol.* **2019**, *8*.
8. Li, X.; Yin, Y.; Li, T.; Zhu, X.; Wang, R. Analytical Study on Reinforced Concrete Columns and Composite Columns under Lateral Impact. *Coatings* **2023**, *13*, 152. [CrossRef]
9. Omran, M.E.; Mollaei, S. Assessment of Empirical Formulas for Estimating Residual Axial Capacity of Blast Damaged RC Columns. *Eur. J. Sustain. Dev.* **2017**, *6*, 383–396. [CrossRef]
10. Momeni, M.; Hadianfard, M.A.; Bedon, C.; Baghlani, A. Damage evaluation of H-section steel columns under impulsive blast loads via gene expression programming. *Eng. Struct.* **2020**, *219*, 110909. [CrossRef]
11. Schneider, S.P. Axially loaded concrete-filled steel tubes. *J. Struct. Eng.* **1998**, *124*, 1125–1138. [CrossRef]
12. Wu, K.; Chen, F.; Zhang, H.; Xu, C.; Lin, S.Q. Experimental Study on the Behavior of Recycled Concrete-Filled Thin-Wall Steel Tube Columns Under Axial Compression. *Arab. J. Sci. Eng.* **2018**, *43*, 5225–5242. [CrossRef]
13. Hu, H.-T.; Asce, M.; Huang, C.-S.; Wu, M.-H.; Wu, Y.-M. Nonlinear Analysis of Axially Loaded Concrete-Filled Tube Columns with Confinement Effect. *J. Struct. Eng.* **2003**, *129*, 1322–1329. [CrossRef]
14. Patel, V.I.; Liang, Q.Q.; Hadi, M.N.S. Nonlinear analysis of circular high strength concrete-filled stainless steel tubular slender beam-columns. *Eng. Struct.* **2017**, *130*, 1–13. [CrossRef]
15. Xu, S.; Wu, C.; Liu, Z.; Shao, R. Experimental investigation on the cyclic behaviors of ultra-high-performance steel fiber reinforced concrete filled thin-walled steel tubular columns. *Thin-Walled Struct.* **2019**, *140*, 1–20. [CrossRef]
16. Lin, S.; Zhao, Y.G.; Lu, Z.H. Fibre beam element models for nonlinear analysis of concentrically loaded circular CFT columns considering the size effect. *Eng. Struct.* **2020**, *210*, 110400. [CrossRef]
17. Radovanovic, M.M.L.; Nikolic, J.Z.; Radovanovic, J.R.; Kostic, S.M. Structural Behaviour of Axially Loaded Concrete-Filled Steel Tube Columns during the Top-Down Construction Method. *Appl. Sci.* **2022**, *12*, 3771. [CrossRef]
18. Giakoumelis, G.; Lam, D. Axial capacity of circular concrete-filled tube columns. *J. Constr. Steel Res.* **2004**, *60*, 1049–1068. [CrossRef]
19. Liang, Q.Q.; Fragomeni, S. Nonlinear analysis of circular concrete-filled steel tubular short columns under axial loading. *J. Constr. Steel Res.* **2009**, *65*, 2186–2196. [CrossRef]

20. Li, G.C.; Chen, B.W.; Yang, Z.J.; Liu, Y.P.; Feng, Y.H. Experimental and numerical behavior of eccentrically loaded square concrete-filled steel tubular long columns made of high-strength steel and concrete. *Thin-Walled Struct.* **2021**, *159*, 107289. [CrossRef]
21. Nikolić, J.; Tošić, N.; Murcia-Delso, J.; Kostić, S.M. Comprehensive review of the structural behaviour and numerical modelling of recycled aggregate concrete-filled steel tubes. *Eng. Struct.* **2024**, *303*, 117514. [CrossRef]
22. Hosseinzadehfard, E.; Mobaraki, B. Investigating concrete durability: The impact of natural pozzolan as a partial substitute for microsilica in concrete mixtures. *Constr. Build. Mater.* **2024**, *419*, 135491. [CrossRef]
23. Mobaraki, B.; Ma, H.; Lozano Galant, J.A.; Turmo, J. Structural health monitoring of 2D plane structures. *Appl. Sci.* **2021**, *11*, 2000. [CrossRef]
24. Sabih, S.M.; Hilo, S.J.; Hamood, M.J.; Salih, S.S.; Faris, M.M.; Yousif, M.A. Numerical Investigation into the Strengthening of Concrete-Filled Steel Tube Composite Columns Using Carbon Fiber-Reinforced Polymers. *Buildings* **2024**, *14*, 441. [CrossRef]
25. Cao, B.; Zhu, L.; Qian, H.; Pan, Z. Mechanical Behavior of Precast Circular Semi-Continuous CFST Columns under a Uniaxial Eccentric Load. *Buildings* **2024**, *14*, 772. [CrossRef]
26. de Oliveira, W.L.A.; De Nardin, S.; de Cresce El Debs, A.L.H.; El Debs, M.K. Influence of concrete strength and length/diameter on the axial capacity of CFT columns. *J. Constr. Steel Res.* **2009**, *65*, 2103–2110. [CrossRef]
27. Liang, Q.Q. High strength circular concrete-filled steel tubular slender beamcolumns, Part II: Fundamental behavior. *J. Constr. Steel Res.* **2011**, *67*, 172–180. [CrossRef]
28. Dundu, M. Compressive strength of circular concrete filled steel tube columns. *Thin-Walled Struct.* **2012**, *56*, 62–70. [CrossRef]
29. Chacón, R.; Mirambell, E.; Real, E. Strength and ductility of concrete-filled tubular piers of integral bridges. *Eng. Struct.* **2013**, *46*, 234–246. [CrossRef]
30. De Oliveira, W.L.A.; De Nardin, S.; El Debs, A.L.H.d.C.; El Debs, M.K. Evaluation of passive confinement in CFT columns. *J. Constr. Steel Res.* **2010**, *66*, 487–495. [CrossRef]
31. Shanmugam, N.E.; Lakshmi, B. State of the art report on steel–concrete composite columns. *J. Constr. Steel Res.* **2001**, *57*, 1041–1080. [CrossRef]
32. Zhang, Y.; Wei, Y.; Bai, J.; Zhang, Y. Stress-strain model of an FRP-confined concrete filled steel tube under axial compression. *Thin-Walled Struct.* **2019**, *142*, 149–159. [CrossRef]
33. Hossain, G.A.; Ahmed, K.F.; Rahman, S. Effect of Geometric and Material Properties on the Behavior of Axially Loaded Concrete-Filled Stainless Steel Tube Columns. *Int. J. Steel Struct.* **2022**, *22*, 1215–1235. [CrossRef]
34. Horváth, A.; Kollár, D.; Kövesdi, B. Behaviour of CFST Stub Columns Subjected to Pure Compression. *Int. J. Steel Struct.* **2022**, *22*, 1175–1188. [CrossRef]
35. Abed, F.; Alhamaydeh, M.; Abdalla, S. Experimental and numerical investigations of the compressive behavior of concrete filled steel tubes (CFSTs). *J. Constr. Steel Res.* **2013**, *80*, 429–439. [CrossRef]
36. Han, L.H.; Yao, G.H.; Tao, Z. Performance of concrete-filled thin-walled steel tubes under pure torsion. *Thin-Walled Struct.* **2007**, *45*, 24–36. [CrossRef]
37. DS SIMULIA Corp. *ABAQUS User Manual 6.9*; DS SIMULIA Corp.: Providence, RI, USA, 2009.
38. *EN 1994-1-1*; Eurocode 4: Design of Composite Steel and Concrete Structures—Part 1-1: General Rules and Rules for Buildings. European Committee for Standardization: Brussels, Belgium, 2004.
39. *ACI 318-11*; Building Code Requirements for Structural Concrete and Commentary. American Concrete Institute: Farmington Hills, MI, USA, 2011.
40. *AS3600*; Australian Standards Concrete Structures. Standards Australia International Ltd.: Sydney, NSW, Australia, 2001.
41. *AS410*; Australian Standard Steel Structures. Standards Australia: Sydney, NSW, Australia, 1998.
42. *ANSI/AISC 360-16: 36010*; Specification for Structural Steel Buildings. American Institute of Steel Construction, Inc.: Chicago, IL, USA, 2005.
43. Architectural Institute of Japan (AIJ). *Recommendations for Design and Construction of Concrete-Filled Steel Tubular Structures*; AIJ: Tokyo, Japan, 2018.
44. *EN 1992-1-1*; Eurocode 2: Design of Concrete Structures—Part 1-1: General Rules and Rules for Buildings. European Committee for Standardization: Brussels, Belgium, 2004.
45. Saenz, L.P. Discussion of Equation for the Stress-Strain Curve of Concrete by Desayi and Krishnan. *J. Am. Concr. Inst.* **1964**, *61*, 1229–1235.
46. Hu, H.; Schnobrich, W.C. Constitutive Modeling of Concrete by Using Nonassociated Plasticity. *J. Mater. Civ. Eng.* **1989**, *1*, 199–216. [CrossRef]
47. Yu, Q.; Tao, Z.; Liu, W.; Chen, Z.B. Analysis and calculations of steel tube confined concrete (STCC) stub columns. *J. Constr. Steel Res.* **2010**, *66*, 53–64. [CrossRef]
48. Gourley, B.C.; Tort, C.; Denavit, M.D.; Schiller, P.H.; Hajjar, J.F. NSEL Report Series A Synopsis of Studies of the A Synopsis of Studies of the Monotonic and Cyclic Behavior of Concrete-Filled Steel Tube Members, Connections, and Frames. 2008. Available online: http://www.terragalleria.com/ (accessed on 8 April 2024).
49. Lee, S.-H.; Uy, B.; Kim, S.-H.; Choi, Y.-H.; Choi, S.-M. Behavior of high-strength circular concrete-filled steel tubular (CFST) column under eccentric loading. *J. Constr. Steel Res.* **2011**, *67*, 1–13. [CrossRef]

50. Mander, J.B.; Priestley, M.J.; Park, R. Theoretical stress-strain model for confined concrete. *J. Struct. Eng.* **1988**, *114*, 1804–1826. [CrossRef]
51. Richart, F.E.; Brandtzæg, A.; Brown, R.L. *A Study of the Failure of Concrete under Combined Compressive Stresses*; University of Illinois Urbana-Champaign: Champaign, IL, USA, 1928.
52. Papanikolaou, V.K.; Kappos, A.J. Confinement-sensitive plasticity constitutive model for concrete in triaxial compression. *Int. J. Solids Struct.* **2007**, *44*, 7021–7048. [CrossRef]
53. Hu, H.T.; Huang, C.S.; Chen, Z.L. Finite element analysis of CFT columns subjected to an axial compressive force and bending moment in combination. *J. Constr. Steel Res.* **2005**, *61*, 1692–1712. [CrossRef]
54. Moon, J.; Roeder, C.W.; Lehman, D.E.; Lee, H.E. Analytical modeling of bending of circular concrete-filled steel tubes. *Eng. Struct.* **2012**, *42*, 349–361. [CrossRef]
55. Baltay, P.; Gjelsvik, A. Coefficient of friction for steel on concrete at high normal stress. *J. Mater. Civ. Eng.* **1990**, *2*, 46–49. [CrossRef]
56. *EN 1993-1-1*; Eurocode 3: Design of Steel Structures—Part 1-1: General Rules and Rules for Buildings. European Committee for Standardization: Brussels, Belgium, 2005.

buildings

MDPI

Article

Push-Out Analysis on the Shear Performance of a New Type of Bellow-Sleeved Stud

Disheng Zou [1], Qingtian Su [1,2], Fei Wu [1,*], Zhiping Lin [3] and Peiran Li [4]

1 Department of Bridge Engineering, Tongji University, Shanghai 200092, China; 2232574@tongji.edu.cn (D.Z.)
2 Shanghai Engineering Research Center of High Performance Composite Bridge, Shanghai 200092, China
3 Fujian Expressway Group Co., Ltd., Fuzhou 350001, China
4 Henan Transportation Planning and Design Research Institute Co., Ltd., Zhengzhou 450008, China
* Correspondence: 1810754@tongji.edu.cn

Abstract: For continuous steel–concrete composite girder bridges based on the post-combined method, the conventional rectangular group studs contribute to the isolation of the steel girder and the concrete slab before prestressing, leading to the majority of prestress forces being introduced to the concrete slab. However, rectangular-group stud holes cause the prestress forces to be unevenly distributed. In this study, a new type of bellow-sleeved stud (BSS) was developed to mitigate the weakening effects of rectangular group stud holes on the slab. A steel corrugated sleeve with a diameter of 60 mm was employed to cover the stud, which served as an internal formwork to prevent the concrete from bonding with the root of the stud. After prestressing was complete, the steel sleeve was filled with ultra-high-performance concrete (UHPC) to create a reliable combination between the concrete slab and the steel girder. To investigate the shear performance of this new type of connection, eight push-out test specimens were designed, and finite-element models were built. This study drew a comparison between the BSS and the ordinary headed stud (OHS). The research findings suggested that the BSS is subjected to less bending–shear coupling and offers a 4.5% increase in shear strength and a 31.9% increase in shear stiffness compared with the OHS. The study also analyzed the structural parameters influencing the shear performance of the BSS. It is found that the steel sleeve of the BSS has a negative effect on shear performance, but this can be mitigated by infusing high-strength material into the sleeve. Furthermore, the study examined the effect of construction quality on shear performance and suggested that sleeve deviation and grout leakage considerably reduced the shear performance of the BSS. Accordingly, strict control over the construction quality of the BSS is necessary.

Keywords: bellow-sleeved stud; shear performance; push-out test; post-combined prestressing method

1. Introduction

The shear connector in the steel–concrete composite girder bridge tightly combines the concrete bridge deck with the steel main girder together to resist the effects of external loads. These connectors not only transmit the horizontal shear force along the steel–concrete interface, but also resist the vertical separation between steel and concrete [1]. Welded studs, i.e., a type of flexible connector commonly used in bridge engineering, exhibit isotropic shear resistance and can be uniformly distributed across distinct regions. Since 1956, a significant number of the push-out tests have been performed, so that several empirical formulas for effective shear capacity and elastic shear stiffness have been derived [2–5]. In 1985, Roik [6] conducted reliability analysis on the shear capacity of studs and recommended the adoption of a partial factor γ_v = 1.25 for stud design, which is an approach that was approved from the EN1990 standard [7] and has since been continuously employed. After decades of development, the design method of the individual shear stud has been clearly specified in the current codal provisions or guidelines across the world [8–10].

After 1990, there have been continual breakthroughs in the new technology of prestressed composite girder bridges [11–14], leading to innovative changes to the construction form and the arrangement of shear studs. In order to enhance the efficiency of prestressing in continuous composite girder bridges, Su Hang introduced the partial-combined method [15] and proposed a new type of rubber-sleeved stud (RSS) [16]. The principle of this method involves utilizing the extrusion of the rubber sleeve to minimize the shear stiffness of the stud, allowing the majority of precompression stress to be transmitted into the slab. Experimental studies indicate that the initial shear stiffness of the RSS decreases to 1% of ordinary headed studs (OHSs), achieving a prestress transfer efficiency of 95%. Simultaneously, Su [17] proposed the post-combined prestressing method, which arranges the shear studs in groups in the negative moment region, ensuring isolation between the slab and the steel girder before prestressing. Initially, prestressed tendons are tensioned to direct all prestress into the concrete bridge slab. Next, high-strength mortar is employed for secondary casting in the reserved group stud holes, such that the concrete bridge slab is combined with the steel girder.

To research the shear behavior of the group studs applied to the post-combined method, Xue [18] conducted push-out tests and found that the group effect reduces the shear performance of the stud, necessitating the introduction of a reduction factor in the design. Yu [19] analyzed the stiffness of the group studs based on the Winkler foundation model, deriving a theoretical calculation formula considering the concrete material, stud diameter, and the number of shear studs. These research indicates that the shear performance of the group studs meets the requirement of the post-combined prestressing method. However, according to onsite testing, Su [20] suggested that the traditional rectangular-group stud holes on the post-combined slab exert a pronounced weakening effect. The precompression stress generated by prestressing tendons cannot be uniformly transferred into the bridge deck, causing stress to be unevenly distributed across the width of the bridge deck. The precompression stress between two consecutive group stud holes was 1.4 MPa, much lower than the average prestress of 10.3 MPa inside the concrete slab.

To ameliorate the uneven distribution of the prestress force, this study proposed a new type of bellow-sleeved stud (BSS). A steel corrugated sleeve with a diameter of 60 mm was employed to cover the stud, which served as an internal formwork to prevent the concrete from bonding with the root of the stud. After prestressing was complete, the steel sleeve was filled with ultra-high-performance concrete (UHPC) to create a reliable combination between the concrete bridge slab and the steel girder. Research found that the BBS increases the precompression stress between two consecutive group stud holes to 7.8 MPa, but its shear performance is still unknown. Differing from the conventional group studs, the steel sleeve is positioned close to the root of the stud, and a significant contrast exists in characteristics between the materials inside and outside the steel sleeve (UHPC inside the sleeve, ordinary concrete outside the sleeve). In this case, the shear performance of this new type of stud has hardly been evaluated by previous research and the current design code. Consequently, push-out tests and numerical studies should be conducted on the BSS to explore its shear mechanism.

2. Push-Out Test

2.1. Specimen Design

In this study, eight pieces of push-out specimens were developed to examine the shear performance of this new type of shear stud, with the characteristics of these specimens listed in Table 1. The type of shear stud was involved in the variation in specimens. OHS and BSS represent the standard push-out specimens welded with ordinary headed studs and bellow-sleeved studs, respectively. Since the construction equality has a considerable influence on the test results, the number of the specimen BSS was set as 5.

Table 1. Characteristics of the test specimen.

Specimen ID	Stud Dimension (Diameter × Height)	Stud Arrangement (Single-Side)	Bellow Dimension (Diameter × Height)	Material Inside Bellow	Number
OHS	22 × 200	1 × 2	NA	NA	3
BSS	22 × 200	1 × 2	60 × 320	UHPC	5

Figure 1 depicts the specimen and the reinforcement layout. Each specimen was developed following the same dimensions, while the only difference in all specimens lay in the type of shear stud. The test piece was cast with C50 concrete characterized by a thickness of 320 mm and then reinforced by HRB400 steel bar possessing a diameter of 20 mm, consistent with the practical bridge slab. The ML15 headed stud, which has been extensively employed in bridge engineering, was selected as the test subject, with a diameter and height of 22 mm and 200 mm, respectively. The bellow buried in BSS was constituted by a steel corrugated sleeve, which exhibited a diameter of 60 mm and a length of 320 mm. Moreover, UHPC served as the grout in the bellow, which was cast after the concrete outside was completed following the casting sequence of the bridge slab based on the post-combined prestressing method.

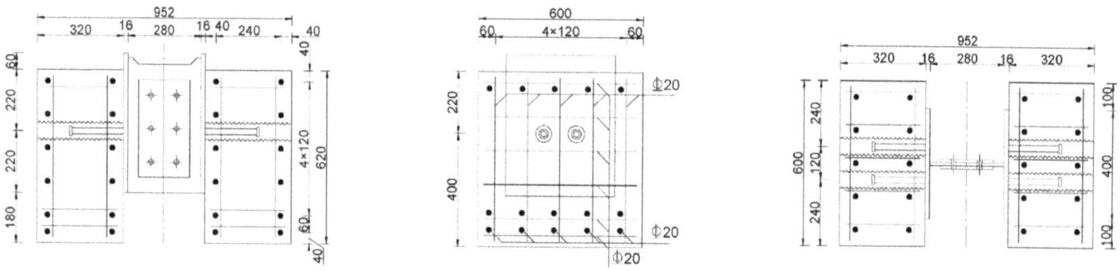

(**a**) Front view (**b**) Side view (**c**) Top view

Figure 1. Specimen and the reinforcement layout of BSS.

The H-shaped steel structure of the specimen was connected by two T-shaped steel plates through bolts. The flange exhibited a thickness of 16 mm, and the respective flange was welded with two headed studs symmetrically centered on the web plate, creating a transverse spacing of 120 mm. Thus, the pushing force was distributed uniformly to four shear studs buried in the specimen, such that this force was not concentrated on the single headed stud.

Figure 2 shows the construction process of the specimen BSS. The respective specimen comprised two identical parts fabricated by lying on the ground. After the shear studs were welded on the outer surface of the steel flange, the interaction surface on the flange was coated with engine oil, with the aim of reducing the friction between steel and concrete. As a result, the test result can be less affected. Subsequently, the framework was erected on the steel flange, laying a basis for constructing the reinforcement of concrete slab. With BSS as an example, the steel corrugated sleeves were employed to cover the shear studs and then fixed by iron wires to the nearby steel bars, serving as an isolation for external C50 concrete. To prevent external concrete leaking into the steel corrugated sleeve, a geotextile was stuffed into the sleeve and then rammed to the stud root with a steel bar for absorbing the cement slurry leaking at the bottom. Moreover, the sponge material was tied by iron wires to absorb the cement slurry leaking at the top. The external C50 concrete was cast after the leak-proof measure was completed. When the concrete was close to the initial set, a rebar with a bending hook was adopted to remove the geotextile out of the sleeve. On that basis, a multitude of small stud holes were created on the concrete slab. After 7 days, the concrete was finally set, and then small stud holes were cast with UHPC. Accordingly, the concrete slab and the steel structure were integrated at a whole. Lastly, two identical parts were connected with bolts.

(**a**) Erect framework (**b**) leak-proof measure

(**c**) Cast stud holes with UHPC (**d**) Connect two parts at a whole

Figure 2. Construction process of BSS.

2.2. Material Properties

The cubic compressive strength, axial compressive strength, elastic modulus, and tensile strength possessed by C50 concrete and UHPC were tested. Moreover, the yield strength, ultimate tensile strength, and elastic modulus of the studs were examined. Table 2 lists the averaged test results.

Table 2. Material test results table (unit: MPa).

Concrete Properties	Cubic Compressive Strength	Axial Compressive Strength	Elastic Modulus ($\times 10^3$)
C50 Concrete	53.2	39.9	35.77
UHPC	136.8	129.4	44.00
Stud Properties	Yield Strength	Ultimate Tensile Strength	Elastic Modulus ($\times 10^3$)
ML15 Stud	367	486	217

2.3. Test Setup and Loading Programs

A total of four displacement transducers were strategically positioned around the steel–concrete interface to examine the relative slip between the stud root and the concrete slab, the arrangement of displacement transducers is shown in Figure 3.

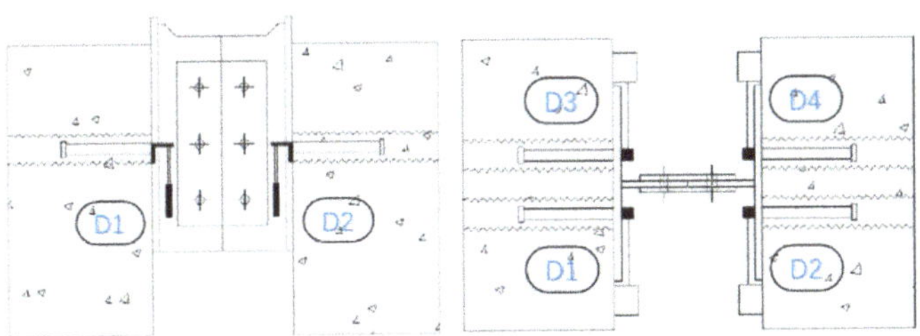

Figure 3. Arrangement of displacement transducers.

Figure 4 shows the experimental setup. The steel structure on the top of the specimens was loaded with a YAJ-10000 micro-controlled electro-hydraulic servo shear test machine, which is capable of exerting a maximum force of 10,000 kN. A steel pad was positioned on the top of the specimen, and a 6 mm rubber pad was arranged between the steel pad and the reaction frame, ensuring that the loading force was evenly distributed to the studs on both sides.

Figure 4. Experimental setup.

The specimens were pre-loaded three times at the estimated ultimate load of 20% before formal loading. The purpose of pre-loading was to examine the symmetry of the readings from the four displacement transducers. If any localized uneven loading was observed, the specimen was unloaded, and the steel plates on the top were re-leveled. In the practical loading process, a loading mode transition was reported at 60% of the estimated load. A force-controlled loading mode was employed, with a loading rate of 5 kN/s, in the first half of loading. In the latter half, a displacement-controlled loading mode was continuously employed with a loading rate of 0.5 mm/min until the specimen reached failure. Sampling was conducted at a frequency of 2 Hz.

3. Experimental Results

3.1. Failure Modes

Figure 5 presents the typical failure modes of the conventional stud push-out specimen OHS and the new bellow-sleeved stud push-out specimen BSS. In all specimens, the studs were sheared at their roots, accompanied by localized crushing of the concrete beneath the studs. As depicted in the figures, the UHPC at the root of the sleeved stud was partially exposed, and the steel sleeve beneath the stud was crushed under the effect of compression. Since the UHPC at the root of the sleeved stud exhibited a higher strength, the concrete failure region of the sleeved stud was slightly smaller than that of the ordinary stud. As indicated by the above results, in the testing process, BSS was primarily subjected to direct shear at its root, whereas OHS was more pronouncedly bent.

(**a**) (**b**) (**c**)

Figure 5. *Cont.*

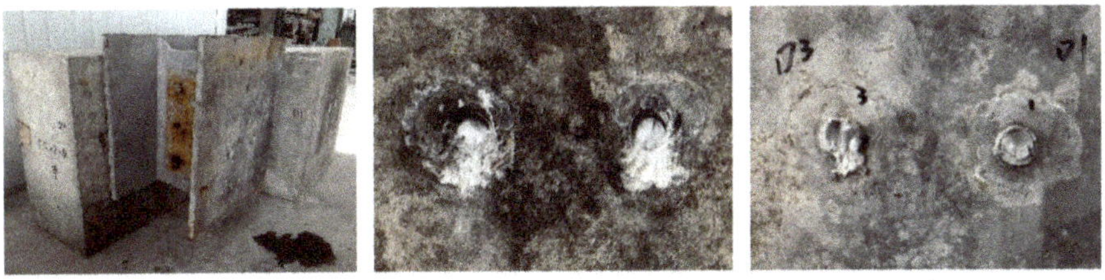

(**d**) (**e**) (**f**)

Figure 5. Failure modes of push-out specimens. (**a**) Failure mode of OHS-a; (**b**) Concrete side of the failure face of OHS-a; (**c**) Steel structure side of the failure face of OHS-a; (**d**) Failure mode of BSS-a; (**e**) Concrete side of the failure face of BSS-a; (**f**) Steel structure side of the failure face of BSS-a.

3.2. Load–Slip Curves and Discussion

Figure 6 illustrates the load–slip curves plotted through the push-out tests. The horizontal axis represents the average slip for the respective specimen, i.e., the mean value of all displacement sensor readings. The vertical axis represents the average shear load per stud for the respective specimen. The load–slip curves with regard to specimens in the identical group exhibited a certain variability. Notably, the shear capacity of OHS-b was significantly higher than that of the other two specimens characterized by identical configurations. The remaining two specimens had relatively consistent load–slip curves. It is noteworthy that the load–slip curves for BSS possessed a slightly greater variability than those of OHS. This variability primarily arose from the effect on the metal corrugated sleeve when the concrete was being poured, such that the studs were off-center in the sleeve.

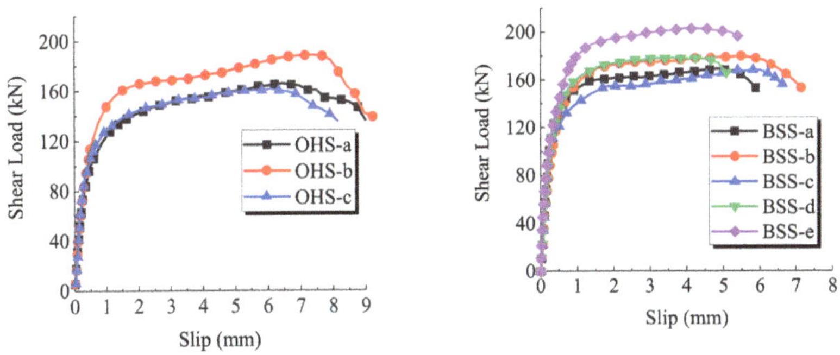

(**a**) Specimen OHS (**b**) Specimen BSS

Figure 6. Push-out test load–slip curve.

Although variability existed among specimens in the identical group, Figure 6 clearly presents the differences in shear performance between the two types of studs. BSS exhibited a higher ultimate shear capacity and shear stiffness than OHS, and a relatively small slip value corresponding to the ultimate load was generated.

Table 3 lists the summarized test results for the respective specimens. Here, V_u is the ultimate shear capacity of the studs, S_p is the slip at the ultimate shear capacity, and $K_{0.2mm}$, $K_{1/3}$, and $K_{1/2}$ denote the shear stiffness of the studs corresponding to the slip values of 0.2 mm, $1/3V_u$, and $1/2V_u$, respectively. This table lists the variations in shear performance for the respective specimens. Compared with OHS, BSS achieved an increase in the ultimate shear capacity by 4.5%, while the ultimate slip decreased by 27.7%. The above-mentioned results suggested that BSS did not confer an advantage over OHS in enhancing shear strength, whereas it exhibited a robust ability to resist slip deformation. Moreover, the shear stiffness of BSS increased by 31.9% compared with OHS.

Table 3. Summary of push-out test results.

Group	ID	Ultimate Shear Capacity V_u (kN)		Slip at Ultimate Shear Capacity S_p (mm)		Shear Stiffness $K_{0.2mm}$ (kN/mm)		Shear Stiffness $K_{1/3}$ (kN/mm)		Shear Stiffness $K_{1/2}$ (kN/mm)	
OHS	a	166.8	Avg: 172.6 0%	6.45	Avg: 6.67 0%	312	Avg: 329 0%	327	Avg: 333 0%	247	Avg: 272 0%
	b	189.7		7.45		331		321		266	
	c	161.3		6.90		342		351		313	
BSS	a	169.7	Avg: 180.3 ↑4.5%	4.59	Avg: 4.82 ↓27.7%	444	Avg: 434 ↑31.9%	422	Avg: 462 ↑38.7%	429	Avg: 402 ↑47.8%
	b	180.0		5.09		370		387		331	
	c	168.7		5.90		455		457		446	
	d	179.1		4.20		424		456		398	
	e	203.8		4.34		477		629		434	

4. Numerical Analysis

4.1. Overview

Finite-element models were built by Abaqus 2020 in terms of the push-out specimens OHS and BSS, and the relevant parameters were analyzed, so as to compare the shear performance of the new bellow-sleeved studs with that of conventional studs. Stud height, sleeve infill material, and sleeve diameter were the parameters analyzed. In accordance with the symmetry of the push-out specimens, only a quarter of the specimen was modeled in the finite-element analysis. The model is shown in Figure 7.

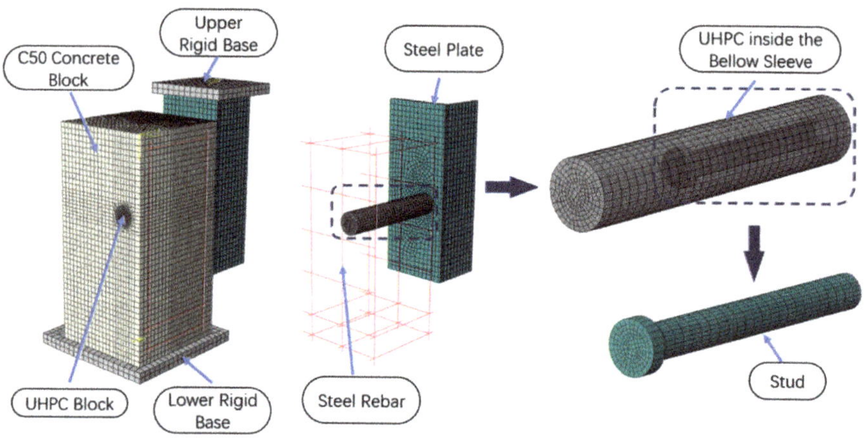

Figure 7. Abaqus finite-element model.

The model was constituted by concrete blocks, steel sleeves, steel plates, reinforcement mesh, and studs. The concrete block was made of two distinct materials, with UHPC arranged inside the sleeve and ordinary C50 concrete placed outside the sleeve. Moreover, the model covered two rigid bases on the top of the steel structure and the bottom of the concrete, and a reference point was placed at the center of the respective base to capture the push-out reaction force in the loading process. In the practical loading process, a displacement of 10 mm was applied to the reference point on the top base of the steel structure.

Concrete blocks, studs, and steel plates were represented using C3D8R solid elements, while steel rebars were modeled with T3D2 truss elements. The overall element size of the specimen was 15 mm. To increase computational accuracy, the mesh was locally refined with a finer resolution of 3 mm around the studs and the steel plates.

4.2. Material Constitutions

(1) Concrete material constitutive model

Table 4 shows the material properties of concrete used in FEM analysis. The stress–strain relationship of the axial compression and tension of concrete can be simulated using the plastic damage model (CDP model) based on Abaqus 2020. This model quantifies the stiffness degradation of concrete in the nonlinear phase using the plastic damage factor d, which conforms to the energy equivalence principle proposed by Sidoroff.

Table 4. The material properties of concrete used in FEM analysis.

Concrete Material	Elastic Modulus E_c (GPa)	Axial Compressive Strength $f_{c,r}$ (MPa)	Ultimate Compressive Strain $\varepsilon_{c,r}$	Axial Tenile Strength $f_{t,r}$ (MPa)	Ultimate Tensile Strain $\varepsilon_{t,r}$
C50 Concrete	35.77	39.9	0.0017	2.7	0.00009
UHPC	44.00	129.4	0.0035	8.0	0.0002~0.002

The calculation of the nonlinear constitutive model was conducted through the following steps. With the compressive constitutive behavior of C50 as an example, the initial elastic modulus E_c = 35.77 GPa and axial compressive strength $f_{c,r}$ = 39.9 MPa were first obtained based on material testing. According to the method specified in GB 50010-2010 [21], the ultimate compressive strain $\varepsilon_{c,r}$ = 0.0017 was computed based on the damage factors d_c for different stages. In accordance with the above computation result, the stress–strain ($\sigma - \varepsilon$) curve was plotted in Figure 8. Subsequently, the process was divided into two segments for input into the software. In the initial linear segment, only the initial elastic modulus and axial compressive strength should be input. With regard to the nonlinear segment, non-elastic strains $\overset{\sim in}{\varepsilon_c}$ for each stage were calculated (tensile cracking strain is expressed as $\overset{\sim in}{\varepsilon_c}$). In accordance with the Sidoroff energy equivalence principle, the corresponding damage factor d was derived. Lastly, the stress σ_c to non-elastic strain $\overset{\sim in}{\varepsilon_c}$ relationship and the damage factor d to non-elastic strain $\overset{\sim in}{\varepsilon_c}$ relationship were imported into Abaqus 2020 software as the constitutive model for concrete.

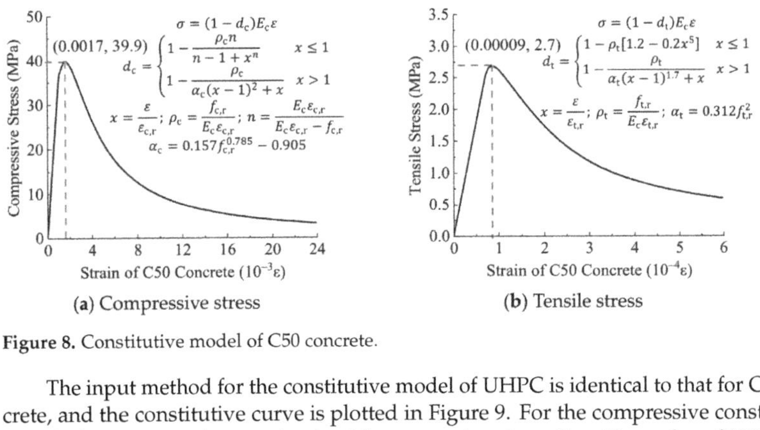

$$\sigma = (1-d_c)E_c\varepsilon \quad d_c = \begin{cases} 1-\dfrac{\rho_c n}{n-1+x^n} & x \le 1 \\ 1-\dfrac{\rho_c}{\alpha_c(x-1)^2+x} & x>1 \end{cases}$$

$$x=\frac{\varepsilon}{\varepsilon_{c,r}};\ \rho_c=\frac{f_{c,r}}{E_c\varepsilon_{c,r}};\ n=\frac{E_c\varepsilon_{c,r}}{E_c\varepsilon_{c,r}-f_{c,r}}$$

$$\alpha_c = 0.157 f_{c,r}^{0.785} - 0.905$$

(**a**) Compressive stress

$$\sigma = (1-d_t)E_c\varepsilon \quad d_t = \begin{cases} 1-\rho_t[1.2-0.2x^5] & x \le 1 \\ 1-\dfrac{\rho_t}{\alpha_t(x-1)^{1.7}+x} & x>1 \end{cases}$$

$$x=\frac{\varepsilon}{\varepsilon_{t,r}};\ \rho_t=\frac{f_{t,r}}{E_c\varepsilon_{t,r}};\ \alpha_t = 0.312 f_{t,r}^2$$

(**b**) Tensile stress

Figure 8. Constitutive model of C50 concrete.

The input method for the constitutive model of UHPC is identical to that for C50 concrete, and the constitutive curve is plotted in Figure 9. For the compressive constitutive model, f_c represents the examined axial compressive strength, with a value of 129.4 MPa. The ultimate compressive strain $\varepsilon_{c.r}$ was taken as 0.0035, and the initial elastic modulus E_c was assigned a value of 44.0 GPa based on the material testing results. E_s represents the secant modulus at the stress peak point. For the tensile constitutive model, f_{ct} was assigned a value of 8.0 MPa, and the strain values for the peak point ε_{ca} and the ultimate point ε_{pc} were computed as 0.0002 and 0.002, respectively. The parameter α principally affected the descending segment, and it was set at 1.106 based on the characteristics of the steel fibers.

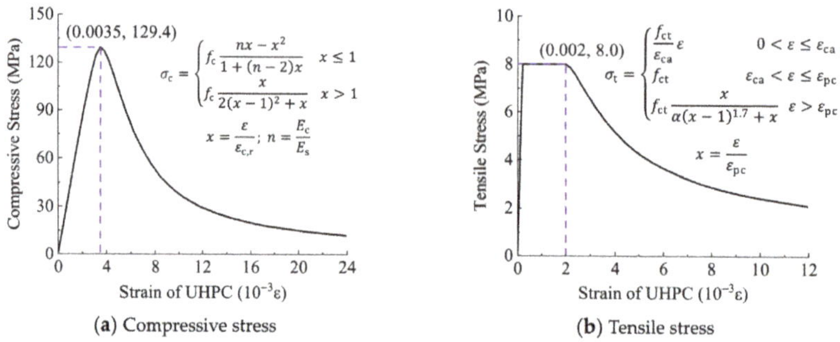

Figure 9. Constitutive model of UHPC.

(2) Steel material constitutive model

Table 5 shows the material properties of steel used in FEM analysis. The stress–strain curve of the stud material is expressed by the three-segment model as illustrated in Figure 10a. The yield stress and ultimate stress of the material were first derived based on the material testing results and then converted into the corresponding true stresses, where were measured at 368 MPa and 498 MPa, respectively. The ultimate strain was set at 2.5%, and the elastic modulus was computed as 207 GPa. The stress–strain curves for steel materials and steel bars were modeled using the dual-segment model, as depicted in Figure 10b. The yield stresses for steel materials and steel bars were obtained as 345 MPa and 400 MPa, respectively, and the elastic modulus was set at 206 GPa.

Table 5. The material properties of steel used in FEM analysis.

Steel Material	Elastic Modulus (GPa)	Yield Strength (MPa)	Ultimate Strength (MPa)
Stud	207	367	486
Steel Plate	206	345	345
Steel Rebar	206	400	400

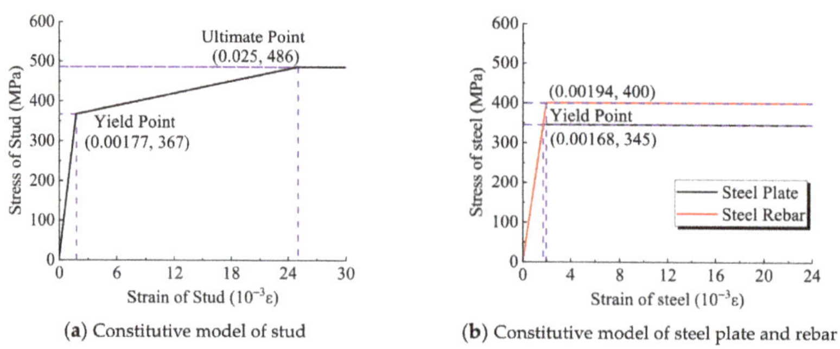

Figure 10. Constitutive model of steel material.

4.3. Boundary Condition

The model with regard to the boundary condition employs surface-to-surface contact to simulate the interactions among the steel–concrete interface, C50–UHPC interface, and

UHPC–stud contact. Tangential contact was simulated with the penalty function friction formula, while normal contact was modeled as hard contact. Since lubricating oil was applied to the steel–concrete interface in the push-out test, the friction coefficient for this interface was set at 0.01. The friction coefficient between UHPC and the stud was set at 0.2. A corrugated metal sleeve with a thickness of 0.6 mm was present between C50 and UHPC, creating a waviness at their interface and generating a higher friction coefficient of 0.6. The interaction between steel reinforcement and concrete was constrained using the embedded command. The upper and lower rigid bases were connected to the surfaces of the specimen with the Tie command.

Figure 11 illustrates the boundary conditions of the model. As the model represents only 1/4 of the practical push-out specimen, appropriate constraints on the surface were applied to the symmetric planes. Furthermore, the frictional interaction between the bottom surface of the concrete block and the test platform was not considered in this model, assuming a complete fixation between the concrete block and the test platform.

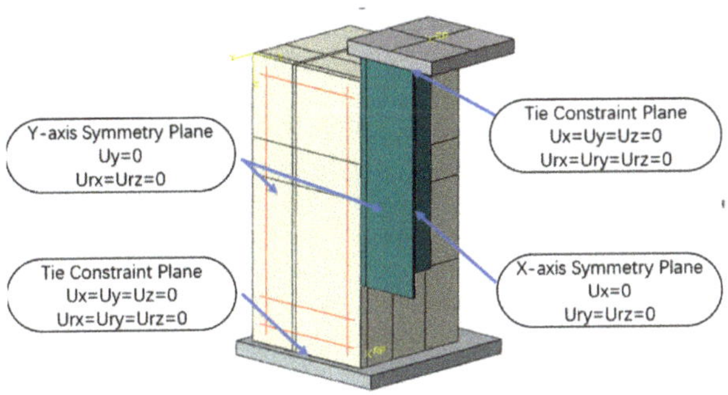

Figure 11. Schematic of model boundary conditions.

4.4. Simulation Verification

(1) Load–Slip curves

Figure 12 presents the comparison between numerically simulated load–slip curves and the results from the push-out tests. The simulated curves are well consistent with the experimental curves. Since the simulation did not explain the stud material damage, the differences in the elastic phase were minor for BSS. However, in the nonlinear phase, the load-bearing capacity was overestimated to a certain extent. Table 6 compares the key shear parameters obtained from the simulation and average test results, revealing discrepancies of 2% to 13%. The above-mentioned differences were primarily attributed to external factors (e.g., sleeve eccentricity, grout leakage at the bottom, and welding quality). In general, the numerical model was validated to have rationality and consistency with the experimental model, effectively capturing the shear characteristics of the studs.

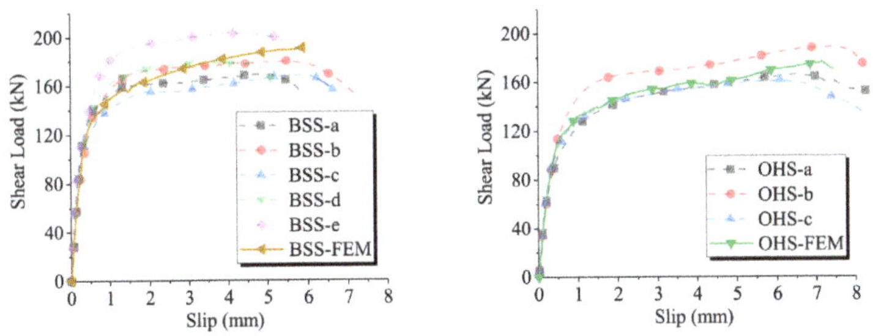

(**a**) BSS (**b**) OHS

Figure 12. Model validation.

Table 6. Comparison of shear parameters between experiments and FEM models.

Group	Ultimate Shear Capacity V_u (kN)		Shear Stiffness $K_{1/3}$ (kN/mm)		Shear Stiffness $K_{1/2}$ (kN/mm)	
BSS-AVG	180.3	↓6%	462	↑13%	402	↑11%
BSS-FEM	191	0%	410	0%	361	0%
OHS-AVG	172.6	↓2%	333	↑3%	272	↓4%
OHS-FEM	176	0%	323	0%	284	0%

(2) Failure mode

The current model cannot simulate the process of stud fracture for not incorporating the metallic damage of the studs. Thus, the moment when the shear capacity reaches its peak serves as the comparative benchmark for the ultimate failure mode. Figure 13 presents the compressive damage distribution of the concrete and Mises stress distribution of the studs that was derived through the finite-element simulation at the ultimate state. As depicted in the figure, the extent of concrete damage in BSS was slightly smaller than that in OHS. Since the damage factor of UHPC was only defined up to 0.7, the internal part of the sleeve, presented by the lighter color, was subjected to greater damage, well consistent with the practical photographs of the failure mode. The comparison of the stress distribution of the studs suggested that the Mises stress contours of BSS were similar to those of OHS. However, the deformation pattern at the root of the studs was pronouncedly distinct. BSS displayed angular deformation at its root, and the overall bending deformation was less pronounced than that of the OHS. The above results suggested that BSS was subjected to less bending–shear coupling, exhibiting a superior shear performance.

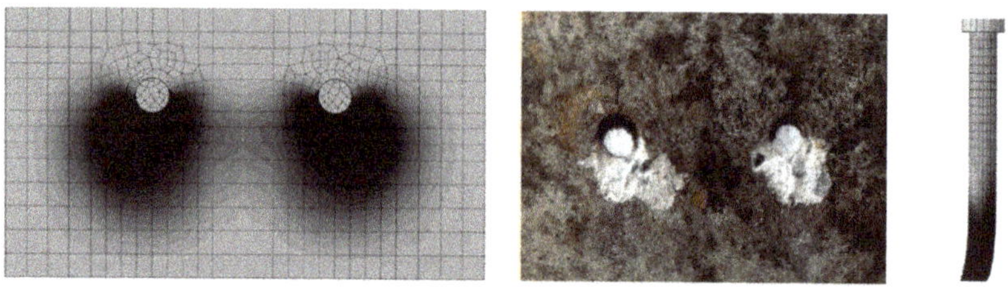

(**a**) OHS

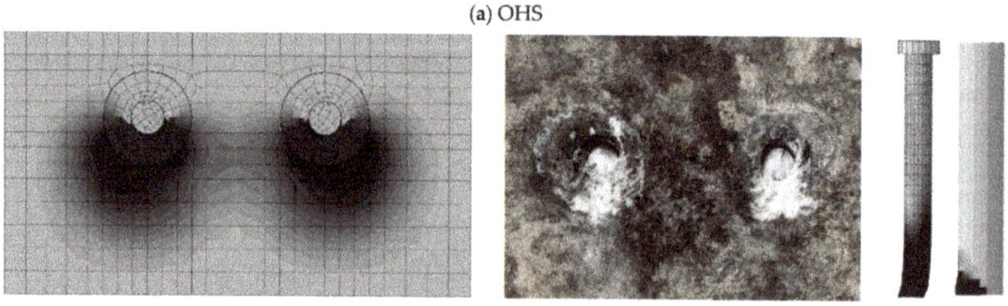

(**b**) BSS

Figure 13. Comparison of failure modes.

4.5. Stress Analysis

(1) Axial normal stress

The axial stresses on the upper and lower edges of the studs in the models for BSS and OHS were extracted, as presented in Figure 14, to investigate the stress distribution along the body of the stud in the loading process. The approximately symmetric axial stresses were imposed on the upper and lower edges of the two studs. With the upper edge of the stud as an example, the upper edge near the root location was subjected to compressive stresses. However, as the height increased, the compressive stress rapidly declined and shifted to tensile stress with a height of nearly 15 mm. At a height of about 35 mm, the stress reached a peak tensile stress and then tended to decline, nearly approaching 0 MPa around 95 mm in height, suggesting a stabilized state. In contrast, the distribution of axial stresses on the lower edge of the stud displayed the opposite pattern. The bending behavior of the stud was indicated by the difference between the upper and lower edge axial stresses. As depicted in the figure, the axial stresses on the upper and lower edges of the stud were almost equal beyond a height of 95 mm, suggesting that the longitudinal loading characteristics of the stud contained bending stresses primarily at a height of 95 mm from the root. However, beyond this height, they were largely subjected to axial tensile forces.

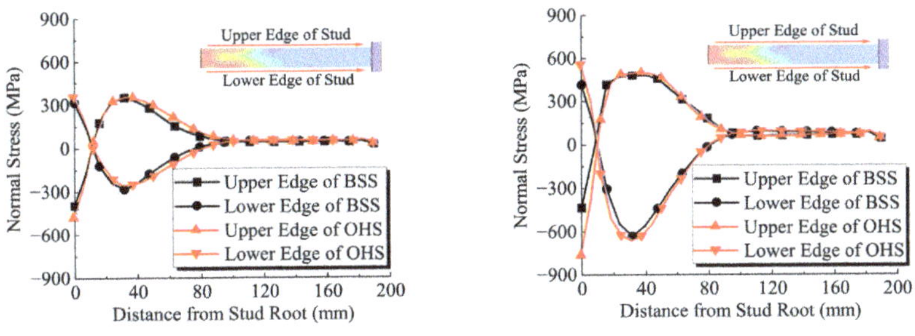

(**a**) 50%V_u (**b**) 90%V_u

Figure 14. Stud normal stress distribution.

As revealed by the further comparison of the axial stress distributions for both types of stud connections at different loading stages, under the shear load of 50%V_u, a minor difference was found in axial stresses between the two types of studs, while BSS had a slightly lower axial stress. However, under the shear load of 90%V_u, the axial stresses of the two studs differed significantly in 15 mm from the root. The axial stress in BSS was approximately 30% to 40% lower than that in OHS, suggesting a weaker degree of bending at the root. The above finding corroborated the conclusions drawn from the observed failure mode. Furthermore, the stress distributions on both types of studs at a height of 15 mm were almost consistent, with similar peak stress values.

(2) Shear stress

The stress displayed a highly complex distribution at the root of the stud during the push-out test. Apart from normal stresses induced by bending and tension, shear stresses were also generated under the effect of direct shear action. Figure 15 presents the numerical extraction of shear stresses along the height direction of the stud's central axis. As depicted in the figure, the shear stress distribution was similar for both types of studs, and peak shear stresses were found at the root of the studs, reaching approximately 280 MPa.

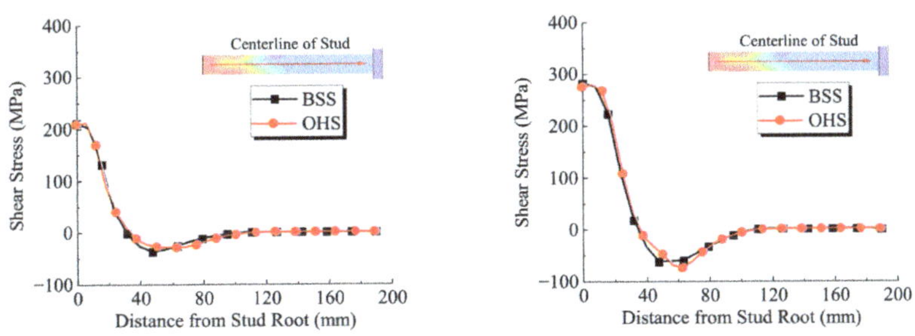

(**a**) 50%V_u (**b**) 90%V_u

Figure 15. Stud shear stress distribution.

Since the shear force was the first derivative of the bending moment, when the distance from the root ranged from 0 mm to 15 mm, the shear stress continued to be high, whereas the normal stresses rapidly varied at this interval. As the distance from the root increased from 15 mm to 35 mm, the shear stress crossed 0 and reversed direction, such that normal stresses achieved reversed peak values that tended to decrease. In the range of 35 mm to 95 mm from the root, the reversed shear stress first increased and then gradually decreased to 0, while the normal stresses also declined gradually and stabilized at a certain level. It can be inferred from the aforementioned analysis that in the range of 35 mm from the root, the stud's lower edge concrete contributed to the shear resistance. Moreover, in the range of 35 mm to 95 mm, the upper edge concrete of the stud contributed to shear resistance. It is noteworthy that the concrete below the stud is critical in providing shear resistance, while it is inevitably subject to compression failure at the ultimate state. When compared with OHS, the reversed shear peak of BSS was closer to the root, suggesting that a more pronounced direct shear effect was generated.

5. Parametric Analysis

5.1. Concrete Material Type for BSS

A total of five finite-element models of the pull-out tests were developed using a wide variety of concrete materials to compare the differences in shear performance between BSS and OHS. The model numbering complied with the rule "base model—sleeve outer material—sleeve inner material". For OHS without sleeves, the concrete materials were comprised of C50 and UHPC. For BSS with sleeves, the computed models consider the economical combination "C50–C50" and the commonly used combination "C50–UHPC", as well as the combination used for UHPC bridge decks "UHPC–UHPC". Figure 16 presents the load–slip curves of different stud connections, and Table 7 summarizes the shear performance indicators. The results of OHS-C were employed as the benchmark for comparison.

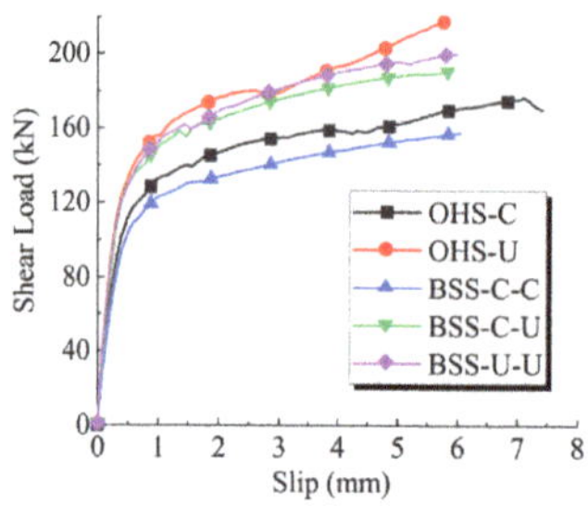

Figure 16. Load–slip curves of stud connectors with different materials.

Table 7. Effect of concrete material type on shear performance of BSS and OHS.

Model Number	Concrete Material Type		Ultimate Shear Capacity V_u (kN)		Shear Stiffness $K_{1/3}$ (kN/mm)		Shear Stiffness $K_{1/2}$ (kN/mm)	
	Outside Sleeve	Inside Sleeve						
OHS-C	C50	N/A	176	0%	323	0%	284	0%
OHS-U	UHPC	N/A	218	↑24%	445	↑38%	367	↑29%
BSS-C-C	C50	C50	152	↓14%	299	↓7%	268	↓6%
BSS-C-U	C50	UHPC	191	↑8%	410	↑27%	361	↑27%
BSS-U-U	UHPC	UHPC	200	↑14%	438	↑36%	372	↑31%

As indicated by the computed results, the shear capacity of BSS-C-C was 24% lower than OHS-C, and the shear stiffness declined by 7%. Likewise, BSS-U-U possessed a shear capacity 10% lower than OHS-U and a shear stiffness decrease of only 3%. The above results suggested that incorporating the steel corrugated sleeve in the concrete led to the reduced overall integrity of the concrete block, reducing the shear performance indicators of the stud. Moreover, the shear strength and stiffness of the concrete materials with lower performance characteristics more pronouncedly decreased. To address the decrease in the shear capacity, BSS-C-U has been generally adopted in engineering practice, which was the bellow-sleeved stud employed in this pull-out test. After UHPC was infused into the sleeve, the shear capacity was improved by approximately 8%, and the shear stiffness was improved by 27%. Superior material performance can be exploited to overcome the shortcomings in construction. Accordingly, in practical bridge design, the steel sleeve should not be filled with ordinary concrete containing fine aggregates. Instead, it is necessary to use a higher modulus of elasticity and better flowability materials (e.g., high-strength mortar or UHPC).

5.2. Stud Height

Due to limitations in the number of specimens, the experimental process only tested the commonly used ML15 studs with a diameter of 22 mm and a height of 200 mm in bridge engineering. There was no comparison made regarding the effect of the aspect ratio (height-to-diameter ratio) on the shear performance of BSS. In this section, finite-element models were established for studs of different heights, ranging from 80 mm to 250 mm. This corresponds to an aspect ratio range of 3.6 to 11.4. The model numbering follows the rule "base model—H stud height". Figure 17 analyzes the variation in shear stiffness of BSS with variations in stud length, and Table 8 presents the computed results of the simulation with varying stud heights. The baseline model is designated as BSS-H200.

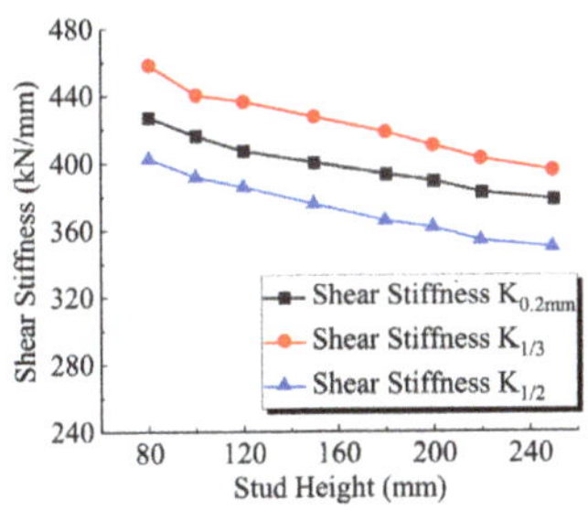

Figure 17. Effect of stud height variation on shear stiffness.

Table 8. Effect of stud height on the shear performance of BSS.

Model Number	Stud Height (mm)	Ultimate Shear Capacity V_u (kN)		Shear Stiffness $K_{1/3}$ (kN/mm)		Shear Stiffness $K_{1/2}$ (kN/mm)	
BSS-H80	80	199	↑4%	458	↑12%	402	↑11%
BSS-H100	100	198	↑4%	440	↑7%	391	↑8%
BSS-H120	120	191	0%	436	↑6%	385	↑7%
BSS-H150	150	192	↑1%	427	↑4%	375	↑4%
BSS-H180	180	192	↑1%	418	↑2%	365	↑1%
BSS-H200	200	191	0%	410	0%	361	0%
BSS-H220	220	190	↓1%	402	↓2%	353	↓2%
BSS-H250	250	189	↓1%	395	↓4%	349	↓3%

The computed results indicate that as the height of the stud increases, the shear strength and shear stiffness show a slight decreasing trend. However, the magnitude of this variation is small, suggesting that the shear behavior of the stud primarily occurs at the root, and its relationship with height is relatively minor. Previous studies have shown that the conventional stud exhibits a coupling of bending and shearing in pull-out tests. As the stud height increases, the bending effect becomes more prominent, leading to a decrease in shear performance. However, in this simulation, UHPC near the root of the sleeved stud has a higher stiffness, resulting in less bending of the stud. Thus, the simulation results are less sensitive to variations in stud height.

Additionally, when the aspect ratio of the stud is less than 4, the shear performance of the stud significantly improves. The shear capacity increases by 4%, and the shear stiffness increases by approximately 11%. When the aspect ratio is between 5 and 11, the shear performance of the stud remains relatively stable, with variations within 4% for all performance indicators. This indirectly indicates that the current design codes have certain limitations on the aspect ratio of studs. In this range, BSS can be uniformly calculated using the prescribed formulas.

5.3. Sleeve Diameter and Stud Diameter

In this study, the commonly used outer diameter 60 mm corrugated metal sleeve was chosen as the steel sleeve for the studs, and no experimental tests were conducted on sleeves of other specifications. Additionally, the effect of the stud diameter and sleeve diameter on the shear performance was not investigated. To address this, finite-element models were developed for sleeves with different diameters (ranging from 50 mm to 90 mm) and varying stud diameters (19 mm, 22 mm, 25 mm). The model numbering convention is "base model—R stud diameter—D sleeve diameter." The computed results are presented in Table 9, and the comparison is based on their respective D60 models.

Table 9. Effect of sleeve diameter on the shear performance of BSS.

Model Number	Ultimate Shear Capacity V_u (kN)		Shear Stiffness $K_{1/3}$ (kN/mm)		Shear Stiffness $K_{1/2}$ (kN/mm)	
BSS-R19-D50	153	↓3%	347	↓1%	302	↓2%
BSS-R19-D60	159	0%	351	0%	307	0%
BSS-R19-D70	164	↑3%	373	↑6%	318	↑4%
BSS-R19-D80	168	↑6%	387	↑10%	318	↑4%
BSS-R19-D90	169	↑6%	403	↑15%	331	↑7%
BSS-R22-D50	186	↓3%	385	↓6%	349	↓3%
BSS-R22-D60	191	0%	410	0%	361	0%
BSS-R22-D70	196	↑3%	415	↑1%	369	↑2%
BSS-R22-D80	206	↑8%	436	↑6%	374	↑4%
BSS-R22-D90	207	↑8%	447	↑9%	380	↑5%
BSS-R25-D50	222	↓3%	443	↓3%	402	↓3%
BSS-R25-D60	228	0%	455	0%	414	0%
BSS-R25-D70	236	↑4%	475	↑4%	426	↑3%
BSS-R25-D80	246	↑8%	474	↑4%	423	↑2%
BSS-R25-D90	250	↑10%	481	↑6%	430	↑4%

Figure 18a,b depict the variations in shear capacity and shear stiffness with variations in steel sleeve diameter under three stud diameters, respectively. It can be observed that as the sleeve diameter increases, the shear strength and shear stiffness of the stud also increase, suggesting that the UHPC inside the sleeve provides a stronger resistance against the bending and shearing of the stud. When comparing models with the same stud diameter, as the sleeve diameter exceeds 80 mm, the rate of increase in shear capacity slows down and approaches a critical point, while the rate of increase in shear stiffness remains unchanged. Although the simulation results demonstrate that larger sleeve diameters lead to a better shear performance of BSS, effectively mitigating the negative effect of the steel sleeve, practical bridge engineering considerations, such as the group arrangement of studs, spatial limitations, and rebar clearances, suggest that the sleeve diameter should not exceed 70 mm. For sleeve diameters of 50 mm to 70 mm, the shear capacity and shear stiffness vary within ±3%. Thus, when designing connections in this range, it may not be necessary to account for the effect of sleeve diameter variation on shear performance.

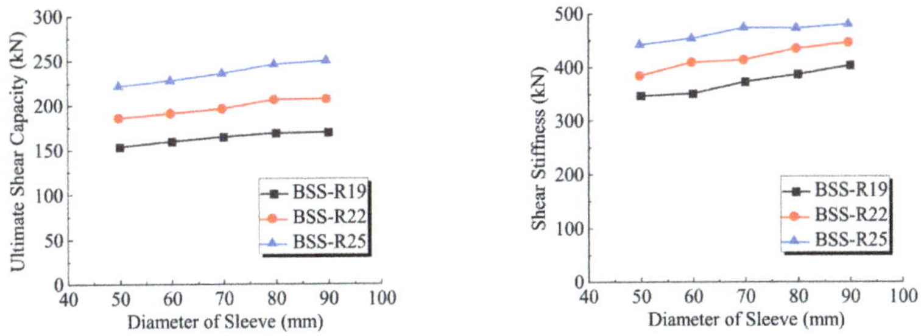

(**a**) ultimate shear capacity (**b**) shear stiffness

Figure 18. Variation curve of shear parameters with sleeve diameter.

Comparing the differences in the shear performance of BSS under various stud diameters, from the perspective of shear capacity, larger stud diameters are more significantly affected by the sleeve diameter. The shear capacity of the stud with a 25 mm diameter fluctuates from −3% to 10%, while that of the 19 mm-diameter stud varies within −3% to 6%. Considering shear stiffness, larger stud diameters are less influenced by sleeve diameter. The shear stiffness of the 25 mm-diameter stud varies within −3% to 6%, while that of the 19 mm-diameter stud even exceeds 10%. From these observations, the shear stiffness of the larger diameter stud is more correlated with the material properties of the stud itself, with the contribution of the concrete material strength to shear stiffness being relatively smaller. As a result, the effect of sleeve diameter variation is relatively minor.

5.4. Deviation of Sleeve and Stud Relative Position

During the course of the experiment, significant variability was observed in the results of BSS. This variation was attributed to certain specimens where the steel sleeve deviated due to the effect of concrete during pouring, causing the stud to be off-center in the sleeve. As a result, the concrete material properties around the root of the stud exhibited considerable differences. To address this issue, finite-element models were developed for specimens with four different positions of the steel sleeve, investigating the effect of a 1 cm deviation in the stud position in the sleeve. The model numbering convention is "base model—position of stud relative to sleeve interior (up—stud positioned higher, down—stud positioned lower, side—stud positioned at the side)", with BSS-middle serving as the comparative baseline.

The load–slip curves of the models are depicted in Figure 19, and shear performance indicators are presented in Table 10. The simulation results indicate that the positioning of the stud inside the sleeve, whether higher or lower, has a significant effect on the shear performance. When the stud is positioned higher, the shear capacity increases by 5%, and the shear stiffness by 7%. In this scenario, there is a thicker UHPC layer below the root of the stud to withstand the compressive stress transmitted from the lower surface of the stud. Conversely, when the stud is positioned lower, the shear capacity decreases by 5%, and the shear stiffness decreases by 13%. This is attributed to the thinner UHPC layer beneath the root of the stud, coupled with its closer proximity to the weakened section near the corrugated sleeve wall, resulting in unfavorable loading conditions. In the case of a sidewise deviation, there is minimal variation in the ultimate shear capacity, and the shear stiffness slightly decreases by 1%. In summary, it can be concluded that the deviation of the sleeve position has a relatively small and symmetric effect on shear capacity, while its influence on shear stiffness is more pronounced.

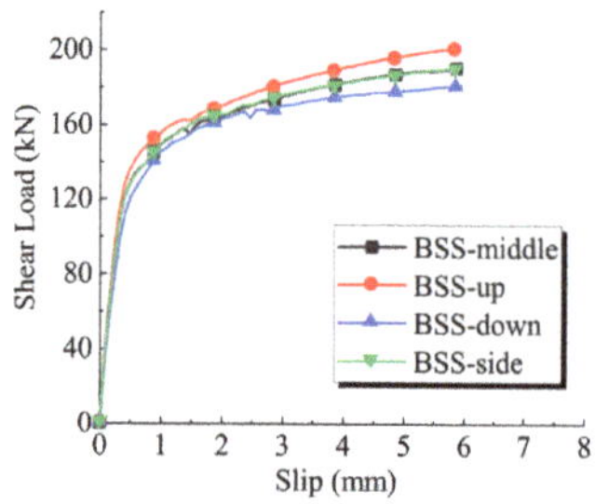

Figure 19. Load–slip curves of BSS with different sleeve positions.

Table 10. Effect of sleeve installation deviation on the shear performance of BSS.

Model Number	Sleeve Installation Deviation	Ultimate Shear Capacity V_u (kN)		Shear Stiffness $K_{1/3}$ (kN/mm)		Shear Stiffness $K_{1/2}$ (kN/mm)	
BSS-middle	Stud Centered	191	0%	410	0%	361	0%
BSS-up	Stud 1 cm Above Center	201	↑5%	435	↑6%	388	↑7%
BSS-down	Stud 1 cm Below Center	181	↓5%	358	↓13%	314	↓13%
BSS-side	Stud 1 cm to the Side	191	0%	406	↓1%	356	↓1%

5.5. Sleeve Grout Leakage Height

During construction, notwithstanding the effect of sleeve deviation, the leakage of grout at the bottom of the sleeve can adversely affect the shear performance of BSS. Grout leakage can reduce the material strength of the concrete surrounding the root of the stud, leading to the premature compressive failure of the concrete. Although measures were taken to prevent leakage in the current experimental tests, since there were no cases of grout leakage, the reality of construction sites with a large number of stud connections cannot ensure the construction quality of leakage-proof measures, inevitably resulting in some degree of grout leakage in the sleeve. Once the grout fills the root of the stud, it can adversely affect the shear performance of the connection. Thus, finite-element models were developed for stud specimens with different degrees of grout leakage in the sleeve. The grout leakage height ranged from 10 mm to 50 mm, assuming that the grout material in the sleeve is C50 concrete, with UHPC remaining above the grout leakage. The model naming convention is "base model—grout leakage height", with BSS-C-U as the baseline model.

Table 11 summarizes the shear performance parameters for each model. Figure 20 depicts the load–slip curves for each model. BSS-C-U represents the model with no grout leakage and essentially serves as the upper limit of the curves, while BSS-C-C represents the model with C50 concrete filling in the sleeve and acts as the lower limit of the curves. The other curves lie between these two scenarios. A comparison of models with grout leakage reveals that grout failure around the root of the studs occurs prematurely, leading to a sharp decrease in shear load. Although the shear load still increases after the grout failure, it cannot exceed the maximum value before failure. Under the premature failure of the grout, a smaller ultimate slip displacement is generated, and the shear capacity develops insufficiently. For instance, with a grout leakage height of 50 mm, the maximum shear strength was reduced by approximately 16%. Figure 21 illustrates the effect of the grout leakage height on shear stiffness, showing a steep decline of about 22% when the leakage height ranged from 0 mm to 50 mm. When the leakage height went beyond 50 mm, the effect gradually declined, with the curve's shape resembling the deformation of the stud.

In brief, grout leakage in the sleeve can reduce the shear strength and shear stiffness of the stud connection. In practice, if proper measures were taken to prevent leakage, with the grout leakage height ensured to be within 10 mm, the effect on the capacity and stiffness was minor. If the grout leakage height exceeded 10 mm, subpar construction quality was indicated, requiring the absorption of cement slurry leaking from the bottom of the sleeve using a geotextile fabric.

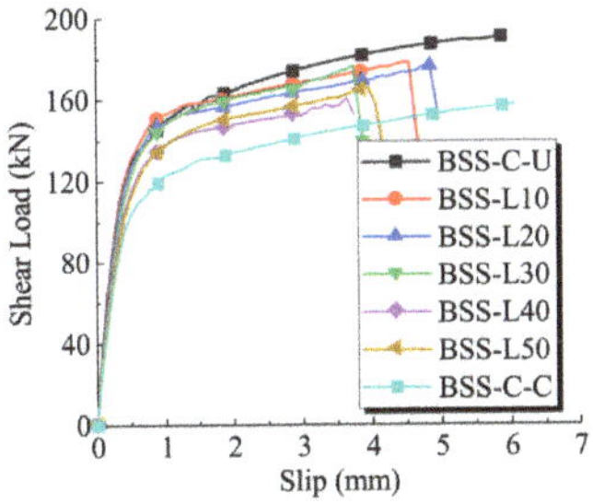

Figure 20. Load–slip curves of BSS with different grout leakage heights.

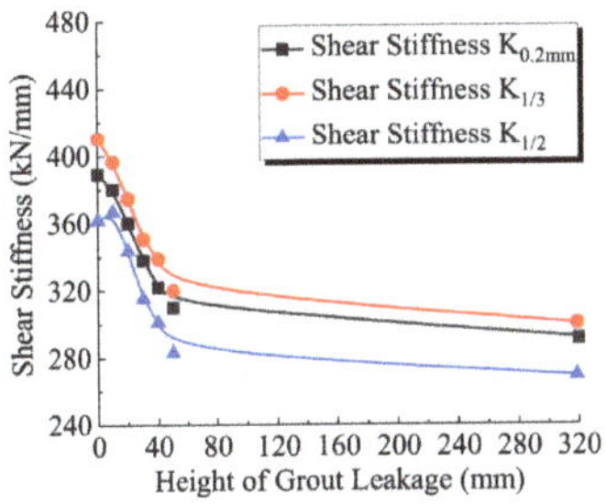

Figure 21. Effect of grout leakage height on shear stiffness.

Table 11. Effect of grout leakage height inside the sleeve on the shear performance of BSS.

Model Number	Grout Height Inside Sleeve (mm)	Ultimate Shear Capacity V_u (kN)		Shear Stiffness $K_{1/3}$ (kN/mm)		Shear Stiffness $K_{1/2}$ (kN/mm)	
BSS-C-U	0	191	0%	410	0%	361	0%
BSS-L10	10	178	↓7%	396	↓3%	366	↑1%
BSS-L20	20	176	↓8%	374	↓9%	343	↓5%
BSS-L30	30	176	↓8%	350	↓15%	314	↓13%
BSS-L40	40	160	↓16%	338	↓18%	300	↓17%
BSS-L50	50	166	↓13%	319	↓22%	282	↓22%
BSS-C-C	320	152	↓20%	299	↓27%	268	↓26%

6. Conclusions

In this study, BSS and OHS were investigated through push-out tests and numerical simulations. Next, the shear performance of these connections was compared. Moreover, the effect of different parameters and construction conditions on the shear performance of BSS was investigated. The conclusions of this study are drawn as follows:

(1) BSS and OSS exhibited shear failure at the root of the studs. The concrete damage region close to the root of BSS was smaller, and the bending deformation turned out to be less pronounced compared with OHS. As indicated by the above result, BSS was subjected to a greater effect of direct shear and a smaller effect of bending–shear coupling, exhibiting a superior shear performance.

(2) It is noteworthy that the steel corrugated sleeve of BSS led to the reduced overall integrity of the concrete block and declining shear performance indicators. Nevertheless, this deficiency in the structural design can be compensated by filling the steel corrugated

sleeve with high-performance concrete materials (e.g., UHPC or high-strength mortar), such that superior material properties are exploited.

(3) The results of the push-out tests and numerical simulations confirmed that BSS exhibited a higher ultimate shear bearing capacity and larger shear stiffness than OHS. Moreover, the slip at the ultimate bearing capacity was minor. BSS did not confer advantages over OHS in enhancing shear strength, with a modest 4.5% increase in the ultimate shear capacity. However, BSS exhibited a strong resistance to slip deformation, with the shear stiffness increasing from 31.9%.

(4) Compared with the stud height, the sleeve diameter and the stud diameter significantly affected the shear capacity and stiffness of BSS. When the sleeve diameter varied from 50 mm to 100 mm, the shear strength and the shear stiffness increased by approximately 8% and 10%, respectively. As indicated by the above results, the UHPC near the root of the stud provided stronger shear support. Under the effect of the reinforcement arrangement in the bridge deck, the diameter of the corrugated sleeve should not exceed 70 mm to ensure sufficient reinforcement spacing.

(5) The sleeve deviation slightly affected the shear bearing capacity and showed symmetric characteristics. However, it significantly affected the shear stiffness, especially when the stud was positioned below the central axis of the sleeve. The leaking grout at the sleeve bottom exerted a more pronounced effect. When the grout leakage height fell into a range of 50 mm, the shear bearing capacity and the shear stiffness decreased by 16% and 22%, respectively. Thus, the installation quality of the corrugated steel sleeves should be tested rigorously.

Author Contributions: Methodology, D.Z.; Software, D.Z.; Validation, Q.S.; Formal analysis, D.Z.; Investigation, F.W.; Resources, F.W. and Z.L.; Data curation, P.L. All authors have read and agreed to the published version of the manuscript.

Funding: This research is sponsored by the Fujian Transportation Technology Project (No. 202126) and Henan Major Research and Development Program for Special Projects (No. 231111241500).

Data Availability Statement: The original contributions presented in the study are included in the article, further inquiries can be directed to the corresponding authors.

Conflicts of Interest: Author Zhiping Lin was employed by the company Fujian Expressway Group Co., Ltd. Author Peiran Li was employed by the company Henan Transportation Planning and Design Research Institute Co., Ltd. The remaining authors declare that the research was conducted in the absence of any commercial or financial relationships that could be construed as a potential conflict of interest.

References

1. Ollgaard, J.G.; Slutter, R.G.; Fisher, J.W. Shear strength of stud connectors in lightweight and normal weight concrete, AISC Eng'g Jr., April 1971 (71-10). *AISC Eng. J.* **1971**, 55–64.
2. Viest, I.M. Investigation of Stud Shear Connectors for Composite Concrete and Steel T-Beams. *J. Am. Concr. Inst.* **1956**, *27*, 875–891. [CrossRef] [PubMed]
3. Oehlers, D.J.; Coughlan, C.G. The shear stiffness of stud shear connections in composite beams. *J. Constr. Steel Res.* **1986**, *6*, 273–284. [CrossRef]
4. Prakash, A.; Anandavalli, N.; Madheswaran, C.K.; Lakshmanan, N. Modified push-out tests for determining shear strength and stiffness of HSS stud connector-experimental study. *Int. J. Compos. Mater.* **2012**, *2*, 22–31. [CrossRef]
5. Oehlers, D.J.; Foley, L. The fatigue strength of stud shear connections in composite beams. *Proc. Inst. Civ. Eng. Part Res. Theory* **2015**, *79*, 349–364. [CrossRef]
6. Roik, K.; Hanswille, G.; Cunze-O Lanna, A. *Harmonisation of the European Construction Codes*; Report of Eurocode; Ministerium fur Raumordnung, Bauwesen und Städtebau: Bonn, Germany, 1989.
7. *EN 1990*; Eurocode—Basis of structural design. Section 6 Verification by the Partial Factor Method. European Committee for Standardization: Brussels, Belgium, 2002.
8. *EN 1994*; Eurocode-4: Design of Composite Steel and Concrete Structures. Part 1-1 General Rules for Buildings. CEN-European Committee for Standardization: Brussels, Belgium, 2004.
9. *AASHTO LRFD Bridge Design Specifications*, 3rd ed.; American Association of State Highway and Transportation Officials: Washington, DC, USA, 2004.

10. *GB50017-2003*; Code for Design of Steel Structures. Ministry of Construction of China, China Planning Press: Beijing, China, 2003.
11. Roik, K. *Methods of Prestressing Continuous Composite Girders*; Proceedings, Conference on Steel, June 1968; BSSA: London, UK, 1969.
12. Ayyub, B.M.; Sohn, Y.G.; Saadatmanesh, H. Prestressed Composite Girders. I: Experimental Study for Negative Moment. *J. Struct. Eng.* **1992**, *118*, 2743–2762. [CrossRef]
13. Ayyub, B.M.; Sohn, Y.G.; Saadatmanesh, H. Prestressed Composite Girders. II: Analytical Study for Negative Moment. *J. Struct. Eng.* **1992**, *118*, 2763–2782. [CrossRef]
14. Chen, S. Experimental study of prestressed steel–concrete composite beams with external tendons for negative moments. *J. Constr. Steel Res.* **2005**, *61*, 1613–1630. [CrossRef]
15. Su, H.; Su, Q.; Xu, C.; Zhang, X.; Lei, D. Shear performance and dimension rationalization study on the rubber sleeved stud connector in continuous composite girder. *Eng. Struct.* **2021**, *240*, 112371. [CrossRef]
16. Xu, X.; Liu, Y.; He, J. Study on mechanical behavior of rubber-sleeved studs for steel and concrete composite structures. *Constr. Build. Mater.* **2014**, *53*, 533–546. [CrossRef]
17. Su, Q.; Yang, G.; Wu, C. Experimental studies on mechanical behavior of continuous composite girder with prefabricate prestressed concrete slab. *J. Tongji Univ.* **2012**, *40*, 996. [CrossRef]
18. Xue, D.; Liu, Y.; Yu, Z.; He, J. Static behavior of multi-stud shear connectors for steel-concrete composite bridge. *J. Constr. Steel Res.* **2012**, *74*, 1–7. [CrossRef]
19. Yu, J.; Wang, Y.H.; Liu, J.-P.; Chen, Y.F.; Xue, S.-L.; Yu, Z. Stiffness analysis of shear stud group connectors in precast concrete deck. *Eng. Struct.* **2023**, *274*, 115148. [CrossRef]
20. Su, Q.; Zou, D.; Zhang, L.; Tao, X.; Huang, C. Field Test and Effect Analysis of Concrete Prestress for the Post-Combined Prestressing Composite Girder Bridge. *J. Tongji Univ.* **2021**, *49*, 1061–1069. [CrossRef]
21. *GB50010-2010*; Code for Design of Concrete Structures. Ministry of Construction of China, China Planning Press: Beijing, China, 2010.

buildings

MDPI

Article

Seismic Performance of Precast Double-Skin Composite Shear Wall with Horizontal Connection Region

Huanqin Liu *, Nuoqi Shi, Xu Fu and Jingjing Zhang

College of Architecture Engineering, North China Institute of Aerospace Engineering, Langfang 065000, China; snq258045@163.com (N.S.); fuxu431@163.com (X.F.); zjj1990@nciae.edu.cn (J.Z.)
* Correspondence: liuhq@nciae.edu.cn

Abstract: This paper proposed a novel, precast double-skin composite (DSC) shear wall, which was composed of two precast parts at the factory and welding and pouring grouting material on site. One monolithic cast-in-place DSC shear wall specimen and two precast DSC shear wall specimens with different axial compression ratios were subjected to reverse cyclic loading tests. The results indicated that the failure mode of both the cast-in-place and precast DSC shear wall shear walls were compression-bending failures, and the damage range of specimens within a height range of 100 mm to 200 mm from the bottom of the DSC shear wall. The load-bearing capacity of the precast specimen was 6.3% higher than that of the monolithic counterpart, but its ductility was reduced by 16%. The precast DSC shear wall with better casting quality and easier site installation exhibited a satisfactory seismic performance on a par with that of the monolithic cast-in-place DSC shear wall. Under higher axial compression ratios, the bearing capacity and energy dissipation of the precast DSC shear wall specimen significantly improved due to the enhanced confinement effect. Finite element (FE) models clarified the stress and deformation mechanisms between the exterior steel plate and the infill concrete. Finally, the key parameters affecting the seismic bearing capacity of the precast DSC shear wall were identified through FE parameter analysis.

Keywords: precast DSC shear wall; seismic performance; post-grouting technology; prefabricated structure

1. Introduction

Shear walls as lateral load-resisting elements have been widely employed in high-rise buildings for decades to withstand the lateral loads induced by earthquakes and wind as well as vertical load induced by gravity [1]. Precast reinforced concrete (RC) shear walls with grouted sleeve connections are predominant in Chinese high-rise prefabricated structural systems [2,3]. As the height of high-rise buildings has increased, the demand for larger cross-sections of RC shear wall has grown due to the amplified base shear, bending moments, and self-weight. The axial compression ratio, as a vital parameter affecting the seismic behavior precast RC shear walls with grouted sleeve connections, must meet the limit value specified in seismic design standards like GB 50011-2010 [4], ACI 318 [5], and Eurocode8 [6] for ensuring sufficient ductility and deformation capacity under seismic action. The increased cross-section of RC shear walls reduces the rentable building space and increases the gravity of buildings that then intensifies the seismic horizontal base shear force [7]. Additionally, precast RC shear walls require grouted sleeve connections with relatively high installation accuracy and highly skilled labor due to their vulnerability to construction errors and uncertainties. Thus, using RC shear walls with grouted sleeve connections in ultra-high buildings is often impractical and uneconomical.

A double-skin composite (DSC) shear wall is an innovative lateral load-resisting system, comprising exterior steel plates, infill concrete, boundary columns, and connectors. On the one hand, the concrete under triaxial compression demonstrates superior strength and ductility due to the confinement effect from the exterior steel plates. The exterior steel plates function as the confinement for the infill concrete and the formwork for casting in situ concrete. On the other hand, the infill concrete can prevent premature local buckling of exterior steel plates. Researchers including Ji et al. [8], Nie et al. [9], Hu et al. [10], Nie et al. [11], Chen et al. [12], Luo et al. [13], Zhang et al. [14], Zhang et al. [15], Ji et al. [16], Guo et al. [17], Huang et al. [18], and Yan et al. [19], have experimentally studied the behavior of DSC shear walls under combined axial loads and reversed cyclic loading. The results reveal that the strength and deformation of DSC shear walls are markedly better than those of RC shear walls even with high axial loads. Thus, the DSC shear walls are a viable substitute for conventional RC shear walls in ultra-high-rise buildings, offering smaller cross-sections, lighter weight, and more rentable building space. DSC shear walls have been utilized in high-rise and super-high-rise buildings like the Yancheng TV Tower, Beijing Fortune Plaza [20], China Zun Tower [21], and Beijing's World Trade Center [22] in China. In practical engineering applications, DSC shear walls typically incorporate boundary columns and connectors. Nie et al. [9], Ji et al. [16,23], and Varma et al. [24] reported on the in-plane seismic response of the proposed DSC shear wall with boundary columns. Furthermore, connectors are also commonly applied to further increase the bonding performance between exterior steel plates and infill concrete. Researchers analyzed the seismic performance of DSC shear walls with different connectors, such as stud connectors, angle connectors, overlapped studs, J-hooks [25], L-connectors [26], tie bars [12,27], tie bolts [28,29], and batten plates [9]. It is worth noting that the head stud connectors in the DSC shear wall can effectively anchor exterior steel plates onto the infill concrete, and tie bolts counteract local buckling of steel plates.

Steel components or precast concrete elements can be prefabricated in factories and directly assembled on site. However, traditional DSC shear walls require welding of steel plates and concrete pouring on-site. Despite eliminating the need for formwork, the construction efficiency of traditional DSC shear walls is still hindered by the in situ concrete pouring process, which prevents synchronization with prefabricated construction. Moreover, on-site concrete pouring has resulted in an intricate construction process and extensive wet work that contribute to environmental pollution. Since the 1990s, scholars have extensively studied precast PC shear walls with a horizontal connection region. Hutchinson [30] proposed the construction form of assembling prefabricated walls with post-tensioned tendons. Soudki [31,32] studied the seismic performance and deformation capacity of precast PC shear walls with various connection regions and proposed design methods in horizontal connection regions in earthquake conditions. Qian [33] utilized sleeve grouting and hole grouting lap-splice methods for connecting vertical reinforcement in the horizontal connection region of precast components. The experiments revealed that the energy dissipation and stiffness of precast PC shear walls matched that of cast-in-place PC shear walls under low reversed cyclic loading. Sun [34] proposed a precast PC shear wall with dry connections, using high-strength bolts to connect the horizontal connection region. Current studies have mainly focused on precast PC shear walls with horizontal connection regions, with fewer studies examining precast DSC shear walls with horizontal connection regions.

This study introduces a precast DSC shear wall, as shown in Figure 1. The precast DSC shear wall is divided into upper and lower sections with boundary columns, featuring precast DSC wall components with factory-cast concrete and a horizontal connection region for site connection. Connection plates are placed within the horizontal connection region for accurate on-site alignment to promote precise alignment. The precast wall components consist of exterior steel plates, infilled concrete, and connectors. During construction, cranes lift the upper precast wall components to align with the lower ones. The upper and lower parts with boundary columns are joined by welding. Subsequently, high-strength grouting material is injected into the horizontal connection region through injected holes to integrate with the internal precast concrete. The innovative precast DSC shear wall using minimal welding and post-grouting technology enhances construction efficiency. Factory casting ensures the component quality, reduces environmental pollution, and minimizes material waste. It is noteworthy that the performance of prefabricated DSC shear walls largely depends on the quality of the grouting material after pouring. Any issues during the grouting process, such as incomplete filling or poor bonding, can significantly compromise structural integrity and seismic performance.

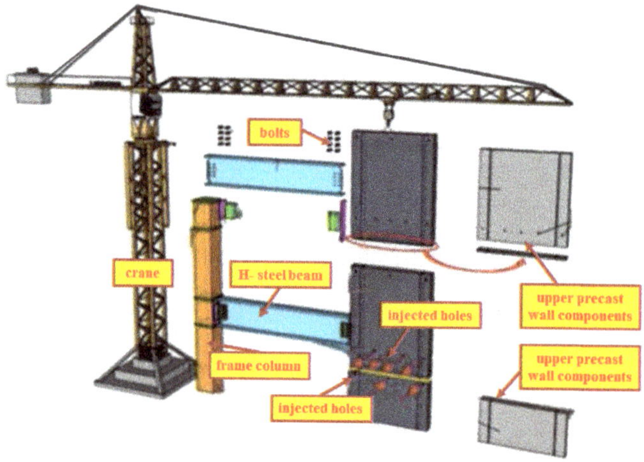

Figure 1. Details of the precast DSC shear wall.

To investigate the seismic performance of the innovative precast DSC shear wall with horizontal connection region, one monolithic cast-in-place DSC shear wall specimen and two precast DSC shear wall specimens with different axial compression ratios were designed. Low reversed cyclic loading tests were conducted to analyze the failure modes, hysteresis curves, load-bearing capacity, energy dissipation capacity, and so on. The buckling behavior of steel plates and the infill concrete was examined through finite element (FE) models to reveal the stress mechanism of precast DSC shear wall under low reversed cyclic loads. Furthermore, a parametric analysis was conducted to investigate the influence of geometric details and material strength on the load-bearing capacity of precast DSC shear walls.

2. Test Program

2.1. Specimen Design and Fabrication

Three half-scale DSC shear wall specimens were designed according to a prototype building to evaluate the seismic behavior of precast shear wall. The variables in the DSC shear wall specimens were the fabrication process (monolithic cast-in-place or precast) and axial compression ratio (0.2 or 0.4). The three test specimens were named SWC-0.4, SWP-0.4, and SWP-0.2, with "SW" indicating shear wall. The letter "C" or "P" indicate the monolithic cast-in-place or precast as the fabrication process of the DSC shear wall specimens, respectively, and the number indicates the axial compression ratio. The details of the DSC shear wall specimens are shown in Figure 2. All test specimens have a total height of 1950 mm, a width of 400 mm, and a thickness of 75 mm. Precast DSC shear wall specimens are comprised of upper and lower precast wall components, with the upper precast walls measuring 900 mm and the lower walls 1050 mm in height. The boundary columns are constructed from cold-formed concrete-filled square steel tubes with dimensions of 100 mm × 3 mm. Box-type loading beams are installed at the top and foundation beams at the bottom of the DSC shear walls, complemented by 20–30 mm steel plates internally as stiffening ribs for increased stiffness. The loading beams feature a 300 mm × 150 mm cross-section, while the foundation beams measure 300 mm × 300 mm.

The exterior steel plates of DSC shear wall were butt-welded to boundary columns with a weld quality of not less than grade 2. To ensure the collaboration between the steel plates and the infilled concrete in DSC shear wall specimens, a mixed connection method utilizing both split bolts and headed studs was employed. Split bolts were staggered with studs interspersed, maintaining a 100 mm interval. The layout of injected holes and exit holes in the upper and lower precast wall components and boundary columns is shown in Figure 2.

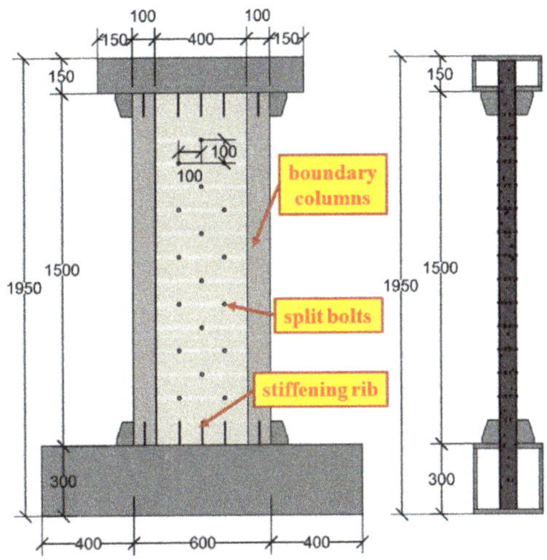

(a) Monolithic cast-in-place DSC shear wall specimens.

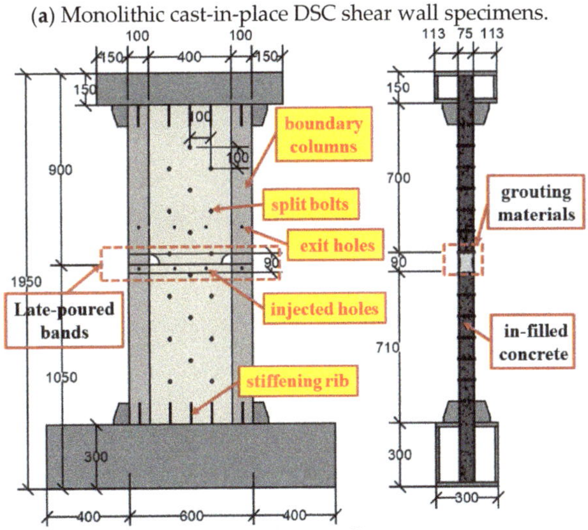

(b) Precast DSC shear wall specimens.

Figure 2. Details of the specimens.

Figure 3 shows the fabrication process of the precast DSC shear wall specimens, taking SWP-0.2 as an example.

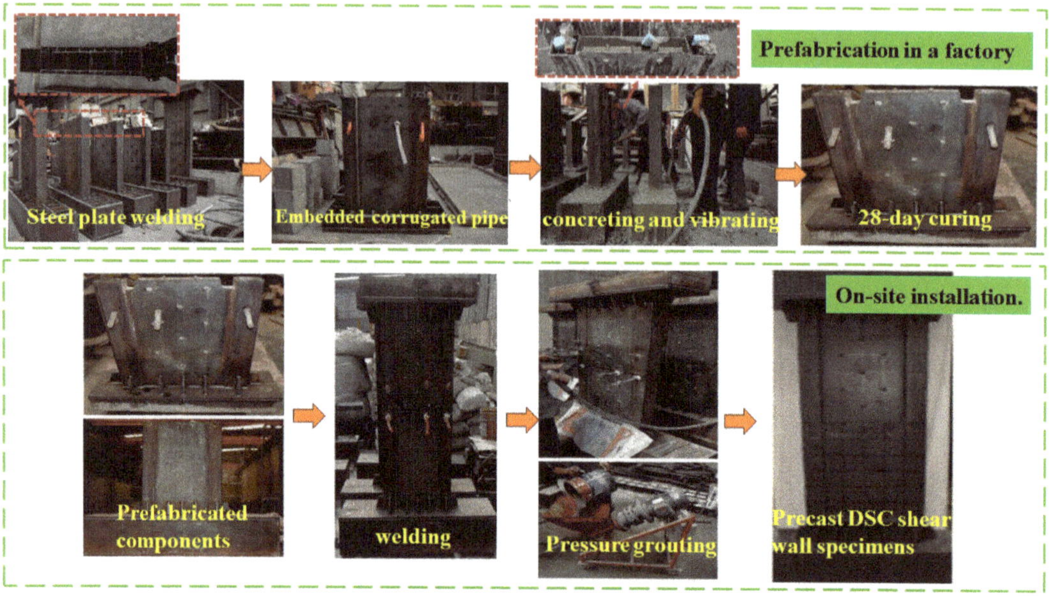

Figure 3. Fabrication process of the precast DSC shear wall specimens.

(1) Weld the steel plates according to the blueprint.
(2) Install corrugated pipes at the injected holes and exit holes of the precast wall components.
(3) Cast concrete and cure for 28 days. After curing, roughen the surface of the infilled concrete to a minimum depth of 6 mm to enhance the bonding performance between the grouting materials and the infilled concrete of the precast wall component.
(4) Hoist the upper precast wall components above the lower precast wall components, and weld them together after alignment.
(5) Inject grouting materials into the corrugated pipes of the lower precast wall components until it flows out from the grouting exit holes of the upper precast wall component. Seal the injected holes and exit holes with wire and execute pressure grouting on boundary columns individually.
(6) After the curing of the grouting materials, cut off the corrugated pipes protruding from the surface of the precast DSC shear wall.
(7) End the process with grinding and painting.

2.2. Material Properties

Three 150 mm cubic blocks for C30 concrete and three 40 mm × 40 mm × 160 mm prismatic blocks for grouting material were prepared and cured in the same conditions as the precast DSC shear wall specimens. After being cured for 28 days, the C30 cubic blocks underwent mechanical testing according to Chinese code GB/T50081-2019 [35] with the cubic compressive strength determined to be 43.2 MPa. The grouting material, assessed base on GB/T17671-1999 [36], showed a compressive strength of 50.7 MPa. According to GB/T 228-2010 [37], the material properties of the steel plates in the same batch as the DSC shear wall are summarized in Table 1.

Table 1. Mechanical parameters of steel plates.

Position	Yield Strength (MPa)	Ultimate Tensile Strength (MPa)	Elasticity Modulus (GPa)
Shear wall	287.5	422.5	211
column	305.2	420.3	213

2.3. Test Setup and Loading Protocol

Figure 4 illustrates the loading setup. Prior to testing, four ground anchors were used to fix the ground beam on the rigid ground, and limit beams at both ends of the ground beam-constrained horizontal displacement to prevent the overall drift of the shear wall specimens during loading. The test beam was connected to a 1000 kN hydraulic actuators on the reaction wall through a threaded rod. The vertical hydraulic actuators, connected to the steel beam via a sliding support, applied a constant vertical load to the distribution beam.

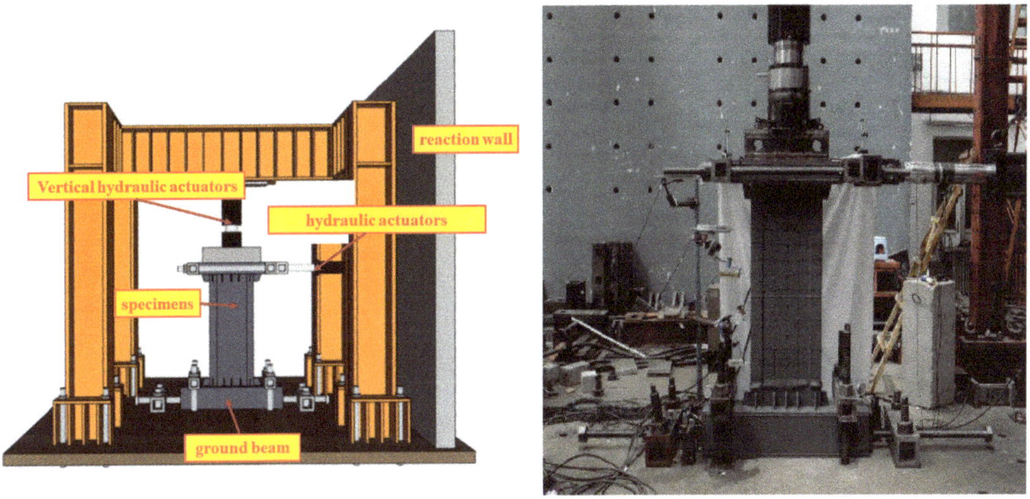

(**a**) Schematic view (**b**) Photography of test setup

Figure 4. Loading setup. (**a**) Schematic view; (**b**) Photography of test setup.

In accordance with JGJ/T 101-2015 [38], the loading protocol in this test was a hybrid load-displacement control with the load-controlled during the elastic phase followed by a displacement-controlled phase, as shown in Figure 5. During the test, a constant vertical load was applied at the top of the distribution beam using different axial compression ratio. The axial compression ratios for SWC-0.4 and SWP-0.4 were 0.4 and 0.2 for SWP-0.2. A load-controlled pattern was adopted, and the load of each stage repeated once until the yielding of steel. Then, a displacement-controlled pattern was applied and loaded at an integer multiple of the yield displacement, which was repeated three times at each stage. The loading was terminated when the force descended to 85% of the peak load.

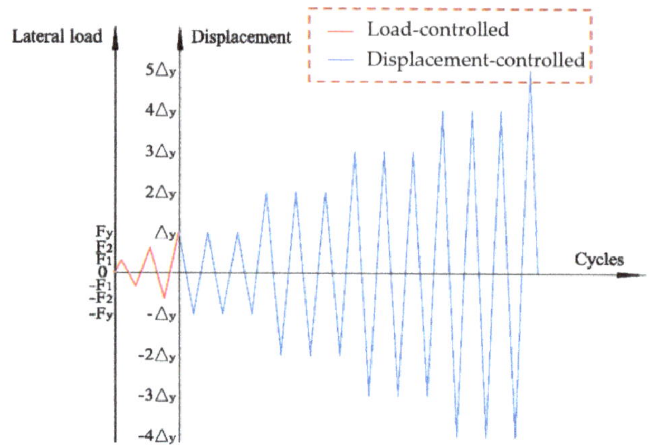

Figure 5. Loading protocol.

The arrangement of measuring points on the specimens is shown in Figure 6, including nine linear variable differential transducers (LVDTs), eight strain gauges, and eighteen strain rosette gauges. L1 to L5 are arranged along the height of the wall to monitor horizontal displacements. L6 to L8 are positioned to correct for the potential drifts and rotations of the foundation beam. Additionally, an out-of-plane LVDT (L9) is installed mid-shear wall to monitor the stability of the DSC shear wall specimens. Diagonal guide rod extensometers G1 and G2 are set at the lower precast wall components to measure shear deformation. Strain rosettes and strain gauges are placed at key locations to record the deformation progression in the late-poured bands and at the base of the wall panel. SWC-0.4 lacks strain and rosette gauges around the late-poured bands, but the other configurations remain consistent.

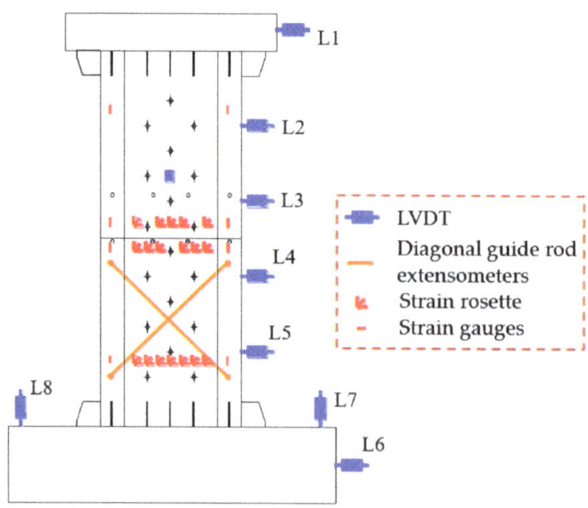

Figure 6. Arrangement of measuring points.

3. Results and Discussions

3.1. Failure Modes

Figure 7 shows the failure modes of all specimens. After vertical loading (axial compression ratio of 0.4), slight buckling occurred in the steel plate of SWC-0.4 specimen.

This may be attributed to the insufficient vibration of the infilled concrete, resulting in a weak bonding performance between steel and concrete. In contrast, the SWP-0.4 and SWP-0.2 specimens with a precast DSC shear wall were no obvious phenomena in the vertical loading period. Hence, the monolithic cast-in-place DSC shear wall with excessive height makes it difficult to ensure adequate consolidation during construction due to the internal connectors.

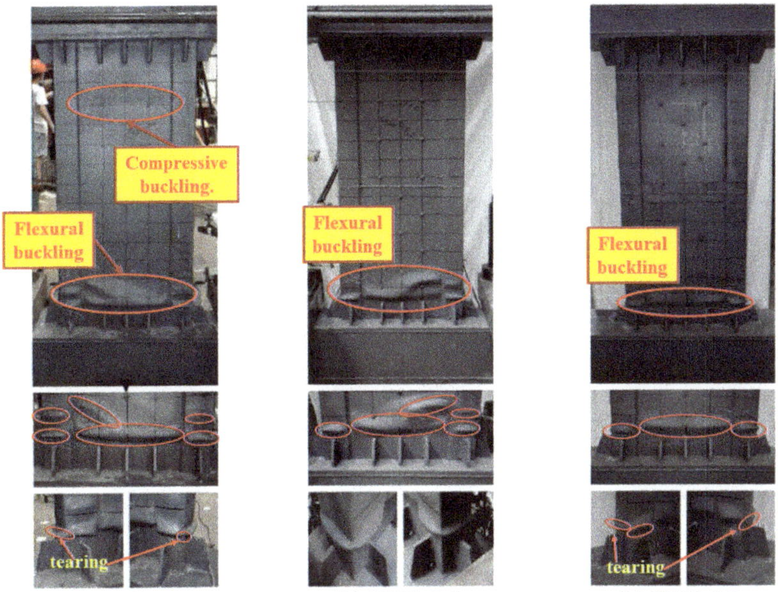

(**a**) SWC-0.4. (**b**) SWP-0.4. (**c**) SWP-0.2.

Figure 7. Failure modes of the specimens.

In the traditional DSC shear wall specimen SWC-0.4, minor buckling began to appear and expanded in the exterior steel plates at a horizontal displacement angle of 1/173. When the horizontal displacement angle increased to 1/55, with the peak load at 334.4 kN, the buckling at the column and DSC shear wall base became more pronounced. Buckling waves were concentrated within 100 mm from the top of the stiffening rib. With increasing displacement, tearing occurred in the base of the column. Then, the concrete was crushed, resulting in a rapid decline in load capacity. The failure mode of SWC-0.4 is a typical flexural–compression failure, with tearing in the exterior steel plate at the base of the boundary column. Buckling waves at the base of the DSC shear wall extend into the boundary column, up to a height of 100 mm from the top of the stiffening rib.

The buckling development of SWP-0.2 and SWP-0.4 specimens were similar to SWC-0.4 at the initial stage. At the same axial compression ratio, buckling in the precast DSC shear wall specimen SWP-0.4 occurred at peak horizontal displacement angles of 1/100, delayed compared to the monolithic cast-in-place wall specimens due to pre-existing damage under the vertical loading phase for SWC-0.4. At a displacement angle of 1/67, the load peaked at 348.9 kN for specimen SWP-0.4, which was higher than that of SWC-0.4. Thereafter, buckling intensified at the base of the boundary columns and steel plates of the DSC shear wall. This occurred without steel plate tearing at the boundary columns in specimen SWP-0.4, attributed to the final loading cycle for CSW1 being conducted only once. Overall, the prefabricated specimen SWP-0.4 exhibits a similar failure process and load-bearing capacity to the monolithic cast-in-place specimens SWC-0.4, yet it offers a simpler construction process and superior concreting quality.

The buckling for SWP-0.2 occurred later, with a horizontal displacement angle of 1/83. This is primarily attributed to withstanding a lesser vertical load for specimen SWP-0.2. Subsequently, buckling in the steel plate and boundary column of SWP-0.2 developed more slowly compared to SWP-0.4, with an axial compression ratio of 0.4. When the horizontal displacement angle reached 1/55, the maximum horizontal for SWP-0.2 was 332.1 kN, less than that of the SWP-0.4 specimen. Overall, the damage severity for the SWP-0.2 specimen with low axial compression ratios was significantly less than that of the SWP-0.4 specimen. The buckling wave of the SWP-0.2 specimen penetrated within a 100 mm height range at the top of the stiffener rib, aligning horizontally with the damage region of boundary columns.

3.2. Hysteretic Behavior

Figure 8 presents the force–displacement relationship for each specimen under low reversed cyclic loading. Initially, loops evolve from spindle to bow shapes, indicating an increase in energy dissipation as the loop area enlarges. Progressive plastic deformation at the bottom of the DSC shear wall and boundary columns leads to increasing residual deformation with each load cycle, causing the loops to assume an S-shape. As displacement increased, yield began in the outer edge of boundary columns, causing concrete crushing and reduced stiffness. The hysteretic curve showed a pinching effect due to rapid plastic deformation. This is attributed to steel–concrete interface slip, steel buckling, and expansion of concrete cracks. After peak load, the specimen showed progressive deformation, reduced load-bearing capacity, and greater residual deformation, with pronounced buckling in the exterior steel plates, culminating in concrete crushing and failure. The hysteresis curves exhibit a full spindle shape, indicating that the designed DSC shear wall has good energy dissipation capacity under low reversed cyclic loading.

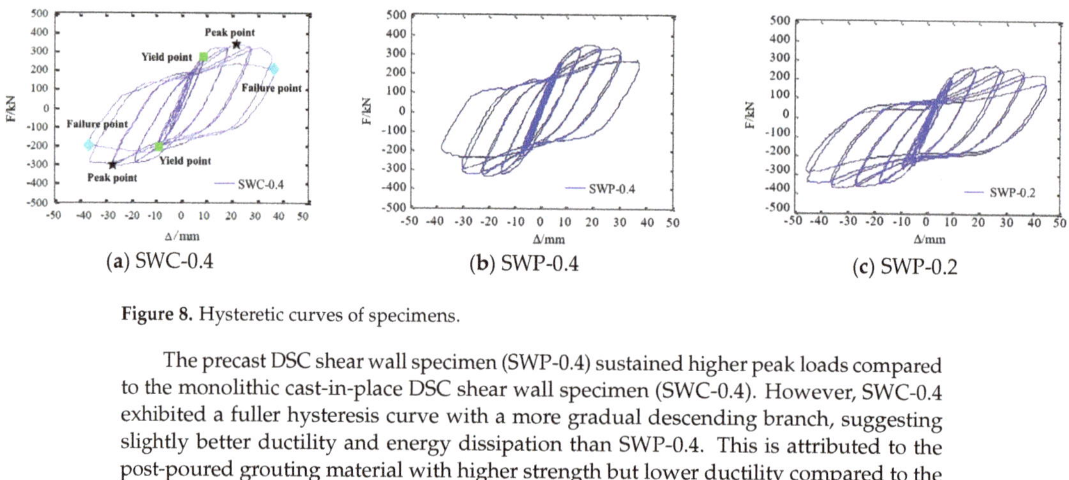

(**a**) SWC-0.4 (**b**) SWP-0.4 (**c**) SWP-0.2

Figure 8. Hysteretic curves of specimens.

The precast DSC shear wall specimen (SWP-0.4) sustained higher peak loads compared to the monolithic cast-in-place DSC shear wall specimen (SWC-0.4). However, SWC-0.4 exhibited a fuller hysteresis curve with a more gradual descending branch, suggesting slightly better ductility and energy dissipation than SWP-0.4. This is attributed to the post-poured grouting material with higher strength but lower ductility compared to the infilled concrete of the precast DSC shear wall. Hence, the proposed precast DSC shear wall exhibits hysteresis performance under low reversed cyclic loading that is essentially equivalent to that of the cast-in-place. Compare SWP-0.4 and SWP-0.2 with different axial compression ratios, SWP-0.2 with a lower axial compression ratio showed a better ductility and energy dissipation capacity than the SWP-0.4 specimen, despite a notable decrease in peak load capacity.

3.3. Skeleton Curves and Ductility

The skeleton curve in Figure 9 is the envelope curve of the hysteresis curve constructed by linking the peak points at each loading stage. The curves exhibit an S-shape and can generally be divided into three parts: elastic stage, elastic–plastic stage, and declining stage. The skeleton curves of three specimens are very close in the elastic stage and diverge upon entering the plastic development stage. Initially, as the steel plate and concrete work together, the curve rises linearly with consistent initial stiffness. As displacement increases, the curve shift and slope decrease signifying stiffness degradation of DSC shear wall due to the yield of exterior steel plates in the bottom of boundary columns and shear wall. Subsequently, the curve falls due to concrete crushing and the irrecoverable deformation of steel plate.

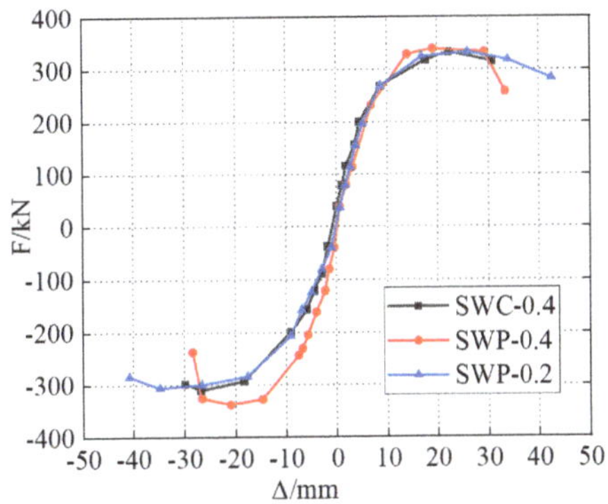

Figure 9. Skeleton curves.

Table 2 displays the characteristic values and ductility for the specimens. Py, Pm, and Pu represent yield, maximum, and ultimate loads, respectively; Δy, Δm, and Δu denote corresponding displacements. θy, θm, and θu are the mean drift angle at yield, peak, and failure stages, averaged from positive and negative loading displacements. The yield point is determined by Park's method [39], and the ultimate point is identified where the load on the descending skeleton curve is 85% of the peak load. The ductility coefficient, μ, is defined as the ratio of ultimate displacement to yield displacement.

Table 2. Characteristic values and ductility.

Specimens	Direction	P_y (kN)	Δ_y (mm)	θ_y	P_m (kN)	Δ_m (mm)	θ_m	P_u (kN)	Δ_u (mm)	θ_u	μ
SWC-0.4	+	250.8	7.649	1/161	334.4	22.53	1/61	284.2	35.04	1/40	4.58
	−	−232.4	−10.88		−309.9	−26.74		−263.4	−39.15		3.60
SWP-0.4	+	261.7	7.787	1/182	348.9	18.93	1/76	296.6	30.07	1/52	3.86
	−	−252.0	−8.610		−252.0	−20.74		−285.6	−27.32		3.17
SWP-0.2	+	249.0	7.852	1/154	332.0	25.91	1/50	282.2	42.94	1/35	5.58
	−	−230.2	−11.70		−306.9	−34.45		−260.9	−42.45		3.63

The peak load for SWC-0.4 is 334.4 kN, which is approximately 95% of that for SWP-0.4. This is attributed to the grouting material with a higher strength used in the late-poured bands of the precast DSC shear wall. The average yield and failure displacement angles for SWC-0.4 are approximately 1/160 and 1/40, respectively, which are 1.13 and 1.25 times those of SWP-0.4, indicating the superior deformation capacity of the precast DSC shear wall specimen. When the axial compression ratio increased from 0.2 to 0.4, the peak load of the precast DSC shear wall improved by 4.8%. Meanwhile, the average yield and failure displacement angles for specimen SWP-0.2 were 0.85 and 0.67 times that of SWP-0.4, respectively. Additionally, with an axial compression ratio of 0.4, the average yield and failure displacement angles of the precast DSC shear wall were approximately 1/180 and 1/50, respectively; at a ratio of 0.2, they were about 1/150 and 1/35, respectively. These values meet the Chinese code JGJ/T 380-2015, which stipulates that the elastic inter-story drift angle for steel plate composite shear walls should not exceed 1/400, and the elastic–plastic inter-story drift angle should not exceed 1/80.

The ductility factor of SWC-0.4 was 16% higher than that of SWP-0.4. While precast DSC shear walls show enhanced load-bearing capacity, their ductility is diminished compared to monolithic cast-in-place walls, a noteworthy consideration in seismic design. Overall, the ductility factors for all specimens exceeded three, indicating relatively good ductility of the monolithic and the precast DSC shear walls. The ductility factor of specimen SWP-0.4 was 0.76 times that of SWP-0.2, showing that the ductility decreases with an increase in axial compression ratio. Jafari et al. [40] also yielded similar results, indicating that excessive vertical load is detrimental to the seismic resistance of shear walls.

3.4. Bearing Capacity Degradation

Bearing capacity degradation is expressed by the bearing capacity degradation coefficient, η, which is the ratio of the last peak load to the first peak load at the same displacement amplitude. Figure 10 illustrates the bearing capacity degradation curves for each specimen. Initially, the bearing capacity degradation coefficient decreases slowly with increasing displacement. After reaching the peak load, the rate of bearing capacity degradation increases significantly, due to irreversible damage. The bearing capacity coefficients of the specimens are mostly between 0.7 and 1.0, indicating stable degradation.

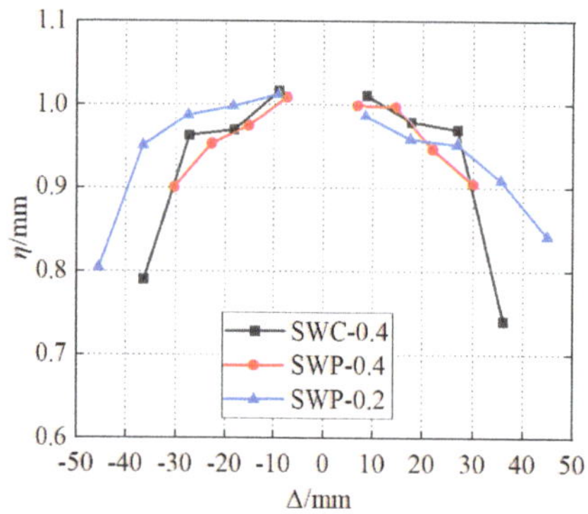

Figure 10. The bearing capacity degradation curves for each specimen.

The monolithic shear wall specimen exhibited less bearing capacity degradation compared to the precast shear wall specimens. The higher the axial compression ratio, the more significant the bearing capacity degradation for precast DSC shear wall. This is due to increased buckling of the square steel tube columns and the edge of the DSC shear wall.

3.5. Stiffness Degradation

Stiffness degradation is characterized by cyclic stiffness, defined as the ratio of peak load to corresponding displacement for the same loading amplitude. Figure 11 depicts the stiffness degradation curves of each specimen. Initially, a rapid decrease in cyclic stiffness occurs due to interface damage between the exterior steel plate and infilled concrete. Entering the plastic phase, the specimen exhibits a slower cyclic stiffness reduction, attributable to concrete crack stabilization and the exterior steel plate's effective confinement of infilled concrete. The SWP-0.4 specimen with rapid construction exhibited a faster decrease in cyclic stiffness compared to SWC-0.4 due to the integrity being slightly inferior to the precast DSC shear wall. The precast DSC shear wall under an axial compression ratio of 0.4 exhibited greater initial stiffness compared to a 0.2 ratio; however, it experienced a faster stiffness reduction during loading, resulting in more significant stiffness degradation.

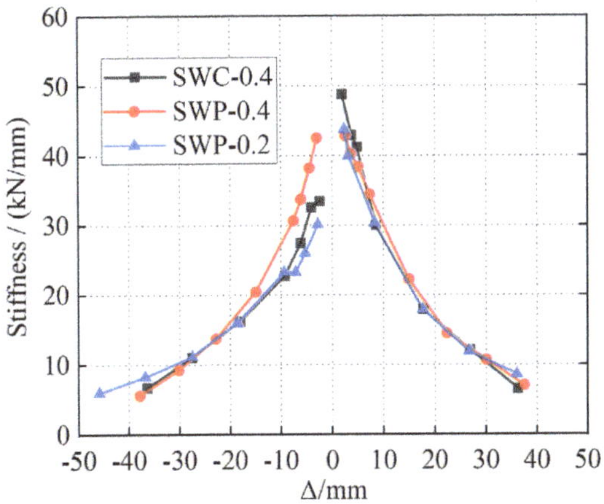

Figure 11. The stiffness degradation curves.

3.6. Energy Consumption

The energy dissipation capabilities of the specimens are quantified using the equivalent viscous damping coefficient, ξ. The equivalent viscous damping is normalized against an equivalent elastic cycle to denote the hysteretic energy dissipation. The equivalent viscous damping ratio for the DSC shear wall specimens is determined using Equation (1).

$$\xi = \frac{S_{ABCD}}{2\pi \cdot (S_{OCF} + S_{OAE})} \quad (1)$$

where SABCD represents the area of the hysteresis loops, while SOCF and SOAE denote the summed triangular areas, as illustrated in Figure 12.

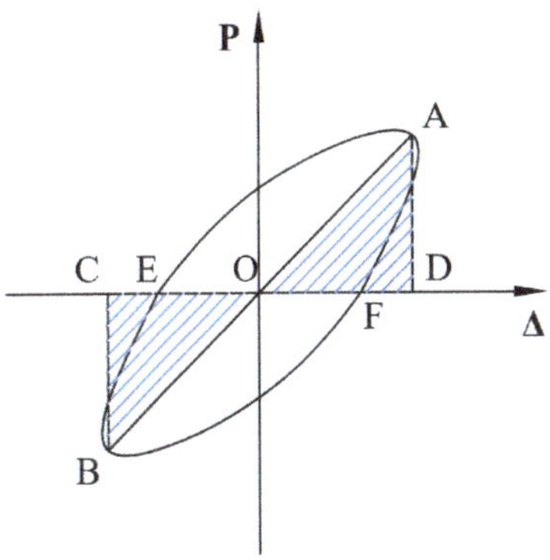

Figure 12. Definitions of equivalent viscous damping ratio.

Figure 13 presents the curves correlating with the equivalent viscous damping coefficient with displacement. Generally, all the specimens exhibit similar damping properties. Initially, the specimens' equivalent viscous damping ratios decreased until reaching a displacement near 10 mm. Subsequently, the equivalent viscous damping coefficient exhibited an increasing trend.

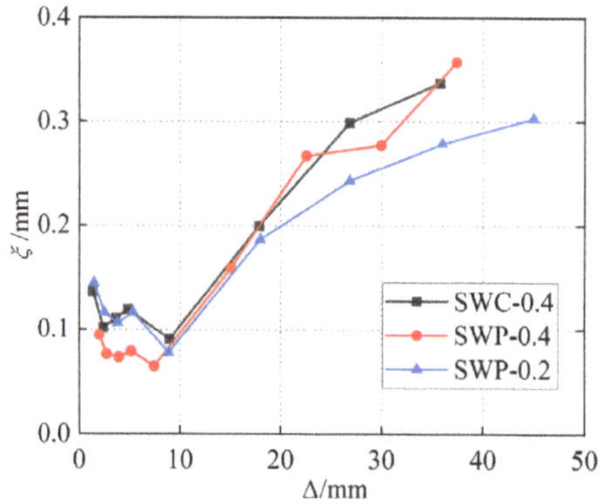

Figure 13. Equivalent viscous damping ratio-displacement curves.

It can be inferred that under the same displacement increment conditions, the precast DSC shear wall with higher equivalent viscous damping coefficient has a superior energy dissipation capability than the monolithic DSC shear wall. Additionally, the equivalent viscous damping coefficient of the precast DSC shear wall increases with the increasing axial compression ratio.

3.7. Analysis of Lateral Deformation

The lateral deformations of shear walls are analyzed using displacement measurements at various heights. L1 to L6 can measure the horizontal displacement of the shear wall at heights of 0 mm, 300 mm, 600 mm, 900 mm, 1200 mm, and 1500 mm. Figure 14 presents the distribution of lateral deformation along the height of the precast DSC shear wall specimen SWP-0.4, where H is the height of SWP-0.4. The graph reveals that the horizontal displacement of the precast DSC shear wall was initially linear. As it entered the plastic phase, the horizontal displacement increased more rapidly and displayed a bent deformation.

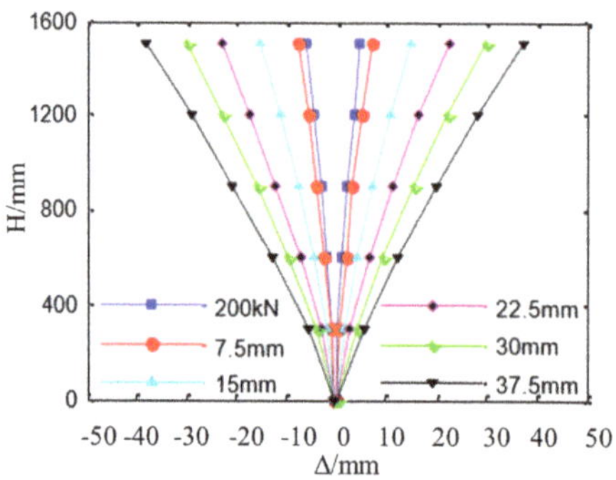

Figure 14. Lateral deformation of SWP-0.4.

4. Finite Element Analysis

4.1. Description of the FE Model

4.1.1. Material Constitutive

To further investigate the behavior of the precast DSC shear wall subjected to low reversed cyclic loading, a finite element (FE) model was developed using ABAQUS. The constitutive relationship of the steel was defined using a tri-linear model, determined by the elastic modulus, yield strength, and ultimate strength measured from material tests discussed in Chapter 2. The stress–strain relationship of the infilled concrete and grouting material was obtained from prior material property tests, and the mechanical properties were simulated by the concrete damage plasticity (CDP) model in the Abaqus with default parameters: Poisson's ratio (0.2), dilation angle (30), eccentricity (0.1), fb0/fc0 (1.16), K (0.667), and viscosity (0.005). Damage factors for tension and compression, based on the energy equivalence principle, were capped at 0.95 to represent the damage progression of concrete [41].

4.1.2. Establishment of FE Model

The FE model with reasonable simplification was established according to the test setup. The model was composed of the precast wall units, the after-casting unit, the loading beam, and the foundation beam, which were all simulated by 8-node solid elements (C3D8R). This study did not consider the loading and foundation beams, which were instead treated as idealized infinite rigid bodies. The interface between the square steel tube columns as well as the exterior steel plate of shear walls and the infilled concrete is defined by hard contact, and tangential interaction uses the Coulomb friction model with a friction coefficient of 0.4, based on studies. All steel components are connected using "tie" constraints. Stress at the load points is mitigated by a central coupling point RPl at the beam's top for vertical loads, and RP2 at the beam's side for horizontal loads. The foundation beam was completely fixed.

A more refined mesh improves accuracy but increases the computational time significantly. To balance accuracy and efficiency, a standard mesh size of 50 mm × 50 mm is used for the precast double-skin composite shear wall models. For buckling and tearing regions near the wall and columns base, a finer 25 mm × 25 mm mesh is applied, as shown in Figure 15. The mesh for foundation and loading beams is coarser, at 100 mm × 100 mm.

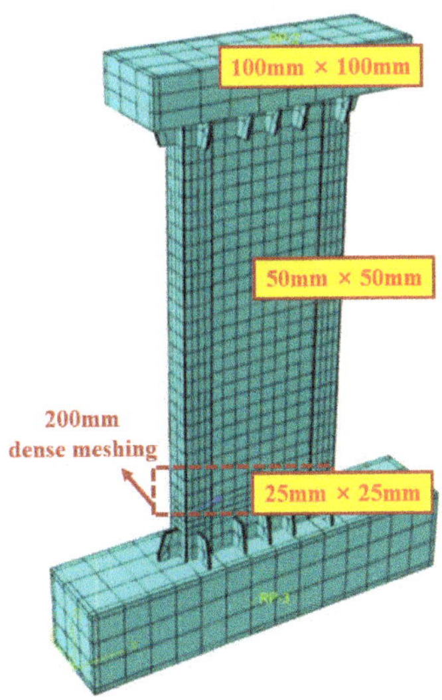

Figure 15. Typical FE mesh of the shear wall system.

4.2. Verification of the FE Model

4.2.1. Comparison of Hysteretic Behavior

Figure 16 illustrates a comparison between the hysteretic curves of three specimens from FE simulations and experimental observations. To improve the working efficiency, FE models used a reduced displacement loading step compared to actual tests. The FE models successfully captured the hysteretic curve characteristics, such as initial stiffness, peak load, stiffness degradation, strengthening degradation, and pinching effects. However, for the precast DSC shear wall specimens, like SWP-0.2, the modeled pinching was less severe than the observed experimental results, attributed to the difficulty in simulating the interaction at the interface between new and old concrete under exterior steel plate confinement. Generally, finite element simulations may not fully align with experimental results, yet their precision can be maintained within an acceptable margin. More precise models are required in subsequent research to improve the accuracy of this study.

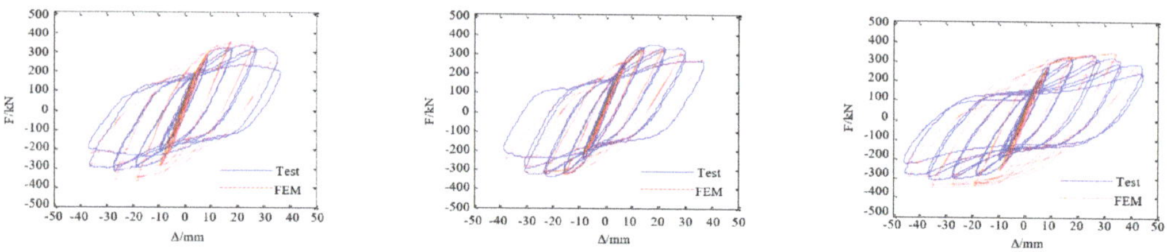

(**a**) SWC-0.4 (**b**) SWP-0.4 (**c**) SWP-0.2

Figure 16. Comparison of hysteretic behavior from experimental results and simulations.

4.2.2. Comparison of Failure Modes

Figure 17 illustrates a comparison between typical failure modes of specimens from FE simulations and experimental observations. Generally, the established FE model under low reversed cyclic loads can reasonably predict the failure modes of monolithic cast-in-place and precast DSC shear walls in the test. For the monolithic cast-in-place DSC shear wall specimens (e.g., SWC-0.4), the buckling deformation occurred and intensified in buckling waves region within a 200 mm region from the base of shear wall. The failure mode of SWC-0.4 is typical flexural compression failure, with tearing in the corners of steel plates of the boundary column. Meanwhile, a high level of von Mises stress was observed in the buckling waves region and the boundary column region where the tears appeared in the tests. For the precast DSC shear wall specimens like SWP-0.2 and SWP-0.4, the buckling deformation in the buckling waves region and severe deformation at the boundary column were well simulated by the FE models. A high level of von Mises stress developed in a smaller range of the exterior steel plate, which was consistent with experimental results.

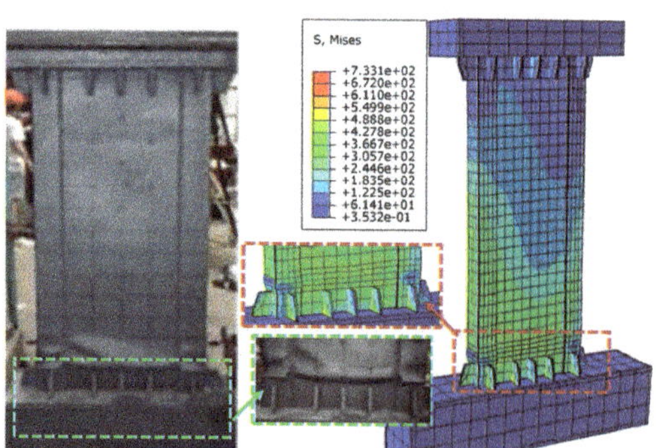

(**a**) SWC-0.4

Figure 17. *Cont.*

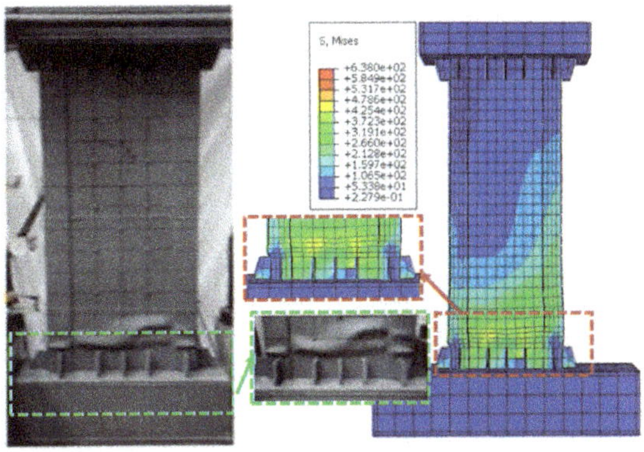

(**b**) SWP-0.4

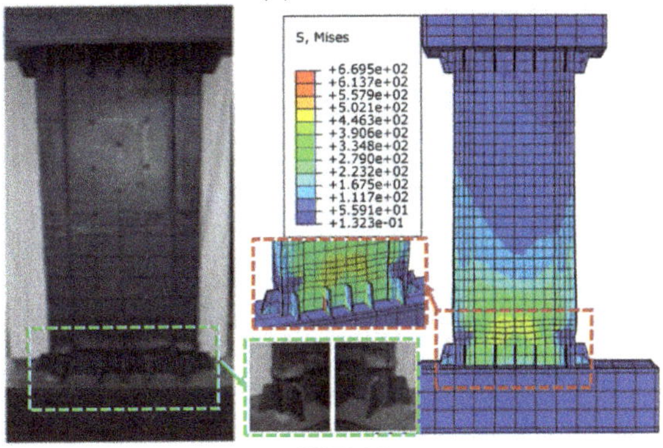

(**c**) SWP-0.2

Figure 17. Comparison of failure modes from experimental results and simulations.

4.3. Stress Maps of Concrete and Steel

Despite the strain gauges on the steel plate, comprehensive monitoring of the deformation and stress of the exterior steel plate and the infilled concrete was challenging. Additionally, the cooperative deformation between the infill C30 concrete and site-cast grouting material was unclear in the experiments. Finite element models of the monolithic cast-in-place DSC shear wall specimens SWC-0.4 and the precast DSC shear wall specimens SWP-0.4 were analyzed at yield, peak, and failure points to assess stress distribution and plastic deformation under low reversed cyclic loading.

4.3.1. SWC-0.4 Specimen

Figures 18 and 19 illustrate the stress–strain distribution of the exterior steel plates and the infilled concrete for the SWC-0.4 specimen at the respective yield, peak, and failure points. Yield points are identified by inflections on the hysteresis curves from FE model analysis. Equivalent plastic strain is used to indicate irreversible deformation. Higher equivalent plastic strain values indicate more extensive cumulative plastic deformation, pointing to a greater likelihood of material damage and failure.

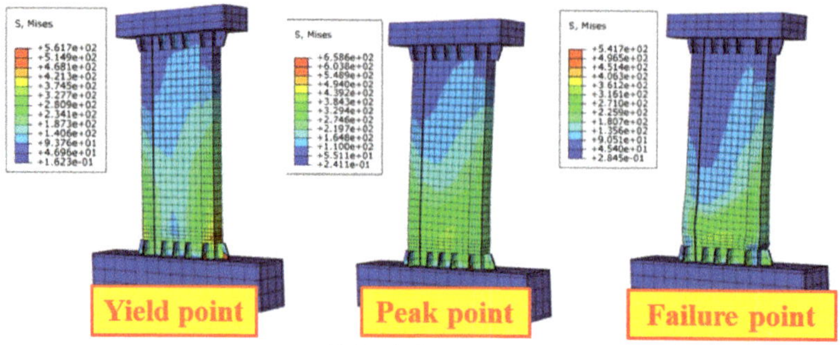

(**a**) Steel plate stress.

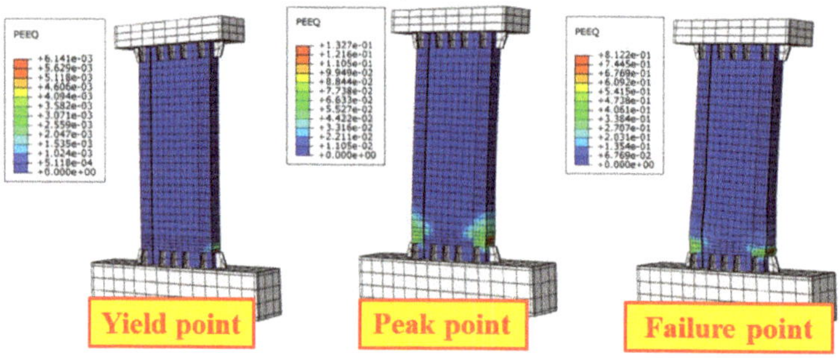

(**b**) Equivalent plastic strain of steel plate.

Figure 18. Stress and strain of exterior steel plates for SWC-0.4.

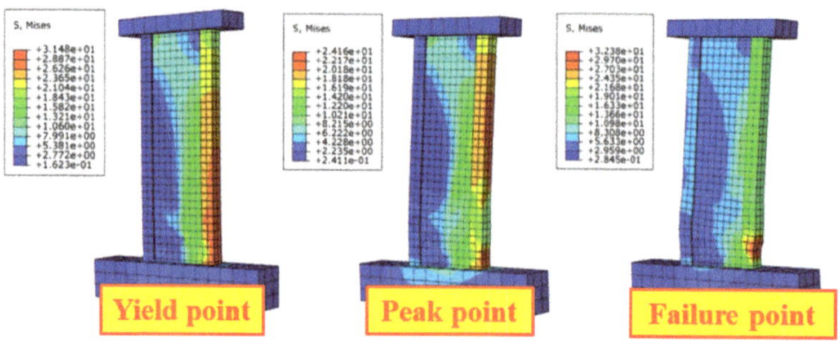

(**a**) Concrete stress.

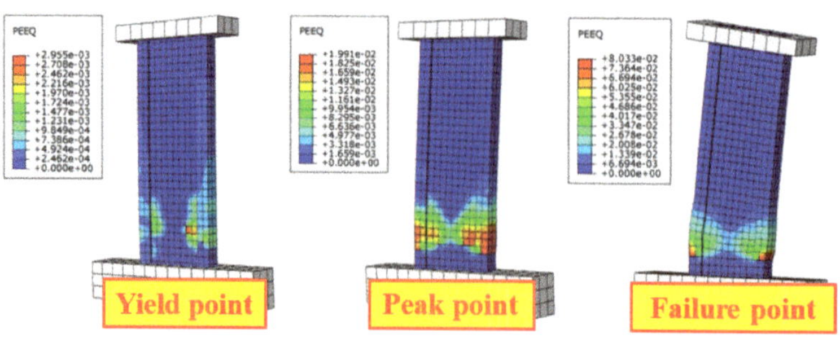

(**b**) Equivalent plastic strain of concrete.

Figure 19. Stress and strain of the infilled concrete for SWC-0.4.

At yield point of the monolithic cast-in-place DSC shear wall model, the von Mises stress on the exterior steel plates of the column base reached 425 MPa, equivalent to the yield stress of Q355B steel, with an equivalent plastic strain of 0.013. At the peak point of the SWC-0.4 specimen, most of the bottom steel plates yielded, experiencing a maximum stress of 486 MPa, which did not reach the ultimate strength of Q355B. Extensive plastic deformation occurred at the bottom of column, with a maximal equivalent plastic strain of 0.153. In the failure phase, the stress of exterior steel plates at the lower part of the column trended towards zero, while the equivalent plastic strain continued to increase, indicating severe damage or tearing for the concrete-filled steel tubular column. The von Mises stress reached yield stress within 100 mm above the top of the shear wall's stiffening rib, aligning with the experimental observations.

For the infilled concrete, the stress gradually intensified as the load increased. At the specimen's yield point, the majority of concrete had not reached its limit state. Meanwhile, minimal plastic strain of concrete was observed at the bottom of the boundary columns and shear wall. At the specimen's peak point, the maximum concrete stress at the column foot was 21.02 MPa, below the concrete's load-bearing capacity, whereas the equivalent plastic strain reached 0.080, exceeding the concrete's ultimate compressive strain. This indicates that concrete crushing occurred before the peak load, but the bearing capacity of the monolithic cast-in-place DSC shear wall continued to increase as the steel entered a reinforcing stage. Compared to the bottom of the wall, the column base showed more pronounced plastic deformation. In the failure stage, concrete damage occurred around 100 mm above the stiffening rib, consistent with the observed experimental phenomenon that the concrete failure. Overall, the exterior steel plates effectively confined the infilled concrete, significantly enhancing the ductility and energy dissipation capacity of the DSC shear wall under seismic loads.

4.3.2. SWP-0.4 Specimen

Figures 20 and 21 illustrate the stress–strain distribution of the exterior steel plates and the infilled concrete for the SWP-0.4 specimen at yield, peak, and failure points. At the yield point of FE model, the maximum von Mises stress in the boundary columns reached the 425 MPa yield stress of Q355B. The maximum equivalent plastic strain of model in the SWP-0.4 specimen was 0.008, less than that in the SWC-0.4 specimen model. Consequently, the maximum von Mises stress borne by the infilled concrete in SWC-0.4 model was also less than that in SWP-0.4 model, indicating that the precast DSC shear wall specimen preceded the monolithic cast-in-place DSC shear wall specimen in reaching yield. At the peak point of the specimen, the maximum von Mises stress at the column edges of the SWP-0.4 specimen reached 489 MPa, higher than SWC-0.4, which corresponds with the skeleton curves discussed in Section 2. At this stage, the maximum equivalent plastic strain of steel plate at the bottom of column peaked at 0.109, below that of SWC-0.4. As the specimen approached failure, the steel plate stress tended towards zero, with a continuously increasing equivalent plastic strain distributed within 200 mm above the top of the shear wall's stiffening rib, indicating severe exterior steel plate damage. This corresponds with the observed buckling on the edges of boundary columns in the experimental phenomena.

As the load increased, the concrete stress gradually intensified. The plastic strain of the concrete appeared at the base of the shear wall and boundary columns, whereas the post-poured grouting material maintained elasticity. At the peak point of the overall model, the maximum concrete stress reached 24.38 MPa, surpassing that of SWC-0.4. The plastic deformation zone expanded to 200 mm above the stiffening rib. The enlargement of the plastic deformation zone enhanced the load-bearing capacity of the prefabricated component under seismic loads. In the failure phase, the equivalent plastic strain of concrete at the column base reached 0.087, exceeding the ultimate compressive strain, while the high-strength grouting materials remained in the elastic stage.

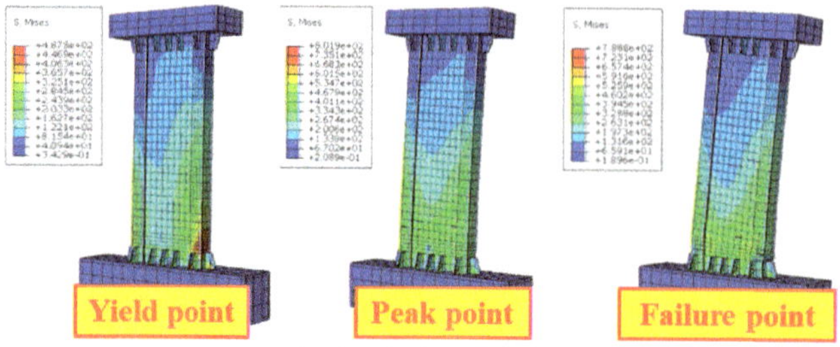

(**a**) Steel plate stress.

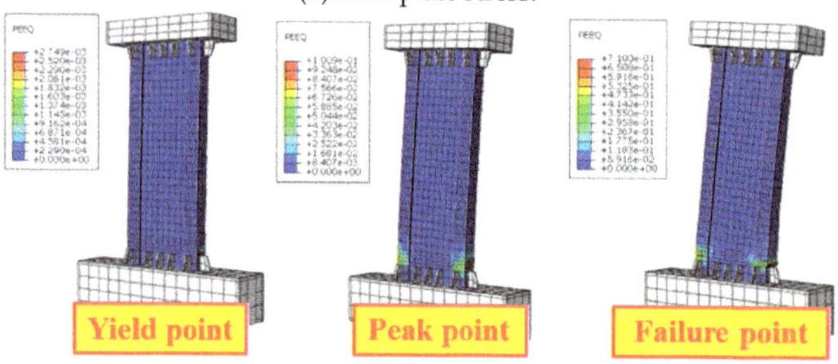

(**b**) Equivalent plastic strain of steel plate.

Figure 20. Stress and strain of exterior steel plates for SWP-0.4.

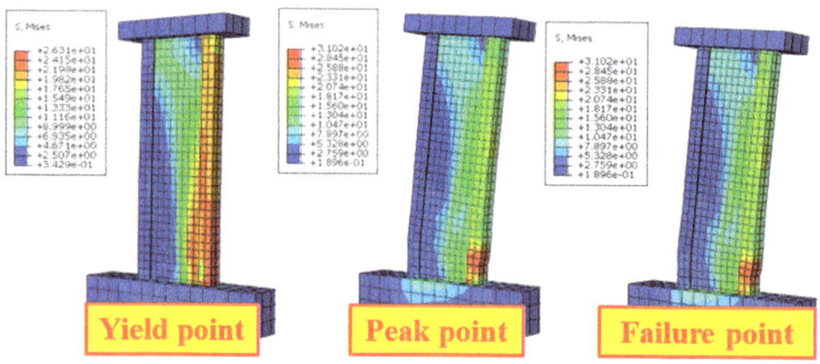

(**a**) Concrete stress.

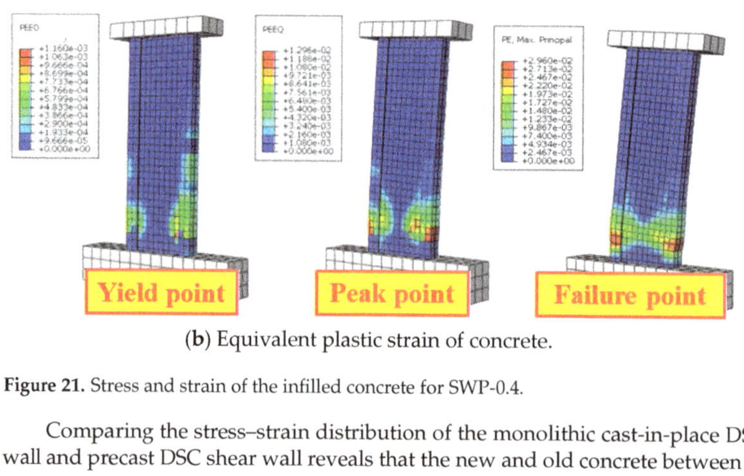

(**b**) Equivalent plastic strain of concrete.

Figure 21. Stress and strain of the infilled concrete for SWP-0.4.

Comparing the stress–strain distribution of the monolithic cast-in-place DSC shear wall and precast DSC shear wall reveals that the new and old concrete between the horizontal connection region and precast components can achieve a cohesive performance for coordinated deformation. The damaged area in precast DSC shear wall is larger than in the monolithic cast-in-place DSC shear wall, enhancing the load-bearing capacity but reducing ductility slightly. Hence, the precast DSC shear wall with guaranteed concrete adhesion and steel plate welding quality demonstrates a seismic performance comparable to that of the cast-in-place counterparts.

4.4. Parametric Analysis

4.4.1. Parameter Settings

This study investigates the influence of geometric details and material strength on the load-bearing capacity of precast DSC shear walls using finite element parametric analysis. The parameters were the overall length (L), overall thickness (T), overall height (H), thickness of exterior steel plates (t), concrete strength, strength of grouting material, and steel strength. The structural configurations of SWP-0.4 corresponding to the general dimensions outlined in Section 2.1 served as the baseline reference models. Fifteen configurations of the precast DSC shear walls were analyzed, with parameter values varying within typical DSC shear wall ranges. Design parameters of individual configurations are shown in Table 3.

Table 3. Design parameters of individual configurations.

Number	L (mm)	H (mm)	T (mm)	t(mm)	Concrete	Grouting Material	Steel
SWP-Base (SWP-0.4)	400	1500	100	3	C40	C50	Q355
SWP-1	600	1500	100	3	C40	C50	Q355
SWP-2	800	1500	100	3	C40	C50	Q355
SWP-3	400	1000	100	3	C40	C50	Q355
SWP-4	400	2000	100	3	C40	C50	Q355
SWP-5	400	1500	70	3	C40	C50	Q355
SWP-6	400	1500	130	3	C40	C50	Q355
SWP-7	400	1500	100	6	C40	C50	Q355
SWP-8	400	1500	100	8	C40	C50	Q355
SWP-9	400	1500	100	3	C30	C50	Q355
SWP-10	400	1500	100	3	C50	C50	Q355
SWP-11	400	1500	100	3	C40	C60	Q355
SWP-12	400	1500	100	3	C40	C80	Q355
SWP-13	400	1500	100	3	C40	C50	Q235
SWP-14	400	1500	100	3	C40	C50	Q390

4.4.2. Influence of Shear Wall Parameters

Figure 22a illustrates the correlation between the overall width of the specimens and their skeleton curves under low reversed cyclic loading. Widening from 400 mm to 600 mm or 800 mm improved the peak load capacity by 64.6% and 110.2%, respectively. This indicated that increasing overall component width would enhance the load-bearing capacity of precast DSC shear walls during an earthquake. However, wider walls lead to a more pronounced decline in post-peak behavior, indicative of higher degradation and less ductility. Concurrently, raising wall heights from 1000 mm to 1500 mm or 2000 mm decreased peak loads by 29.3% and 47.8%, respectively, as shown in Figure 22b. Higher walls also had a gentler post-peak slope, indicating improved ductility for the high precast DSC shear walls. Thus, it is essential to appropriately configure the height-to-width ratio of the precast DSC shear walls in seismic design. Figure 22c shows the effect of the overall thickness of precast DSC shear walls on the load–displacement curve. The data indicate a clear increase in peak load with greater wall thickness.

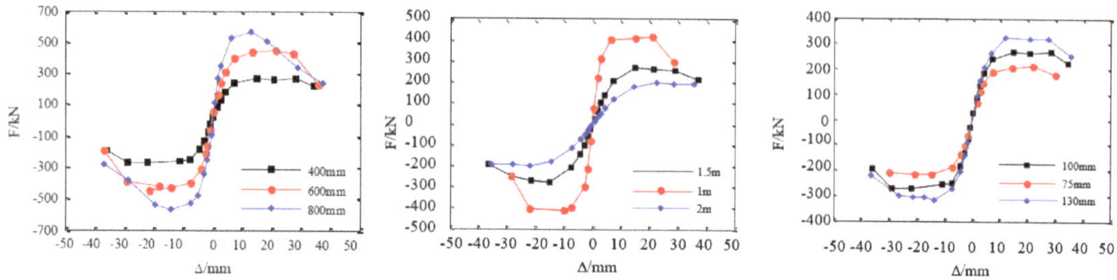

(a) Width (b) Height (c) Thickness

Figure 22. Influence of geometric details of shear wall.

4.4.3. The Effect of Exterior Steel Plates

Increasing the exterior steel plate thickness significantly enhanced the peak load capacity, with a 72.5% increase for 6 mm and 118.6% for 8 mm, compared to 3 mm plates, as shown in Figure 23a. This was attributed to the enhanced confinement effect on the infilled concrete with increased exterior steel plate thickness. Additionally, excessively thick steel plates not only waste resources but also reduce seismic ductility. Upgrading the steel grade of exterior steel plates from Q235 to Q345 or Q390 steel yields peak load gains of 28.21% and 40.66%, showing a correlation between steel grade and load performance, as shown in Figure 23b. However, based on the design size of SWP-0.4, upgrading the steel grade does not significantly enhance the seismic resistance capability of the precast DSC shear walls.

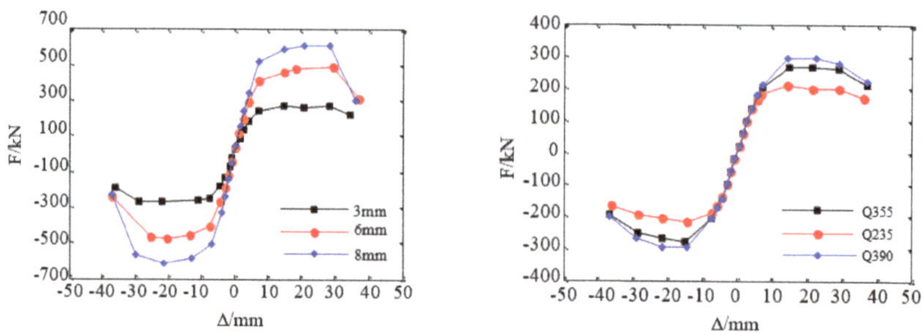

(a) Thickness (b) Strength

Figure 23. The effect of exterior steel plates.

4.4.4. The Effect of the Infilled Concrete

The peak load increases with the higher infilled concrete strength grades, showing a 6.0% and 11.3% rise when upgrading from C30 to C40 and C50, respectively, as shown in Figure 24a. The strength grade of the grouting material appears to have no effect on horizontal load, attributed to the grouting material remaining within the elastic range without plastic deformation. The seismic performance of the novel precast DSC shear walls is essentially equivalent to that of monolithic cast-in-place DSC shear walls. Enhancing the bonding performance between infilled concrete and grouting material is more effective for seismic enhancement than increasing concrete strength.

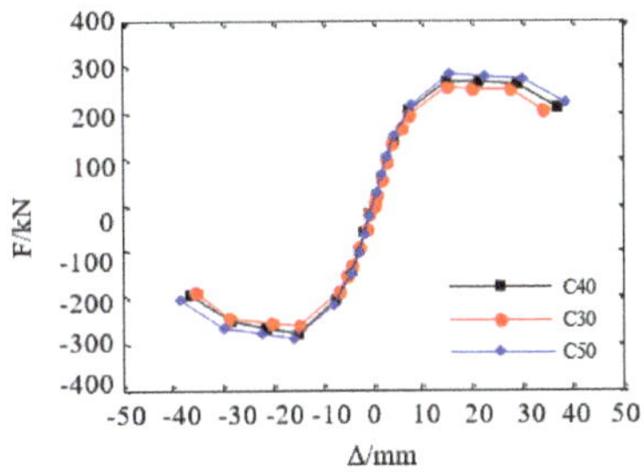

(**a**) The precast concrete.

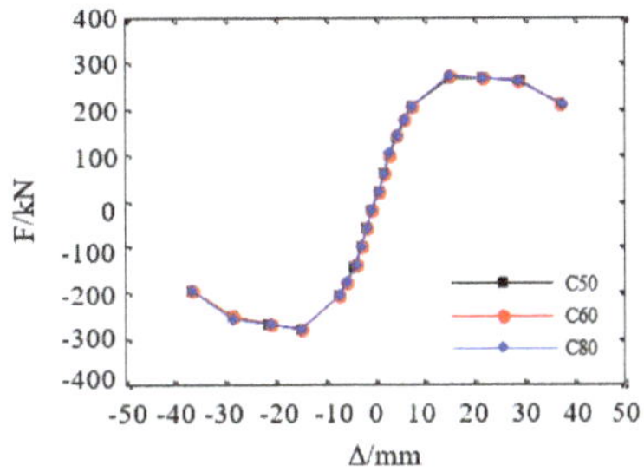

(**b**) The grouting material.

Figure 24. The effect of the infilled concrete.

5. Conclusions

This paper studied the effect on seismic performance of precast DSC shear walls with on-site welding and cast grouting. Three half-scale DSC shear wall specimens including two precast DSC shear walls with horizontal connection region specimens and monolithic cast-in-place DSC shear wall specimens were tested under reversed cyclic loading. Finite element models were used for stress mechanism analysis and parametric study. The conclusions are as follows:

All specimens demonstrated satisfactory seismic performance and cooperative deformation capacity. The precast DSC shear wall with horizontal connection region specimens and on seismic performance monolithic cast-in-place DSC shear wall specimen exhibited compression-bending failure. The damage range of specimens was within a height range of 100 mm to 200 mm from the bottom of the shear wall.

With guaranteed welding quality and concrete bonding performance, precast DSC shear wall specimens with a horizontal connection region offered high construction efficiency and standardized components on-site. The SWP-0.4 specimen showed superior load-bearing capacity, deformation capability, and energy dissipation compared to SWC-0.4, with slower stiffness degradation due to improved component quality. However, on the monolithic cast-in-place DSC shear wall specimen SWC-0.4 exhibited better ductility under seismic loads.

An increased axial compression ratio enhanced the load-bearing and energy dissipation capacities of precast DSC shear walls by improving the constraint effect of the exterior steel plates on the infilled concrete. However, the SWP-0.4 specimen with a higher axial compression ratio resulted in more significant stiffness and strength degradation and poorer ductility compared to SWP-0.2.

Parametric analysis under low reversed cyclic loading revealed a positive correlation between the width, thickness, steel plate thickness, infilled concrete strength, steel strength, and peak load-bearing capacity of prefabricated double steel plate shear walls, with the shear wall height negatively correlated. The strength of the post-cast grouting material had a negligible impact on the load-bearing capacity.

Author Contributions: Conceptualization, H.L.; methodology, H.L., N.S. and X.F.; software, N.S.; formal analysis, H.L.; investigation, X.F. and J.Z.; supervision, H.L., N.S. and X.F.; funding acquisition, H.L. All authors have read and agreed to the published version of the manuscript.

Funding: This research was funded by the Higher Education Natural Science Research Project of Hebei Province, China (QN2024290).

Data Availability Statement: All relevant data are within the paper.

Conflicts of Interest: The authors declare no conflicts of interest.

References

1. Yang, J.; Yang, Y.; Deng, L.; Sun, B.; Gu, Z.; Zeng, L.; Zhao, S. Seismic Behaviors of Prefabricated Reinforced Concrete Shear Walls Assembled with a Cast-in-Place Vertical Joint. *Buildings* **2023**, *13*, 3013. [CrossRef]
2. Xu, G.; Wang, Z.; Wu, B.; Bursi, O.S.; Tan, X.; Yang, Q.; Wen, L. Seismic performance of precast shear wall with sleeves connection based on experimental and numerical studies. *Eng. Struct.* **2017**, *150*, 346–358. [CrossRef]
3. Zheng, Y.; Guo, Z.; Guan, D.; Zhang, X. Parametric study on a novel grouted rolling pipe splice for precast concrete construction. *Constr. Build. Mater.* **2018**, *166*, 452–463. [CrossRef]
4. *GB 50011-2010*; Code for Seismic Design of Buildings. Ministry of Housing and Urban-Rural Development of the People's Republic of China: Beijing, China, 2008. (In Chinese)
5. *CI. ACI 318-19*; Building Code Requirements for Structural Concrete and Commentary. American Concrete Institute: Farmington Hills, MI, USA, 2019.
6. *EN. Eurocode 8*; Design of Structures for Earthquake Resistance–Part 1: General Rules, Seismic Actions and Rules for Buildings. European Committee for Standardization: Brussels, Belgium, 2004.
7. Qian, K.; He, P.; Deng, N.; Li, H. Impact of concrete strength on seismic behavior of T-shaped double-skin composite walls. *Eng. Struct.* **2021**, *236*, 112039. [CrossRef]
8. Ji, X.; Jiang, F.; Qian, J. Seismic behavior of steel tube–double steel plate–concrete composite walls: Experimental tests. *J. Constr. Steel Res.* **2013**, *86*, 17–30. [CrossRef]
9. Nie, J.-G.; Hu, H.-S.; Fan, J.-S.; Tao, M.-X.; Li, S.-Y.; Liu, F.-J. Experimental study on seismic behavior of high-strength concrete filled double-steel-plate composite walls. *J. Constr. Steel Res.* **2013**, *88*, 206–219. [CrossRef]
10. Hu, H.-S.; Nie, J.-G.; Eatherton, M.R. Deformation capacity of concrete-filled steel plate composite shear walls. *J. Constr. Steel Res.* **2014**, *103*, 148–158. [CrossRef]
11. Nie, J.-G.; Ma, X.-W.; Tao, M.-X.; Fan, J.-S.; Bu, F.-M. Effective stiffness of composite shear wall with double plates and filled concrete. *J. Constr. Steel Res.* **2014**, *99*, 140–148. [CrossRef]
12. Chen, L.; Mahmoud, H.; Tong, S.-M.; Zhou, Y. Seismic behavior of double steel plate–HSC composite walls. *Eng. Struct.* **2015**, *102*, 1–12. [CrossRef]
13. Luo, Y.; Guo, X.; Li, J.; Xiong, Z.; Meng, L.; Dong, N.; Zhang, J. Experimental Research on Seismic Behaviour of the Concrete-Filled Double-Steel-Plate Composite Wall. *Adv. Struct. Eng.* **2015**, *18*, 1845–1858. [CrossRef]
14. Zhang, X.; Qin, Y.; Chen, Z.; Jie, L. Experimental behavior of innovative T-shaped composite shear walls under in-plane cyclic loading. *J. Constr. Steel Res.* **2016**, *120*, 143–159. [CrossRef]
15. Zhang, X.; Qin, Y.; Chen, Z. Experimental seismic behavior of innovative composite shear walls. *J. Constr. Steel Res.* **2016**, *116*, 218–232. [CrossRef]
16. Ji, X.; Cheng, X.; Jia, X.; Varma, A.H. Cyclic In-Plane Shear Behavior of Double-Skin Composite Walls in High-Rise Buildings. *J. Struct. Eng.* **2017**, *143*, 04017025. [CrossRef]
17. Guo, L.; Wang, Y.; Zhang, S. Experimental study of rectangular multi-partition steel-concrete composite shear walls. *Thin-Walled Struct.* **2018**, *130*, 577–592. [CrossRef]
18. Huang, S.-T.; Huang, Y.-S.; He, A.; Tang, X.-L.; Chen, Q.-J.; Liu, X.; Cai, J. Experimental study on seismic behaviour of an innovative composite shear wall. *J. Constr. Steel Res.* **2018**, *148*, 165–179. [CrossRef]
19. Yan, J.-B.; Li, Z.-X.; Wang, T. Seismic behaviour of double skin composite shear walls with overlapped headed studs. *Constr. Build. Mater.* **2018**, *191*, 590–607. [CrossRef]
20. Goman, H.; Liu, P.; Yin, C.; Ching, R.; Liu, M.; Zhu, L. Design and study on super high-rise structural system of Beijing Fortune Plaza 2 office tower. *Build. Struct.* **2009**, *39*, 1–8.
21. Liu, P.; Yin, C.; Cheng, Y.; Zhu, Y.S.; Liu, Y.B.; Wu, H.; Li, X.; Yang, M.L.; Liu, H.; Li, Z.X. Structural design and research of Beijing CBD Core Area Z15 plot China Zun Tower. *Build. Struct.* **2014**, *44*, 1–8.
22. Ge, D.; Zhang, B.; Wei, J.; Zhou, W. Welding of Main Tower at Stage a of World Trade Center Phase III. *Archit. Technol.* **2008**, *39*, 265–269.
23. Lv, X.; Dong, Y.; Ding, Z. Study on seismic behavior of steel reinforced concrete wall. *Earthq. Eng. Eng. Vib.* **2006**, *26*, 101–107.
24. Varma, A.H.; Malushte, S.R.; Sener, K.C.; Lai, Z. Steel-plate composite (SC) walls for safety related nuclear facilities: Design for in-plane forces and out-of-plane moments. *Nucl. Eng. Des.* **2014**, *269*, 240–249. [CrossRef]
25. Sohel, K.; Liew, J.R.; Yan, J.; Zhang, M.; Chia, K. Behavior of Steel–Concrete–Steel sandwich structures with lightweight cement composite and novel shear connectors. *Compos. Struct.* **2012**, *94*, 3500–3509. [CrossRef]
26. Chen, L.; Xia, D.; Liu, W.; Zhang, X. Experimental study on seismic behavior of double steel plates and concrete composite shear wall. *China Civ. Eng. J.* **2017**, *50*, 10–19.
27. Mydin, M.; Wang, Y.C. Structural performance of lightweight steel-foamed concrete-steel composite walling system under compression. *Thin-Walled Struct.* **2011**, *49*, 66–76. [CrossRef]
28. Zhu, J.-S.; Guo, Y.-L.; Wang, M.-Z.; Yang, X.; Zhu, B.-L. Strength design of concrete-infilled double steel corrugated-plate walls under uniform compressions. *Thin-Walled Struct.* **2019**, *141*, 153–174. [CrossRef]
29. Zhu, J.-S.; Guo, Y.-L.; Wang, M.-Z.; Yang, X.; Pi, Y.-L. Seismic performance of concrete-infilled double steel corrugated-plate walls: Experimental research. *Eng. Struct.* **2020**, *215*, 110601. [CrossRef]

30. Hutchinson, R.L.; Rizkalla, S.H.; Lau, M.; Heuvel, S. Horizontal Post-Tensioned Connections for Precast Concrete Load bearing Shear Wall Panels. *PCI J.* **1991**, *36*, 64–76. [CrossRef]
31. Soudki, K.A.; Rizkalla, S.H.; Leblanc, B. Horizontal Connections for Precast Concrete Shear Walls Subjected to Cyclic Deformations Part 1: Mild Steel Connections. *PCI J.* **1995**, *40*, 78–96. [CrossRef]
32. Soudki, K.A.; West, J.S.; Rizkalla, S.H.; Blackett, B. Horizontal Connections for Precast Concrete Shear Wall Panels Under Cyclic Shear Loading. *PCI J.* **1996**, *41*, 64–80. [CrossRef]
33. Qian, J.; Peng, Y.; Qin, H. Tests on seismic behavior of pre-cast shear walls with vertical reinforcements grouted in holes and spliced indirectly. *Build. Struct.* **2011**, *41*, 7–11.
34. Sun, J.; Qiu, H.; Lu, Y. Experimental study and associated numerical simulation of horizontally connected precast shear wall assembly. *Struct. Des. Tall Spéc. Build.* **2016**, *25*, 659–678. [CrossRef]
35. *GB/T 50081-2019*; Standard for Test Methods of Concrete Physical and Mechanical Properties. China Building Industry Press: Beijing, China, 2019. (In Chinese)
36. *GB/T17671-1999*; SBTS, Analysis Methods for Strength of Cement Mastic and Sand (ISO Methods). China Standards Press: Beijing, China, 1999. (In Chinese)
37. *GB/T 228.1-2010*; Metallic Materials-Tensile Testing at Ambient Temperature. China Standards Press: Beijing, China, 2010. (In Chinese)
38. *JGJ/T 101—2015*; Specification for Seismic Test of Buildings. China Architecture & Building Press: Beijing, China, 2015. (In Chinese)
39. Park, R.; Priestley, M.J.N.; Gill, W.D. Ductility of Square-Confined Concrete Columns. *J. Struct. Div.* **1982**, *108*, 929–950. [CrossRef]
40. Jafari, A.; Beheshti, M.; Shahmansouri, A.A.; Bengar, H.A. Cyclic response and damage status of coupled and hybrid-coupled shear walls. *Structures* **2024**, *61*, 106010. [CrossRef]
41. Wang, X.; Ma, S.; Wu, W.; Fang, J.; Chen, Z. Tensile damage development and constitutive relationship establishment of recycled coarse aggregate concrete based on a double–interface numerical model. *Constr. Build. Mater.* **2023**, *392*, 131908. [CrossRef]

buildings

MDPI

Article

Finite-Element Analysis of Temperature Field and Effect on Steel-Concrete Composite Pylon of Cable-Stayed Bridge without Backstays

Boxu Gong [1], Lianjun Feng [2], Jiang Liu [1,*], Shiming Liu [3], Zhuang Wang [1] and Yongjian Liu [1,4,*]

1 School of Highway, Chang'an University, Xi'an 710064, China; 2023021038@chd.edu.cn (B.G.); 2019021028@chd.edu.cn (Z.W.)
2 Xiamen Company of China First Highway Engineering Co., Ltd., Xiamen 361000, China; feng18689566987@hotmail.com
3 School of Civil Engineering and Transportation, North China University of Water Resources and Electric Power, Zhengzhou 450046, China; liushm@ncwu.edu.cn
4 School of Civil Engineering, Chongqing University, Chongqing 400044, China
* Correspondence: liu-jiang@chd.edu.cn (J.L.); liuyongjian@chd.edu.cn (Y.L.)

Abstract: The backless cable-stayed bridge has the advantages of beautiful shape and reasonable force, but due to the low overall stiffness of the bridge pylon during cantilever construction, it is susceptible to the effect of solar temperature. To reveal the temperature deformation laws and achieve accurate alignment prediction during the installation process of steel–concrete composite pylons in complex environments, a refined numerical simulation model for the 3D bridge temperature field was established based on the proposed automatic sunshine-shadow recognition method. Subsequently, the optimal time periods for construction control are provided. The results of the study show that, during the cantilever construction of the bridge pylon, one pylon column will shade the other pylon column, resulting in asynchronous deformation that can reach 7.6 mm. The effect of solar temperature on the displacement of the bridge pylon is significant, where the maximum daily change in transverse displacement in the cantilevered state of the pylon can reach 33.6 mm, and the maximum change in cable force value can reach 52 kN. In order to mitigate the effect of solar radiation, the best construction time for the bridge pylon is 19:30~9:30, while the tensioning and measurement of the cable should be avoided from 6:00~18:00. This strategy ensures that the control of the pylon top displacement is maintained within 1/4000 of the pylon height, and the error in cable force is kept within 5%.

Keywords: bridge engineering; cable-stayed bridge without backstays; steel-concrete composite pylon; bridge pylon cantilever construction stage; thermal effect; refined simulation

1. Introduction

Bridges are generally exposed to the open-air environment, resulting in an inhomogeneous temperature field under the influence of solar radiation, atmospheric temperature, and ambient wind speed [1,2]. This temperature effect will cause additional temperature stresses on the structure, in addition to the deformation of the bridge structure. In the area of extreme solar-radiation conditions, the effect of solar temperature will even exceed the constant load and live load to become the first control role, which brings a series of difficulties to the bridge construction, including elevation control and internal force control [3–8].

A cable-stayed bridge without backstays typically has cables on only one side, with the force of the main girder balanced by the weight of the pylon. This type of bridge is often used in the construction of urban-landscape bridges because of its unique structural form. During the construction of a traditional cable-stayed bridge without backstays, bracing is typically employed to support the construction pylon. However, due to the long period and high cost of bracing, an increasing number of cable-stayed bridges without backstays are transitioning to using steel–concrete composite pylons and employing the bridge pylon cantilever assembly construction method. The bridge pylon adopts the structural form of a steel box and concrete, first lifting and welding the steel box of the bridge pylon, then pouring the concrete inside the box, and gradually constructing until the bridge pylon is merged. Compared to the traditional bracket method, the temperature effect on the alignment of this construction method will be more pronounced.

For instance, in open-air environments, the welding seams of large steel structures need to be welded in specific sequences. Upon the completion of welding on one side of the structure, there will be significant differences between the structural temperature field and the temperature field at the beginning of welding due to the influence of solar and environmental temperature. As the structure is constrained by the welded seam on one side, adjacent welding plates cannot freely expand or contract, resulting in an asynchronous deformation between welding segments. Therefore, if appropriate constant temperature measures are not taken during welding, forced welding and fusion between adjacent segments may occur, forming wedge-shaped weld seams, thereby significantly affecting the linearity and internal forces of the steel structure [9–13]. Especially in the process of bridge pylon construction, one side of the pylon column will produce shadow blocking to the other side of the pylon column at some moments. As a result, the temperature of the shaded area decreases significantly [14], leading to a reduced temperature gradient in the pylon column. This reduced temperature gradient can reduce the bending deformation, causing the deformation of the two sides of the pylon columns to be unsynchronized, which will further increase the difficulty of the construction control [15,16].

Previous research results are based on the computational assumption that the longitudinal temperature difference of the bridge pylon is negligible. The three-dimensional temperature field in space is simplified to a two-dimensional temperature field in the plane, and the temperature-stress coupling method is used to analyze the temperature effect of the cross-section of the concrete bridge pylon [17–20]. However, the method cannot take into account the shading effect between pylon columns. The applicability to inclined bridge pylons with sunshine-shadow shading characteristics is debatable, and the calculation results of the temperature effect cannot reflect the phenomenon of unsynchronized deformation of bridge pylons. In order to more accurately grasp the real temperature field and temperature effect of the bridge structure, research scholars propose a numerical simulation method of a three-dimensional solar temperature field that can take real-time shadows into account based on the shadow identification principle of ray-tracing calculation [21,22].

This article focuses on the Longgun River Bridge, utilizing thermal conduction theory and a three-dimensional temperature field calculation method to precisely simulate the pylon temperature fields and temperature effects during construction. Finally, it provides the optimal time periods for controlling bridge pylon construction. The method employed in this article can be applied to the precise control of construction alignment for large-span complex structures, the selection of construction hoisting windows, temperature response separation in health monitoring data, and temperature modal frequency subtraction for structural damage identification, among other applications.

2. Three-Dimensional Solar Temperature Field Analysis of Bridge

2.1. Differential Equation of Heat Conduction

Based on the law that the rate of heat conduction in an object is proportional to the temperature gradient, in 1822, Fourier [23] proposed the partial differential equation for heat conduction:

$$\rho c \frac{\partial T}{\partial t} = \frac{\partial}{\partial x}\left(k_x \frac{\partial T}{\partial x}\right) + \frac{\partial}{\partial y}\left(k_y \frac{\partial T}{\partial y}\right) + \frac{\partial}{\partial z}\left(k_z \frac{\partial T}{\partial z}\right) \quad (1)$$

where ρ and c are density (kg/m^3) and specific heat capacity (J/kg·°C); k_x, k_y, and k_z are the thermal conductivity (W/m·°C) of the material in three coordinate axes x, y, and z.

2.2. Boundary Conditions under the Solar Effect

In actual engineering, the heat exchange between the bridge and the surrounding environment mainly includes three types, namely solar radiation, radiation heat transfer, and convection, as presented in Figure 1, and the total heat flux absorbed by structural surfaces is expressed as:

$$q = q_s + q_c + q_r \tag{2}$$

where q_s is the solar-radiation heat flux absorbed by the structure surface; q_c and q_r are the heat fluxes of convection and radiation heat transfer between the structure's surface and outside, respectively.

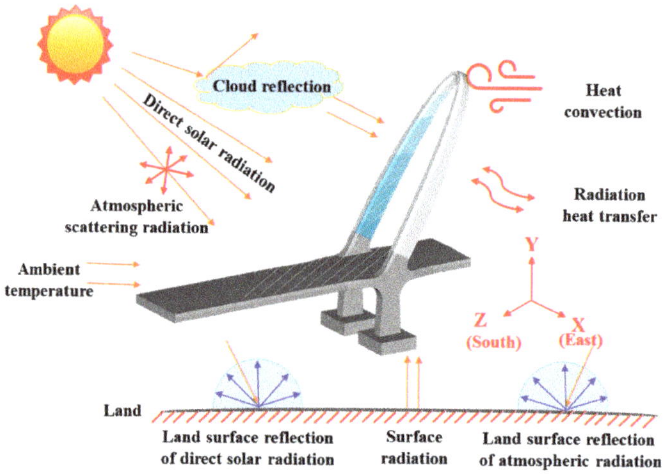

Figure 1. Thermal boundary condition of bridge pylon under solar effect.

2.2.1. Solar Radiation

Solar radiation includes direct solar radiation that passes directly through the atmosphere, diffuse radiation that has been scattered by atmospheric constituents and clouds, and reflected radiation from the ground or water [24]. It is calculated with respect to factors such as the normal direction of the surface of the structure, the solar altitude angle, the solar azimuth angle, and the composition of the ground surface, as well as the material properties of the structure [25,26]:

$$q_s = \xi\left(I_{cb} \times R_b + I_{cd} \times R_d + I_{cr} \times R_\rho\right) \tag{3}$$

where ξ is the absorptivity of the structural surface to solar radiation; I_{cb}, I_{cd}, and I_{cr} are the intensities of direct solar radiation, diffuse radiation, and ground-reflected radiation intensity in the horizontal plane. R_b, R_d, and R_ρ are correction factors of direct radiation, diffuse radiation, and ground-reflected radiation.

2.2.2. Radiation Heat Transfer

The radiative heat transfer between the surface of the structure and its surroundings can be approximated as [27]:

$$q_r = eC_s\left[(T_a + 273)^4 - (T + 273)^4\right] \tag{4}$$

where e is the surface emissivity of the structure; C_s is the Stefan-Boltzmann constant. T_a and T are the temperature of the surrounding environment and bridge structure, respectively.

2.2.3. Convection

The convective heat transfer between the surface of the structure and the outside atmosphere follows the Newtonian cooling equation, and the calculation method is:

$$q_c = h_c(T_a - T) \tag{5}$$

where h_c is the convective heat-transfer coefficient, whose value is related to the surface roughness of the structure, wind speed, and atmospheric temperature. The specific calculation can be referenced from [28,29].

2.3. *Calculation Method of 3D Solar Temperature Field*

The solar radiation received on the surface of the structure is closely related to the distribution range of sunshine shadows. The bridge can be divided into light areas, mutual-sheltering areas, and self-shelter areas under sunlight. The shadows can be determined by using a ray-tracing algorithm, as shown in Figure 2. First, from the point to be judged to send a ray parallel to the sun's rays, determine whether the ray and the structure of the intersection point, if the intersection point exists, is the point to be judged for the shadow areas, and vice versa, the point to be judged for the light areas. For example, in Figure 2, point A is the light areas, point B is his mutual-shelter areas, and point C is the self-shelter areas.

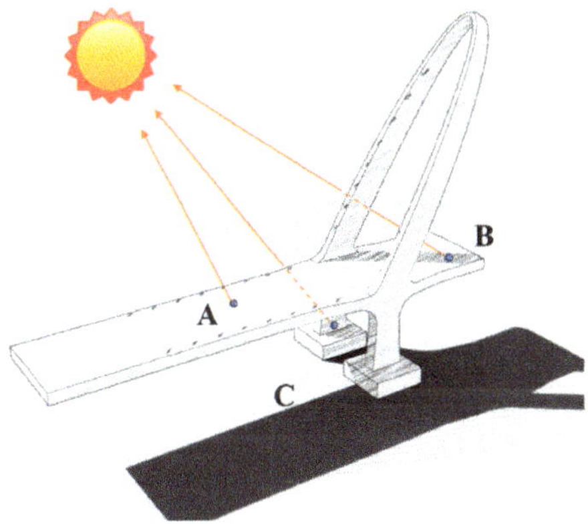

Figure 2. Sunshine-shadow recognition.

This paper is based on the Hottel clear-sky modeling algorithm [26], discretizing the structural surface into a mesh using finite-element techniques and identifying different areas through ray-tracing technology. However, traditional identification methods involve a large number of invalid calculations, which greatly wastes computational resources. Therefore, this paper adopts the triangular center coordinate method based on a raster accelerated structure to effectively and conveniently calculate the solar temperature field. The method realizes the accurate and efficient identification of the shadows of complex structural bridges and the accurate calculation of the structural three-dimensional temperature field by efficient intersection algorithms and reasonable raster divisions [15,21].

3. Finite-Element Numerical Simulation

3.1. *Engineering Background*

This paper is based on the engineering of the Longgun River Bridge, which is a cable-stayed bridge without backstays, with a span arrangement of 80 + 30 m. The bridge pylon is a composite structure of a steel box filled with counterweight concrete, standing at a height of 55 m and forming a 60° angle with the bridge deck. The cross-section of the bridge gradually transitions in size from the bottom to the top, with a width of 3.0 m × 6.2 m at the base and tapering to 2.0 m × 3.0 m at the top. The tension cables are arranged in a spatial double-corded vertical lute shape, comprising a total of 16 cables. The bridge stands in a north-south direction, with geographical coordinates of 110.5° E longitude and 19.0° N latitude. Figure 3 presents a structural diagram of the bridge pylon.

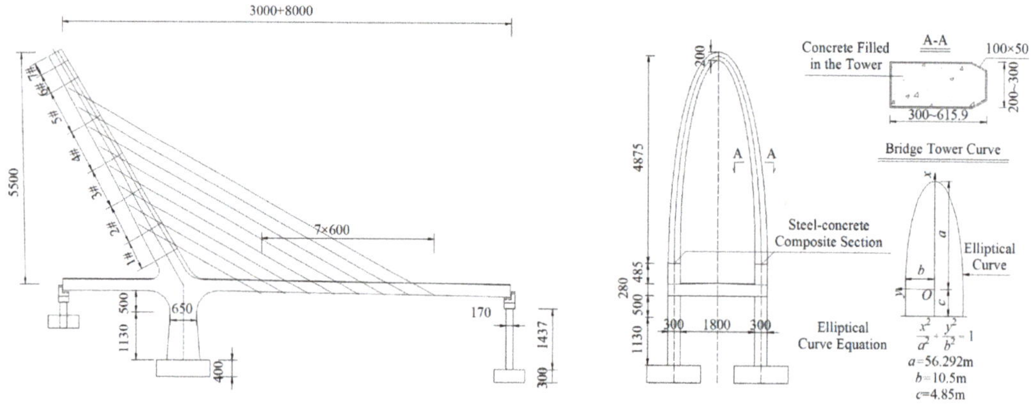

Figure 3. Schematic diagram of bridge structure (unit: cm): (**a**) elevation diagram; (**b**) bridge pylon section.

The bridge pylon of the Longgun River Bridge adopts the cantilever assembly construction method, with the construction process illustrated in Figure 4a. The bridge pylon is divided into seven sections, with two pairs of cables on each of the 2#~5# segments, and the cables are tensioned after lifting completion. Figure 4b shows the bridge pylon lifting date (*DL*) for the year 2023, the weight per segment (*WS*), and the segment length (*LS*). Because the overall stiffness of the bridge pylon in the cantilever state is lower and more sensitive to the temperature effect of solar radiation, this paper only analyzes the solar temperature field and temperature effect during the construction of the bridge pylon.

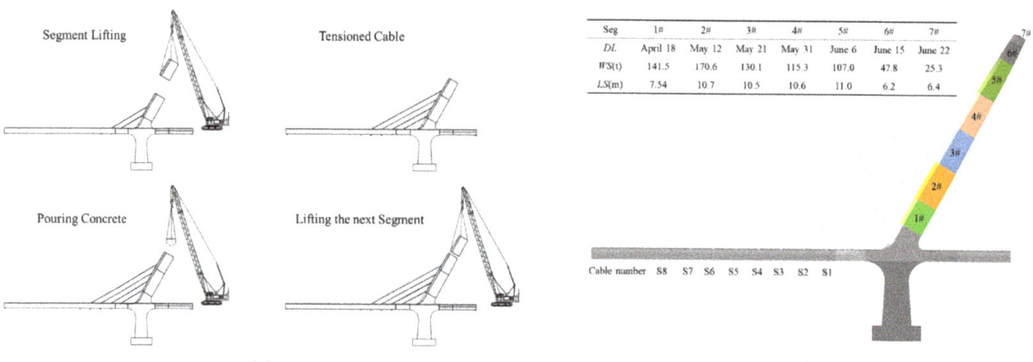

Seg	1#	2#	3#	4#	5#	6#	7#
DL	April 18	May 12	May 21	May 31	June 6	June 15	June 22
WS(t)	141.5	170.6	130.1	115.3	107.0	47.8	25.3
LS(m)	7.54	10.7	10.5	10.6	11.0	6.2	6.4

Figure 4. Construction information: (**a**) construction process; (**b**) segment information.

3.2. Finite-Element Model

The analysis model is established by ABAQUS 2019 software, employing DS3/DS4 heat-transfer elements to simulate the steel structure and DC3D8/DC3D6 heat-transfer elements to simulate the concrete structure. The S3/S4 and C3D8/C3D6 elements are used to simulate the mechanical calculations of shell and solid elements. The mesh division of the bridge pylon is illustrated in Figure 5, and the material thermal parameters are provided in Table 1.

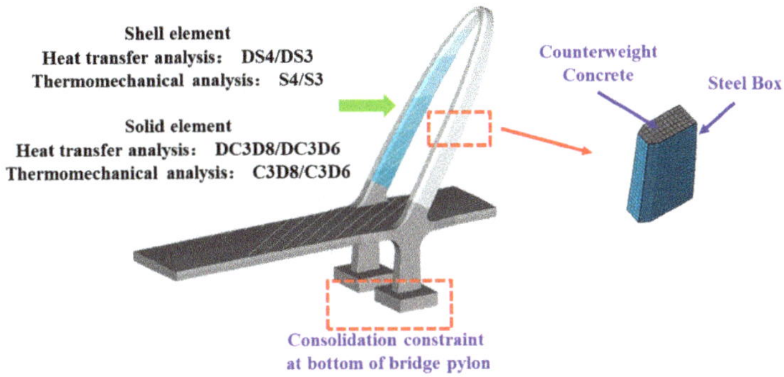

Figure 5. Schematic diagram of bridge model.

Table 1. Thermal parameter.

Material Properties	Counterweight Concrete	Concrete	Steel
Density $\rho/[\mathrm{kg{\cdot}m^{-3}}]$	3800	2592	7850
Specific heat capacity $c/[\mathrm{J/(kg{\cdot}°C)}]$	820	898	475
Thermal conductivity $\lambda/[\mathrm{W/(m{\cdot}°C)}]$	3.49	2.94	55
Absorptivity ξ	0.6	0.6	0.3
Emissivity ε	0.85	0.85	0.8
Elastic modulus MPa	3.25×10^4	3.6×10^4	2.06×10^5
Thermal expansion coefficient 1/°C	1.0×10^{-5}	1.0×10^{-5}	1.2×10^{-5}

Prior to the temperature effect analysis, the relationship between the temperature field, temperature effects, and the mesh size under maximum cantilever conditions was calculated. The temperature and displacement of the pylon top are presented in Figure 6, in which the convergence results are obtained. The correlation between the temperature field of the bridge and the mesh size is minimal. However, there is a certain correlation between the size of the bridge displacement and the mesh size. When the mesh size exceeds 600 mm, there is a significant deviation in bridge displacement, while displacement is more consistent when the mesh size is less than 400 mm. Considering the computational cost and accuracy, a model element size of 400 mm was chosen.

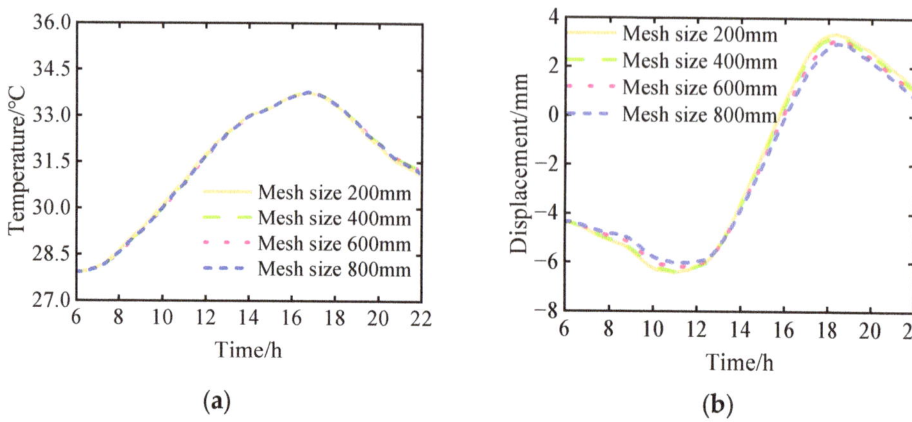

Figure 6. Model calculation results under different mesh sizes: (**a**) temperature variation; (**b**) longitudinal displacement of pylon top.

3.3. Heat-Transfer Boundary Conditions

In this paper, the meteorological conditions measured at the bridge location during the crane construction of the bridge pylon from May to June 2023 are utilized as the boundary conditions for calculation. The calculation time is advanced by 2 days to eliminate the effect of the initial temperature field. The ambient temperature on the day of each construction section is shown in Figure 7a, and the convective heat-transfer coefficient is shown in Figure 7b. The temperature effect during the construction of the steel box in segment #1 is not significant, primarily due to the stiffness of the concrete lower pylon column beneath section #1. Therefore, this paper selects the 2#~6# segment bridge pylon to analyze the temperature field and temperature effect of the bridge pylon during the construction process.

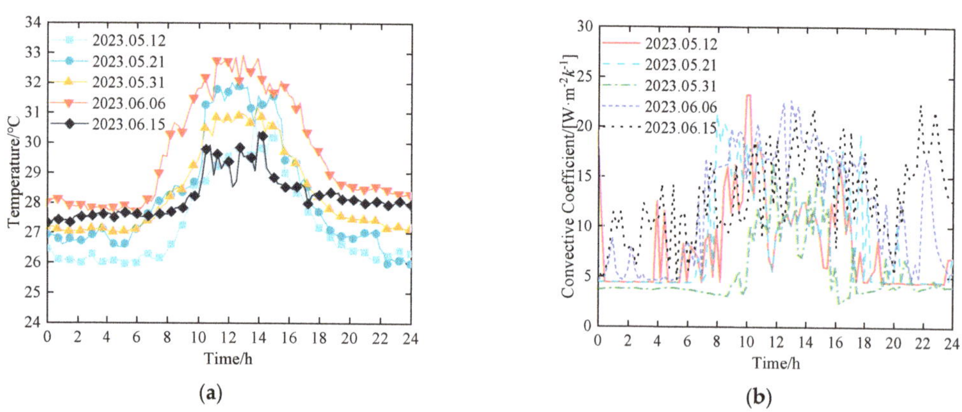

Figure 7. Boundary condition: (**a**) atmospheric temperature; (**b**) bridge pylon surface convection coefficient.

3.4. Model Validity Analysis

The solar temperature field of the bridge pylon structure will be affected by shading between the bridge components. Therefore, accurately identifying the sunshine shadow of the bridge structure is essential for accurately calculating the temperature field of the bridge structure. In this paper, 24 and 25 April 2023, were selected for verifying the accuracy of the calculation through a comparison between the measured temperature field of the main girder and the simulated temperature field derived from the finite-element analysis. Figure 8 shows the arrangement of measurement points, and Figure 9 illustrates the measured and simulated results of the main beam temperature field. 'LY-S' denotes the simulated results, and 'LY-M' represents the measured results. From Figure 9, it is evident that the calculated simulated values of the main beams are largely consistent with the actual temperature variations, with a maximum error of no more than 1.5 °C. This indicates that the method adopted in this paper can realize the accurate simulation of the three-dimensional solar temperature field of the complex structure.

(**a**) (**b**)

Figure 8. Measurement point layout: (**a**) main beam roof; (**b**) main girder floor.

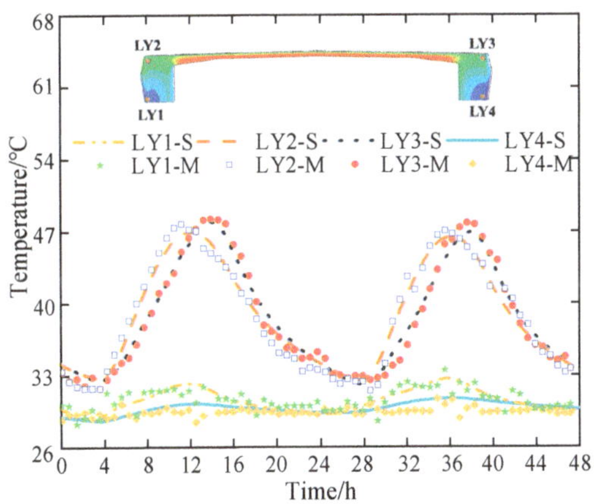

Figure 9. Bridge temperature field comparison results.

4. Analysis of Temperature Field

To illustrate the variations in the bridge temperature field during the construction of the Longgun River Bridge, the bridge structure was analyzed after the completion of the first diagonal cable tensioning. Figure 10 depicts the current stage of the bridge calculation model, indicating that the 1# steel box has been encased in counterweight concrete, and the tensioning of the first and second pairs of diagonal cables has been completed.

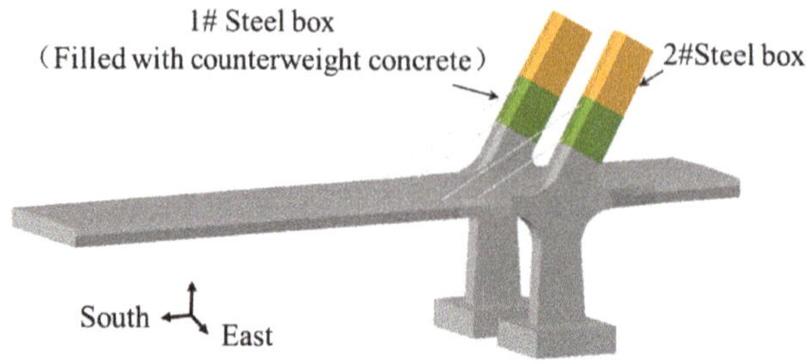

Figure 10. Model diagram after the first cable tension is completed.

Figure 11 displays a cloud representation of the temperature field of the bridge structure as of 19 May 2023. It can be observed that the temperature between the bridge components continued to rise after 8:00. Due to the significantly higher thermal conductivity of steel compared to concrete, the surface temperature of the steel box on the bridge pylon increases rapidly under sunlight, remaining consistently higher than that of the concrete throughout the day. The maximum surface temperature of the steel box is recorded at 16:00, reaching 51.2 °C. Owing to the larger specific heat capacity of the concrete in the pylon, the temperature difference between the poured and unpoured sections is more obvious, with the maximum temperature difference reaching 10.6 °C at 16:00. At this time, the temperature of the concrete main girder inner deck plate can reach up to 47.2 °C, and the two sides of the solid section are 44.3 °C.

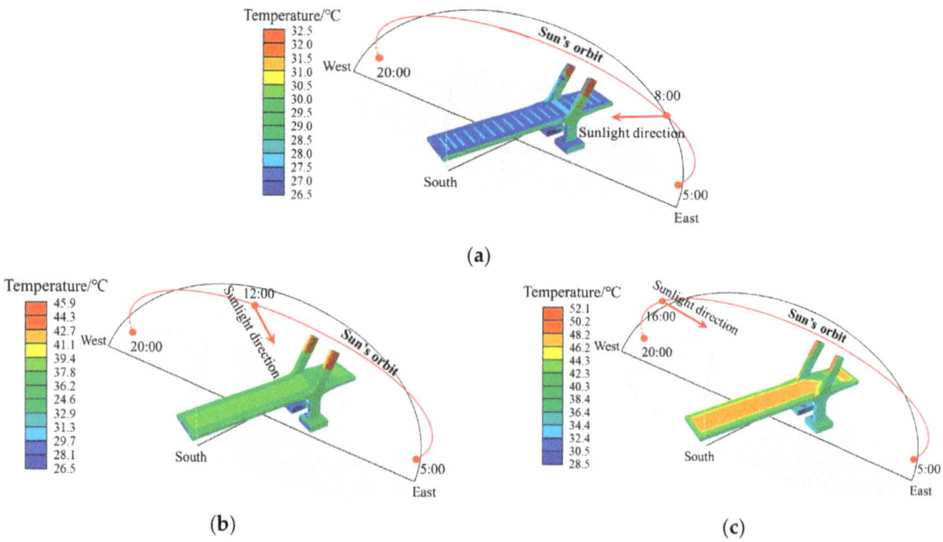

Figure 11. Bridge temperature cloud at typical moments: (**a**) 8:00; (**b**) 12:00; (**c**) 16:00.

Figure 12 shows the surface-temperature variation curves for the 1# and 2# sections of the east and west bridge pylons, where 'E-E', 'E-S', 'E-W', and 'E-N' represent the east, south, west, and north side walls of the east bridge pylon, respectively. Similarly, 'W-E', 'W-S', 'W-W', and 'W-N' denote the east, south, west, and north side walls of the west bridge pylon.

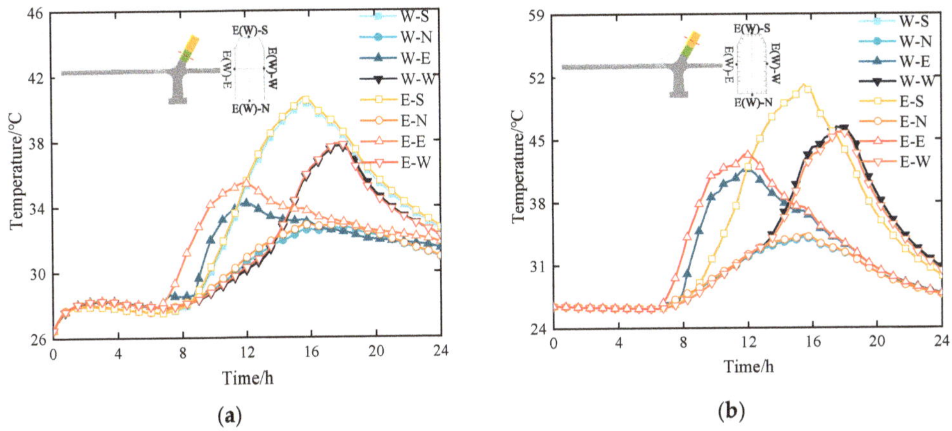

Figure 12. Bridge pylon temperature variation curve: (a) 1# bridge pylon steel box; (b) 2# bridge pylon steel box.

It can be seen from Figure 12b that the highest temperature on the surface of the steel box of the bridge pylon appears on the south pylon wall, reaching 52 °C. Between 7:00 and 8:00, the shading effect of the east side bridge pylon on the west side bridge pylon results in significant temperature differences between the eastern walls of the two pylons, with a maximum temperature difference of 4.3 °C observed at 8:00. Figure 12a illustrates the surface-temperature variation curve of the 1# bridge pylon segment. Due to the heat transfer between the steel box and the concrete in the pylon, the surface temperature of the 1# bridge pylon segment will be significantly lower than that of the 2# steel box, although the trends are similar. At 16:00, the maximum temperature difference between the upper and lower segments of the 1# and 2# sections on the south side of the pylon wall can reach 10.6 °C.

5. Analysis of Temperature Effects on Bridge Pylon

Through the analysis of the temperature field of the bridge pylon in the previous subsection, it is evident that, at different times of the day, one side of the bridge pylon casts shade on the other side, resulting in a sudden drop in temperature in the shaded area. The temperature difference will cause asynchronous deformation on both sides of the bridge pylon, thereby imposing higher requirements on the control of the construction alignment of the bridge pylon. Since the bridge pylon is approximately in a single cantilever state, the base stress of the bridge pylon and the displacement change at the pylon top become more noticeable under solar temperature.

5.1. Pylon Top Displacement

The overall stiffness of the bridge pylon in the cantilevered construction state is weak, and the effect of temperature on the pylon displacement is more significant. Figure 13a illustrates the variation in pylon girder displacement during the construction of segment 5# under temperature effect, where $\delta_{\text{east(west)}}$-$x$ represents the longitudinal displacement of the east (west) bridge pylon, $\delta_{\text{east(west)}}$-$y$ represents the transverse displacement of the east (west) bridge pylon. After sunrise, due to the shading from the east bridge pylon to the west bridge pylon, significant asynchronous vertical deformation occurs. At 12:00, the vertical displacement difference between the two sides of the bridge pylon can reach 7.7 mm, while the magnitude of vertical displacement change in the bridge pylon during the whole day was only 10.8 mm. Both the longitudinal and transverse displacements of the bridge pylons exhibit similar trends, with a longitudinal variation range of 10.1 mm and a transverse variation range of 29.3 mm over the course of the entire day. It can be analyzed that the transverse displacement should be the key factor of the alignment control of the bridge pylon construction of the cable-stayed bridge without backstays.

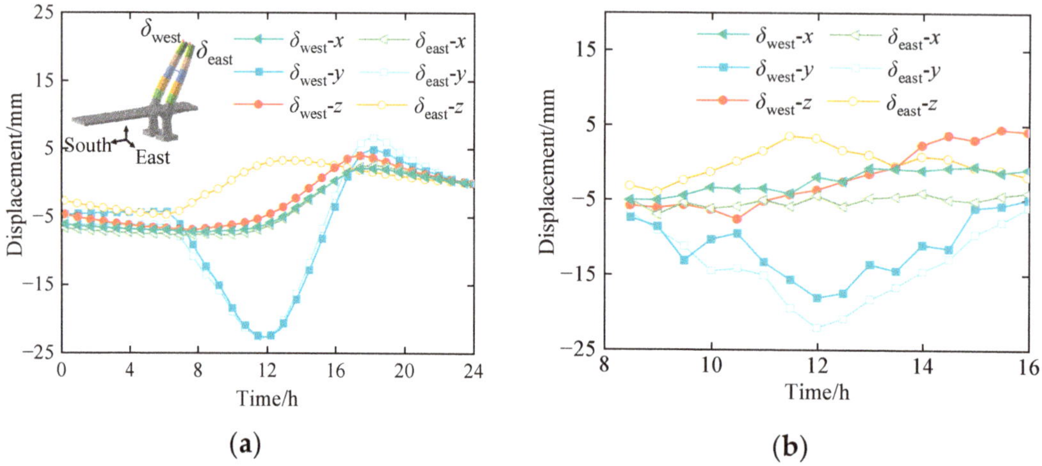

Figure 13. Bridge pylon displacement: (**a**) calculated displacement of bridge pylon; (**b**) measured displacement of bridge pylon.

Figure 13b shows the measured displacement of the bridge pylon from 8:00 to 16:00 during the current construction phase. As can be seen from the figure, the method used in this paper is similar to the measured results. At 12:00, the minimum displacement in the transverse measurement was −17.4 mm, while the calculated result was −22.3 mm, with a difference of only 4.9 mm between the two. The maximum deviation throughout the day is only 3.7 mm in the vertical direction and 3.8 mm in the longitudinal direction of the bridge, indicating the accuracy of the simulation.

5.2. Bridge Pylon Base Stress

Figure 14 shows the concrete stress change at the base of the bridge pylon under the construction phase of the 5# bridge pylon. Under the influence of solar temperature, the bridge pylon moved to the south side before 18:00. The south side of the pylon wall exhibited a gentle transition, with the bridge's surface stress showing less sensitivity to sunlight temperature, while the north side of the pylon wall exhibited the opposite trend. The tensile stress on the north side of the pylon wall gradually increased from −1.4 MPa to 0.4 MPa before 18:00, followed by a gradual transition to compressive stress. The stress change trend of the east and west side pylon walls of different bridge pylons was relatively consistent. However, due to the shading influence of the east side pylon, a stress difference gradually appears on the east pylon wall of the east and west pylon after 8:00, reaching a maximum of 0.44 MPa at 17:30.

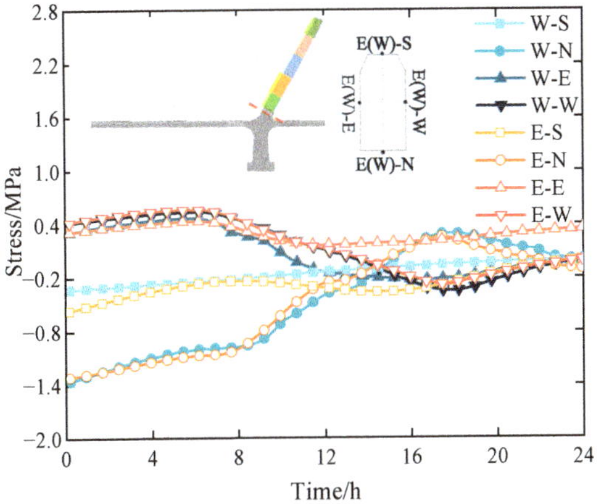

Figure 14. Bridge pylon surface stress change.

6. Construction Control Recommendations

This paper is based on the engineering of the Longgun River Bridge; the vertical height of the pylon is 55 m. According to the requirements of the construction stage control accuracy in the Chinese code 'Technical Specifications for Construction Monitoring and Control of Highway Bridges' (JTG/T 3650-01-2022) [30], the steel structure installation of cable-stayed bridge pylons should be controlled within an accuracy of $H/4000$ (where H is the height of the bridge pylon), with an elevation error not exceeding $\pm$ 2 mm $\times$ n (where n is the number of bridge pylon segments), and the allowable relative error for cable-stayed cable force should not exceed 5%. To ensure the aforementioned construction control accuracy target, the construction error caused by temperature can be mitigated by reasonably arranging the lifting time of the bridge pylon and tensioning time of the cable-stayed cable, ensuring that the alignment shape and cable force after the bridge is completed meet the design target.

The temperature effect caused by solar radiation in each construction stage is depicted in Figure 15. Through analysis, it is found that, as the height of the bridge pylon increases during the construction process, the impact of solar temperature on the bridge displacement also increases. The maximum daily change in transverse displacement of the bridge pylon in the cantilever state can reach up to 33.6 mm, while the effects on longitudinal and vertical displacement are relatively smaller, remaining within specification limitations.

Table 2 provides a list of recommended construction control times for the bridge pylon construction process. During certain moments, one side of the bridge tower will be shaded by the other, and the temperature in the shaded area will drop significantly. This leads to asynchronous displacement between the two bridge towers. Therefore, the construction control time for the two bridge towers also differs. From the table, it can be seen that, when the pylon height is low, the bridge pylon's stiffness is high, allowing for most of the day to meet construction requirements. However, as the pylon's height increases, the bridge pylon's stiffness weakens, and the positioning of the bridge pylon and welding time should be controlled between 19:30~9:30 to weaken the impact of sunlight temperature on the bridge pylon alignment.

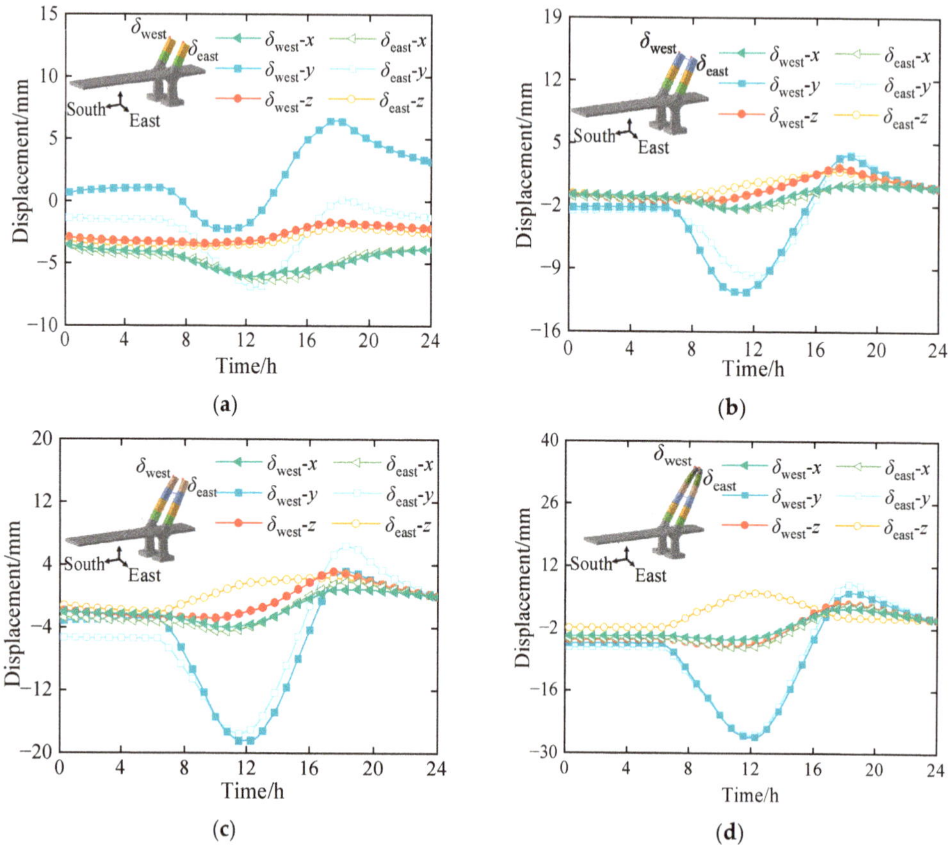

Figure 15. Displacement changes of bridge pylon under solar radiation at different construction stages: (**a**) 2# bridge pylon segment; (**b**) 3# bridge pylon segment; (**c**) 4# bridge pylon segment; (**d**) 6# bridge pylon segment.

Table 2. Reasonable construction control time for bridge pylons.

Segments	Pylon Height/m	East Bridge Pylon	West Bridge Pylon
2#	22.658	whole day	18:30~17:00
3#	31.566	whole day	13:30~9:00
4#	40.432	19:30~9:30 13:00~17:30	14:30~9:00
5#	49.361	19:00~9:30 13:30~17:00	14:00~9:30
6#	53.924	19:30~9:00 14:00~17:30	14:30~9:00

Figure 16 shows the change in cable force for each cable under the influence of solar temperature after the tension of cable S1~S8. Table 3 shows the maximum value of cable force change in each construction stage; the most significant changes in cable force occur between 6:00 to 18:00. The change in cable force of the cable-stayed bridge without backstays increases with the height of the pylon and the length of the cable. In the maximum cantilevered state of the bridge pylon, the maximum increase in cable force of cable S8 can reach up to 52 kN, which accounts for 1.4% of the design force. In order to ensure the quality of construction, the cable tensioning and measurement should be selected between 18:00~6:00, when the influence of solar temperature is small.

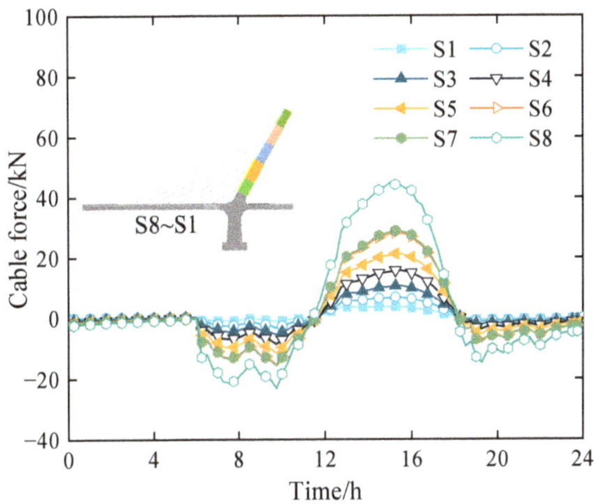

Figure 16. Cable force under the influence of solar temperature.

Table 3. Maximum position and value of cable force change in different construction stages.

Construction Stage	Maximum Change Position of Cable Force	Maximum Value of the Cable Force Change/kN
2# bridge pylon segment	S2	8.8
3# bridge pylon segment	S4	10.7
4# bridge pylon segment	S6	35.3
5# bridge pylon segment	S8	52

According to the solar temperature effect on the Longgun River Bridge, construction activities are conducted in adherence to the predetermined timelines. Through a comparative analysis of measured deviations in pylon top and cable tensions against their theoretical counterparts, it was observed that the deviation in pylon top is 11 mm, while the average relative error in cable tension is 2.1%. These discrepancies fall well within the permissible construction tolerances, thus ensuring the construction quality of the Longgun River Bridge amidst the challenging conditions of high temperature and humidity prevalent in Hainan.

7. Conclusions

This paper accurately analyses the temperature distribution patterns of the curved special-shaped pylon cable-stayed bridge without backstays by employing the shadow recognition algorithm proposed in previous studies. Based on this, the temperature-induced stress and displacement of the structure were calculated, the deformation characteristics of the bridge pylon were summarized, and the optimal construction times for bridge pylon lifting and cable tensioning were predicted. The following conclusions were drawn:

(1) During the construction process, one side of the bridge pylon will produce shadow shading on the other side; the maximum temperature difference between the light area and the shadow area of the bridge pylon can reach 4.3 °C. Additionally, the temperature difference between the upper and lower sections of the poured and unpoured concrete steel box surfaces can be up to 10.6 °C, due to the presence of the concrete in the pylon during the construction process;

(2) Due to shadow shading between the bridge pylons, asynchronous deformation occurs between the two sides of the bridge pylons. Moreover, the direction of these

asynchronous deformation changes during the construction process. When the height of the bridge pylon is low, asynchronous deformation in the transverse direction is more pronounced. Meanwhile, as the height of the bridge pylon increases, asynchronous deformation in the vertical direction becomes more apparent. The stresses at the base of the pylons are also affected by the shadow between the pylons, resulting in stress differences. During the construction of segment 5# of the bridge pylons, the transverse stress difference can reach 0.44 MPa;

(3) Under the influence of temperature, the transverse displacement of the bridge pylon is particularly sensitive to solar temperature. As the height of the bridge pylon increases during the construction process, the effect of solar temperature on bridge displacement and cable-stayed cable force also increases. The maximum daily variation in the transverse displacement of the bridge pylon in the maximum cantilever condition can reach up to 33.6 mm, and the maximum variation in cable force is 52 kN for cable S8;

(4) To mitigate the impact of solar effects, the construction schedule for bridge pylons in the cable-stayed bridge without backstays should be limited to between 19:30 and 9:30. The tensioning and measurement of cables should be avoided between 6:00 and 18:00. Furthermore, the daily variation pattern of each control parameter should be taken into consideration to correct for the temperature effect.

Author Contributions: Conceptualization, J.L. and B.G.; methodology, Z.W., Y.L. and J.L.; software, B.G.; validation, B.G. and Z.W.; formal analysis, S.L. and L.F.; investigation, B.G. and J.L.; supervision, J.L.; project administration, J.L.; writing—original draft preparation, B.G. and J.L.; writing—review and editing, B.G., L.F., J.L., Y.L., S.L. and Z.W. All authors have read and agreed to the published version of the manuscript.

Funding: This research was funded by the Key Research and Technologies for the Construction of the Main Pylon of Curved Special-shaped Pylon Cable-stayed Bridge without Backstays (Grant No. T-GL-XMGS(J)-HAN-HDLYGL-02-JS-034).

Data Availability Statement: All datasets are publicly available.

Conflicts of Interest: Author Lianjun Feng is employed by the Xiamen Company of China First Highway Engineering Co., Ltd. The remaining authors declare that the research was conducted in the absence of any commercial or financial relationships that could be construed as a potential conflict of interest.

References

1. Liu, J.; Liu, Y.J.; Zhang, C.; Zhao, Q.; Lyu, Y.; Jiang, L. Temperature action and effect of concrete-filled steel tubular bridges: A review. *J. Traffic Transp. Eng.* **2020**, *7*, 174–191. [CrossRef]
2. Meng, Q.L.; Zhu, J.S. Fine temperature effect analysis–based time-varying dynamic properties evaluation of long-span suspension bridges in natural environments. *J. Bridge Eng.* **2018**, *23*, 04018075. [CrossRef]
3. Tong, M.; Tham, L.G.; Au, F.T.K.; Lee, P.K.K. Numerical modelling for temperature distribution in steel bridges. *Comput. Struct.* **2001**, *79*, 583–593. [CrossRef]
4. Bai, Y.X.; Liu, Y.J.; Liu, J.; Wang, Z.; Yan, X.K. Research on solar radiation-caused radial temperature difference and interface separation of CFST. *Structures* **2024**, *62*, 106151. [CrossRef]
5. Tu, G.Y.; Yan, D.H.; Zhang, K.N. Calculation method and countermeasure for closure temperature influence in trussed CFST arch bridge. *Eng. Mech.* **2008**, *25*, 236–240.
6. Li, X.G.; Luo, H.R.; Ding, P.; Chen, X.H.; Tan, S.L. Prediction Study on the Alignment of a Steel-Concrete Composite Beam Track Cable-Stayed Bridge. *Buildings* **2023**, *13*, 882. [CrossRef]
7. Lou, P.; Zhu, J.P.; Dai, G.L.; Yan, B. Experimental study on bridge—Track system temperature actions for Chinese high-speed railway. *Arch. Civ. Mech. Eng.* **2018**, *18*, 451–464. [CrossRef]
8. Hagedorn, R.; Martí-Vargas, J.R.; Dang, C.N.; Hale, W.M.; Floyd, R.W. Temperature gradients in bridge concrete I-girders under heat wave. *J. Bridge Eng.* **2019**, *24*, 04019077. [CrossRef]
9. Mirambell, E.; Aguado, A. Temperature and stress distributions in concrete box girder bridges. *J. Struc. Eng.* **1990**, *116*, 2388–2409. [CrossRef]
10. Long, G.; Yang, Y.H.; Zhu, W.Q.; Liu, Y.J. Analysis and handing of upwarping accident of main girder of a steel box girder cable-stayed bridge. *Bridge Constr.* **2018**, *48*, 113–117.
11. Zhou, L.; Xia, Y.; Brownjohn, J.M.W.; Koo, K.Y. Temperature analysis of a long-span suspension bridge based on field monitoring and numerical simulation. *J. Bridge Eng.* **2016**, *21*, 04015027. [CrossRef]

12. Wang, J.F.; Xu, Z.Y.; Fan, X.L.; Lin, J.P. Thermal effects on curved steel box girder bridges and their countermeasures. *J. Perf. Constr. Facil.* **2017**, *31*, 04016091. [CrossRef]
13. Mosavi, A.A.; Seracino, R.; Rizkalla, S. Effect of temperature on daily modal variability of a steel-concrete composite bridge. *J. Bridge Eng.* **2012**, *17*, 979–983. [CrossRef]
14. Liu, H.B.; Chen, Z.H.; Zhou, T. Theoretical and Experimental Study on the Temperature Distribution of H-shaped. *Appl. Therm. Eng.* **2012**, *37*, 329–335. [CrossRef]
15. Liu, Y.; Han, S.; Gong, B.X.; Wang, Z.; Liu, J.; Shen, Z. Refined 3D Solar Temperature Field and Effect Simulation of Ultra-High Steel Bridge Pylon. *Appl. Sci.* **2023**, *13*, 4400. [CrossRef]
16. Hu, W.; Liu, J.; Liu, Y.J.; Lyu, Y.; Li, H.T.; Xian, J.P.; Ma, Y.P. Construction alignment and closure control of CFST truss arch bridges based on temperature effect. *Structures* **2024**, *63*, 106471. [CrossRef]
17. Zhu, J.S.; Meng, Q.L. Effective and fine analysis for temperature effect of bridges in natural environments. *J. Bridge Eng.* **2017**, *22*, 04017017. [CrossRef]
18. Kennedy, J.B.; Soliman, M.H. Temperature distribution in composite bridges. *J. Struc. Eng.* **1987**, *113*, 475–482. [CrossRef]
19. Dai, P.; Qian, Y.J. Short-term temperature characteristics of H-shaped section concrete pylon of cable-stayed bridge. *J. Southwest Jiaotong Univ.* **2014**, *49*, 59–65.
20. Zhang, H.L.; Li, J.; Liu, C.G.; Jiang, T.H.; Wei, J. Thermal Effect of the Cable-Stayed Bridge Tower. *Wuhan Univ. J. Nat. Sci.* **2003**, *04*, 1121–1125.
21. Wang, Z.; Liu, Y.J.; Tang, Z.W.; Zhang, G.J.; Liu, J. Three-dimensional temperature field simulation method of truss arch rib based on sunshine shadow recognition. *China J. Highw. Transp.* **2022**, *35*, 91–105.
22. Gu, B.; Xie, P.Z.; Qian, H.; Lei, L.H. 3D-temperature field and its effect of inverted Y-shaped concrete pylon in lower Yangtze region. *J. Cent. South Univ.* **2020**, *50*, 248–260.
23. Elbadry, M.M.; Ghali, A. Temperature Variations in Concrete Bridges. *J. Struct. Eng.* **1983**, *109*, 2355–2374. [CrossRef]
24. Chen, D.S.; Wang, H.J.; Qian, H.L.; Li, X.Y.; Fan, F.; Shen, S.Z. Experimental and Numerical Investigation of Temperature Effects on Steel Members Due to Solar Radiation. *Appl. Therm. Eng.* **2017**, *127*, 696–704. [CrossRef]
25. Liu, B.Y.H.; Jordan, R.C. The Interrelationship and Characteristic Distribution of Direct, Diffuse and Total Solar Radiation. *Sol. Energy* **1960**, *4*, 1–19. [CrossRef]
26. Hottel, H.C. A Simple Model for Estimating the Transmittance of Direct Solar Radiation Through Clear Atmospheres. *Sol. Energy* **1976**, *18*, 129–134. [CrossRef]
27. Zhao, R.D.; Wang, Y.B. Studies on Temperature Field Boundary Conditions for Concrete Box-girder Bridges Solar Radiation. *Chin. J. Highw.* **2016**, *29*, 52–61.
28. Kehlbeck, F. *Einfluss der Sonnenstrahlung bei Brückenbauwerken*; Werner-Verlag: Bielefeld, Germany, 1975.
29. ASHRAE. *Handbook of Fundamentals*; American Society of Heating, Refrigerating, and Air Conditioning Engineers: Peachtree Corners, GA, USA, 1972.
30. *JTG/T 3650-01-2022*; Technical Specifications for Construction Monitoring and Control of Highway Bridges. Ministry of Transport of China: Beijing, China, 2022. (In Chinese)

buildings

MDPI

Article

Evaluation Methods and Influence Factors of Blisters Disease in Concrete Composite Bridges

Chen Yu [1,*] and Mengya Zhang [2]

[1] Engineering and Technology Research Institute, First Highway Consultants Co., Ltd., Xi'an 710075, China
[2] College of Management, Xi'an University of Finance and Economics, Xi'an 710100, China; zhangmy@xaufe.edu.cn
* Correspondence: vincentrainy@hotmail.com

Abstract: The decks of steel–concrete composite bridges are constantly exposed to severe environmental conditions, which frequently give rise to significant issues, including cracks and holes. These problems occur due to the formation of blisters under the paving layer with waterproofing membranes. This paper aims to delve into the characteristics of blisters during their expansion and propagation stages. Additionally, it proposes a rating index and a simplified calculation formula to assess the interface propagation performance of bridge deck pavement. To achieve this, the research group developed a simulated blister test device and employed the digital image correlation (DIC) technique. The study investigated the impact of pavement structure, waterproofing layer, and air voids on blister propagation behavior. It was discovered that the pavement blister test encompassed two distinct stages: expansion and propagation. Furthermore, the SMA-13 asphalt mixture exhibited slightly superior resistance to blistering compared to AC-13. It was also observed that when the mixture void ratio is less than 3.5%, it becomes more susceptible to blistering deformation, ultimately leading to debonding damage. Among the waterproofing materials tested, SBS-modified emulsified asphalt demonstrated the weakest adhesion to cement concrete substrates, while SBS-modified asphalt performed slightly better than rubberized asphalt.

Keywords: concrete bridge deck pavement; blister test; asphalt blisting; asphalt deflection; evaluation of blister propagation

1. Introduction

Blistering commonly occurs on bridge decks, and it is primarily caused by the accumulation of air and water vapor pressure beneath the pavement of steel–concrete composite bridge [1–6]. The existence of blister will change the stress state of the pavement, leading to the redistribution of the stress field. Cracks, breakages, and even holes in the pavement caused by blisters under traffic load can significantly impact the service life and safety of bridge structures [7,8]. Generally, blisters can be treated through post-processing methods such as drilling drainage and slotting drainage, among others. Although these methods have a mitigating effect, they cannot fundamentally solve the problem. Numerous studies have been conducted on the formation mechanism of blisters [9,10], while there is a lack of research on the propagation mechanism of blisters. Therefore, it is urgent to determine the propagation criterion and influencing factors of blisters in order to implement effective measures and minimize or eliminate problems related to asphalt overlays.

Dannenberg [11] initially introduced the idea of the blister tests to assess the bonding strength between organic coatings and metals. They have also found significant application in studying the propagation of debonding, particularly the separation of a thin layer of adhesive from the substrate [12]. Several researchers have contributed valuable insights and techniques in this field. Steven [13] utilized the Hencky series equation to examine blister propagation. Utilizing an atomic force microscope (AFM), the shape of bulged graphene membranes at the nanoscale could be precisely measured, with the maximum deflection serving as a crucial parameter in such analyses. Kaimin [14] emphasized that adhesive interactions are encompassed within an energy balance equation, which considers both strain and adhesion energies. This provides a comprehensive understanding of the forces governing debonding processes. In assessing blister propagation, Liao [15] implemented a displacement sensor that accurately measured the expansion of blisters. This enabled the quantification and analysis of blister behavior over time. Furthermore, the blister test holds the potential to simulate crack propagation along the interface between the paving layer and substrate. By subjecting these systems to blister tests, researchers gain insights into the dynamics of crack propagation and the underlying mechanisms. Harvey [16] established a mechanical model for the adhesion toughness of monolayer films under pressure load and point load, namely mode I and II. Williams [17] developed a linear elastic fracture mechanics model to analyze a blistering plate under bending. Jensen further enhanced the model by considering stress intensity factors. Cotterell and Chen [18] solved the Von Karman equation to study blistering films with a wide range of thicknesses. They calculated the constitutive relation and derived the fracture energy (G) through an energy balance. Tan [19] utilized digital speckle correlation techniques to explore the tensile strain rate of asphalt beams across varying stress levels, revealing their deformation patterns over time. In 2013, Yuan [20] combined these techniques with four-point bending fatigue tests, providing a microscopic view of asphalt mixture fatigue deformation. Then, in 2015, Xing Yongming [21] applied the same methods to semi-circular bending tests on UV-aged and modified asphalt mixtures, offering insights into crack initiation and propagation patterns.

In this paper, we will outline the criteria for blister propagation by deriving the fracture energy (G) through a comprehensive analysis of energy balance from first principles. The procedure is detailed in the following steps: (1) In the first section, the preparation process of the investigated sample and the experimental device is introduced. (2) Then, the deformation of the blister is presented and its failure pattern is described. (3) Consequently, the propagation criteria based on energy change are proposed and the expansion mechanism is analyzed. (4) Finally, the influence of pavement structure, waterproofing layer and air voids on blister propagation is discussed.

2. Materials and Methods

2.1. Investigated Samples

The main objective of this article was to determine the propagation behavior of blistering. Six different types of deck pavement were used for this investigation (see Table 1). Considering that the common blister disease of bridge deck pavement is mainly caused by asphalt mixtures with low air voids, AC-13 (optimal asphalt content is 5.0%) and SMA-13 (optimal asphalt content is 6.15%) were used in the experiment.

Table 1. Different types of blister test samples.

Sample	Mixture	Waterproofing Layer	Bottom Plate
1	AC-13	SBS-modified asphalt	Cement concrete substrate
2	AC-13	Asphalt rubber	Cement concrete substrate
3	AC-13	SBS-modified emulsion asphalt	Cement concrete substrate
4	SMA-13	SBS-modified asphalt	Cement concrete substrate
5	SMA-13	Asphalt rubber	Cement concrete substrate
6	SMA-13	SBS-modified emulsion asphalt	Cement concrete substrate

The sample preparation process is as follows:

- Cement concrete plates with dimensions of 300 mm width, 300 mm length, and 50 mm height were prepared as substrates. Each plate included a centrally positioned hole with a diameter of 10 mm that serves as a hydraulic inlet at the bottom. The orifice (filled with fine sand) in the substrate is sealed using asphalt release agent on the surface.

- The waterproofing membrane was set in an interlayer between the concrete and the asphalt mixture to prevent water intrusion and increase the interfacial bonding force.
- Next, 5 cm thick asphalt mixture was laid over the substrates, and the total test mold was rolled into shape under the wheel mill.
- Finally, the well-formed specimen was conditioned at room temperature for 48 h before removing the mold. Figure 1a shows the prepared sample.
- To facilitate the observation of blister behavior in experiments, a paint spray technique was employed on the asphalt mixture, resulting in a spackle-like pattern as illustrated in Figure 1b. Subsequently, a layer of white primer was applied and allowed to thoroughly dry. To ensure clear visibility and enhance contrast, black spots were manually sprayed onto the surface, creating a distinct gray differentiation.

(a) (b)

Figure 1. (**a**) The virgin sample and (**b**) the speckled sample.

2.2. Blister Test

The blister loading device and data acquisition device used in this experiment are mainly composed of a hydraulic loading system, a sample bearing platform, a sample fixing device, an environment box developed by the research group and an MTI-3D non-contact displacement measurement system. The blister test method uses an electro-hydraulic servo fatigue machine to control the flow rate of injected oil.

The sample bearing platform is a high-strength steel plate (80 cm × 80 cm × 1 cm) with a stainless-steel bolt at each corner (for sample securing), as is shown in Figure 2a. The platform contains a central hole with a size of 1 cm diameter as oil pressure inlet from the hydraulic oil chamber below. In order to prevent the hydraulic oil from leaking from the bottom of the sample during the loading process, three O-shaped rubber seal ring are arranged on the outside of the orifice, as presented in Figure 2b.

To ensure the sample was in close contact with the O-ring seal and no slippage occurred under liquid pressure, a sample fixing device was designed as indicated in Figure 3. The device is a high-strength steel plate of thickness 1 cm with a central hole size of 28 cm × 28 cm. The device is tightened on the sample bearing platform by bolts to form a fixed reaction frame. Therefore, it not only ensures the sealing, measuring precision, but also convenient handling and simple operation.

The transparent environmental box was connected to the bearing platform, which was used to control the test temperature and humidity. As shown in Figure 4, the environment box consisted of a heating and humidifying system, temperature and humidity control system, lighting, and other components. The four sides of the environmental box were made up of stainless-steel plate and an insulation layer to ensure the constant temperature and humidity of the internal gas. The top of it was made of low-refractive-index toughened glass, which is convenient for the non-contact displacement measurement system for observing blister behavior.

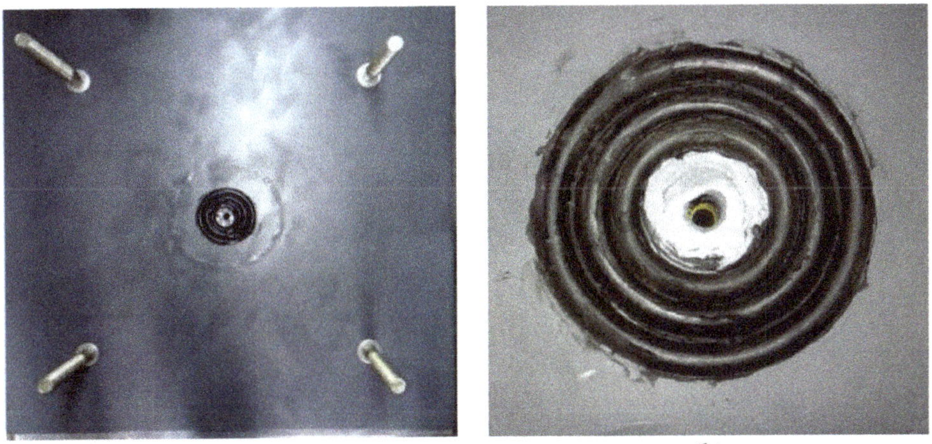

(a) (b)

Figure 2. (**a**) The sample bearing platform and (**b**) the seal ring.

Figure 3. The sample fixing device.

In order to overcome the limitations of a traditional deformation measurement device, such as single point realization testing, complicated optical path testing and difficult equipment control, etc., the Digital Speckle Correlation Method (DSCM) was used to collect the deformation data of blister in this paper. The evolution of the blisters was documented by the MTI-3D structural deformation analysis system, comprising both hardware and software components. The hardware subsystem included two high-precision industrial CCD cameras, two high-performance lenses with a focal length of 25 mm and 50 mm, respectively, a signal synchronous collector, a lighting system, and a workstation. The MIT-3D analysis calculation flow is shown in Figure 5. The pressure simulation test was carried out at 60 °C until the specimen was damaged.

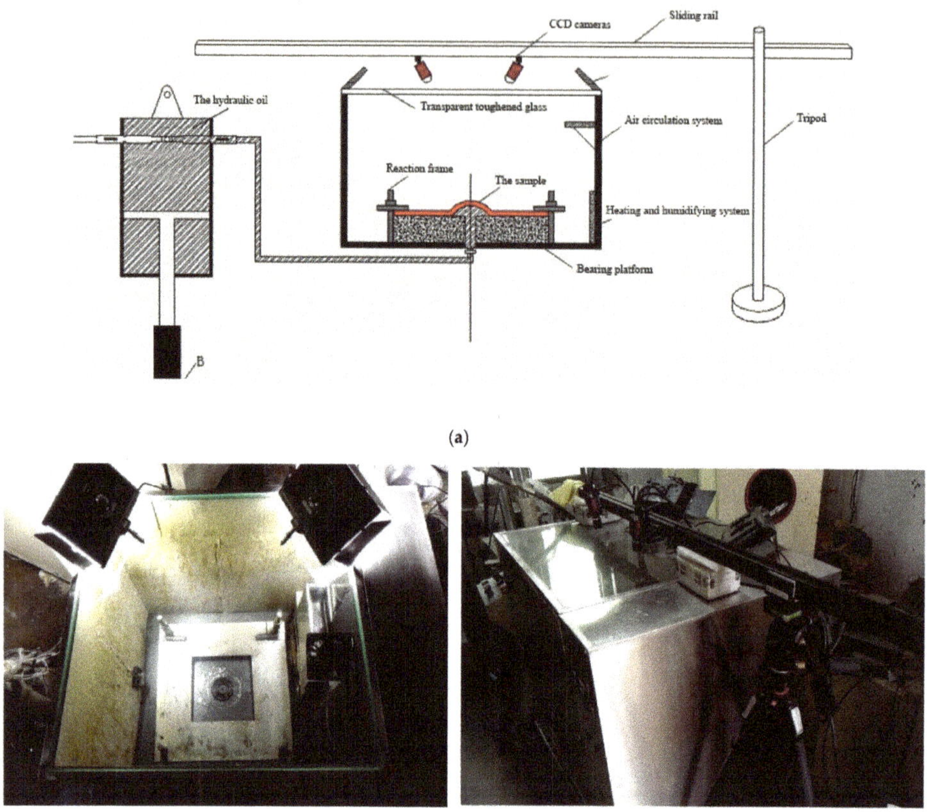

(a)

(b)

(c)

Figure 4. The environmental box (Note: (**a**) the schematic diagram of test equipment; (**b**) the top of the environmental box; (**c**) data acquisition device).

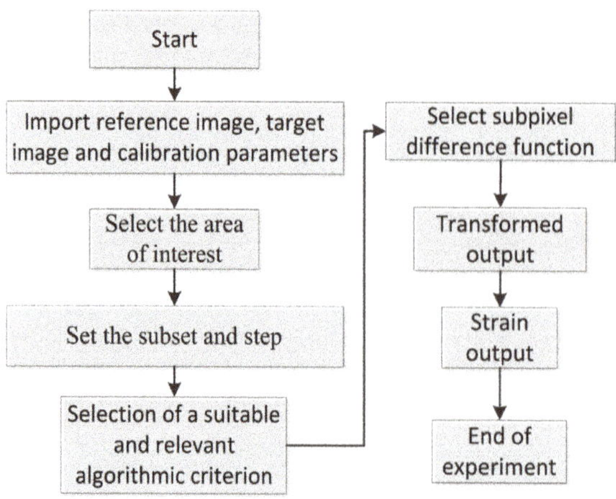

Figure 5. MIT-3D calculation flowchart.

3. Results and Discussion

3.1. The Blister Propagation and Failure Patterns

Under uniform load, the failure surface of the blister on asphalt mixture pavement is not regular. The deflection at middle points of blister is relatively large (about 4.5 mm). This irregular deformation is mainly due to the uneven paving layer. The final failure of the blister is caused by the interfacial debonding between the asphalt mixture paving layer and the cement concrete substrates with the increasing pressure. Meanwhile, the pressurized liquid seeps from the side of the sample. The expansion of the blister along the bonding interface indicates that the bonding surface between the deck pavement and the concrete plates is its area of weakness.

It is found that the deformation of the blister is divided into two stages: expansion and propagation. In the expansion stage, the obvious deflection occurs after the load reaches a certain value due to the high rigidity of the pavement layer, and increases sharply with the load (see Figure 6a). Afterwards, the blister begins to propagate and the deflection decreases when the load reaches the critical load of blister expansion (see Figure 6b). This observation suggests that the crack initiation and propagation of blister may be the reason for the decrease in the overall stiffness of the sample, which will affect the mechanical behavior of the material under the load.

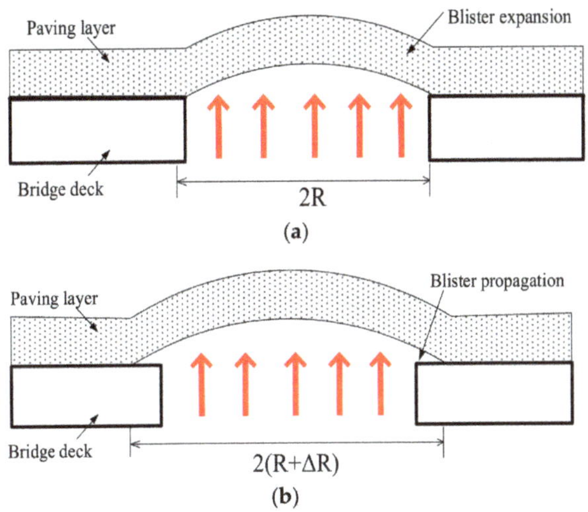

Figure 6. A pressurized circular blister (**a**) the expansion stage and (**b**) the propagation stage.

3.2. The Propagation Criteria of Blister

The center deflection ω of the paving layer changes with the pressure P. Because the edge of the paving layer is clamped on substrate, the boundary conditions are:

$$\begin{aligned} \omega = 0 \text{ and } \tfrac{d\omega}{dr} = 0, \text{ for } r = R \\ \omega = \omega_0 \text{ and } \tfrac{d\omega}{dr} = 0, \text{ for } r = 0 \end{aligned} \quad (1)$$

If residual stress in a paving layer can be ignored, the relationship between pressure and deflection can be written as

$$\omega(r) = \frac{P}{64D}\left(R^2 - r^2\right)^2 = \frac{P}{64D}a^2 \quad (2)$$

where the stiffness D is determined by the following expression:

$$D = \frac{Eh^2}{12(1 - v^2)} \tag{3}$$

where a is the radius of a blister at propagation and h is the pavement thickness, P is the critical pressure at propagation, E is the elastic modulus, ω is the largest deflection of the blister center, and v is the Poisson ratio.

The elastic modulus can be quantitatively determined from the deflection of the blister and the applied pressure, and given by Equation (4)

$$E = \frac{p}{w}\left[\frac{(1 - v^2)a^4}{5.3h^3} + \frac{(1 + v)a^2}{1.7h}\right] \tag{4}$$

Equation (2) indicates that the blister deflection is related to the pressure, the radius of debonding, the pavement thickness and the stiffness modulus. Under certain deflection, the applied force of the blister is inversely proportional to the debonding radius, which is proportional to the stiffness modulus and pavement thickness. Therefore, the choice of the appropriate loading radius is critical for the effective acquisition of blister cracking loads.

Propagation criterion is a basis for determining whether a crack will expand and in which direction it will expand. The initial crack direction for a blister was predicted using the maximum energy release rate criterion in this study.

From the perspective of fault theory, crack expansion occurs because the strain energy is greater than the energy required for the material to resist crack development. Thus, the critical condition of blister propagation is expressed as:

$$G = G_{cr} \tag{5}$$

The energy release rate, denoted as G, represents the force required to detach the membrane from the substrate. G is calculated per unit width situated along the elliptical failure boundary separating the asphalt paving layer and the underlying concrete. It is expressed in N·m^{-1}, as illustrated in Equation (6).

$$G = \frac{\Delta W - \Delta W_2}{2\pi R \Delta R} = \frac{q(V|_{R+\Delta R} - V|_R) - \Delta W_2}{2\pi R \Delta R} = \frac{3q^2R^4(1 - v^2)}{32Eh^3} + \frac{3q^2R^2(1 + v)}{10Eh} \tag{6}$$

The cumulative energy involved in the spread of the adhesive blister is composed of two components: the potential energy resulting from an externally applied load and the elastic strain energy stored within the membrane. The magnitude of this energy is influenced by the geometry of the adhesive failure region [22,23], as described by Equation (7).

$$\begin{gathered}\Delta W = \Delta W_1 + \Delta W_2;\ \Delta W = q\Delta V = q(V|_{R+\Delta R} - V|_R) = q\left(\tfrac{\partial V}{\partial R}\right)_q \Delta R \\ \Delta W_1 = 2\pi R G \Delta R; \\ \Delta W_2 = \int \sigma d\varepsilon|_{R+\Delta R} - \int \sigma d\varepsilon|_R\end{gathered} \tag{7}$$

where ΔW is the blister propagation energy, ΔW_1 is the energy consumed for blister propagation and ΔW_2 is the elastic strain energy, and q is the uniform load.

The critical pressure p_{cr} required for a blister to expand from an initially flat state to a certain height can be determined when the strain energy release rate G reaches the critical release rate G_{cr}, as shown in Equations (8) and (9).

$$p_{cr} = \sqrt{\frac{512Eh^3G_{cr}}{3(1 - v)}} \times \frac{1}{(2R)^2} = \frac{k_1}{(2R)^2} \tag{8}$$

$$k_1 = \sqrt{\frac{512Eh^3G_{cr}}{3(1 - v)}} \tag{9}$$

3.3. The Mechanism of Blister Propagation

A specimen was placed in the test setup, and the pressure inside the blister was gradually increased until it reached its highest point, resulting in the separation of the blister. Each sample had three parallel groups. The changes in the radius and deflection of the blister over time are illustrated in Figure 7. Within the initial 60 s of loading, the blister started to bulge without any change in its radius, while its height increased proportionally to the pressure, as shown in Figure 7a. The deflection of the blister increased after 20 s and expanded rapidly after 60 s (see Figure 7b). Finally, the radius of the blister stabilized at 80 s, and its deflection showed the tendency to decrease slightly first and then increase gradually. This was the generation phase of the next blister propagation. Subsequently, the specimen experienced a cycle of unloading and reloading. This process was repeated until the asphalt paving layer completely separated from the substrate.

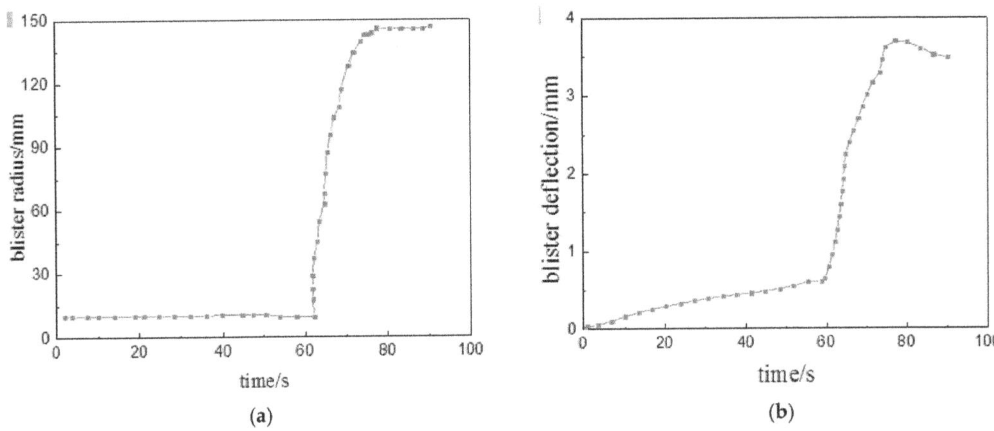

Figure 7. (**a**) The blister radius with time and (**b**) measured blister deflection.

Boyle's law can be stated as Equation (10)

$$V\Delta p + p\Delta V = 0 \tag{10}$$

The product of pressure and volume is equal to a constant value, denoted as k. In this equation, P denotes the pressure of the gas, $\triangle p$ is the variable quantity of the gas, V denotes the volume of the gas, and $\triangle V$ is the variable quantity of the volume, which can be written as:

$$V = \int_0^R \omega(r)2\pi r dr = \frac{\pi q R^6}{192D} \tag{11}$$

$$\Delta V = \frac{\pi R^5 p}{32D}\Delta R \tag{12}$$

When G is equal to G_{cr}, the critical pressure for blister expansion can be calculated from Equations (8) to (12):

$$\Delta p = -\frac{3k_1}{2R^3}\Delta R \tag{13}$$

The above Equation (13) indicates that the blister pressure decreases $\frac{3k_1}{2R^3}\Delta R$ when the blister radius increases gradually from R to $R + \Delta R$ during the propagation process.

Differentiating the above equation in Equation (8) and solving the change in blister radius ΔR to the change in blister critical pressure Δp_{cr} results in

$$\Delta p_{cr} = -\frac{k_1}{2R^3}\Delta R \tag{14}$$

It can be seen from Equations (13) and (14) that the pressure and critical pressure of a blister decreases with the growth of the blister radius. This means that the propagation will stop when the pressure in the blister is reduced to less than the critical pressure. When the blister radius remains stable, the volume of gas in it is unchanged. At this point, the blister pressure increases gradually with the load. Eventually the pressure will be increased to exceed the critical pressure, Δp_{cr}, and an interface separation will occur between the asphalt paving layer and concrete substrate, at which point a new propagation cycle with the growth in blister radius begins until final failure.

3.4. The Influence of Pavement Structure on Blister Propagation

Figure 8 shows the curves indicating the variation in blister deflection with time under different pavement layer structures. Each sample has three parallel groups. Bulging of the blister on AC-13 asphalt pavement was observed after the first 38 s loading period and increased with the load. Then the blister height reached up to 4.5 mm (the maximum) after 77 s and the deflection gradually decreased, while the deflection of SMA-13 occurred 20 s later than that of AC-13. This indicated that the blister propagation of SMA-13 was relatively slow, but the deflection of both pavements was basically the same. Thus, the degree of blister propagation of SMA asphalt pavement is lower than that of AC-13 asphalt pavement.

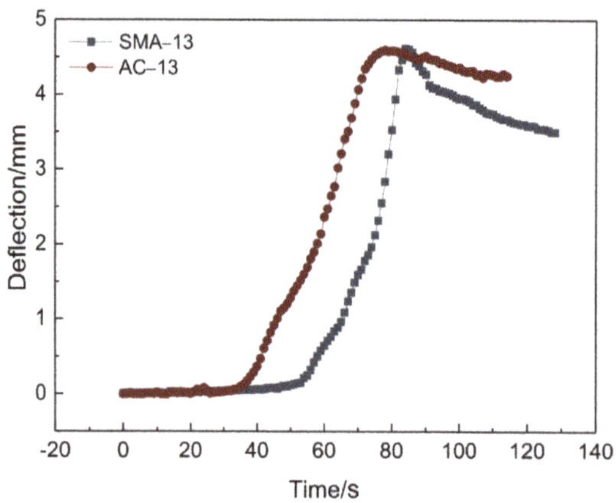

Figure 8. The variation in blister deflection with time under different pavement layer structure.

The blister critical propagation loads and critical propagation deflections of the SMA-13 and AC-13 obtained in the tests were substituted into Equations (4) and (6) to obtain the critical energy release rates G for both SMA-13 and AC-13 samples, as shown in Table 2. The critical propagation load and the critical propagation deflection on the SMA-13 asphalt paving layer are larger than those of AC-13. The results prove that the ability of SMA-13 to resist blister deformation is better than that of AC-13.

Table 2. The blister propagation energy release rate G under different pavement layer structure.

Mixture	Waterproofing Layer	Load (at Peak) /N	Deflection (at Peak) /mm	Propagation Time /s	Critical Energy Release Rate/N·m^{-1}
SMA-13	SBS-modified asphalt	1550	4.65	85	179
AC-13	SBS-modified asphalt	1432	4.58	77	163

3.5. The Influence of Waterproofing Layer on Blister Propagation

For comparing and ranking the influence of membranes on blister propagation, the deflection of blister with SBS-modified asphalt, asphalt rubber and SBS-modified emulsion asphalt waterproofing membranes were determined in a blister test. The results of each blister central deflection are shown in Figure 9. Blister tests were performed at 60 °C, increasing the pressure for 140 s during blister growth.

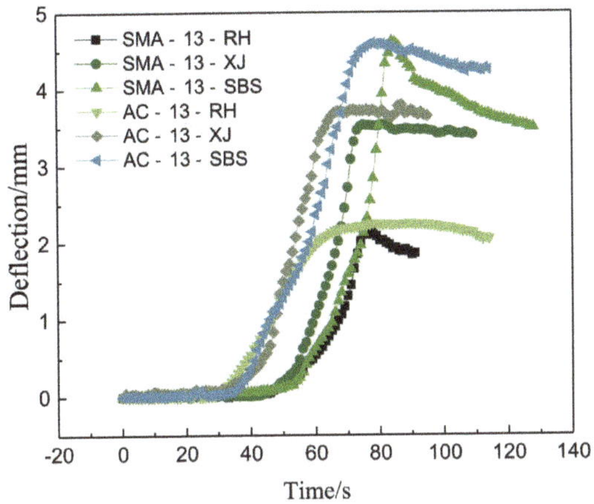

Figure 9. Deflection curves of the blister with different waterproofing layers. (Note: SBS—SBS-modified asphalt; XJ—asphalt rubber; RH—SBS-modified emulsion asphalt).

When comparing the deflection results of blister tests for the SBS (SBS-modified asphalt), XJ (asphalt rubber), and RH (SBS-modified emulsion asphalt) membranes in the same pavement structure, it can be observed that the SBS membrane has a greater deflection value compared to the other types. For the same paving layer, the bulging of the blister on the AC-13 pavement (waterproofing adhesive is SBS-modified asphalt) was observed earliest (at 73 s), with a maximum deflection of 4.6 mm; the bulging of the blister on the AC-13 pavement (waterproofing adhesive is SBS-modified emulsion asphalt) was observed earliest (at 62 s), with a maximum deflection of 3.7 mm; the bulging of the blister on the AC-13 pavement (waterproofing adhesive is asphalt rubber) was observed earliest (at 57 s), with a maximum deflection of 2.0 mm. It can be seen that SBS-modified emulsified asphalt has the worst adhesion to cement concrete bridge deck decks, rubber asphalt is better, and SBS-modified asphalt is the best. The reason for the difference in blister propagation time can be attributed to the different membranes' properties.

For the same waterproofing layer, the AC-13 pavement structure has lower deflection and a shorter propagation time compared to SMA-13. For example, the bulging of the blister on SMA-13 pavement (waterproofing adhesive is SBS-modified asphalt) was detected at the latest time point, at about 85 s, with a maximum deflection of 4.7 mm. In contrast, the bulging of the blister on AC-13 pavement (waterproofing adhesive is SBS-modified asphalt) was observed at the earliest time point (at 73 s), with a maximum deflection of 4.6 mm. This shows that the SMA-13 asphalt mixture bulge deformation is smaller under the same load level condition.

3.6. The Influence of Air Voids on Blister Propagation

The influence of air voids on blister propagation behavior was investigated in this paper. The samples were obtained by coring from the central debonding site. Prior to the experiments, the bottom cement concrete of the samples was removed. In order to avoid the influence of the speckle on the air voids, the sample surfaces were polished. Then they were dried to a stable weight by putting them in a drying oven. Each sample had three parallel groups. Finally, their air voids were measured, as shown in Figure 10.

(a) (b) (c)

Figure 10. The sample with different air voids (**a**) coring sample (**b**) original core sample (**c**) polished sample.

It is possible to use the energy release rate as an index of blister propagation to evaluate the propagation behavior. As shown in Table 3, the critical energy release rate of the blister decreased as the air voids increased. The failure mode of the pavement layer changed from deboning to the seepage of the asphalt mixture. For the sample with an air void of less than 3.5%, the blister occurred under uniform pressure and eventually experienced propagation to destruction. However, for the sample with an air void greater than 3.5%, the hydraulic oil flowed through the micro voids under uniform pressure. When the internal pressure of a sample is greater than the oil seepage pressure, the hydraulic oil will eventually penetrate the surface of the mixture; the reason for this can be the increasing connectivity of micro voids from a higher air void, which decreases the oil seepage pressure. Therefore, the low air voids of asphalt mixture are an important cause of blisters in bridge deck pavement.

Table 3. The blister propagation energy release rate G under different air voids.

Mixture	Waterproofing Layer	Air Voids/%	Load (at Peak) /N	Deflection (at Peak) /mm	Critical Energy Release Rate/N·m^{-1}	Failure Pattern
AC-13	SBS-modified asphalt	3.21	1135.2	2.71	90.6	Asphalt mixture seepage + bonding course destruction
AC-13	SBS-modified asphalt	3.44	1035.3	2.53	77.1	Asphalt mixture seepage + bonding course destruction
AC-13	SBS-modified asphalt	4.67	878.9	1.59	41.5	Asphalt mixture seepage

4. Conclusions

In this study, blister propagation in concrete bridge deck pavement was investigated. A detailed analysis of the factors influencing propagation behavior, such as pavement structure and air voids, was conducted by calculating the critical energy release rate. The following conclusions were drawn from this study.

(1) Blister deformation is divided into an expansion phase and a propagation phase. In the expansion phase, the blister radius remains stable, and the deflection of the pavement layer gradually increases; in the propagation phase, the blister radius increases and the deflection rate of pavement layer also increases. This is because

propagation leads to a decrease in the overall stiffness of the specimen. Furthermore, catastrophic debonding often occurs once a critical pressure is reached.

(2) For the same waterproof bonding layer, a blister in the AC-13 pavement structure is more likely to propagate than one in the SMA-13 pavement structure. For the same pavement structure, pavement with SBS-modified asphalt membranes is the most difficult to propagate, followed by asphalt rubber. Blister propagation on pavement with SBS-modified emulsified asphalt membranes is the most likely. In addition to this, the low air voids of asphalt mixture are a crucial reason for blisters forming in bridge deck pavement.

(3) In this study, using the critical energy release rate as an evaluation index for the interface expansion performance of bridge deck pavement is recommended.

Author Contributions: Conceptualization, C.Y.; investigation, C.Y.; supervision, C.Y.; writing—review and editing, C.Y.; writing—original draft preparation, M.Z.; formal analysis, M.Z. All authors have read and agreed to the published version of the manuscript.

Funding: This research was funded by Natural Science Foundation of Shaanxi Province of China, grant number 2024JC-YBQN-0429; the Shaanxi Provincial Innovation Capacity Support Program Project, grant number 2022TD-16. And the APC was funded by Shaanxi Provincial Innovation Capacity Support Program Project.

Data Availability Statement: The original contributions presented in the study are included in the article material, further inquiries can be directed to the corresponding authors.

Conflicts of Interest: Author C.Y. was employed by First Highway Consultants Co., Ltd., Engineering and Technology Research Institute. The remaining authors declare that the research was conducted in the absence of any commercial or financial relationships that could be construed as a potential conflict of interest.

References

1. Hailesilassie, B.W.; Partl, M.N. Mechanisms of asphalt blistering on concrete bridges. *J. ASTM Int.* **2012**, *9*, 104135. [CrossRef]
2. Partl, M.N.; Gubler, R.; Hean, S. Belagsschäden infolge Blasenbildung auf einer neuen Hängebrücke in Hongkong. In *Nachhaltige Material-und Systemtechnik*; Walter, M., Kurt, S., Eds.; Swiss Federal Laboratories for Material Science and Technology: Dubendorf und St. Gallen, Switzerland, 2001; pp. 337–343.
3. Hean, S.; Partl, M.N. Determination of the decisive factors influencing the behaviour of bridge waterproofing systems with polymer-bitumen membranes. Federal Department of the Environment, Transport. *Energy Commun.* **2006**, *597*, 1–14.
4. Warlow, W.J.; Pye, P.W. Osmosis as a cause of blistering of in situ resin flooring on wet concrete. *Mag. Concr. Res.* **1978**, *30*, 152–156. [CrossRef]
5. Sasaki, I.; Moriyoshi, A.; Tsunekawa, M. Water accumulation and behavior of surfactant associated with moisture permeation in bituminous pavement on concrete deck bridge. *J. Jpn. Pet. Inst.* **2006**, *49*, 315–320. [CrossRef]
6. Zhou, C.L.; Zheng, Y.R.; Hua, Z.L.; Mou, W.J.; Liu, X.H. Recent insights into hydrogen-induced blister fracture of rubber sealing materials: An in-depth examination. *Polym. Degrad. Stabil.* **2024**, *224*, 110747. [CrossRef]
7. Budziński, B.; Mieczkowski, P. Application of Innovative SMA-MA Mixtures on Bridges. *Appl. Sci.* **2020**, *10*, 6958. [CrossRef]
8. Liu, H.; Yuan, G.; Zhang, Q.; Hao, P.; Dong, S.; Zhang, H. Study on influence factors of asphalt mixtures pavement blistering on Portland cement concrete bridge deck. *Int. J. Pavement. Eng.* **2018**, *22*, 249–256. [CrossRef]
9. Liu, H.; Li, Y.; Zhang, Q.; Hao, P. Deformation characteristic and mechanism of blisters in cement concrete bridge deck pavement. *Constr. Build. Mater.* **2018**, *172*, 358–369. [CrossRef]
10. Wang, Z.; Hao, J.; Sun, Z.; Ma, B.; Xia, S.; Li, X. Blistering Mechanism Analysis of Hydraulic Asphalt Concrete Facing. *Appl. Sci.* **2019**, *9*, 2903. [CrossRef]
11. Dannenberg, H. Measurement of adhesion by a blister method. *J. Appl. Polym. Sci.* **1961**, *5*, 125–134. [CrossRef]
12. Fini, E.H.; Al-Qadi, I.L. Development of a Pressurized Blister Test for Interface Characterization of Aggregate Highly Polymerized Bituminous Materials. *J. Mater. Civil. Eng.* **2011**, *23*, 656–663. [CrossRef]
13. Hailesilassie, B.W.; Hean, S.; Partl, M.N. Testing of blister propagation and peeling of orthotropic bituminous waterproofing membranes. *Mater Struct.* **2015**, *48*, 1095–1108. [CrossRef]
14. Yue, K.; Gao, W.; Huang, R.; Liechti, K.M. Analytical methods for the mechanics of graphene bubbles. *J. Appl. Phys.* **2012**, *112*, 083512. [CrossRef]
15. Liao, M.; Luo, R. Influencing factors and mechanisms of blistering in epoxy asphalt mixtures for steel deck pavements. *Constr. Build. Mater.* **2022**, *316*, 126036. [CrossRef]

16. Harvey, C.M.; Wang, S.; Yuan, B.; Thomson, R.C.; Critchlow, G.W. Determination of mode I and II adhesion toughness of monolayer thin films by circular blister tests. *Theor. Appl. Fract. Mech.* **2018**, *94*, 34–39. [CrossRef] [PubMed]
17. Williams, M.L. The continuum interpretation for fracture and adhesion. *J. Appl. Polym. Sci.* **1969**, *13*, 29–40. [CrossRef]
18. Cotterell, B.; Chen, Z. The blister test–Transition from plate to membrane behaviour for an elastic material. *Int. J. Fract.* **1997**, *86*, 191. [CrossRef]
19. Tan, Y.Q.; Zhang, X.N.; Sun, L.J. Study on strain rate of asphalt by digital speckle gram processing technology. *J. Build. Mater.* **2003**, *3*, 321–324. (In Chinese)
20. Yuan, M.M. *Study on Fatigue Damage Mechanism of Asphalt Mixture Based on Digital Speckle Gram Processing Technology*; South China University of Technology: Guangzhou, China, 2013.
21. Wang, L.; Gong, N.; Xing, Y. Characteristics of interface cracking for ultraviolet aged hot-mix asphalt based on digital speckle correlation method. *Fuhe Cailiao Xuebao/Acta Mater. Compos. Sin.* **2016**, *33*, 732–740. [CrossRef]
22. Robert, J.H. Bulge and Blister Testing of the Films and Their Interfaces. Ph.D. Thesis, Materials Science and Engineering Stanford University Stanford, Stanford, CA, USA, 1999.
23. Kai-Tak, W.; Yiu-Wing, M. Fracture mechanics of a new blister test with stable crack growth. *Acta Metall. Matter.* **1995**, *43*, 4109–4115. [CrossRef]

buildings

MDPI

Article

Fatigue Evaluation of CFST Arch Bridge Based on Vehicle–Bridge Coupling Vibration Analysis

Wenliang Hu [1,*], Bin Zhou [2] and Xiaobo Zheng [3,4,*]

1 School of Astronautics, Northwestern Polytechnical University, Xi'an 710064, China
2 School of Civil Engineering, Tianjin University, Tianjin 300072, China; chdzb5168@163.com
3 School of Highway, Chang'an University, Xi'an 710064, China
4 Shaanxi Institute of Teacher Development, Shaanxi Normal University, Xi'an 710062, China
* Correspondence: leonwen_hu@nwpu.edu.cn (W.H.); zhengxiaobo@chd.edu.cn (X.Z.)

Abstract: This study proposes a fatigue life analysis method for long-span CFST arch bridges based on a vehicle–bridge coupled vibration analysis model, which can analyze the structural dynamic effects and the excessive fatigue damage caused by the passage of vehicles. In situ test analysis of bridge dynamic characteristics is carried out, and a numerical model considering the vehicle–bridge coupled system is validated according to the measured vibration modes, frequency, and displacement time history. The results indicate that the proposed vehicle–bridge coupled vibration numerical model can be used to simulate the dynamic response of the bridge under various conditions. The factors of vehicle speed, vehicle weight, and road surface condition are further selected to analyze the vehicle–bridge coupled vibration effect, and it is found that the response time history is more sensitive to the vehicle weight factor. In addition, the fatigue life of suspenders at different positions is compared, which is found to decrease significantly with a reduction in suspender length. Due to damage to the suspender caused by environmental erosion, the cross-sectional area decreases and the stress amplitude changes, resulting in a decrease in the fatigue reliability of the suspender under different conditions.

Keywords: CFST arch bridge; vehicle–bridge coupling vibration; dynamic property; fatigue evaluation; suspender corrosion

1. Introduction

There has been an increasingly fast rate of development in the application of concrete-filled steel tubular (CFST) arch bridges in the construction industry, due to their certain distinctive benefits including adequate cost, proper appearance, ease of construction, and powerful cross performance [1,2]. However, the vertical and lateral stiffnesses are relatively weak for long-span concrete-filled steel tubular arch bridges, and the structural vibration caused by vehicle loads is prominent. As crucial components of CFST arch bridges, CFST members play a key role in their structural mechanical performance. A previous study has explored a data-driven optimization approach for the torsion design of CFRP-CFST columns, while another study has focused on the vibration mechanics of concrete-filled steel tubular composite columns with ultra-high-performance concrete plates [3,4]. Additionally, the trend towards heavier and faster transportation of vehicles exacerbates fatigue damage to the bridge structure [5,6], resulting in severe damage to the bridge before it reaches its designed service life; as such, some damping measures need to be provided to control vibration-induced fatigue damage [7].

Bridge finite element models aimed at fatigue analysis can effectively reflect local hot spot stresses in the bridge. During the modeling process, vehicle loads are typically simulated as moving concentrated forces or simplified by considering an impact factor for fatigue damage analysis [8–11]. While these analysis methods can partially depict the dynamic response of the bridge, they neglect the interaction between the bridge and vehicles and fail to consider the influence of factors such as road surface roughness on the bridge's dynamic response. The accuracy of these methods is insufficient to meet the requirements of fatigue assessment for existing bridges. Additionally, some researchers have analyzed the fatigue performance of bridges using normal stress, hot spot stress, and notch stress methods [12,13]. The vehicle–bridge coupled dynamic analysis technology effectively addresses these issues [14–16]. For highway bridges, due to factors such as road surface roughness and expansion joints, fatigue vehicles passing through the bridge at certain speeds will inevitably impact the bridge, and the dynamic effects of vehicle loads accelerate bridge fatigue damage [17]. Previous studies have conducted analyses of bridge fatigue performance based on dynamic response numerical results, including the effects of vehicle speed, axle weight, road surface roughness, and lane configuration on bridge fatigue stress [18–20]. These studies indicate that dynamic vehicle loads increase with increasing vehicle weight and worsening road surface roughness, leading to the accumulation of fatigue damage [21,22].

As the primary load-bearing components of steel pipe concrete arch bridges, suspenders are inherently susceptible to fatigue failure under dynamic loads, significantly impacting the safety of long-span arch bridges. In particular, the alternating stresses generated by vehicle loads and wind-induced vibrations can cause fatigue failure of the suspenders well below their static strength [23]. However, limited studies have addressed fatigue life assessments for arch bridge suspenders. Moreover, aside from vehicle loading, which induces fatigue failure in suspenders due to varying stress amplitudes, environmental corrosion also emerges as a critical factor contributing to the damage and degradation of suspender components, and studies have demonstrated the significant impact of environmental corrosion on suspender fracture, where short suspenders are particularly susceptible to the severe effects of fatigue on long-span arch bridges, compared to other suspenders [24,25].

In this study, a vehicle–bridge coupled numerical analysis model is established for a five-hole half-through concrete-filled steel tube tied-arch bridge, comprehensively analyzing the structural dynamic effects and excessive fatigue damage caused by the passage of vehicles. The model considers the effects of various parameters, such as vehicle weight, speed, and road surface roughness, on the dynamic performance of the bridge structure. The S-N curve and Miner linear cumulative damage theory are applied to develop a vehicle–bridge coupled dynamic analysis method in order to assess the fatigue life of suspenders in large-span bridge structures. The impacts of different factors on the fatigue life of suspenders at different positions on the subside span and the central span are separately conducted. This study provides theoretical support for the fatigue design of similar bridge structures.

2. Description of the Bridge

The Yuehai bridge is located on the northwest section of the Yinchuan Ring Expressway in Ningxia Hui Autonomous Region, which runs east–west across Yuehai, connecting Yuehai Bay and the Binhe Ecological Park. The bridge adopts the five-hole half-through concrete-filled steel tube tied-arch structure, the span combination is 30 m + 3 × 80 m + 30 m (i.e., the central span and the subside span are 80 m long, the side span is 30 m long), and the width of the bridge deck is 26 m. The arch rib is a concrete-filled steel tube with a dumbbell-shaped section. The rise span ratios of the central hole, the subside hole, and the side hole are 1/2.5, 1/3, and 1/5, respectively, with the axis of a quadratic parabola. The bridge adopts a flexible tied bar of epoxy-sprayed strands, which is anchored to the beams at both ends of the flying swallow part to balance the horizontal thrust of the arch. The suspender is anchored to the upper chord of the arch rib and the lower edge of the crossbeam with a horizontal distance of 5 m. The crossbeam used to anchor the suspender is made of precast box-shaped prestressed concrete members. The columns on the arch rib and the arch seats are concrete-filled steel tubular members, where the diameter of the columns on the arch seat is 80 cm and the diameter of the columns on the arch rib is 60 cm. The slab of the bridge is a precast reinforced concrete trough plate, and the bridge deck is paved with 18 cm deep pavement. The trough plate and bridge deck pavement were originally designed to be continuous, with no expansion joints in the middle. The abutment and pile caps use reinforced concrete, and each pile cap arranges six bored piles with a diameter of 160 cm. The layout of the bridge is shown in Figure 1.

(**a**) Overview of bridge

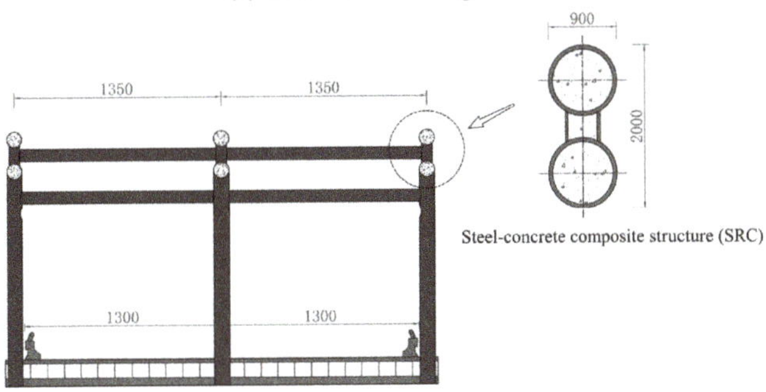

(**b**) Cross-section of the bridge (unit: cm)

Figure 1. The layout of the bridge.

3. Bridge–Vehicle Dynamic Interaction Model

The vibration of a bridge is a cyclical process in which the deformation energy and the kinetic energy of the structural system are mutually converted under an external input. Vehicle dynamic load is a kind of external input that produces a coupling response with the bridge structure, further reflecting the structure's mechanical properties [26,27]. Therefore, the dynamic performance of a structure can be used to evaluate its structural performance. In the analysis process, the Newmark-β numerical integration method was used to solve non-linear dynamic equilibrium equations, which discretizes the continuous time domain into a series of equally spaced time steps. Each time step corresponds to a specific moment, at which the system's displacement, velocity, and acceleration are calculated.

3.1. Vehicle Model Parameters

A five-axle vehicle was modeled as a multiple-degree-of-freedom (DOF) system with 16 independent DOFs, which contributes to reflecting the dynamic response of the bridge. The model proposed by Deng et al. [28], based on the design vehicle in Chinese bridge specification, was adopted, which has already been validated for obtaining the dynamic response of a bridge. The vehicle is a rigid body, the constraint between masses M_1 and M_2 is a rigid connection, and a spring damping system is formed between the vehicle body, the wheels, and the bridge deck. The vehicle analysis model is shown in Figure 2, and the parameters of the vehicle model are listed in Table 1.

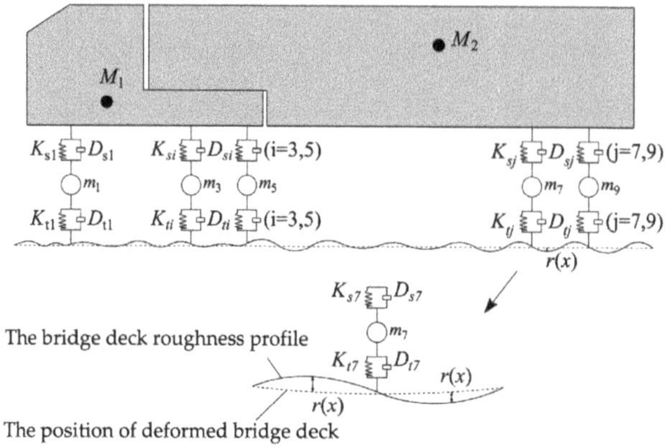

Figure 2. Five-axle vehicle analysis model.

Table 1. Vehicle model analysis parameters.

Parameters	Values (kN·m^{-1})	Parameters	Values (kg)	Parameters	Values (kN·s·m^{-1})
Suspension stiffness K_{s1}, K_{s2}	300.0	M_1	2276.5	Suspension damping D_{s1}, D_{s2}	10.0
Suspension stiffness K_{s3}–K_{s6}	500.0	M_2	45,246.0	Suspension damping D_{s3}–D_{s10}	53.0
Suspension stiffness K_{s7}–K_{s10}	1250.0	m_1	700.0	Suspension damping D_{t1}–D_{t10}	3.0
Tire stiffness K_{t1}, K_{t2}	1500	m_3, m_5	1000.0		
Tire stiffness K_{t3}–K_{t10}	3000.0	m_7, m_9	800.0		

3.2. Bridge Model Parameters

3.2.1. Geometric Model

The FE software ABAQUS 2022 was used to establish a three-dimensional model with the geometry provided in the bridge design. Two-node linear beam elements (B31) were used to model the main beams, columns, arch ribs, suspender cross beams, bent caps, and foundations, while two-node linear truss elements (T3D2) were used for the suspenders and tie bars; additionally, the bridge deck was modeled using shell elements (S4R) with four-node and reduced integration. A dumbbell concrete-filled steel tube section (SRC) was adopted for the arch ring of the main bridge, assuming that there was no relative slip between the steel pipe and the concrete, and the steel pipe wrapping the concrete was converted into the corresponding area of concrete during the calculation. The cross bracings adopted steel tubes, and concrete was not considered to have been poured inside. There were 121 suspenders and four tie bars, which were composed of high-strength and low-relaxation prestressed strands, and the tie bars were embedded within the main beam. The crossbeams were simulated by box-shaped variable-section elements, the arch columns were simulated by the cross-section of the concrete-filled steel tubes, and the combined properties were the same as those of the arch ribs. The bridge deck pavement consisted of 8 cm C40 concrete and 10 cm asphalt concrete. In addition, the model's boundary conditions included fixed constraints at the base of skewback, while there were roller supports at both ends according to the types of bearings, and the connections between the columns, arch ribs, and crossbeams were of a rigid type. The contact properties between the main beam and the bridge deck were as follows: no relative slip in tangential behavior and no interference in normal behavior. The global FE model is partially shown in Figure 3.

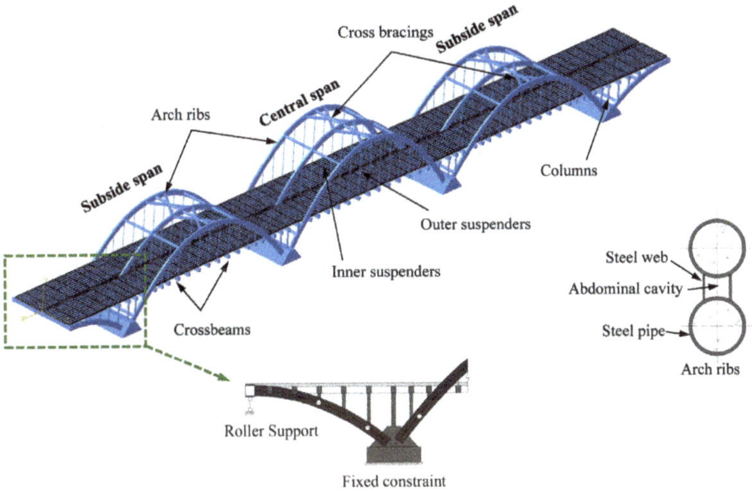

Figure 3. The global FE model.

3.2.2. Material Parameters

The material parameters in the model were based on the Code for Design of Highway Reinforced Concrete and Prestressed Concrete Bridges and Culverts [29]; in particular, C50 concrete was used for the prefabricated trough-shaped plates, end crossbeams, and concrete in the main arch rings, while C40 concrete was used for the boom crossbeams, cover beams, and abutments. The anchor underlay and arch ring steel pipe adopted Q345D, the suspenders were made of PES (FD) 7–121 galvanized low-relaxation prestressed steel wire rope, and the tie rod adopted a φ15.2 prestressed steel strand. Both the concrete and steel materials behaved in the linear elastic stage, the elastoplastic process and strain rate of the material were not considered [30], and all bridge components were considered to be damage-free during vehicle loading and unloading. The concrete and steel material parameters are shown in Tables 2 and 3.

Table 2. Concrete material parameters.

Materials	Elasticity Modulus (MPa)	Unit Weight ($kN{\cdot}m^{-3}$)	Poisson's Ratio	Compression Strength (MPa)	Tensile Strength (MPa)	Coefficient of Linear Expansion
C50	34,500	25.0	0.167	32.4	2.65	0.000010
C40	32,500	25.0	0.167	26.8	2.40	0.000010

Table 3. Steel material parameters.

Materials	Elasticity Modulus (MPa)	Unit Weight ($kN{\cdot}m^{-3}$)	Poisson's Ratio	Yield Strength (MPa)	Coefficient of Linear Expansion
Q345D	205,000	7850	0.3	315	0.000012
Prestressed strand	195,000	7850	0.3	1860	0.000012
Parallel wire	205,000	7850	0.3	1670	0.000012

3.2.3. Loading Condition

The grooved slab of the bridge deck was simplified as a shell for the calculation. The weight of the bridge deck pavement was equivalent to a surface load of 4.3 $kN{\cdot}m^{-2}$, while the weight of the anti-collision guardrail on the unilateral deck was a linear load of 6.8 kN·m^{-1}. In addition, the influence of temperature on the conventional concrete structures, such as the main beam of bridge decks, suspender beams, bent caps, and end floor beams, was calculated according to General Specifications for the Design of Highway Bridges and Culverts [31].

3.2.4. Road Roughness Simulation

The road roughness was assumed to be a stationary Gaussian random process with zero mean value, and the triangular series method was used to simulate the road roughness [32].

$$r(x) = \sum_{k=1}^{n} G_d(n_k)\cos(2\pi n_k x + \varphi_k) \quad (1)$$

where x is the distance between a roughness point on the bridge deck and the starting point of the bridge; $G_d(n_k)$ is the road surface power spectral density; $n_k = n_1 + (k - 1/2)\Delta n$, where $k = 1,2,3\cdots$, $\Delta n = (n_u - n_1)/N$, n_u and n_1 are the upper and lower limits of spatial frequency, respectively; N is the number of frequency bands dividing spatial frequency; and φ_k is the random phase angle.

The above method was used for simulation to obtain the sample functions of different grades of bridge deck roughness. In this study, after comparing it with the test value, class C pavement was taken as an example, and one sample function curve is shown in Figure 4.

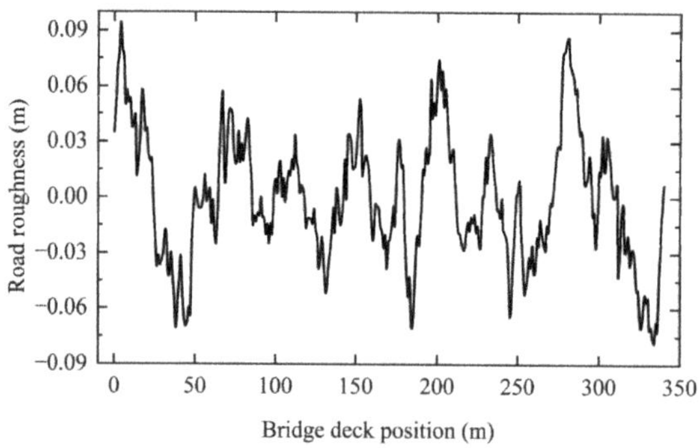

Figure 4. Road roughness sample function.

3.3. *Vehicle–Bridge Coupled Vibration Analysis Model*

The vibration of a CFST arch can be considered as the vibration of a multi-degree-of-freedom system. Its vibration balance equation is shown in Equation (2). The assumption of Equation (2) is based on the Timoshenko beam theory [33–36], which is suitable for modeling deep beams where shear deformation has a significant impact. When the object of calculation is a slender beam, the results are similar to the Euler–Bernoulli beam theory.

$$\boldsymbol{M}_b\{\ddot{q}_b\} + \boldsymbol{C}_b\{\dot{q}_b\} + \boldsymbol{K}_b\{q_b\} = \{\boldsymbol{F}_b\} \quad (2)$$

where $\boldsymbol{M}_b$ is the mass matrix, $\boldsymbol{C}_b$ is the damping matrix, $\boldsymbol{K}_b$ is the stiffness matrix, $\{q_b\}$ is the generalized coordinates of different mode shapes, and $\{\boldsymbol{F}_b\}$ is the force vector.

The vehicle dynamic equilibrium equation is shown in Equation (3), and the lateral position of the vehicle remained unchanged during driving.

$$\boldsymbol{M}_v\{\ddot{u}_v\} + \boldsymbol{C}_v\{\dot{u}_v\} + \boldsymbol{K}_v\{u_v\} = \{\boldsymbol{F}_v\} \quad (3)$$

where $\{u_b\}$ is the displacement matrix, $\{\dot{u}_b\}$ is the velocity matrix, $\{\ddot{u}_b\}$ is the acceleration vector, and the remaining letters have the same meaning as in Equation (2).

The equation of the vehicle–bridge coupled vibration system was established by combining the above equations and considering displacement coordination and force equilibrium, as shown in Equation (4) [37]:

$$\begin{bmatrix} M_v & 0 \\ 0 & M_b \end{bmatrix}\begin{Bmatrix} \ddot{u}_v \\ \ddot{q}_b \end{Bmatrix} + \begin{bmatrix} C_v & C_{vB} \\ C_{bv} & C_b + C_b^v \end{bmatrix}\begin{Bmatrix} \dot{u}_v \\ \dot{q}_b \end{Bmatrix} + \begin{bmatrix} K_v & K_{vb} \\ K_{bv} & K_b + K_b^v \end{bmatrix}\begin{Bmatrix} u_v \\ q_b \end{Bmatrix} = \begin{Bmatrix} F_v \\ F_b \end{Bmatrix} \quad (4)$$

where $\boldsymbol{M}$, $\boldsymbol{C}$, $\boldsymbol{K}$, and $\boldsymbol{F}$ are mass, damping, stiffness, and force terms, respectively, and the subscripts $\boldsymbol{bv}$ and $\boldsymbol{vb}$ express vehicle–bridge coupled terms.

Through the establishment of the vehicle model, bridge model, and the coupling model of their interaction, the dynamic analysis process based on the vehicle–bridge coupling system was obtained, as shown in Figure 5. The uncontacted mode refers to the state where the moving vehicle is not in contact with the base beam, and the contact force is zero during its horizontal travel [38].

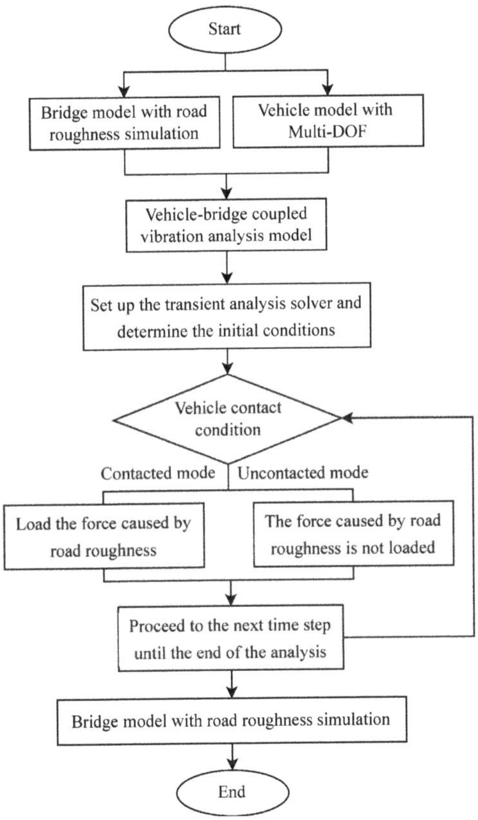

Figure 5. Flowchart of dynamical analysis of bridge structure acted upon by moving vehicles.

3.4. Fatigue Assessment Method

The rainflow counting algorithm [39] processes the time history of the tension stress σ. Then, with the extracted stress cycles, the fatigue damage of suspension bridge hangers can be estimated using the Miner rule [40]:

$$D = \sum_{i=1}^{k} \frac{n_i}{N_i} \quad (5)$$

where n_i is the cycle number of the stress range S_i, N_i is the fatigue life when the stress range is S_i, and D is the fatigue damage. According to the Basquin relation and standard for the design of steel structures [41], the total cycle number N of the allowable stress amplitude of the steel strand is specified as 2×10^6 times, and the corresponding fatigue stress range S is 328 MPa. The S-N curves of the suspenders are obtained using Equation (6).

$$\log N = 15.1 - 3.5 \log S_m \tag{6}$$

The effects of mean stress on fatigue damage were considered using the Goodman diagram. The modified stress range S_m is shown in Equation (7) [11], where σ_m and σ_u are the mean stress and ultimate tension strength, respectively:

$$S_m = \frac{S}{1 - \frac{\sigma_m}{\sigma_u}} \tag{7}$$

4. Dynamic Parameter Analysis

To check the actual dynamic working status of the bridge structure, dynamic tests were conducted to evaluate the performance of the bridge structure under the ambient and vehicle load, and numerical simulation was conducted according to the actual bridge test conditions.

4.1. Verification of Numerical Analysis

The dynamic test included ambient vibration and a moving-vehicle loading test. The mode parameters were identified using the eigensystem realization algorithm (ERA) [42] based on the measured results. The ambient stochastic excitation method was used to collect the accelerometer data using an INV3062W (China Orient Institute of Noise & Vibration, Beijing, China) data acquisition system with a 100 Hz sampling frequency. Vertical sensors were arranged at mid-span sections, quarter-span sections, and the top of the piers, and horizontal sensors were arranged according to the corresponding vertical sensor positions. The vibration mode points are shown in Figure 6.

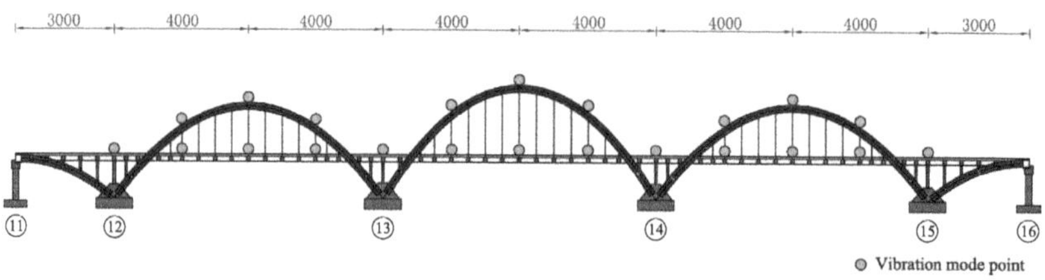

Figure 6. Layout of the vibration mode points (unit: cm).

The first four modal frequencies of each span were obtained (as shown in Figure 7) through modal identification of the measured natural excitation response data. In addition, a moving-vehicle loading test with vehicles passing at speeds ranging from 20 km/h to 50 km/h was carried out on the bridge. The quality of the vehicle model (Figure 2) was the same as that of the test vehicle. The dynamic response data were recorded at the measuring points on the bridge when the vehicle passed.

Based on the in situ test values of the subside and middle spans of the bridge, a finite element numerical model was established to obtain the corresponding modal and frequency numerical values. The analysis results were based on consideration of the axial prestress effect of the tie bars on the main girder, as shown in Figure 8. It can be seen that the modal shapes of the eastern and western subside spans and the central span were basically consistent with those of the numerical model. As the in situ tests only focused on the main beam, the modal shape of the first-order arch ribs was obtained only through numerical analysis. The second to fourth orders corresponded to antisymmetric vertical bending, torsion, and symmetric vertical bending, respectively, and were consistent with the numerical values.

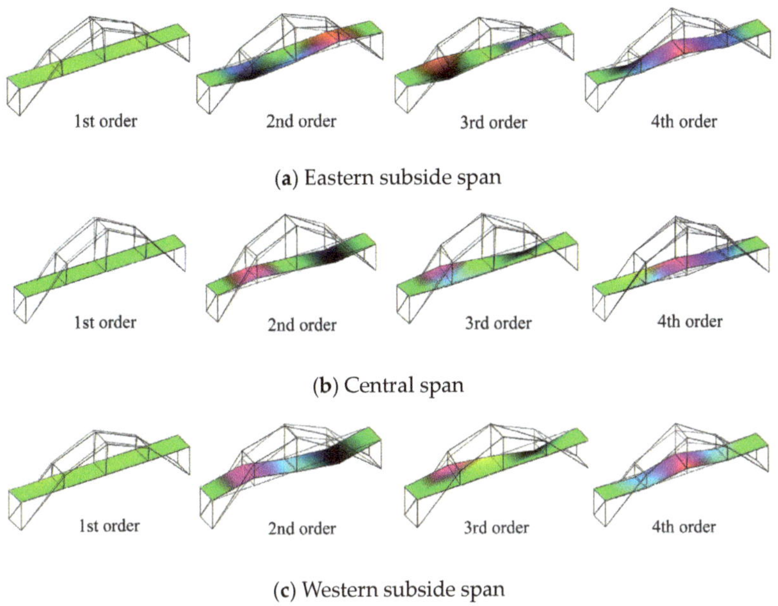

Figure 7. In situ test of bridge vibration mode in different spans.

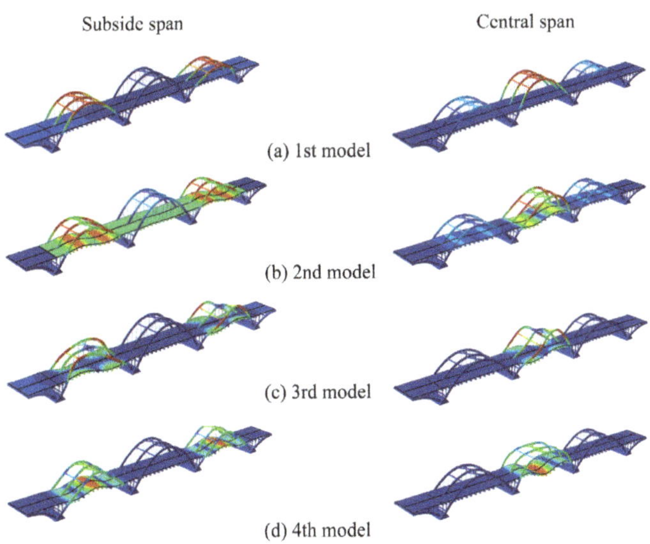

Figure 8. FE analysis of bridge vibration mode.

Figure 9 shows the frequency comparison of the experimental and FE analysis values; the frequency values and relative errors for the first four measured orders are given for the eastern subside span (E), the central span (C), and the western subside span (W). The graph illustrates that the maximum errors for the three spans were 6.75%, 8.93%, and 8.82%, respectively, from which it can be concluded that the FE analysis results in this study were in good agreement with the experimental values, thus verifying the reliability of the numerical model.

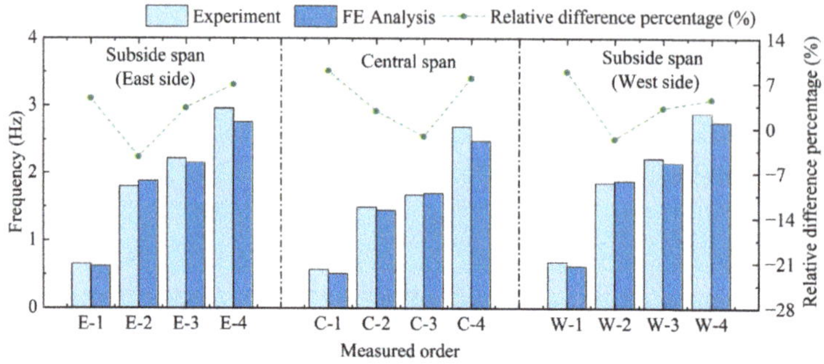

Figure 9. Frequency comparison of experimental and FE analysis data.

The deflection variation of the arch rib at the quarter point on the west side of the main span was recorded, using a test vehicle with a weight of 400 kN driving at 30 km/h. The comparison between the displacement predicted in the FE analysis and that obtained from the experiment is shown in Figure 10. From the figure, it is evident that the variation trend of the test and simulation strain values was in good agreement. There was a small degree of reverse deflection phenomenon in the bridge structure model between 30 s and 36 s, which indicates that the interaction of each span of the model is more prominent, and the relative error values at the maximum and minimum displacements were 4.9% and 1.9%, respectively. Therefore, it can be concluded that the variation law of the results presented in this study verifies the numerical analysis model based on the five-axle vehicle and the road roughness, from which reliable vehicle–bridge coupling dynamic response analysis can be obtained. Thus, the model was further utilized for subsequent analysis.

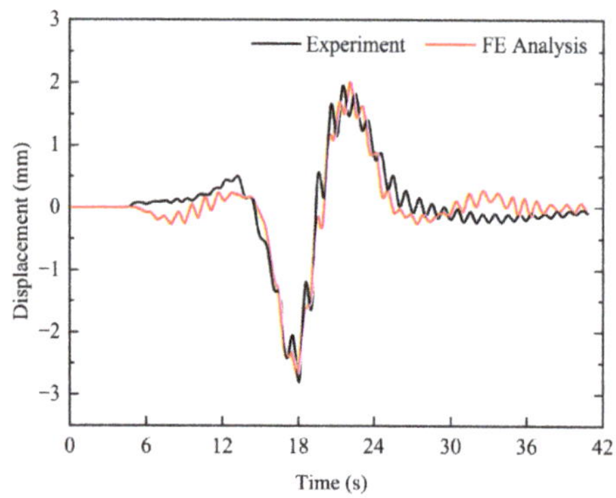

Figure 10. Comparison between the displacement predicted by the FE analysis and that obtained experimentally.

In the numerical analysis of structures, mesh size significantly impacts the accuracy of the results. Therefore, selecting an optimal mesh size is crucial for balancing accuracy and computational efficiency. Figure 11 presents the displacement time history curves under the three mesh sizes of 2 m, 1 m, and 0.5 m, respectively. The analysis results were similar under the different mesh sizes. It should be noted that the mesh sizes shown in the figure are specifically for the bridge deck, with other components having been adjusted proportionally. Considering both accuracy and efficiency, a 1 m mesh size was selected for the numerical model.

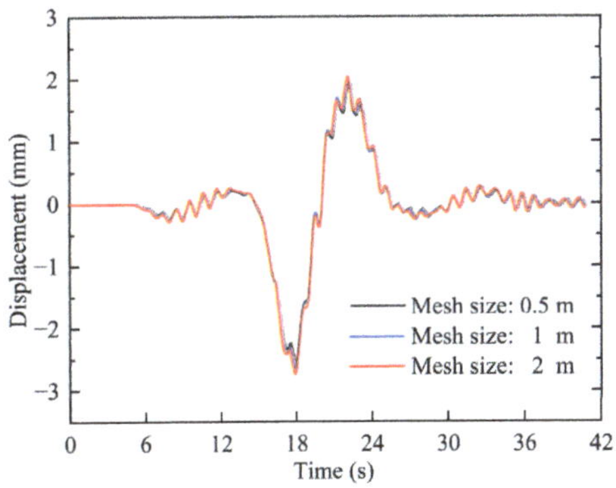

Figure 11. Comparison of numerical calculation results under different mesh sizes.

4.2. Dynamic Parameter Analysis

In the vehicle–bridge coupling system in Section 3.3, the vehicle acts as the vibration source, and its velocity affects the dynamic response value of the bridge structure. In this study, an analysis of vehicle speed, vehicle weight, and road surface conditions was conducted to evaluate the dynamic characteristics of the bridge. The dynamic performance was evaluated through time history analysis of the main span's vertical displacement and acceleration. The basic parameters used in the analysis were a vehicle speed of 30 km/h, a vehicle weight of 400 kN, and a road surface condition of grade C. The vehicle and bridge parameters are shown in Table 4.

Table 4. Effect factors in vehicle–bridge coupled vibration analysis model.

Subsystem Type	Effect Factor	Parameter Range
Vehicle	Speed (km/h) Weight (kN)	20, 30, 40, 50 300, 400, 500, 600
Bridge	Road surface condition	A, C, E

4.2.1. Vehicle Speed

The effect of vehicle speed on the dynamic performance of the structure was analyzed using speed values ranging from 20 km/h to 50 km/h. As shown in Figure 12, with an increase in the vehicle speed, the peak displacement of mid-span and quarter-span did not change much, with a maximum change of 13.9%. Relative to displacement, the acceleration time history curves were more sensitive to vehicle speed, as shown in Figure 13. It is worth noting that the quarter-span acceleration time history curve oscillated significantly at 40 km/h, with a time interval of 13–19 s, which is a critical speed. This phenomenon arises from the periodic loading caused by the spaced-out wheel loads of the vehicle. As a vehicle passes across the bridge at a speed of V, its axle load creates a periodic dynamic force on the bridge with a loading period. When this period matches the natural frequency of the bridge's nth vibration mode, resonance occurs. Based on calculations from the literature [43], the critical vehicle speed is 38.62 km/h. The numerical analysis results differed from the theoretical calculations in the literature by only 3.6%, and the results can be used to identify conditions that cause resonance. Avoiding resonance frequencies can prevent excessive structural stress, fatigue damage, and material failure, thereby ensuring the bridge's safety, durability, and comfort.

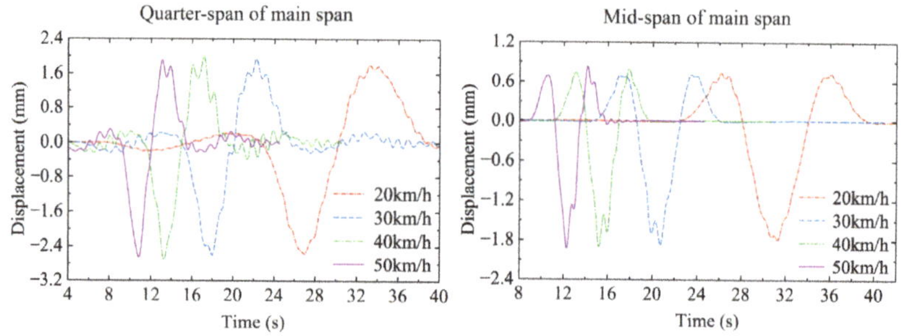

Figure 12. Displacement time history under different vehicle speeds.

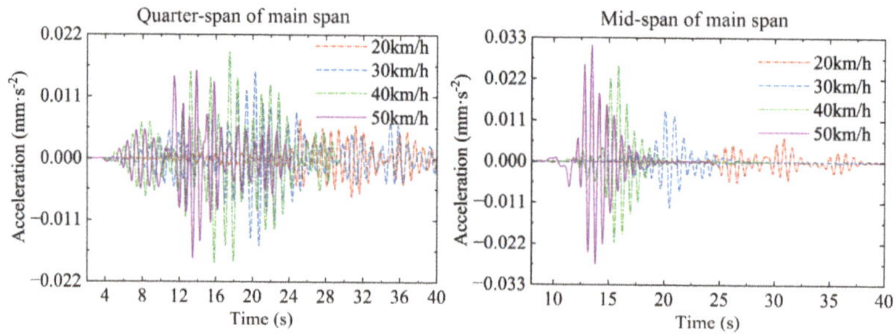

Figure 13. Acceleration time history under different vehicle speeds.

4.2.2. Vehicle Weight

The vehicle weight, as an additional aspect of the vibration source, further affects the dynamic response of the bridge structure. This study analyzed vehicle weights ranging from 300 kN to 600 kN. When compared with the vehicle speed, it can be seen that time history results were more sensitive to vehicle weight, as shown in Figure 14. The peak displacement of the mid-span and the quarter-span of the central span increased with an increase in vehicle weight, with the maximum change being 95.3%. Unlike the vehicle speed, the shape of the time curve generated by the vehicle weight was the same, which indicates that the response levels at the same point remain consistent under different vehicle weights. It is worth noting that the relative residual displacement of the quarter-span was larger than that of the mid-span of the main span, which indicates that different positions have different sensitivities to vehicle weight parameters. Figure 15 shows that the vibration degree of the bridge increased with increasing vehicle weight, while the shape of the acceleration time history curve under different vehicle weights was basically the same; this was the same as the time history law of displacement, and no resonance phenomenon was observed.

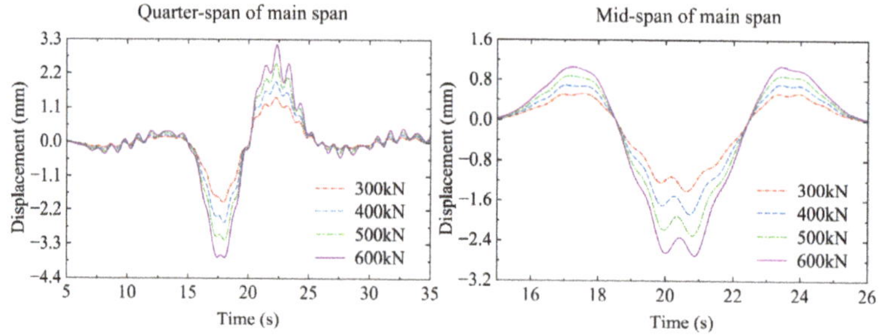

Figure 14. Displacement time history under different vehicle weights.

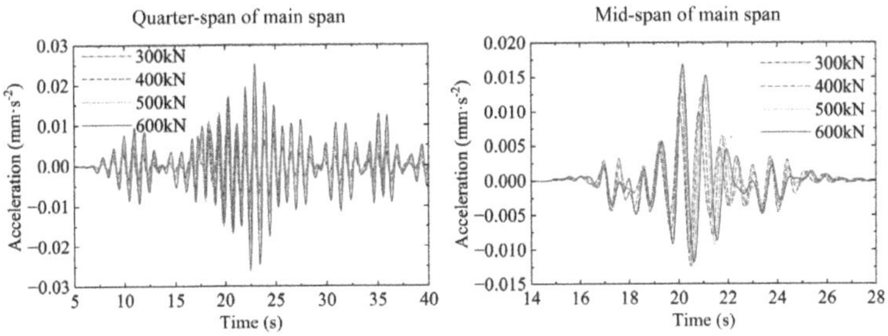

Figure 15. Acceleration time history under different vehicle weights.

4.2.3. Road Surface Condition

As a parameter of the bridge structure, the road surface condition (RSC) also plays a key role in the response of the vehicle–bridge coupling system. Therefore, three grades of A (very good), C (average), and E (very poor) were used for road surface condition analysis. It can be seen, from Figure 16, that with an increase in the road surface condition level, the oscillation degree of the displacement curve of the mid-span and the quarter-span increases; this is due to the road surface roughness indirectly leading to an increase in vehicle excitation, which, in turn, increases the damage state of the bridge and leads to an increase in displacement. The oscillation amplitude of the curve is more evident in the acceleration time history diagram in Figure 17, from which different road surface conditions can be clearly distinguished; notably, grade E was much higher than the others, and the road surface condition is an essential factor that must be considered in vehicle–bridge coupling analysis.

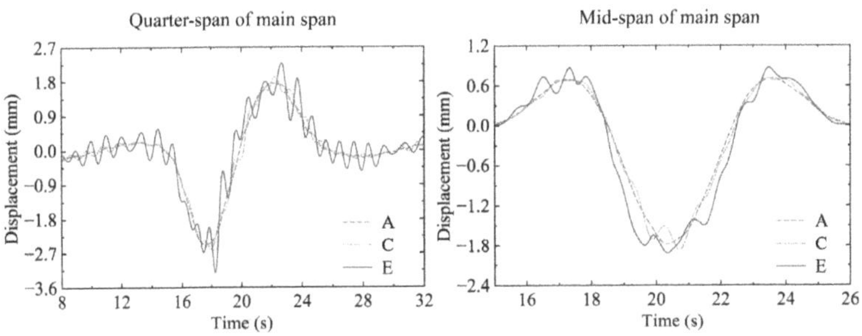

Figure 16. Displacement time history under different road surface conditions.

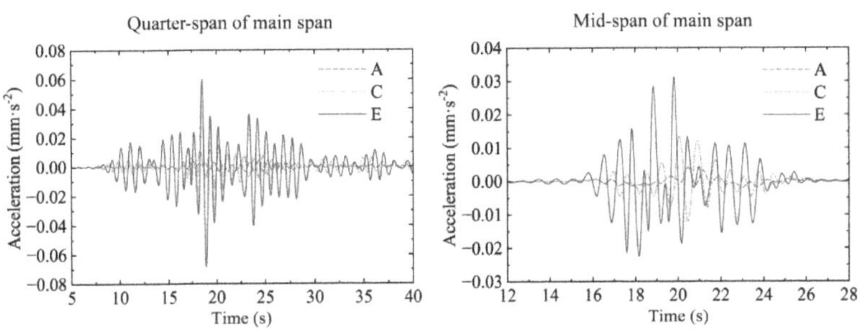

Figure 17. Acceleration time history under different road surface conditions.

4.2.4. Results of Key Position

Based on the analysis of various dynamic parameters such as vehicle speed, vehicle weight, and road surface roughness, the stress and displacement results at the mid-span (a key position of the bridge) were evaluated, as shown in Table 5. The results demonstrate that, as vehicle speed and weight increased and when road conditions worsened, the stress and displacement at the key position increased. Compared to vehicle speed and road surface conditions, the results were more sensitive to vehicle weight. Additionally, to further describe the differences in the actual working conditions of the bridge under dynamic loads, compared to those under the same static loads, the variation in the impact coefficient for the key position in the central span under different effect factors was analyzed. The results indicate that, as the vehicle speed increased and the road conditions deteriorated, the impact coefficient at the key position increased, reflecting the enhanced dynamic effects. However, as the vehicle weight increased, the impact coefficient gradually decreased. This is mainly because an increase in vehicle weight leads to a greater static response at the key position of the bridge, while the increase in dynamic response is relatively small, resulting in smoother vehicle passage and reduced dynamic impact on the bridge. According to the specification [31], the impact coefficient calculated based on the frequency was 0.05. The numerical analysis results indicate that it is unsafe to consider only the fundamental frequency of the bridge when calculating the vehicle impact coefficient for long-span concrete-filled steel tube arch bridges.

Table 5. The stress and deflection results at the mid-span under different factors.

Subsystem Type	Effect Factor	Parameter Range	Stress (MPa)	Deflection (mm)	Impact Coefficient
Vehicle	Vehicle speed (km/h)	20	0.96	−1.80	0.11
		30	1.06	−1.94	0.20
		40	1.10	−2.03	0.25
		50	1.12	−2.05	0.26
	Vehicle weight (kN)	300	0.74	−1.42	0.28
		400	1.06	−1.94	0.20
		500	1.37	−2.33	0.14
		600	1.65	−2.63	0.08
Bridge	Road surface condition	A	0.98	−1.82	0.12
		C	1.06	−1.94	0.20
		E	1.07	−1.96	0.21

5. Fatigue Analysis

5.1. Fatigue Life Estimation

An in-depth analysis was conducted to examine the impacts of factors on the fatigue stress cycle counts of the suspenders. Specifically, the inner and outer suspenders located at the subside span and central span of the tied-arch bridge were selected for fatigue analysis. The suspenders (A1, A4, A7, B1, B4, B7) are indicated in Figure 18.

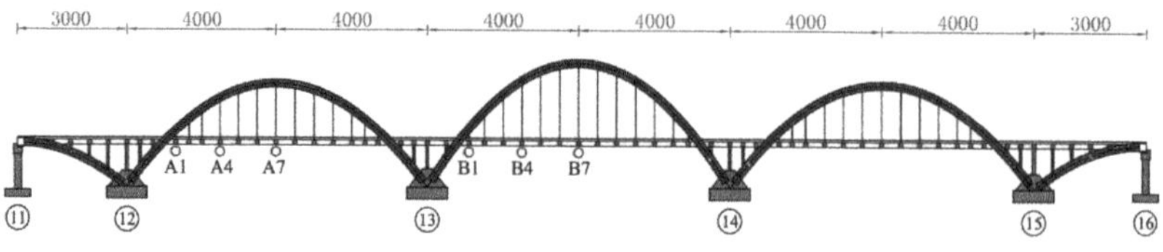

Figure 18. Layout of the points for fatigue analysis (unit: cm).

To investigate the effect of vehicle speed, fatigue stress cycle counts during the operational period were calculated for vehicle speeds ranging from 20 km/h to 50 km/h. According to Figure 19, the fatigue stress cycle number of the suspenders decreased slightly with increasing vehicle speed and the maximum value changed by 7.6%, which occurred at the inner suspender A1 in the subside span, while the inner suspender B1 in the central span changed by 6.6%, indicating insensitivity to vehicle speed parameters. Compared to the central span suspenders, the fatigue cycle number of the bridge's subside suspenders was relatively smaller.

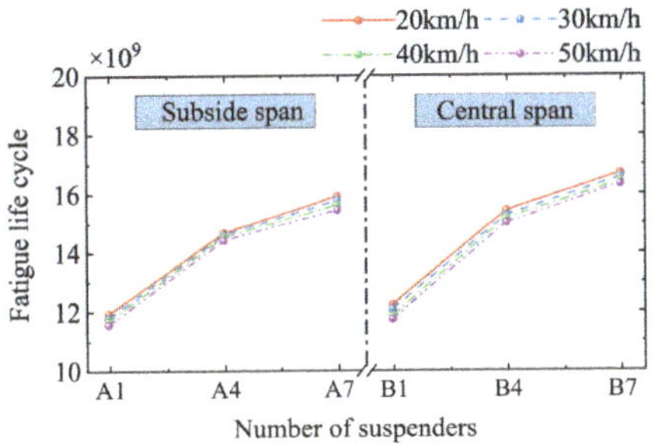

(**a**) Outer suspenders

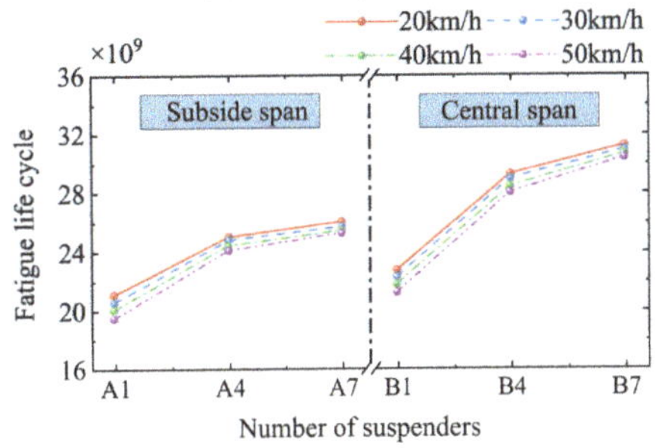

(**b**) Inner suspenders

Figure 19. Fatigue life at different vehicle speeds.

To study the effect of vehicle weight, fatigue stress cycle counts during the operational period were calculated for vehicle weights ranging from 300 kN to 600 kN. The fatigue impact of vehicle weight on each suspender is illustrated in Figure 20. Analysis of the figure reveals significant variations in the fatigue stress cycle counts of bridge suspenders with different vehicle weights. Overloaded vehicles notably impacted the structural fatigue life, as increased overloading accelerates damage accumulation rates, consequently reducing fatigue life [44]. When the vehicle weight was reduced, the variation in fatigue stress cycles for different span suspenders ranged from 63.0% to 92.5%. The fatigue life of the suspenders increased non-linearly with decreasing vehicle weight. Considering the actual operational conditions of the bridge, imposing a weight limit of 400 kN during bridge operations would effectively enhance the fatigue life of the suspenders. Additionally, the analysis indicates that the fatigue stress cycle count of outer suspenders was significantly lower than that of inner suspenders, and generally increased with suspender length. Consequently, special attention should be given to the shorter outer suspenders during the operation of the bridge.

The road surface condition of the bridge deck directly affects the interaction between vehicles and the bridge, significantly impacting fatigue stress [21]. To study the influence of bridge deck unevenness levels, the fatigue cycle counts of suspenders were calculated for levels A, C, and E, where the analysis employed multiple random road surface mean values to eliminate the effects of randomness. Figure 21 illustrates the impact of bridge deck unevenness on the fatigue cycle counts of suspenders, from which it is evident that as the bridge deck deteriorated, the fatigue cycle counts of the suspenders gradually decreased. Moreover, as the bridge deck level changed from A to E, the rate of change in fatigue cycle counts of suspenders gradually increased, indicating a decrease in the fatigue life of the bridge due to severe bridge deck deterioration.

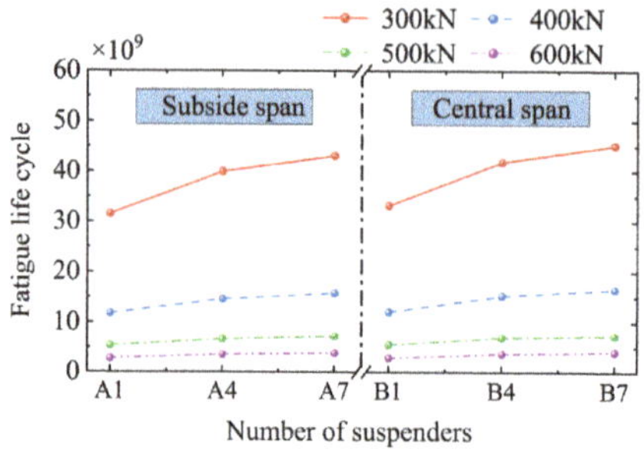

(**a**) Outer suspenders

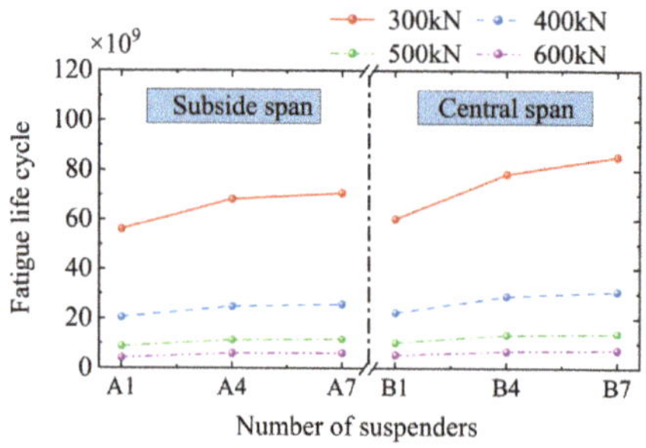

(**b**) Inner suspenders

Figure 20. Fatigue life at different vehicle weights.

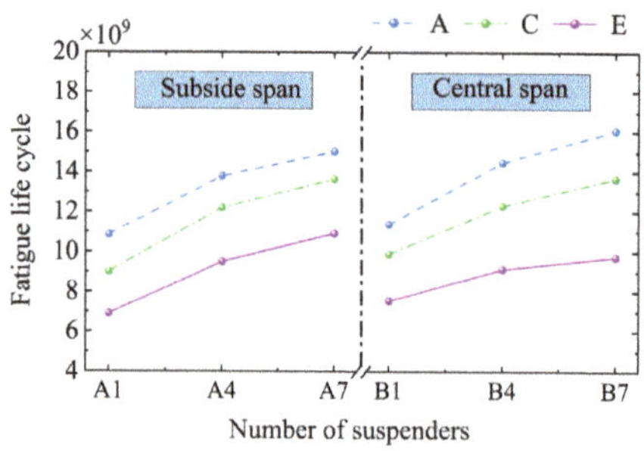

(**a**) Outer suspenders

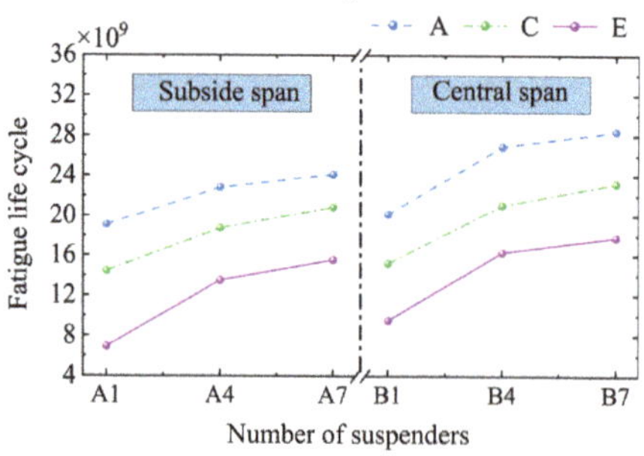

(**b**) Inner suspenders

Figure 21. Fatigue life under different road surface conditions.

5.2. Fatigue Reliability under Corrosion

A fatigue reliability study was conducted on the suspenders before and after corrosion. The suspender corrosion type was uniform corrosion, which was simulated by reducing the cross-sectional area of the suspender [45]. This method was used to analyze the variations in fatigue reliability under the influence of vehicle speed, vehicle weight, and road surface condition parameters, as shown in Table 6. From the results, it can be observed that, compared to the reliability indicators without corrosion, the mean reliability indicators considering corrosion exhibited decreases ranging from 15.1% to 23.5%, and the decrease was primarily evident in terms of vehicle weight, indicating that changes in vehicle weight have a more significant effect on fatigue under the effect of corrosion. According to the relevant regulations regarding the fatigue limit state of steel components [46], the fatigue reliability index β for the cable was taken as 3.5; as such, the fatigue reliability of the suspenders was greater than the safety standard under non-corrosive conditions. However, under the corrosion effect, the fatigue reliability was less than the safety standard when the weight of the car was 600 kN. As for road surface conditions, the difference between level A and level E was 1.6%. Compared to changes in vehicle weight and road surface condition, the change before and after corrosion was relatively small at different speeds.

Table 6. The fatigue reliability of the suspenders.

Vehicle Speed (km/h)	Non-Corrosion	Corrosion	Error Percentage (%)
20	4.99	4.12	17.6
30	4.99	4.11	17.6
40	4.98	4.10	17.6
50	4.98	4.10	17.7
Vehicle Weight (kN)	**Non-Corrosion**	**Corrosion**	**Error Percentage (%)**
300	5.36	4.55	15.1
400	4.99	4.11	17.6
500	4.68	3.73	20.2
600	4.43	3.39	23.5
Road Surface Condition	**Non-Corrosion**	**Corrosion**	**Error Percentage (%)**
A	4.96	4.08	17.8
C	4.89	3.99	18.4
E	4.76	3.84	19.4

6. Conclusions

The fatigue life of a CFST arch bridge was evaluated based on vehicle–bridge coupling vibration analysis. Through establishing a vehicle–bridge coupling model and employing the triangular series method to simulate road roughness, the actual working state of the bridge when a vehicle passes was reflected. The reliability of the numerical model was verified by comparing its predicted results with in situ measured results. The dynamic performance of the considered structure under different vehicle and road surface parameters was analyzed, and discussions regarding the structure's fatigue life were provided. Corrosion factors were further considered, in order to calculate the change rate of fatigue life of the bridge under different conditions. The main conclusions derived from the performed calculations can be summarized as follows:

(1) While the effect of vehicle speed on the dynamic performance of the bridge is relatively small, it should be noted that the frequency caused by a change in vehicle speed gradually approaches a certain order of the bridge's vertical bending frequency, leading to a resonance phenomenon. Compared to vehicle speed and road surface condition, the response time history results are more sensitive to vehicle weight. The increased proportion between vehicle weight response and displacement time history is basically the same. With an increase in the road surface condition, the

oscillation degree increases, and this phenomenon is more evident in the acceleration time history.

(2) Under vehicle loads, the fatigue cycle number of the bridge's subside span suspenders is relatively smaller, compared to the central span suspenders; that of outer suspenders is significantly lower than that of inner suspenders from the cross-sectional direction of the bridge; and the fatigue life of short suspenders is significantly less than that of longer suspenders.

(3) Environmental corrosion reduces the effective cross-sectional areas of suspenders, substantially raising their cyclic stresses when subjected to vehicle loads, especially for bridges with heavy vehicle loads and poor road conditions. Thus, specific consideration is required during the evaluation and analysis of suspender fatigue life.

(4) Improving structural dynamic performance is necessary for a long-serving CFST arch bridge. The proposed fatigue evaluation method based on the vehicle–bridge coupled vibration analysis system reflects the structure's dynamic performance well.

Author Contributions: Methodology, W.H.; software, B.Z.; validation, B.Z.; formal analysis, X.Z.; data curation, X.Z.; supervision, W.H.; funding acquisition, X.Z. All authors have read and agreed to the published version of the manuscript.

Funding: Key research and development program of Shaanxi (2024SF-YBXM-616), Shaanxi Teacher Development Research Project (SJS2022ZQ019).

Data Availability Statement: Data are contained within the article.

Conflicts of Interest: The authors declare no conflicts of interest.

References

1. Chen, B.; Wang, T. Overview of Concrete Filled Steel Tube Arch Bridges in China. *Pract. Period. Struct. Des. Constr.* **2009**, *14*, 70–80. [CrossRef]
2. Zheng, J.; Wang, J. Concrete-Filled Steel Tube Arch Bridges in China. *Engineering* **2018**, *4*, 143–155. [CrossRef]
3. Huang, H.; Xue, C.; Zhang, W.; Guo, M. Torsion Design of CFRP-CFST Columns Using a Data-Driven Optimization Approach. *Eng. Struct.* **2022**, *251*, 113479. [CrossRef]
4. Wei, J.; Ying, H.; Yang, Y.; Zhang, W.; Yuan, H.; Zhou, J. Seismic Performance of Concrete-Filled Steel Tubular Composite Columns with Ultra High Performance Concrete Plates. *Eng. Struct.* **2023**, *278*, 115500. [CrossRef]
5. Deng, L.; Nie, L.; Zhong, W.; Wang, W. Developing Fatigue Vehicle Models for Bridge Fatigue Assessment under Different Traffic Conditions. *J. Bridge Eng.* **2021**, *26*, 04020122. [CrossRef]
6. Zhang, J.; Song, Z.; Han, B.; Xie, H.; Li, P. Research Progress on Fatigue of Highway Concrete Bridges under Vehicle Loading. *China Civ. Eng. J.* **2022**, *55*, 65–79. [CrossRef]
7. Zhang, C. The Active Rotary Inertia Driver System for Flutter Vibration Control of Bridges and Various Promising Applications. *Sci. China Technol. Sci.* **2023**, *66*, 390–405. [CrossRef]
8. Chan, T.H.T.; Guo, L.; Li, Z.X. Finite Element Modelling for Fatigue Stress Analysis of Large Suspension Bridges. *J. Sound Vib.* **2003**, *261*, 443–464. [CrossRef]
9. Zhou, H.; Liu, K.; Shi, G.; Wang, Y.Q.; Shi, Y.J.; De Roeck, G. Fatigue Assessment of a Composite Railway Bridge for High Speed Trains. Part I: Modeling and Fatigue Critical Details. *J. Constr. Steel Res.* **2013**, *82*, 234–245. [CrossRef]
10. Imam, B.M.; Righiniotis, T.D. Fatigue Evaluation of Riveted Railway Bridges through Global and Local Analysis. *J. Constr. Steel Res.* **2010**, *66*, 1411–1421. [CrossRef]
11. Basso, P.; Casciati, S.; Faravelli, L. Fatigue Reliability Assessment of a Historic Railway Bridge Designed by Gustave Eiffel. *Struct. Infrastruct. Eng.* **2015**, *11*, 27–37. [CrossRef]
12. Yang, H.; Wang, P.; Qian, H. Fatigue Behavior of Typical Details of Orthotropic Steel Bridges in Multiaxial Stress States Using Traction Structural Stress. *Int. J. Fatigue* **2020**, *141*, 105862. [CrossRef]
13. Huang, Y.; Zhang, Q.; Bao, Y.; Bu, Y. Fatigue Assessment of Longitudinal Rib-to-Crossbeam Welded Joints in Orthotropic Steel Bridge Decks. *J. Constr. Steel Res.* **2019**, *159*, 53–66. [CrossRef]
14. Xiao, X.; Yan, Y.; Chen, B. Stochastic Dynamic Analysis for Vehicle-Track-Bridge System Based on Probability Density Evolution Method. *Eng. Struct.* **2019**, *188*, 745–761. [CrossRef]
15. Shamsi, M.; Zakerinejad, M.; Vakili, A.H. Seismic Analysis of Soil-Pile-Bridge-Train Interaction for Isolated Monorail and Railway Bridges under Coupled Lateral-Vertical Ground Motions. *Eng. Struct.* **2021**, *248*, 113258. [CrossRef]

16. Lou, P.; Yau, J.D.; Au, F.T.K.; Urushadze, S. Shifted Resonance of Railway Bridges Under Trains Passing by Each Other. *Int. J. Str. Stab. Dyn.* **2022**, *22*, 2271001. [CrossRef]
17. Zhang, W.; Cai, C.S.; Pan, F. Nonlinear Fatigue Damage Assessment of Existing Bridges Considering Progressively Deteriorated Road Conditions. *Eng. Struct.* **2013**, *56*, 1922–1932. [CrossRef]
18. Cui, C.; Chen, A.; Ma, R.; Wang, B.; Xu, S. Fatigue Life Estimation for Suspenders of a Three-Pylon Suspension Bridge Based on Vehicle–Bridge-Interaction Analysis. *Mater.* **2019**, *12*, 2617. [CrossRef] [PubMed]
19. Bhetuwal, U.; Shrestha, J.K.; Pradhananga, R. Fatigue Life Analysis of Steel–Concrete Composite Bridge Considering Road Surface Conditions. *Innov. Infrastruct. Solut.* **2022**, *7*, 124. [CrossRef]
20. Wang, C.; Zhang, J.; Tu, Y.; Sabourova, N.; Grip, N.; Blanksvärd, T.; Elfgren, L. Fatigue Assessment of a Reinforced Concrete Railway Bridge Based on a Coupled Dynamic System. *Struct. Infrastruct. Eng.* **2020**, *16*, 861–879. [CrossRef]
21. Zhang, W.; Cai, C.S. Fatigue Reliability Assessment for Existing Bridges Considering Vehicle Speed and Road Surface Conditions. *J. Bridge Eng.* **2012**, *17*, 443–453. [CrossRef]
22. Wang, W.; Deng, L. Impact Factors for Fatigue Design of Steel I-Girder Bridges Considering the Deterioration of Road Surface Condition. *J. Bridge Eng.* **2016**, *21*, 04016011. [CrossRef]
23. Zhu, J.; Yi, Q. Study of Safety of New Type Suspenders and Their Impacton Static and Dynamic Performance of Arch Bridge. *Bridge Const.* **2011**, *5*, 39–52+51.
24. Liu, Z.; Guo, T.; Hebdon, M.H.; Zhang, Z. Corrosion Fatigue Analysis and Reliability Assessment of Short Suspenders in Suspension and Arch Bridges. *J. Perform. Constr. Facil.* **2018**, *32*, 04018060. [CrossRef]
25. Nakamura, S.; Miyachi, K. Ultimate Strength and Chain-Reaction Failure of Hangers in Tied-Arch Bridges. *Struct. Eng. Int.* **2021**, *31*, 136–146. [CrossRef]
26. Tan, Y.; He, X.; Shi, L.; Zheng, S.; Zhang, Z.; Wang, X. Reinforcement Effect Evaluation on Dynamic Characteristics of an Arch Bridge Based on Vehicle-Bridge Coupled Vibration Analysis. *Comput. Model. Eng. Sci.* **2022**, *131*, 1041–1061. [CrossRef]
27. Chen, L.; He, W.; Li, Q.; Cao, D.; Wang, K.; Ling, L.; Zhu, S.; Li, Y.; Luo, J. Assessment of the Strengthening of Existing Benchmark Monorail Railway Bridge: A Case Study of a Generic Method for Lateral Reinforcement. *Struct. Eng. Int.* **2019**, *29*, 362–369. [CrossRef]
28. Deng, L.; Duan, L.; He, W.; Ji, W. Study on Vehicle Model for Vehicle-bridge Coupling Vibration of Highway Bridges in China. *China J. Highw. Transp.* **2018**, *31*, 92–100. [CrossRef]
29. *JTG D62-2004*; Code for Design of Highway Reinforced Concrete and Prestressed Concrete Bridges and Culverts. MOT, (Ministry of Transportation): Beijing, China, 2004. (In Chinese)
30. Lu, D.; Meng, F.; Zhou, X.; Zhuo, Y.; Gao, Z.; Du, X. A Dynamic Elastoplastic Model of Concrete Based on a Modeling Method with Environmental Factors as Constitutive Variables. *J. Eng. Mech.* **2023**, *149*, 04023102. [CrossRef]
31. *JTG D60-2015*; General Specifications for Design of Highway Bridges and Culverts. MOT, (Ministry of Transportation): Beijing, China, 2015. (In Chinese)
32. Wang, G.; Li, W. Analysis on the vehicle ride comfort based on vehicle-bridge coupled vibration. *J. Vib. Shock* **2016**, *35*, 224–230. [CrossRef]
33. Kiani, K.; Roshan, M. Nonlocal Dynamic Response of Double-Nanotube-Systems for Delivery of Lagged-Inertial-Nanoparticles. *Int. J. Mech. Sci.* **2019**, *152*, 576–595. [CrossRef]
34. Sapountzakis, E.J.; Kampitsis, A.E. Nonlinear Response of Shear Deformable Beams on Tensionless Nonlinear Viscoelastic Foundation under Moving Loads. *J. Sound Vib.* **2011**, *330*, 5410–5426. [CrossRef]
35. Yu, G.; Kiani, K.; Roshan, M. Dynamic Analysis of Multiple-Nanobeam-Systems Acted upon by Multiple Moving Nanoparticles Accounting for Nonlocality, Lag, and Lateral Inertia. *Appl. Math. Model* **2022**, *108*, 326–354. [CrossRef]
36. Ma, X.; Roshan, M.; Kiani, K.; Nikkhoo, A. Dynamic Response of an Elastic Tube-like Nanostructure Embedded in a Vibrating Medium and under the Action of Moving Nano-Objects. *Symmetry* **2023**, *15*, 1827. [CrossRef]
37. Li, Y.; Cai, C.S.; Liu, Y.; Chen, Y.; Liu, J. Dynamic Analysis of a Large Span Specially Shaped Hybrid Girder Bridge with Concrete-Filled Steel Tube Arches. *Eng. Struct.* **2016**, *106*, 243–260. [CrossRef]
38. Stăncioiu, D.; Ouyang, H.; Mottershead, J.E. Vibration of a Beam Excited by a Moving Oscillator Considering Separation and Reattachment. *J. Sound Vib.* **2008**, *310*, 1128–1140. [CrossRef]
39. Matsuishi, M.; Endo, T. Fatigue of Metals Subjected to Varying Stress. *Jpn. Soc. Mech. Eng.* **1968**, *68*, 37–40.
40. Miner, M.A. Cumulative Damage in Fatigue. *J. Appl. Mech.* **1945**, *67*, 159–164. [CrossRef]
41. *GB 50017-2017*; Standard for Design of Steel Structures. MOT (Ministry of Housing and Urban-Rural Development): Beijing, China, 2017. (In Chinese)
42. Juang, J.-N.; Pappa, R.S. An Eigensystem Realization Algorithm for Modal Parameter Identification and Model Reduction. *J. Guid. Control Dyn.* **1985**, *8*, 620–627. [CrossRef]
43. Guo, W.; Xia, H.; De Roeck, G. Dynamic Characteristics and Coupling Vehicle-Bridge Vibration Analysis of the High Speed Railway Sesia Bridge in Italy. *J. Vib. Shock* **2013**, *32*, 82–88. [CrossRef]
44. Zhu, Z.; Huang, Y.; Xiang, Z.; Mai, P. Fatigue Performance of Floorbeam Cutout Detail of Orthotropic Steel Bridge on Heavy Freight Transportation Highway. *China J. Highw. Transp.* **2017**, *30*, 104–112. [CrossRef]

45. Zhu, J.; Xiao, R. Damage Identification of Long-Span Concrete Cable-Stayed Bridge Based on Periodic Detection and Genetic Algorithms. *China Civ. Eng. J.* **2006**, *39*, 85–89. [CrossRef]
46. Li, Y.; Lu, D.; Sheng, H. Fatigue Reliability Analysis on Cable of Cable-Stayed Bridge Under Random Vehicle Load and Wind Oad. *China J. Highw. Transp.* **2012**, *25*, 60–66. [CrossRef]

buildings

Article

Accelerated Bridge Construction Case: A Novel Low-Carbon and Assembled Composite Bridge Scheme

Ling Kang [1], Jinhua Xu [2], Tingmin Mu [1], Huan Wang [1,*] and Ping Zhao [2]

[1] Sichuan Highway Planning, Survey, Design and Research Institute Ltd., Chengdu 610041, China; kangling@schdri.com (L.K.); moutm@vip.sina.com (T.M.)
[2] Sichuan Chengle Expressway Co., Ltd., Chengdu 610041, China; clgs01@cygs.com (J.X.); clgs02@cygs.com (P.Z.)
* Correspondence: wh_chd@163.com

Abstract: Modern bridge construction towards a higher degree of low carbonization and assembly has been the general trend, while developing and broadening the low-carbon and assembled-oriented Accelerated Bridge Construction (ABC) technology can better realize the trade-offs between construction quality, efficiency, cost and sustainability. In the current mainstream ABC technologies such as precast-assembled concrete bridge and assembled steel bridge schemes, it is difficult to achieve an excellent balance between the above multicriterion trade-offs. To this end, this paper proposes a novel low-carbon and assembled composite bridge scheme as an innovative case of ABC technology based on a 26.7 km-length urban viaduct project in China with urgent environmental protection and assembly demands. Construction sustainability, the comprehensive economy and low-carbon performance are well balanced by the collaborative application of new steel–concrete composite structures, the rapid assembly interface design and low-carbon material technologies. The proposed scheme has been applied to a completed real-scale bridge, and the whole construction process only experienced 105 days of effective time, accompanied with slight environmental interference and construction noise and a small amount of labor and equipment input. In addition, the safety of the bridge, the rationality of the design concept and the calculation method have been verified by the static and dynamic loading tests of the real-scale bridge.

Keywords: accelerated bridge construction; low-carbon bridge; assembly construction; sustainability; composite structure

1. Introduction

In recent years, under the background of the increasing traffic demand in China, a new expressway for urban traffic and reconstruction and extension projects for existing highways have gradually spread [1,2]. Urban standard-span bridges (20–40 m spans) are numerous and commonly found in densely populated, traffic-concentrated, and environmentally sensitive areas. The onsite bridge construction activities can have significant impacts on the environment, mobility and safety [3,4]. Noise, dust, lane closures and other adverse effects caused by the traditional bridge construction process will most likely determine the feasibility of the bridge scheme [5]. With the continuous implementation of the carbon peaking and carbon neutrality goals in China, the environmental protection requirements of urban bridge schemes have been referred to an unprecedented decision-making priority. Minimizing the impact of bridge construction activities on the existing traffic, environment and surrounding residents has become a new requirement for bridge engineering in the development of a harmonious society [6].

In order to solve the above environmental protection needs, accelerated bridge construction (ABC) [7–9] has rapidly developed and become the mainstream technology of urban bridge construction. The basic connotation of ABC is to adopt highly prefabricated and rapid assembly methods to minimize the adverse impact of bridge construction on the existing environment and traffic, as well as speeding up construction and reducing engineering risks and investment [10]. At present, ABC gradually develops in the direction of low carbonization and assembly. The industrialization-oriented ABC characterized by standardized design, industrial prefabrication and assembly construction is ushering in unprecedented development opportunities [11,12] and it is the general trend for modern bridge construction to move towards a higher level of assembly. The large-scale application of low-carbon and assembled (LA) bridge technology [13,14] with multiple performance attributes, such as an excellent structural quality, weak environmental impact, less traffic interference, low construction cost and rapid on-site assembly, can open up a new situation for the integration and development of digital technology, intelligent manufacturing, rapid assembly and intelligent equipment industry and promote the development of the industrial chain of the process, equipment, management and maintenance [15–20].

To realize the LA goal, assembled bridge schemes such as the precast-assembled concrete bridge (P-ACB) and assembled steel bridge (ASB) have been developed and widely used. The P-ACB was originally proposed to solve a series of problems of on-site-cast concrete bridges such as high carbon emissions, a high input of labor and equipment, a long construction period and a strong environmental impact [21,22]. At present, the research frontier of P-ACB focuses on the prefabrication and installation of substructures such as capping beams and piers. The components are prefabricated in the factory with an overall or segmented prefabrication scheme, and transported to the site through large-scale transport equipment for modular assembly. The pre-embedded holes, steel bars or specific connecting materials are designed at the assembly interfaces between capping beams and piers to realize the connection. The structural form and performance improvement of the connection are the key factors affecting the reliability of P-ACB. Zhen et al. [23] proposed a novel connection adopting lap-spliced large-diameter bars and ultra-high-performance concrete grout to accelerate the assemble progress of precast bridge piers. Shafieifar et al. [24] envisioned a new connection detail, where the capping beam and the pier are spliced in the pier and joined with a closure joint of ultra-high-performance concrete. Li et al. [25] proposed precast segmental ultra-high-performance concrete piers with post-tensioned tendons to realize the LA goal. Some scholars [26–28] also proposed to adopt a high-performance grouted sleeve to realize the connection of segmented columns or capping beams.

The P-ACB technology can greatly accelerate the construction, especially for urban bridges in scenarios of environmental protection requirements, small traffic impact requirements, limited on-site areas, difficult construction organization and complex construction conditions. However, for the overall prefabrication scheme, the large size of the prefabricated components will bring strict requirements on the trafficability of large-scale transport lines and the capacity of hoisting equipment. For the segmented prefabrication scheme, the large number of assembly interfaces will bring the problems of the high demand of assembly accuracy, complex connection details and the low reliability of the connection quality. Therefore, the application of the P-ACB is limited to developed coastal cities. For instance, the maximum mass and height of the prefabricated concrete piers of Hangzhou Bay Bridge [29] in the Zhejiang province, China, are 458 t and 17.38 m, respectively, while the prefabricated piers of the Hong Kong–Zhuhai–Macao Bridge [30], China, need to be transported and installed by large floating cranes at sea.

The ASB is a typical type of LA bridge, since almost no concrete material is used, accompanied with fewer carbon emissions from the cement production and lightweight steel components which facilitate on-site assembly [31,32]. Thus, the ASB embodies a considerably higher degree of LA performance than P-ACB. The main structural forms of ASB are a steel box, steel plate and steel truss girder, where a large number of stiffeners are arranged to maintain stability, which leads to high steel consumption and large welding work. To this end, many scholars performed a structural optimization design with the goal of reducing steel consumption, carbon emissions and cost. D. Martínez-Muñoz et al. [33] carried out an optimization of a box-girder steel bridge, with cost and emissions as objective functions, through applying the unsupervised learning technique of the k-means and the continuous swarm intelligence metaheuristics. Fé van Lookeren Campagne [34] generated a steel truss bridge using only reclaimed steel, by optimizing the material usage in relation to the function and avoiding cut-off waste, which achieves a 63% reduction in carbon emission compared to a traditionally designed new steel truss bridges. Zekeriya Aydın et al. [35] performed a size, layout and tendon profile optimization of a prestressed steel-trusses bridge using the Jaya algorithm to minimize carbon emissions and cost under the constraint conditions of ensuring the stability of the steel structure. Nevertheless, to realize the large-scale popularization and application of the ASB, some key technical problems must be overcome, such as the high investment consumption of construction and life-cycle maintenance, redundant multi-links of processing, manufacturing, coating and transportation, high industrialization requirements of assembly, welding and bolting and difficult quality control [36].

To this end, a composite bridge combines the double advantages of steel and concrete materials, and eliminates the problems of the high cost and poor stability performance of ASB and the problems of the heavy self-weight and high material consumption of P-ACB [37–39]. Therefore, based on the sustainable needs of a specified engineering project, this paper proposes a novel LA composite bridge scheme with less material consumption, lower carbon emissions, lighter components, reliable structural safety and an acceptable engineering cost. The advanced design concept, new structure and material application of the LA composite bridge structural system are elaborated in detail, and the comprehensive sustainable effects on economic and social indicators are provided through an engineering application.

2. Engineering Project Overview

2.1. Project Information

As shown in Figure 1, the Chengle Highway Expansion Project (CHEP), a highway to connect Chengdu and Leshan in the Sichuan province, China, is the first case of a large-scale highway reconstruction and expansion project from four lanes to eight lanes and its completion will greatly promote the urban development of Chengdu, Deyang, Meishan and Ziyang in the Sichuan Province, China. This paper studies the CHEP viaducts within Chengdu, located between the Third-ring Road and the Second-ring Highway. The total length of the viaducts is about 26.7 km, which is composed of five main-line bridges and four interchange bridges, involving 16.7 km main-line bridges and 10 km ramp-line bridges. The total investment of the project is about 4.37 billion RMB (£0.49 billion), which is expected to be completed and opened to traffic in 2025.

The initial design criteria included the following:

- Traffic speed: 120 km/h.
- Traffic load: Chinese highway-I load.
- Road layout: dual eight-lane highway with a standard width of 42 m.
- Service life: 100 years.
- Earthquake: seismic intensity VII, peak ground acceleration 0.10 g.
- Standard span of main girder: mostly 30 m (accounting for 75%).

2.2. Environmental Protection and Assembly Demands

Considering the environmental sensitivity and high traffic density of the project area, specific measures for environmental protection and efficient assembly are crucial. As shown in Figure 1, the viaducts project within the Chengdu urban area is located in the urban built-up area and planning area, passing through the Wuhou District, the Shuangliu District and the Xinjin District of Chengdu, with a dense population, concentrated traffic and sensitive environment. The whole design range involves 34 environmental sensitive points, which need to cross multiple underlying roads, operating railways and municipal pipe networks. The houses and factories on both sides of the route line are dense, accompanied by complex construction boundary conditions. The bridge design scheme needs to be forced to adopt a low-carbon structural system and ABC technology with performances of low carbon emissions and rapid assembly so as to reduce the impacts on existing traffic, the surrounding environment and residents.

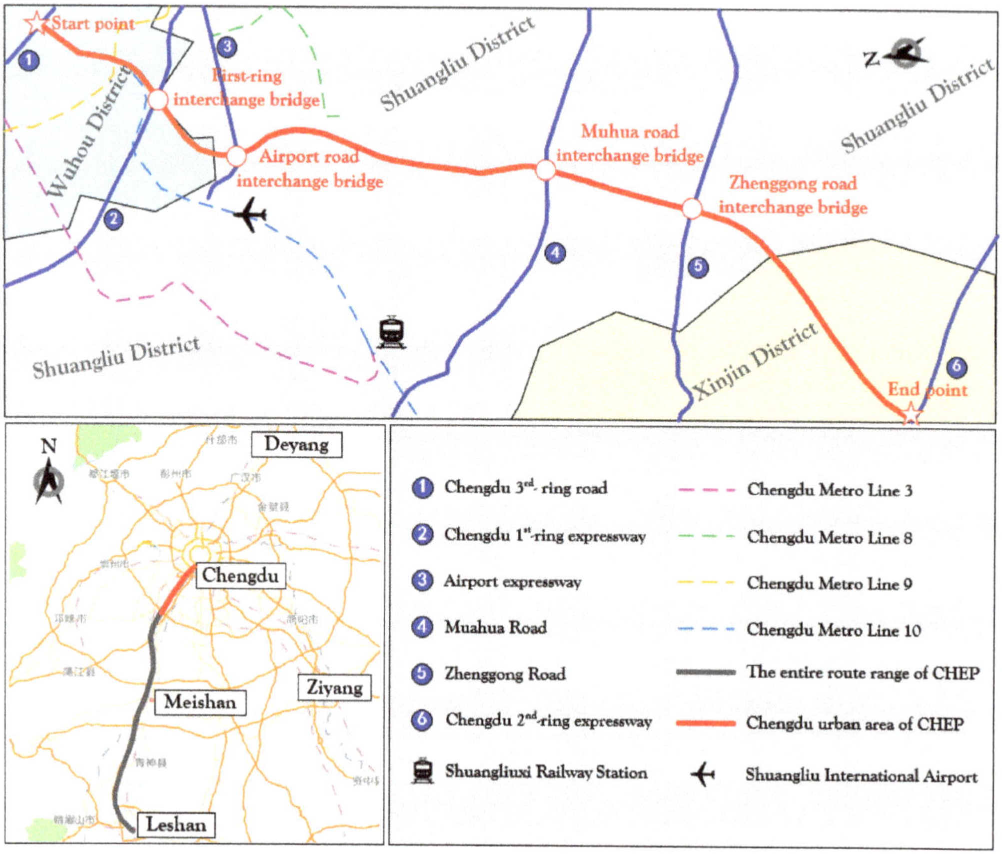

Figure 1. Design range of viaducts within Chengdu urban area. (Remark: the terms in Chinese in the figure refer to the names of various cities near the CHEP in Sichuan province, China).

3. ABC-Oriented LA Composite Bridge Scheme

3.1. Overall Scheme Comparison

For the construction requirements of CHEP, overall, feasible bridge schemes include concrete bridges, P-ACB and ASB. As shown in Figure 2, for concrete bridges with a span of 20 m~60 m, the superstructure generally adopts a prefabricated box girder or T-shape girder and the superstructure adopts cast-in-place capping beams and piers. This scheme owns good structural integrity and a mature construction procedure, but there are the following problems in using this kind of scheme in CHEP: (1) The concrete consumption is 1.41 million m^3; the large consumption of cement and gravel leads to high carbon emissions. The construction period is long due to the low efficiency, accompanied by serious construction noise and dust pollution. (2) A multi-pile foundation has to be used in this scheme to withstand the heavy structural weight from the concrete girders, capping beams and piers and a total of 5388 pile foundations and 1596 pile caps are used, which means more construction links and a longer time consumption. In addition, 3×10^6 m^3 of waste soil stacking has a great impact on the environment. (3) To ensure the traffic of the bottom road under the bridge, partial capping beams have to adopt the integral structure with a span of 42 m, a maximum weight of 700 t and a concrete consumption of nearly

300 m^3. Building the superstructure using the temporary support method needs much equipment and labor and inevitably causes serious safety risk, environmental impacts and traffic disruption. (4) There are too many assembly interfaces between prefabricated girders, including 60 transverse connections and 70 wet joints, which will lead to a large amount of on-site work and an insufficient assembly degree.

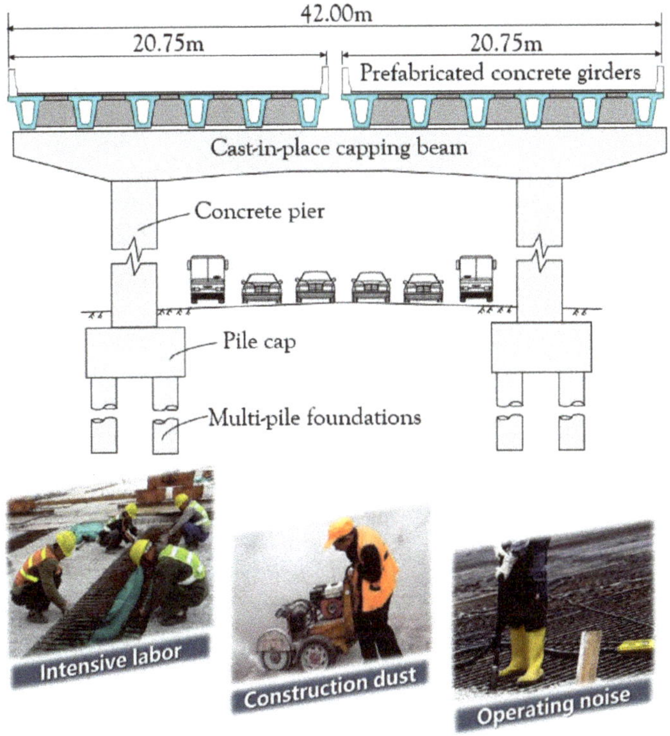

Figure 2. Concrete bridge scheme.

As shown in Figure 3, for the P-ACB scheme, the superstructure generally adopts a prefabricated box girder or T-shape girder and the superstructure adopts precast-assembled concrete capping beams and piers. The components are prefabricated and staged in the factory and transported to the site using large-scale transport equipment for modular assembly. The pre-embedded holes, steel bars or specific connecting devices are designed at the assembly interfaces between the capping beams and piers to realize the connection. This technology can greatly promote ABC, but there are still the following problems for this project: (1) The standard size of the prefabricated capping beam is 42.3 m and the overall weight is 700 t. The lifting weight is more than 200 t even if it is divided into three segments for assembly and the requirements for the trafficability of large-scale transport routes and the hoisting capacity of the equipment are extremely high. (2) The large number of assembly interfaces will bring the problems of the high demand of assembly accuracy, complex connection details and the low reliability of the connection quality. (3) The small number of superstructures and the low degree of standardization will lead to low production efficiency even if using the smart precast field. Moreover, the structure is complex and the construction process is numerous, which means a high construction cost for LA goals.

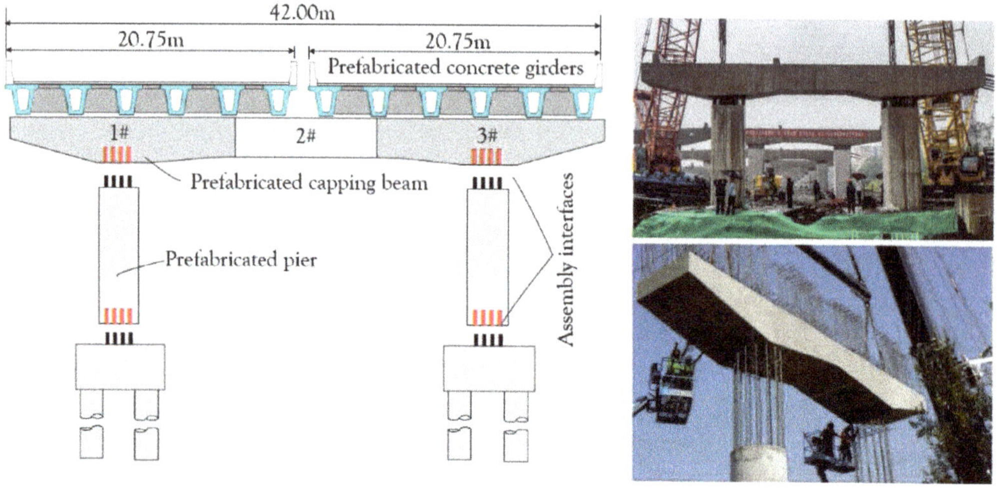

Figure 3. P-ACB scheme.

As shown in Figure 4, the ASB scheme generally adopts steel box girder or steel plate girder as the bridge superstructure. The production of the steel girder has to go through multiple complex processes, such as plate unit manufacturing, welding, test pre-assembly and coating, and the production cycle is long, with a high demand for advanced labors and equipment. The steel material consumption of the steel girder is usually greater than 300 kg/m^2 and the cost only for processing and manufacturing exceeds 30% of the material cost. Therefore, the cost of the steel bridge is usually unacceptable to the government or investors compared with the concrete scheme and is limited to the application cross-line and cross-obstacles scenes. Furthermore, there is a wide variety of components involving steel boxes, steel plates, stiffened ribs and other components with different sizes and weights, which are of low standardization, and it is difficult to control the accuracy of on-site assembly.

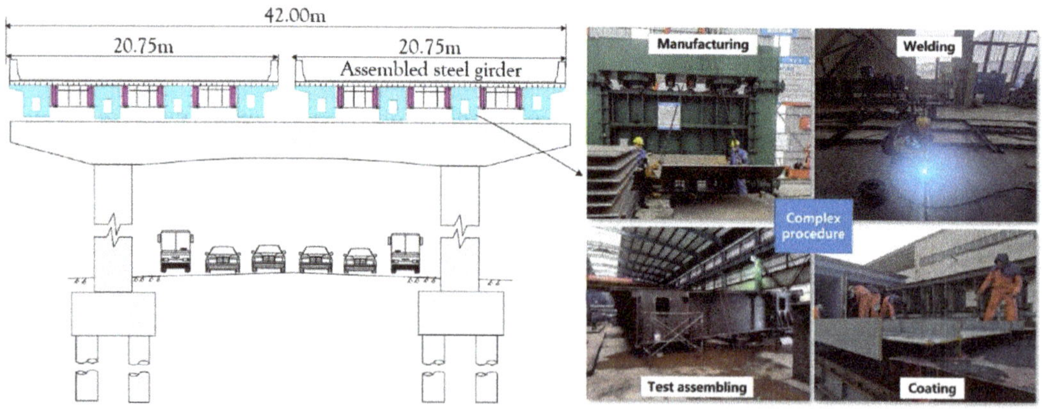

Figure 4. ASB scheme.

To sum up, the above bridge schemes cannot meet the high requirements of the low-carbon, intelligent and assembled construction for the CHEP. Accordingly, this paper proposes a novel LA composite bridge scheme for ABC, which minimizes the amount of on-site work and reduces the impacts of construction activities on the environment through standardized designs, industrialized prefabrication, assembly construction and the sustainability concept. As shown in Figure 5, the LA composite bridge's structural system consists of dense-girder–H-shaped (DGH) composite girders, a prestress-assisted concrete-filled steel box (PCSB) capping beam, a concrete-filled steel tube (CFST) pier and a single-pile reinforced concrete (SPRC) pile foundation. The detailed description of the structural system and the application of low-carbon materials are shown in Sections 3.2 and 3.3, respectively.

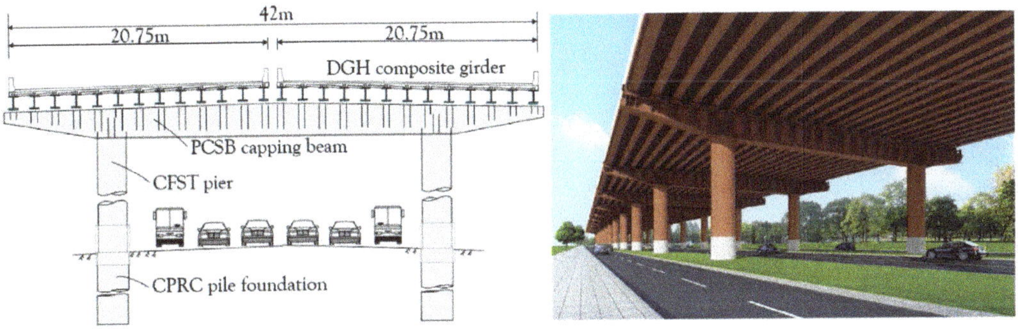

Figure 5. LA composite bridge scheme.

3.2. A Novel LA Composite Bridge Structural System

3.2.1. Dense-Girder–H-Shape Composite Girder

At present, the commonly applied forms of steel plate composite girders in the world are multi-girder and few-girder systems, according to the number of girders. The few-girder system greatly reduces the consumption of steel for a better economy, but requires dense transverse connecting members between girders to maintain structural integrity and stability, and even the need to stretch transverse prestress in the bridge deck brings complex and cumbersome processes. In addition, its safety redundancy under extreme loads is not excessive, and the dynamic performance needs to be focused on. Oppositely, the transverse load distribution coefficient of the multi-girder system is small, showing better stress uniformity for each girder, while the safety redundancy and disaster resilience under the scenario of local component failure can be guaranteed. In addition, its small girder height and light lifting weight make it more suitable for ABC. However, the problem of excessive steel consumption needs to be overcome under today's low-carbon and sustainable development concept.

Based on the multi-girder design concept, this paper proposes the DGH composite girder that adapts to the transverse force transmission of vehicle loads, so that any wheel load is borne by at least two girders. Meanwhile, a large number of transverse connection members and stiffened steel plates are innovatively eliminated through structural optimization, which greatly overcomes the problem of the high consumption of steel structures in a multi-girder system. As shown in Figure 6, the DGH composite girder is composed of multiple H-shaped steel girders and a corrugated steel–concrete composite bridge deck, with a girder height of 1.0 m and a deck thickness of 0.3 m. The applicable span of the DGH composite girder is 20–35 m. Specifically, there is no stiffened steel plate arranged on the girders, as well as no intermediate transverse connecting members within the span. Only the corrugated steel–concrete composite end beams filled with concrete are set at both ends, which greatly eliminates the stiffener structure and the number of welds. Only the concrete-filled corrugated steel composite girder-end crossbeams are designed at both ends, which significantly simplifies the stiffener structure and the number of welds and greatly reduces the steel consumption.

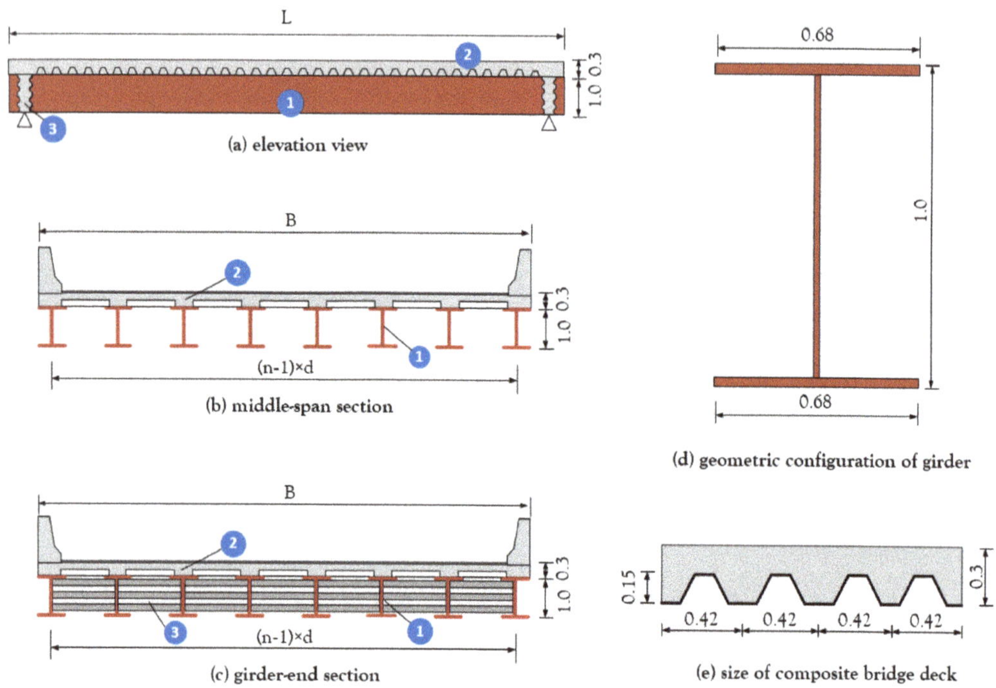

Figure 6. Structural description of the DGH composite girder (unit: m). (Remark: ①: H-shape steel girder, ②: corrugated steel–concrete composite bridge deck, ③: girder-end crossbeam, L and B refer to the span of girder and the width of the deck and n and d refer to the number and spacing of girders).

The elimination of transverse connecting members and stiffeners may imply a need to focus on the integrity, stability and lateral force transmission performance of the structure. The above concerns are ensured by the following multiple design strategies: (1) a girder spacing of less than 1.8 m (wheel spacing of standard vehicles in China) ensures a uniform structural response from an individual girder even under the worst-case scenario and the dense-girder system adaptively increases the lateral stiffness of the superstructure owing to the sectional contribution of multiple girders. (2) The optimal thickness parameter of 0.3 m for the composite bridge deck greatly improves the integrity and stiffness of the deck, with the assistance of the corrugated steel base plate which owns large axial stiffness and out-of-plane stiffness relative to flat steel plates. (3) The optimum width design value of 0.68 m for the top plate and bottom plate of the H-shape steel girder makes it have a greater torsional stiffness than the traditional I-shape steel girder to resist the local instability risk. (4) The girder-end crossbeams are designed using the combination of a corrugated steel shell and concrete, and a larger thickness of 0.46 m (usually about 0.3 m for a concrete box girder or a T-shape girder) improves the stiffness of the girder-end crossbeams to provide strong boundary constraints for the DGH girder, so as to magnify the integrity and stability of the bridge superstructure.

In addition, the dense-girder arrangement increases the transverse force-transmission paths of the structural system and improves the redundancy and robustness of the system when dealing with extreme loads or disaster situations [40]. When local members fail, the stress of the structure be can easily redistributed due to the strong composite bridge deck and girder-end crossbeams and the further spread of damage can be significantly limited by using alternate transmission paths to distribute the loads to other H-shaped girders. The local stability, load lateral distribution and other structural responses of the DGH composite girder have been verified by loading tests. As shown in Figure 7, two testing conditions of two girders and four girders are respectively designed to explore the transverse force-transmission performance of the DGH girder without transverse connecting members. Firstly, the loading test of four girders under Condition 1 (loads are transferred laterally through the composite bridge deck) is carried out, after which the structure is cut in half for the loading test of four girders under Condition 2 (no transverse force-transmission path).

(a) Condition 1: loading test of 4 girders **(b) Condition 2: loading test of 2 girders**

Figure 7. Mechanical behavior experiment.

Figure 8 shows the loading-displacement curve of Condition 1 and the ultimate bearing capacity of the structure is 9360 kN. After unloading, the loading test is repeated four times and the bearing capacity remains at the same level, which verifies the structural safety. Figure 9 shows the loading-displacement curve of Condition 2 and the ultimate bearing capacity of the structure is 7050 kN. By comparing the testing results of four girders and two girders, 2310 kN of the loads is transversely transferred through the composite bridge deck, although no transverse connecting members exists. It proves that the transverse force-transmission performance of the composite bridge deck is reliable and the basis for canceling the transverse connecting members between girders is sufficient.

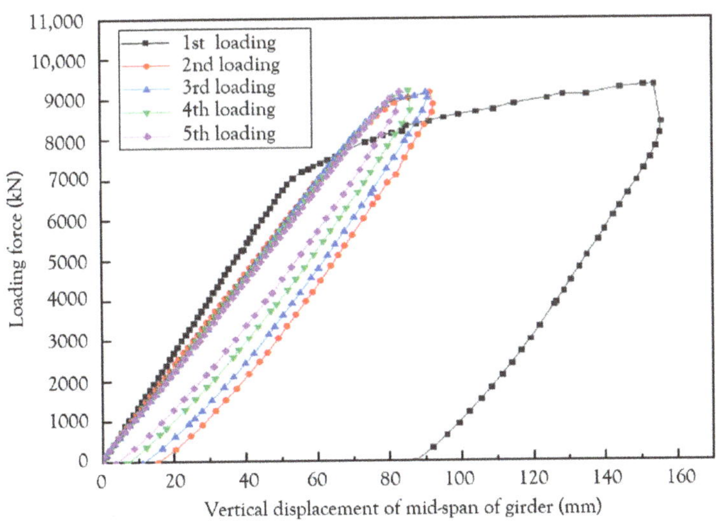

Figure 8. Loading-displacement curve of the loading test of 4 girders.

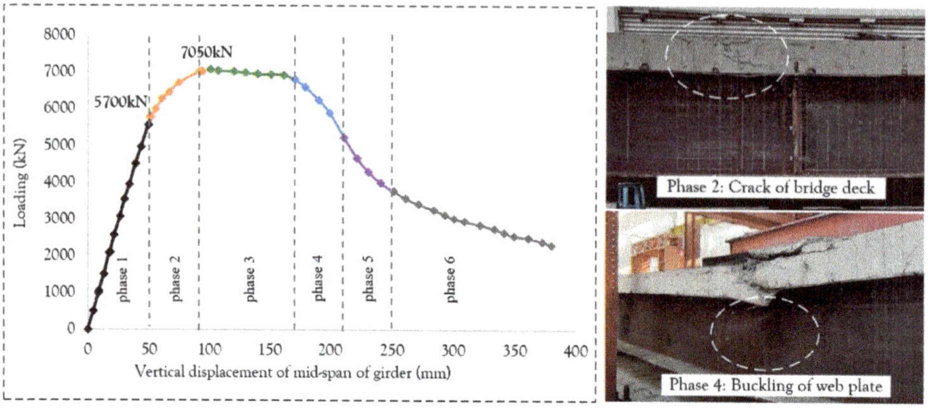

Figure 9. Loading-displacement curve of the loading test of 2 girders.

In addition, from Figure 9, the loading-displacement curve of Condition 2 can be divided into six phases: phase 1 (the elastic working stage), phase 2 (the cracking stage of the bridge deck), phase 3 (the bridge deck is crushed), phase 4 (the buckling of the web plate of the H-shaped girder), phase 5 (the deformation development of the web plate) and phase 6 (the large deformation of the web plate). It can be seen that the first-order failure of the structure is the crushing failure of the bridge deck and the second-order failure is when the local buckling of the web plate occurs. In other words, the failure of the web plate buckling occurs during a long loading process after the elastic working stage, which proves that the web plate without stiffeners has no safety problems in the objective service stage.

3.2.2. Prestress-Assisted Concrete-Filled Steel Box Capping Beam

The standard width of the capping beam of the CHEP is 42.3 m, which involves technological challenges in the structural design of the extra-large span capping beam. The construction of the extra-large-span capping beam usually adopts the P-ACB scheme (as shown in Figure 10) and its technological limitations are described in detail in Section 3.1. To this end, this paper proposes the PCSB capping beam as one of the solution schemes for the technological challenges in the extra-large-span capping beam design, which is a composite structure formed by stretching prestress on the steel box and then pouring self-compacting concrete (SCC) into the box. The combination of a steel box and concrete fully integrates the advantages of the two materials, that is, the strength of the concrete is significantly improved owing to the constraints from the steel box and the steel box is laterally supported by the concrete, which reduces the instability risk of the steel structure. Meanwhile, the application of prestressed wires reduces the tensile stress level of the steel box and concrete under the objective service stage and improves the service performance during the whole life-cycle. Compared with the concrete capping beam, the toughness is improved while the weight is reduced. Compared with the steel capping beam, less steel consumption and greater stiffness and stability make it more suitable for the application scenarios of extra-large-span capping beams.

As shown in Figure 11, for the CHEP, the standard width of the capping beam is 42.3 m, which meets the traffic demand of eight lanes in two lines. The spacing between the two piers is 25.8 m and the middle-span section height of the capping beam is 2.6 m. A total of 36 prestressed wires with a horizontal straight line arrangement are installed on the capping beam and the amount of steel structure is 70 t and the amount of concrete is 35.6 m^3. The typical construction sequence of the PCSB capping beam is as follows: (1) stretching the prestressed bundles and anchoring them at both ends of the steel box; (2) hoisting the capping beam with the anchored prestressed bundles and welding the bottom plate of the capping beam to the top surface of the pier steel pipe; (3) the SCC is poured into the steel box from one end of the capping beam; and (4) after the concrete strength meets the design requirements, the prestressed bundles are loosened to transfer the prestress to the steel box and concrete, so as to form a composite structure.

Figure 10. P-ACB scheme for capping beam.

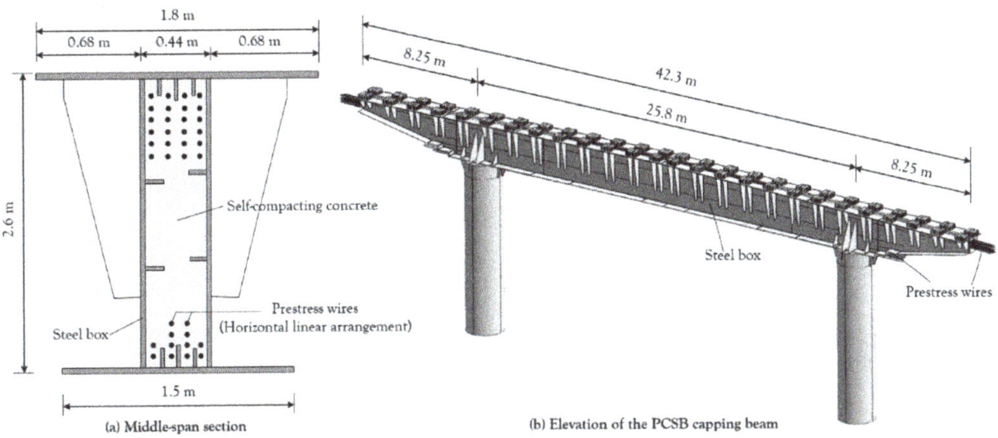

Figure 11. Structural detail of the PCSB capping beam.

The PCSB capping beam is very suitable for the construction scenarios with the ABC requirements and environmentally friendly requirements. The steel box lifting weight of the PCSB capping beam is only 70 t, which shows excellent lightweight characteristics compared with the concrete capping beam, which can accelerate the construction and promote the level of industrial construction, relying on the steel box as the template for concrete pouring with no brackets and template engineering, which can minimize the impacts of the construction process on traffic and the environment. The use of SCC without vibration operation can greatly reduce the impact of construction noise on the environment.

3.2.3. Pier and Pile Structure

As shown in Figure 10, the superstructure of the LA composite bridge structural system adopts CFST piers, and each pier is connected to a SPRC pile foundation through a pier–pile transition connector. As the total weight of the DGH composite girders, PCSB capping beam and CFST piers of the novel bridge scheme is reduced by about 40% compared with that of the traditional concrete bridge scheme, only a single pile can meet the safety demand without the need to design a pile-group foundation, which thus eliminates the huge pile cap structure, greatly saves the amount of concrete, speeds up the construction and also significantly reduces the noise and mud pollution associated with multi-pile drilling operations.

Meanwhile, for the CFST pier, the compressive strength of the concrete is increased by 1.8~2.5 times due to the hoop confinement effect of the steel tube on concrete and the stability of the steel tube is also improved by the support effect of the concrete in the tube. The mutual achievement of the steel and concrete materials makes the CFST pier have better ductility and crashworthiness. During the construction, for the piers with a length of 12 to 20 m, the lifting weight of a single steel pipe is only 10.3 to 17.2 t so that the ordinary lifting equipment can achieve rapid construction. In addition, the steel tube can be used as a support template for concrete pouring without the need to set up any temporary scaffolding, which avoids the interference to the traffic under the bridge.

As shown in Figure 12, a variety of anchorage structures are used to realize the structural safety of the pier–pile transition connector so that the force flow transmission between the two structures of the CFST and reinforced concrete is smooth. The shear holes are set at the bottom of the CFST pier and the PBLs are welded on the inner wall of the steel tube so that the concrete and steel tube in the pier–pile transition area can work together with a good bonding performance and the opening holes on the PBLs can transform the shear force of the steel tube into the concrete. The outer-layer protective sleeve provides an additional hoop confinement effect and increases the cross-sectional area in the pier–pile transition area, thereby improving the bearing capacity and bending performance under the earthquake action. The annular stirrups and vertical reinforcements in the pile foundation extend upward to the CFST pier area and weld with the steel tube, which enhances the anchorage reliability and force transmission uniformity between the two structures. The above measures make the pier–pile transition connector very strong and improve the seismic performance, which can eliminate the tied beam between piles and obtain the beneficial effects of saving materials, simplifying the process and speeding up the construction period.

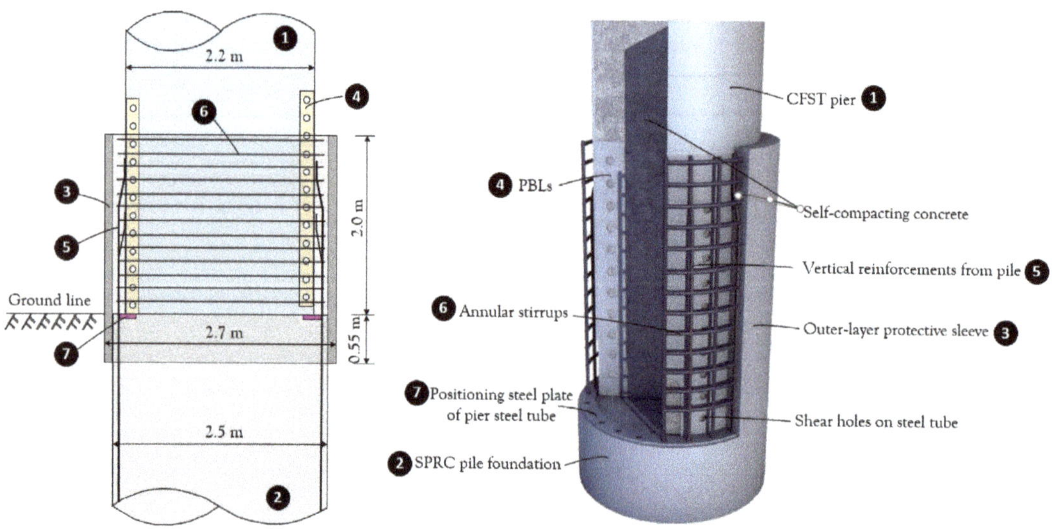

Figure 12. Structural detail of the pier–pile transition connector.

3.3. Large-Scale Application of Low-Carbon Materials and Technologies

3.3.1. Self-Compacting Concrete

SCC is widely used in all the composite structures of the LA composite bridge scheme and the concrete can be quickly poured and compacted by its own gravity (as shown in Table 1 and Figure 13), so the labor intensity of workers can be greatly reduced owing to the elimination of the use of vibrators and the noise pollution to the surrounding environment can be minimized. The multi-objective mixture design of self-compacting concrete is proposed to minimize the amount of slurry material, while satisfying the strength and working performances. Compared with the same-grade concrete, the amount of cement can be reduced by about 25 kg/m^3, which can reduce carbon emissions by about 12,500 t for the CHEP.

Table 1. Working performance of SCC.

Item	Slump Index (mm)	Flow Index (mm)	Slump-Flow Time (s)	V-Funnel Emptying Time (s)
C40 SCC in PCSB capping beam	265	650	10	15
C30 SCC in CFST pier	270	650	10	16

Remarks: The evaluation indexes of SCC in the table above are derived from China code [41], where the Slump and the Flow index shows the filling ability, while the Slump-flow time and the V-funnel emptying time indicate the passing ability and flow velocity.

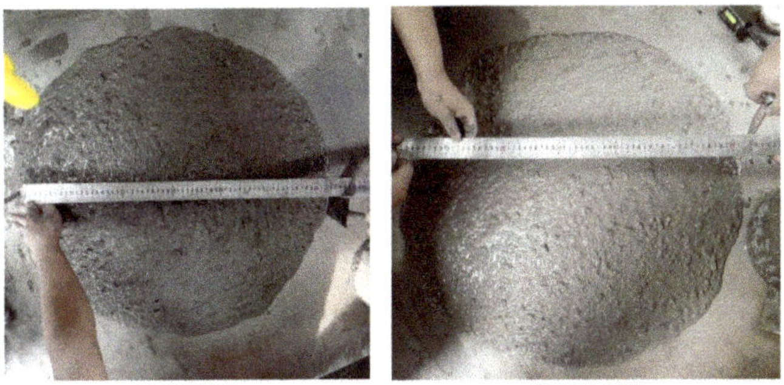

(**a**) C40 SCC (**b**) C30 SCC

Figure 13. Working condition of SCC.

3.3.2. Uncoated Weathering Steel

All steel components in contact with atmospheric environment in CHEP are made of uncoated weathering steel (UWS) and the amount of UWS is 327,000 t. Table 2 shows the list of UWS bridges built in recent years and the CHEP is the first large-scale application of UWS in China; it is also the largest case of UWS consumption in a single project around the world.

The LA composite bridge scheme using UWS integrates the following advantages: (1) The coating process in the factory is eliminated, so the cost is reduced, the construction speed is accelerated and the labor intensity is reduced. (2) No secondary coating is required, so the maintenance frequency as well as the cost in the life-cycle is observably reduced. (3) The corrosion resistance of the bridge components is improved owing to the atmospheric corrosion resistance of the UWS, which greatly improves the life-cycle durability of the structure.

It should be noted that the use of UWS in wet areas and particularly in contact with rain water may cause the removal of the protective rust layer and then the corrosion can progress over time, reducing the resistance and durability of steel sections [42]. The corrosion resistance performance of UWS is greatly related to the composition of alloying elements and their compounds and the protective effect of the rust layer is closely related to environmental conditions (temperature, humidity, sunlight, salt content, etc.) [43]. Therefore, the designer should choose the alloying element composition of the UWS according to the bridge operating environment (defined by the atmospheric corrosion resistance index I, a larger I means a better weather resistance). For the CHEP, considering the climate environment of Chengdu, China, the value of I is 6.0. In addition, the formation of the protective rust layer on the surface of UWS generally takes about three years, thus yellow rust liquid appears on the surface of UWS in the early stage of use. It is difficult to form a stable rust layer on the surface of UWS under the climate condition with a high salt content [44], which reduces the performance advantage of UWS. Therefore, in the LA composite bridge scheme, the side H-shape girder, the PCSB capping beam and the CFST pier are required to perform the rapid stabilization treatment of the rust layer on the surface of UWS.

Table 2. UWS bridges built in recent years.

Project	Year	Country	Bridge Length (m)	Bridge Type	Amount of UWS (t)
The CHEP (under construction)	2023	China	26,700	Beam bridge	327,000
Linyi Huanghe River Bridge	2021	China	5427	Beam bridge	56,000
Fuzhou Hongtang Bridge	2021	China	1055	Cable-stayed bridge	24,000
Shaxi Bridge of Putian–Yanling Expressway	2021	China	1408	Rigid frame bridge	22,000
Zangmu Brahmaputra River bridge	2021	China	525.1	Arch bridge	12,800
China–Russia Amur River Bridge	2022	China Russia	1274	Cable-stayed bridge	6600
Guanting Reservoir Grand Bridge	2016	China	1140	Truss-beam bridge	7200
Dalian Puwan New District Cross-sea Bridge	2015	China	2882	Beam bridge	20,334
Tanana River Railway Bridge	2012	America	1006	Beam bridge	15,000
New Pattullo Bridge (under construction)	2023	Canada	/	Cable-stayed bridge	12,725
China–Maldives Friendship Bridge	2018	Maldives	760	Beam bridge	/
Suez Canal El Ferdan Railway Bridge	2021	Egypt	640	Truss-beam bridge	14,000

In addition, designers are advised to consider a corrosion allowance of 0.5~1.5 mm in the thickness design of the steel plate to reduce the impact of the lost rust layer on the durability of steel sections. More importantly, from the disease analysis of UWS bridges in the United States, a good drainage detail and ventilation structure design are necessary. As shown in Figure 14, an inclined water baffle is arranged on the top flange of the side H-shape girder to avoid the hydrops on the surface of the UWS that affects its weatherability. Meanwhile, the water baffle is also the construction template of the concrete of the bridge deck, as well as the fact that it improves the stability of the top flange of the side H-shape girder. In addition, two beam-end water baffles are arranged on both ends of the PCSB copping beam to prevent rainwater from entering the inclined bottom plate of the PCSB copping beam.

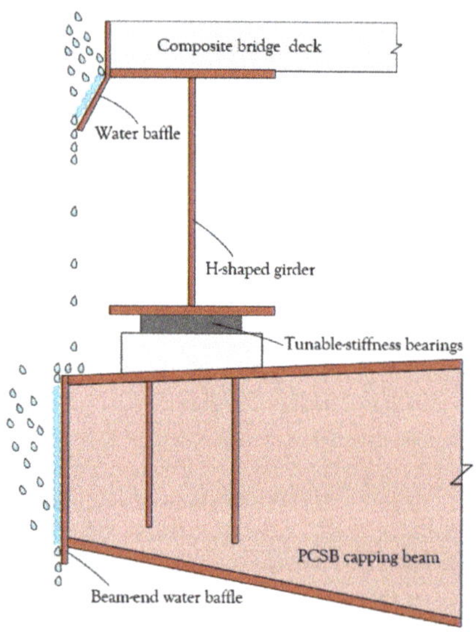

Figure 14. Structural details of the water baffle.

3.3.3. Tunable-Stiffness Bearings Technology

Equal-stiffness bearings are generally adopted in a multi-span girder bridge, where the horizontal stiffness of each bearing between two expansion joints is the same regardless of the pier height, resulting in significant differences in the mechanical response of the piers with different heights, uneven stress and damage to the bearings. In addition, the length of a single multi-span girder bridge between two expansion joints is less than 150 m and too many expansion joints are designed, resulting in poor driving comfort and a high level of driving noise.

To this end, the tunable-stiffness bearings technology, which has been formed to a local standard [45] in the Sichuan Province, China, is proposed to achieve a long-connected and few-joint structural system through the following strategies:

(1) A coordinated matching strategy between the horizontal stiffness of the bearing and the piers is proposed to enhance the total stiffness of the main girder within a long-connected bridge, as well as reduce the horizontal displacement of the main girder under the actions of temperature, braking force and earthquake. As shown in Figure 15, the calculation principle for the aggregative horizontal stiffness of a multi-span girder bridge is as follows:

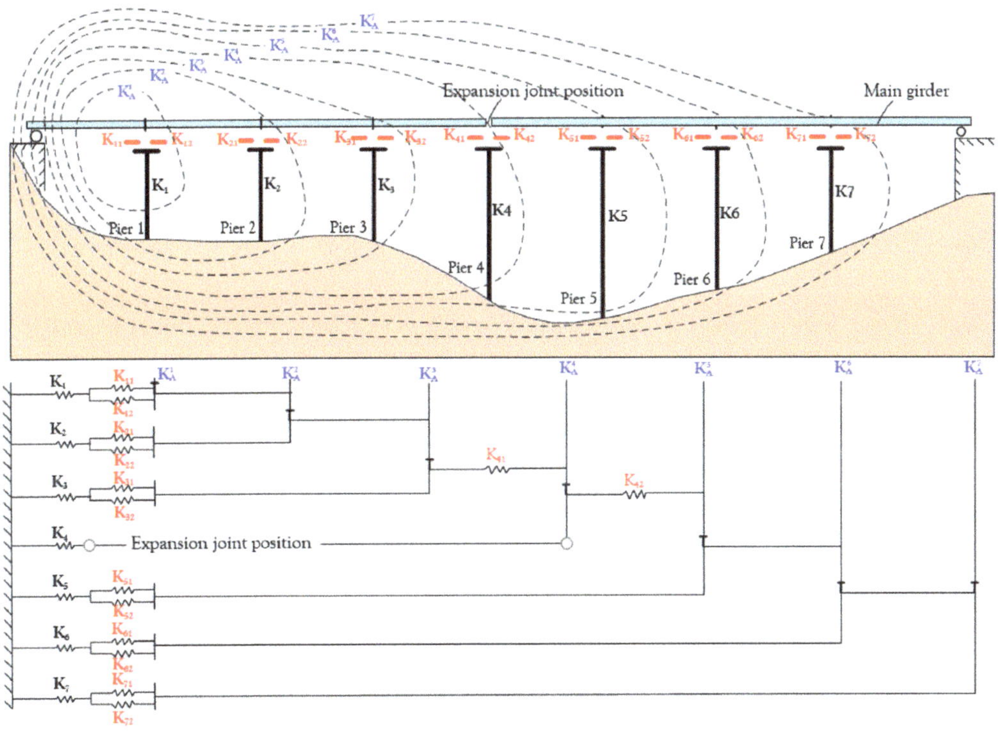

Figure 15. Aggregative horizontal stiffness of multi-span girder bridge.

Step 1: the horizontal stiffness of all bearings on the same pier K_{bi} is superimposed

$$K_{bi} = K_{i1} + K_{i2} \tag{1}$$

where K_{i1} and K_{i2} are the horizontal stiffness of the bearings on both sides of the i-th pier, respectively.

Step 2: The horizontal stiffness of the bearings and the i-th pier are integrated into the aggregative horizontal stiffness of the i-th pier K_A^i

$$K_A^i = \frac{K_{bi}K_i}{(K_{bi} + K_i)} \tag{2}$$

where K_i is the horizontal stiffness of the i-th pier, which is related to parameters such as the pier height and section size. Then, the total horizontal stiffness of a length-connected bridge composed of n piers is

$$K_{total} = \sum_{i=1}^{n} K_A^i = f(K_{bi}, n,\ K_i) \tag{3}$$

Equation (3) relates the total horizontal stiffness of the main girder K_{total} to the horizontal stiffness of the bearings K_{bi}, the horizontal stiffness of the bridge piers K_i and the number of piers n . By adjusting K_{bi} and n, the group piers can participate in the stiffness contribution to improve the K_{total}. Moreover, the adjustable horizontal stiffness of the bearings can change the boundary conditions of the piers, thereby regulating the internal forces of each pier and making the mechanical response of all piers uniform regardless of their height.

(2) The calculation method for the optimal length of the main girder between two expansion joints is proposed, where the horizontal stiffness of bearings K_{bi} and the number of piers n are adopted as the optimization design variables D to carry out multi-objective trade-offs between the stress optimality, driving comfort and life-cycle sustainability. The optimization model is established as follows:

$$\begin{array}{ll} find & D = [K_{bi}, n]^T \\ obj. & f_1(D) = max\{n\} \\ & f_2(D) = max\{K_{total}\} \\ & f_3(D) = min\left\{ \sum\limits_{i=1}^{n} (M_i - \overline{M})^2 \right\} \\ s.t. & g(D) = f \geq 0.2\ \text{Hz} \end{array} \tag{4}$$

where M_i is the bottom bending moment of the i-th pier and $\overline{M}$ is the average bottom bending moment of all piers. The above model ensures structural stiffness and dynamic performance by satisfying the constraint condition that the natural frequency of vibration is greater than 0.2 Hz. f_1 is designed to maximize the number of piers to extend the length of the main girder, so as to minimize the number of expansion joints, improve driving comfort and reduce noise pollution during the operation stage. f_2 is designed for maximizing the total longitudinal stiffness of the main girder in pursuit of better stiffness and a dynamic performance for better driving comfort. f_3 is designed to achieve a uniform stress distribution on bearings and piers in order to enhance the durability and sustainability during the entire bridge life-cycle.

Specifically, for the CHEP, based on the proposed tunable-stiffness bearings technology, the length of the main girder between the two expansion joints is increased to 300 m and the number of expansion joints reduces to 50% of traditional bridges, which can greatly reduce the vehicle operation noise of bridges in urban areas.

4. Sustainability Assessment on the LA Composite Bridge Scheme

4.1. Construction Sustainability Performance

Considering the various influencing factors on construction sustainability, this paper uses an index system by considering the following six aspects: carbon emissions, environmental impacts, traffic disruption, construction time, construction cost and assembly degree. The construction sustainability performances of the concrete bridge scheme, P-ACB scheme, ASB scheme and LA composite bridge scheme are compared in Table 3. The carbon emissions are calculated according to the carbon emissions of various materials in the production process. The construction time is determined according to the construction schedule of various bridge schemes. The construction cost is calculated according to the project budget of various bridge schemes. The assembly degree is measured according to the proportion of materials processed in the factory and assembled on site.

Table 3. Primary construction sustainability performances' comparison between several bridge schemes.

Performance \ Schemes	Concrete Bridge Scheme	P-ACB Scheme	ASB Scheme	LA Composite Bridge Scheme
Carbon emission	**High:** 11.6 × 10^5 t Huge material consumption	**High:** 11.4 × 10^5 t Huge material consumption	**Low**: 10.8 × 10^5 t Small amounts of concrete	**Low**: 10.0 × 10^5 t Small amounts of material 160,000 t
Environment impact	**Serious:** Noise, dust, temporary supports, multi-pile foundation	**Middle:** Damage by multi-pile construction	**Middle:** Damage by multi-pile construction	**Almost none**
Traffic disruption	**Serious:** Traffic under the bridge is blocked by temporary supports	**Almost none**	**Almost none**	**Almost none**
Construction time	**Long**: 36 months Loss of time consumption of pouring concrete, erecting and removing supports	**Short**: 24 months Component erection and connection	**Middle**: 30 months The production cycle of steel girder is time-consuming	**Short**: 20 months Fast component erection and connection
Construction cost	**Low:** 4.08 billion RMB	**Middle:** 4.16 billion Additional costs from large-scale transporta-tion equipment	**High:** 8.63 billion RMB Extensive use of steel materials; expensive manufacturing costs	**Middle:** 4.37 billion RMB Small amounts of steel material
Assembly degree	**Low**: 15% Lots of on-site cast work	**High**: 70% Lots of precast-assembled structures	**Middle**: 60% Many kinds of components and low degree of standardization	**High**: 80% Lightweight components, simple structural form, few welds

Then, the hierarchical evaluation method based on a rose chart is employed to compare sustainability associated with different bridge schemes. As shown in Figure 16, each performance is scored with 1–4 points according to the merit ranking of the various bridge schemes and the area S_i enclosed by connecting lines is taken as the analysis index of the sustainability associated with the bridge schemes. Considering the LA goal, this paper designs the performance enhancement factor α_1 and α_2 to increase the focus on low-carbon emissions and the assembly degree. The final sustainability index S_i' of each scheme is:

$$S_i' = \alpha_1 \alpha_2 S_i \tag{5}$$

where α_1 and α_2 refer to the performance enhancement factors considering low-carbon emissions and the assembly degree, respectively. When the carbon emission performance of each scheme is scored 4 points, $\alpha_1 = 1.1$, then the score is 3 points, $\alpha_1 = 1.05$, otherwise $\alpha_1 = 1.0$. Similarly, when the assembly degree performance of each scheme is scored 4 points, $\alpha_2 = 1.1$, then the score is 3 points, $\alpha_2 = 1.05$, otherwise $\alpha_2 = 1.0$.

As seen in Figure 16, the traditional concrete bridge scheme receives the lowest grade ($S' = 5.20$), and it results in not only a serious environment impact and traffic disruption, but also a high onsite construction cost, time and risk. Even if the P-ACB and ASB schemes overcome the problems of the concrete bridge scheme and promote the ABC to a certain extent, there are still many ABC performances that need to be further improved for the

LA goals. The LA composite bridge scheme receives the highest grade ($S' = 41.92$) since it achieves an excellent balance between the various performance indicators such as low carbon and assembly.

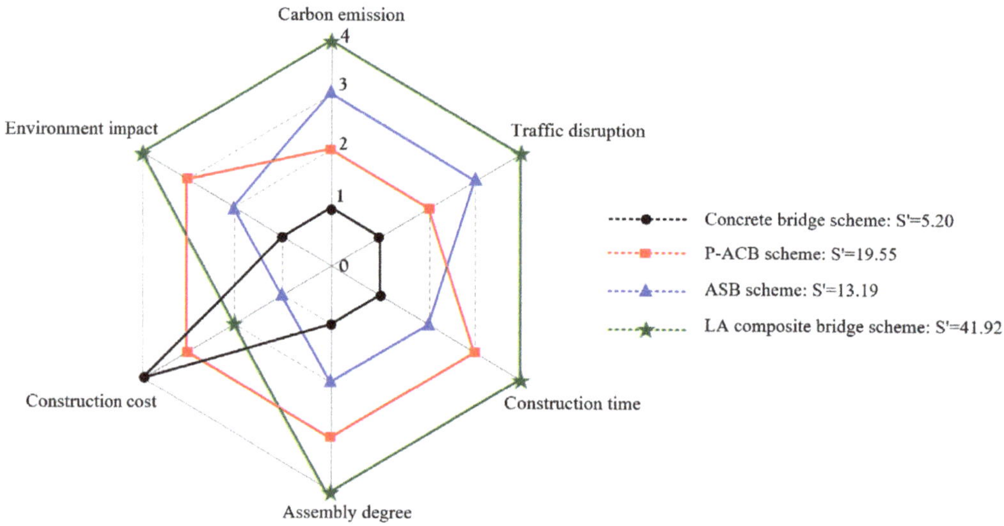

Figure 16. Construction sustainability performance assessment of bridge schemes.

4.2. *Comprehensive Economic Performance*

4.2.1. Material and Construction Costs

The comparison of the material consumption and construction cost between the concrete bridge scheme and the LA composite bridge scheme is shown in Table 4. Compared with the concrete bridge scheme, although the amount of steel (including uncoated weathering steel, rebar and prestressed wires) increased by 11.79×10^4 t with an increase of about 38.2%, the amount of concrete reduced by about 8.02×10^5 m^3 with a great reduction of 56.8% thanks to the lightweight design concept of the LA composite bridge scheme. For the construction cost composed of direct and indirect expenses caused by the installation and disassembly of the support template, equipment investment, hoisting construction and labor input, the LA composite bridge scheme can reduce 123.5 million RMB with a great reduction of 52.4% compared to the concrete bridge scheme, owing to the excellent comprehensive sustainable performance of the LA composite bridge scheme.

Table 4. Material and construction cost comparison.

Item	Material	Concrete Bridge Scheme	LA Composite Bridge Scheme
Material consumption	Steel (t)	59,036	327,281
	Rebar (t)	221,450	97,433
	Prestressed wires (t)	27512	1272
	Concrete (m^3)	1,410,802	608,880
Construction cost		235.7 million RMB	112.2 million RMB
Project investment cost		4.08 billion RMB	4.37 billion RMB

Although the amount of steel for the LA composite bridge scheme is increased, the comprehensive unit price is lower than that of the traditional steel bridge because of the simple structure, convenient processing and manufacturing and less welding of the LA composite bridge scheme. Finally, the project investment cost of the composite bridge scheme is 4.37 billion RMB, which is only 7.1% more than that of the concrete bridge scheme.

4.2.2. Life-Cycle Cost

When the project investment cost is further extended to the life-cycle range of the bridge, the life-cycle cost of the two schemes is almost equivalent. Firstly, the tunable-stiffness bearings are adopted in the CHEP, which have the same life as the bridge and do not need to be replaced every 15~20 years like the traditional bearings. Therefore, the tunable-stiffness bearings technology can save a rear bearing replacement cost of about 86.4 million RMB within the 100-year life-cycle of the bridge (as shown in Table 5). Secondly, the uncoated weathering steel is so adopted for all the components of the Chengle project, which does not need to be re-coated every 15~20 years like the traditional steel bridge, so the coating cost can be saved, which is about 66.2 million RMB during the life-cycle of the bridge (as shown in Table 6). Furthermore, at the end of the bridge's life, the steel material of the LA composite bridge scheme can be reused again and has an additional residual value compared to the concrete bridge scheme. Hence, the LA composite bridge scheme replacing the concrete bridge scheme has good economy and has the potential of large-scale popularization and application.

Table 5. The cost of replacing bearings during the life-cycle of the bridge.

Bridge Scheme	Number of Bearings	Replacement Times during Life-Cycle	Cost (RMB)
Concrete bridge scheme	1.80×10^4	6	86.4 million
LA composite bridge scheme	3.25×10^4	0	0

Table 6. The cost of re-coating during the life-cycle of the bridge.

Bridge Scheme	Coating Area (m^2)	Re-Coating Times during Life-Cycle	Cost (RMB)
Concrete bridge scheme	57.56×10^4	6	66.2 million
LA composite bridge scheme	0	0	0

4.3. Social Benefit Contribution

The main social benefit contributions of the LA composite bridge scheme are as follows:

- The LA composite bridge scheme is proposed to provide a feasible path for the development of low-carbon bridge technology. Compared with the concrete bridge scheme, the concrete consumption of the LA composite bridge scheme is reduced by 80.2×10^4 m^3, and the steel consumption including rebars and prestressed wires increases only by 11.79×10^4 t. The final carbon emissions considering the superposition of various materials are reduced by about 16×10^4 t (without accounting for other energy consumption reductions), which is equivalent to the carbon dioxide absorption of 1 km^2 of forest for 4.3 years.
- The LA composite bridge scheme minimizes the interference of bridge construction activities on the surrounding environment. Any supports and formworks do not need to be erected during the construction, so as not to interrupt traffic and occupy land. The whole bridge adopts self-compacting concrete without vibration operation, reducing the noise during construction. And the number of expansion joints is halved compared with the traditional bridge, meaning the driving comfort associated with driving noise are greatly improved.

- The standardized and simple structure form is applied to the LA composite bridge scheme, which is especially convenient for intelligent processing and manufacturing and can promote the industrialization development of steel-structure bridges. Eighty percent of the workload is performed in the factory, which can liberate the intensive labor input and improve the working environment and work intensity of the workers.

5. Engineering Application

5.1. Structural Description

Before the large-scale application of the LA composite bridge scheme in the CHEP, its technical feasibility was tested and verified by the construction of the testing bridge, the Pengshan Qinglong Bridge in Meishan, Sichuan, China. As shown in Figures 17 and 18, in order to achieve a highly standardized design, the span of all the main girders is 30 m, with a total length of 638 m and a standard deck width of 13 m. The maximum length of the main girder between two expansion joints reaches 270 m. The steel structure of the whole bridge is made of UWS and the concrete used for the whole bridge is SCC. The main design parameters of various components are as follows:

- DGH composite girder: the girder height of 1 m, the deck thickness of 0.3 m and the girder spacing of 1.76 m.
- PCSB capping beam: the length of 13.6 m and the girder height of 1 m.
- CFST pier: the diameter of 1.0 m and the average height of 9 m.
- SPRC pile: the diameter of 1.3 and the average length of 27 m.

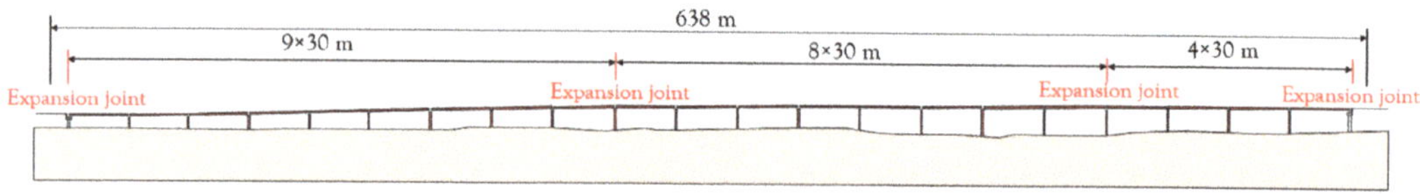

Figure 17. Elevation view of the Pengshan Qinglong Bridge.

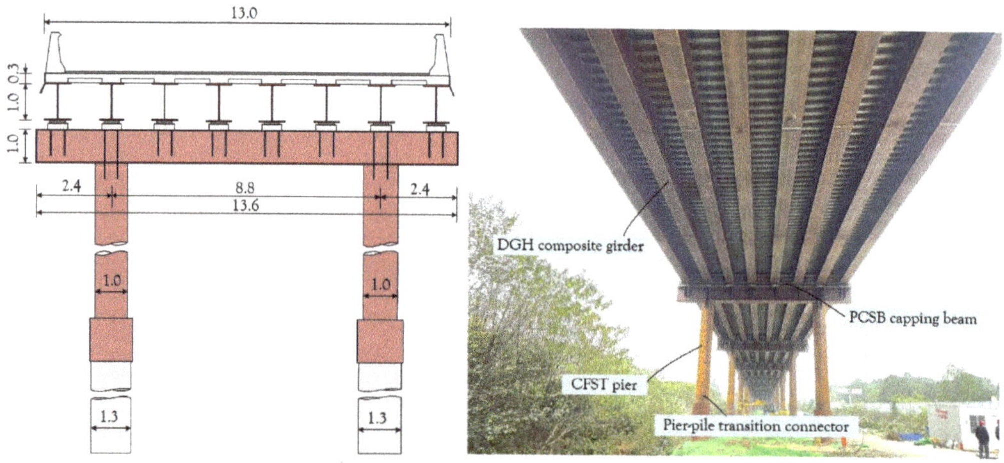

Figure 18. Cross-section view of the Pengshan Qinglong Bridge (unit: m).

5.2. Construction Procedure

The main components of the LA composite bridge scheme are manufactured in the steel structure factory and only the connection of the assembly interfaces between the components and a small amount of SCC need to be completed on site, which minimizes the amount of on-site work for ABC and shortens the interference time to the surrounding environment and traffic. The Pengshan Qinglong Bridge began construction in July 2022, completed in December 2022 and opened to traffic in February 2023. Excluding the impacts of no labor to carry out construction activities during COVID-19 and the local government's requirement to suspend construction during periods of heavy air pollution, the effective construction time is only 105 days, which saves more than 30% of the construction time compared with the traditional concrete bridge scheme. As shown in Figures 19 and 20, the main construction procedures of the Pengshan Qinglong Bridge are as follows:

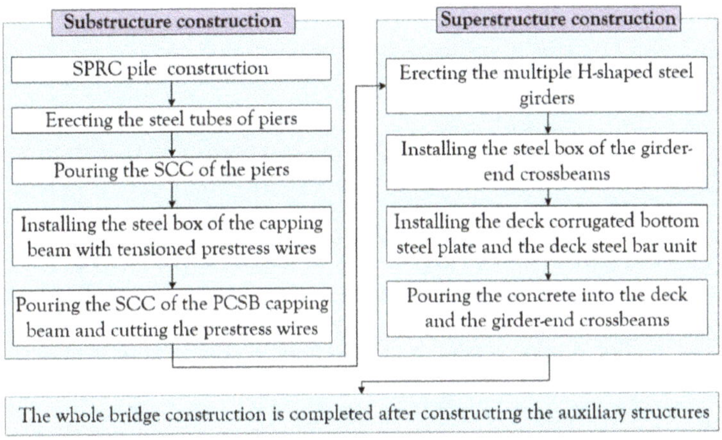

Figure 19. The main construction procedure of the Pengshan Qinglong Bridge.

Figure 20. The onsite construction process of the Pengshan Qinglong Bridge.

Step 1: After the construction of the SPRC pile foundation, the steel tube of the CFST pier is lifted to the top of the pile foundation and aligned with the pile foundation and then the pier–pile transition connector is constructed. Finally, the SCC is poured from the top of the pier into the pier–pile transition connector and the interior of the CFST pier until it is poured to 0.6 m away from the top of the pier.

Step 2: The whole segment of the PCSB capping beam is transported from the factory to the site and is lifted to the pier-top area after stretching the prestressed wires on the ground under the bridge so that the PBLs arranged at the bottom of the capping beam can be accurately positioned and inserted into the range of 0.6 m at the top of the pier. Then, the connection between the capping beam and the pier is realized by welding the steel tube of the pier and the bottom plate of the capping beam.

Step 3: The concrete inside the capping beam and within 0.6 m of the pier top is poured. After the concrete strength reaches the design requirements, the prestressed wires are cut to form a composite capping beam structure with the superposition of the steel box, concrete and prestresses.

Step 4: After multiple H-shaped steel main girders are lifted and placed on the PCSB capping beam, the steel boxes of the girder-end crossbeams are installed to realize the temporary transverse connection between the multiple main girders. Then, the deck corrugated bottom steel plates are hoisted and placed on the top surface of the main girder and welded with the main girder. After the prefabricated units of the steel bar of the bridge deck are installed, the concrete for the composite bridge deck and the girder-end crossbeams is poured to complete the construction of the bridge superstructure.

Step 5: the whole bridge construction is completed after constructing the bridge auxiliary structures.

5.3. *Real-Scale Bridge Testing*

After the completion of the Pengshan Qinglong bridge, the real-scale bridge testing including static and dynamic testing was carried out to verify the safety of the bridge and the rationality of the design and calculation method, mainly involving the strength, stiffness and dynamic performance of the main girder.

5.3.1. Static Loading Test

The static loading test was carried out on the 10th span of the Qinglong Bridge and two testing conditions of central loading and eccentric loading considering the overload scenario were designed, respectively (as shown in Figure 21). Figure 22 shows the longitudinal stress of the bottom plate of the H-shaped girder under the mid-span section, which reflects the structural strength. Under Static Testing Condition 1 (the worst-case condition), the stress of the 1#-8# girders gradually decreased and the maximum stress of 90.64 MPa existed in the 1# side girder. Meanwhile, the maximum stress of 49.03 MPa existed in the 1# side girder under Static Testing Condition 2. Even under the overload scenario, the maximum stress is less than the allowable stress of the steel after superimposing the dead load. In addition, the measured stress values are less than the calculated values and no residual stress was found in the structure after unloading, which shows that the structural strength meets the relevant specifications.

Figure 23 shows the vertical deformation of the H-shaped girder under the mid-span section, which reflects the structural stiffness. The maximum vertical deformation of 24.35 mm exists in the 1# side girder under Static Testing Condition 1, which is less than 1/600 of the span of the girder. Similarly, the measured deformation values are less than the calculated values, which means there is adequate structural stiffness.

5.3.2. Dynamic Testing

The dynamic testing mainly includes structural natural vibration frequency testing and all of the spans of the Qinglong Bridge are tested. As shown in Table 7, the measured natural vibration frequency of each span is obtained by the spectrum curve of the pulsating signal.

The minimum measured frequency is 3.32 Hz, which is 1.10 times that of the calculation value. The average measured frequency is 3.57 Hz, which is 1.19 times that of the calculation value, which means a good structural stiffness and dynamic performance. In addition, to evaluate the dynamic performance objectively, the natural vibration frequencies of the steel truss beam, the T-shaped, I-shaped and box-shaped concrete beams, the continuous concrete beam and the steel–concrete composite box beam are also measured and collected for comparative analysis. As shown in Figure 24, the natural vibration frequency of the proposed LA composite bridge is consistent with the concrete girder bridge and the steel girder bridge of the same scale.

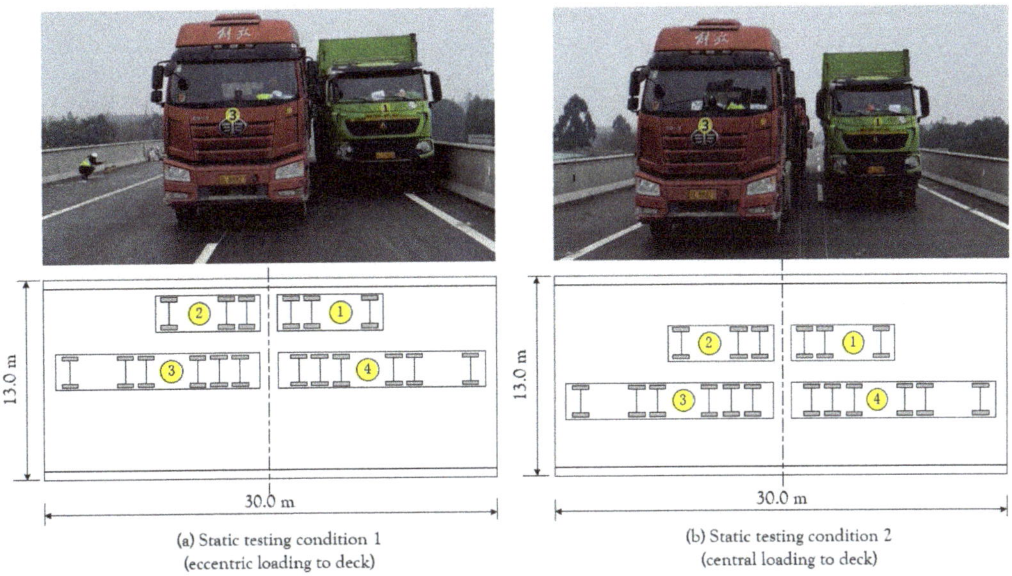

Figure 21. Condition design of static testing. (Remarks: each of 1# and 2# vehicle weighs 40 t, while each of 3# and 4# vehicle weighs 55 t).

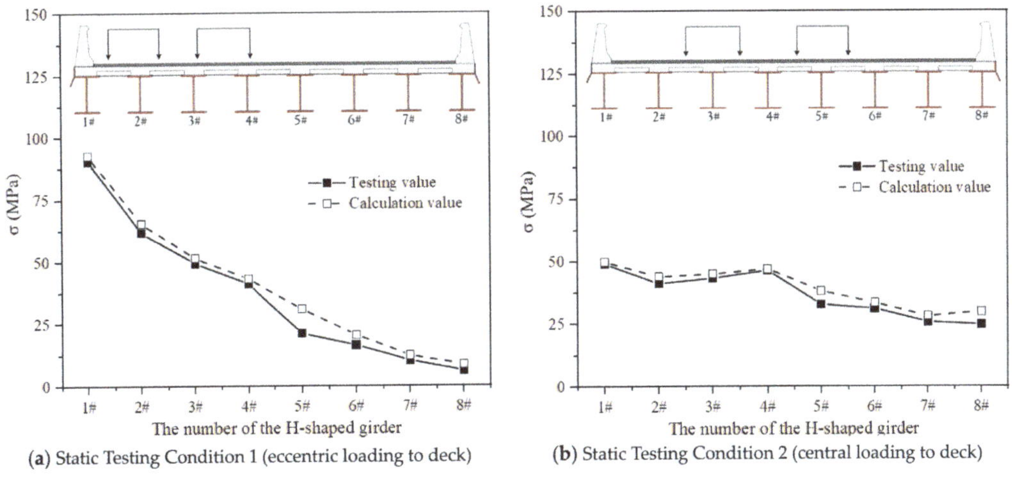

Figure 22. Longitudinal stress of the bottom plate of the H-shaped girder under mid-span section.

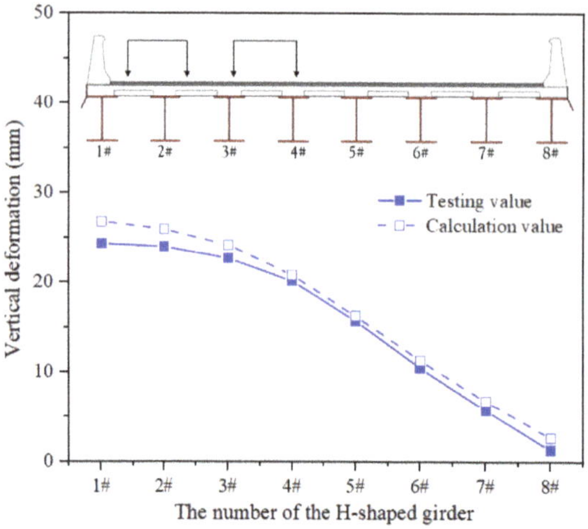

(**a**) Static testing condition 1 (eccentric loading to deck)

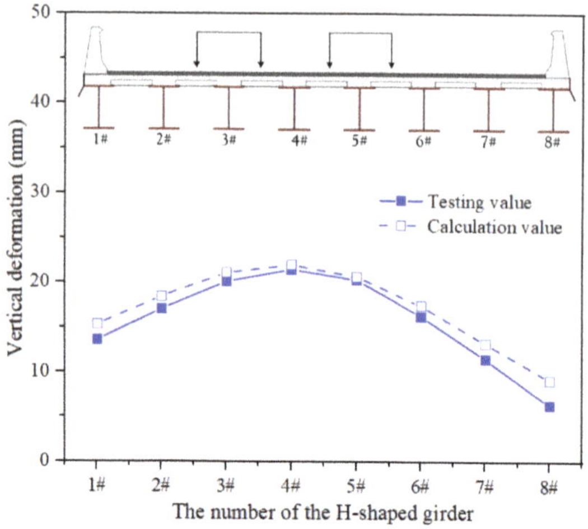

(**b**) Static testing condition 2 (central loading to deck)

Figure 23. Vertical deformation of the H-shaped girder under mid-span section.

Table 7. Results of natural vibration frequency testing.

Span No.	The Measured Value f_1 (Hz)	The Calculation Value f_2 (Hz)	f_1/f_2
1–3, 8–9	3.32	3.01	1.10
4–5, 7, 10–11, 14–15, 17	3.52		1.17
6, 12–13, 20–21	3.71		1.23
16, 18–19	3.91		1.30

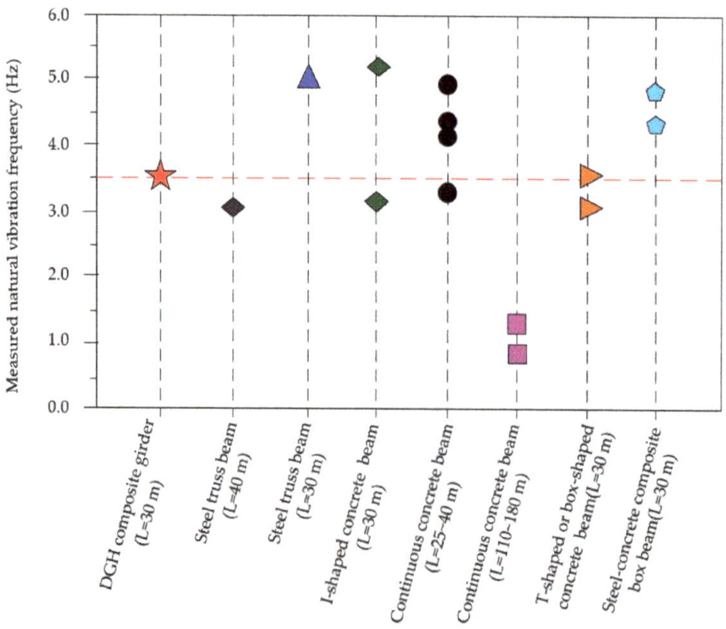

Figure 24. Testing results of natural vibration frequency for various bridge structural systems.

5.3.3. Long-Term Performance Observation

As of February 2024, the Pengshan Qinglong Bridge has been in normal service for one year. The researchers went to the site for a total of 15 times to inspect and monitor the structural performance of the bridge and no abnormalities were found in the bridge appearance. In addition, the structural strength and stiffness performance were not attenuated. The successful construction of the Pengshan Qinglong Bridge provides strong design, construction and management experience for the construction of the viaducts within the Chengdu urban area of the CHEP.

6. Conclusions

This study proposes a novel LA composite bridge scheme for ABC through the collaborative application of new steel–concrete composite structures, rapid-assembly interface designs and low-carbon material technologies. Compared with the traditional ABC scheme, the proposed scheme has less material consumption, lower carbon emissions, lighter components, reliable structural safety and an acceptable engineering cost.

(1) Based on the urban viaduct project of the CHEP, this paper comprehensively compares the advantages and disadvantages of the concrete bridge scheme, the P-ACB scheme, the ASB scheme and the LA composite bridge scheme in terms of the construction speed, environmental interference, equipment input and labor input and, finally, the LA composite bridge scheme is recommended. In this scheme, the structural design, stress and installation characteristics of the DGH composite girder, PCSB capping beam and CFST pier are described in detail and large-scale low-carbon materials and technologies such as the SCC, the UWS and the tunable-stiffness bearings technology are introduced into the scheme for a better LA goal.

(2) The proposed scheme is evaluated comprehensively from the aspects of the construction sustainability performance, comprehensive economic performance and social benefit contribution. The LA composite bridge scheme achieves an excellent balance between various performance indicators such as low carbon and assembly. In addition, the proposed scheme has a good economic performance owing to less material consumption and no need for coating and bearing replacements during the life-cycle.

(3) The proposed scheme is applied to a real-scale bridge and the whole construction process only experienced 105 days of effective time, accompanied with slight environmental interference and construction noise and a small amount of labor and equipment inputs. The safety of the bridge, the rationality of the design concept and the calculation method are verified by performing the static loading test, dynamic loading tests and long-term performance monitoring.

Author Contributions: Conceptualization: L.K., T.M. and H.W.; Methodology: L.K., T.M. and H.W.; Software: J.X. and P.Z.; Validation: P.Z. and H.W.; Formal analysis: J.X., T.M. and P.Z.; Investigation: L.K. and H.W.; Resources: T.M.; Data curation: J.X.; Writing—original draft: L.K.; Writing—review and editing: T.M., H.W., J.X. and P.Z.; Visualization: L.K., H.W. and P.Z.; Supervision: L.K.; Project administration: J.X.; Funding acquisition: J.X. All authors have read and agreed to the published version of the manuscript.

Funding: This research was funded by the Transportation Scientific and Technology Project of the Sichuan Province, China [Grant No. 2022-ZL-07, 2022-ZL-08], and the Sichuan Science and Technology Program, China [Grant No. 23JBGOV0009].

Data Availability Statement: Data is contained within the article.

Conflicts of Interest: Authors Ling Kang, Tingmin Mu and Huan Wang are employed by the Sichuan Highway Planning, Survey, Design and Research Institute Ltd. Authors Jinhua Xu and Ping Zhao are employed by the Sichuan Chengle Expressway Co., Ltd.

Abbreviations

ABC	Accelerated Bridge Construction
LA	Low-carbon and Assembled
P-ACB	Precast-Assembled Concrete Bridge
ASB	Assembled Steel Bridge
CHEP	Chengle Highway Expansion Project
DGH	Dense-Girders–H-shape
PCSB	Prestress-assisted Concrete-filled Steel Box
CFST	Concrete-Filled Steel Tube
SPCR	Single-Pile Reinforced Concrete
SCC	Self-Compacting Concrete
UWS	Uncoated Weathering Steel

References

1. Bai, Y.; Wang, X.; Wang, X.; Wang, H.; Tchuente, N.F.C.; Wu, W. Pavement Strategy Optimization of Cable-Stayed Bridges against the Negative Reaction Risks of Auxiliary Piers. *Appl. Sci.* **2023**, *13*, 4877. [CrossRef]
2. Li, H.; Tao, Z.; Han, D. Finite Element Analysis of Axial Compression Behavior of L-Shaped Concrete-Filled Steel Tubular Columns with Different Combinations. *Buildings* **2024**, *14*, 730. [CrossRef]
3. Wang, X.; Fei, P.; Dong, Y.; Wang, C. Accelerated construction of self-anchored suspension bridge using novel tower-girder anchorage technique. *J. Bridge Eng.* **2019**, *24*, 05019006. [CrossRef]
4. Shu, J.; Ma, H.; Ding, W.; Jin, Z. Improved Evidence Fusion Theory for the Safety Assessment of Prestressed Concrete Bridges. *Buildings* **2024**, *14*, 1144. [CrossRef]
5. Wang, X.; Wang, H.; Sun, Y.; Liu, Y.; Liu, Y.; Liang, P.; Bai, Y. Fault-tolerant interval inversion for accelerated bridge construction based on geometric nonlinear redundancy of cable system. *Automat. Constr.* **2022**, *134*, 104093. [CrossRef]
6. Wang, X.; Wang, F.; Zhao, J.; Qi, Z.; Wang, H.; Li, P.; Tao, P. Machine learning-based assembly fault-tolerant interval inversion method for stay cables. *J. Harbin Inst. Technol.* **2023**, *55*, 60–69.
7. Tazarv, M.; Saiidi, M.S. UHPC-filled duct connections for accelerated bridge construction of RC columns in high seismic zones. *Eng. Struct.* **2015**, *99*, 413–422. [CrossRef]
8. Kim, D.; Kwak, Y.; Sohn, H. Accelerated cable-stayed bridge construction using terrestrial laser scanning. *Automat. Constr.* **2020**, *117*, 103269. [CrossRef]
9. Zhang, G.; Su, S.; Han, Q.; Xu, K.; Li, Z.; Du, X. Experimental and numerical investigation of seismic performance of prefabricated double-column piers used in accelerated bridge construction. *Eng. Struct.* **2023**, *293*, 116688. [CrossRef]
10. Kagioglou, P.; Katakalos, K.; Mitoulis, S.A. Resilient connection for accelerated bridge constructions. *Structures* **2021**, *33*, 3025–3039. [CrossRef]
11. Xu, Y.; Wei, K.; Zhang, Z.; Wang, J.; Mao, H.; Fu, W. Seismic Behavior of Precast Bridge Column–Cap Beam Joints with Grouted Corrugated Duct Connections: Experimental and Numerical Study. *J. Bridge Eng.* **2022**, *28*, 04022153. [CrossRef]
12. Jiao, C.; Liu, W.; Li, Y.; Shi, W.; Long, P. Experimental study of CFST embedded precast concrete bridge column-foundation connection with studs. *Soil Dyn. Earthq. Eng.* **2023**, *168*, 107826. [CrossRef]
13. Xia, B.; Xiao, J.; Ding, T.; Guan, X.; Chen, J. Life cycle assessment of carbon emissions for bridge renewal decision and its application for Maogang Bridge in Shanghai. *J. Clean. Prod.* **2024**, *448*, 141724. [CrossRef]
14. Yu, F.; Ma, K.; Yuan, B.; Liu, Z.; Guo, Y.; Wang, Y. Experimental study of a new assembled integral concrete–steel open-web sandwich plate composite bridge. *Eng. Struct.* **2022**, *272*, 115018. [CrossRef]
15. Zhang, C.; Yin, Z.; Qin, R. Attention-Enhanced Co-Interactive Fusion Network (AECIF-Net) for automated structural condition assessment in visual inspection. *Automat. Constr.* **2024**, *159*, 105292. [CrossRef]
16. Nguyen, D.-C.; Jeon, C.-H.; Roh, G.; Shim, C.-S. BIM-based preassembly analysis for design for manufacturing and assembly of prefabricated bridges. *Automat. Constr.* **2024**, *160*, 105338. [CrossRef]
17. Xiao, F.; Hulsey, J.L.; Balasubramanian, R. Fiber optic health monitoring and temperature behavior of bridge in cold region. *Struct. Control Health Monit.* **2017**, *24*, e2020. [CrossRef]
18. Xiao, F.; Hulsey, J.L.; Chen, G.S.; Xiang, Y. Optimal static strain sensor placement for truss bridges. *Int. J. Distrib. Sens. N.* **2017**, *13*, 155014771770792. [CrossRef]
19. Xiao, F.; Chen, G.S.; Hulsey, J.L.; Dolan, J.D.; Dong, Y. Ambient loading and modal parameters for the Chulitna River Bridge. *Adv. Struct. Eng.* **2016**, *19*, 660–670. [CrossRef]
20. Xiao, F.; Chen, G.S.; Hulsey, J.L. Monitoring Bridge Dynamic Responses Using Fiber Bragg Grating Tiltmeters. *Sensors* **2017**, *17*, 2390. [CrossRef]
21. Zheng, H.; Gong, Z.; Cui, C.; Chen, Z.; Ma, X.; Li, H.; Li, Y. Experimental study on oblique-shear performance of splicing joint for precast ultrahigh-performance concrete segmental bridges. *Case Stud. Constr. Mat.* **2023**, *19*, e02336. [CrossRef]
22. Li, X.-X.; Deng, E.; Wang, Y.-W.; Ni, Y.-Q. 3D laser scanning for predicting the alignment of large-span segmental precast assembled concrete cable-stayed bridges. *Automat. Constr.* **2023**, *155*, 105056. [CrossRef]

23. Wang, Z.; Wang, J.; Zhao, G.; Zhang, J. Numerical study on seismic behavior of precast bridge columns with large-diameter bars and UHPC grout considering the bar-slip effect. *Bull. Earthq. Eng.* **2020**, *18*, 4963–4984. [CrossRef]
24. Shafieifar, M.; Farzad, M.; Azizinamini, A. Investigation of a Detail for Connecting Precast Columns to Precast Cap Beams Using Ultrahigh-Performance Concrete. *J. Bridge Eng.* **2020**, *25*, 04020001. [CrossRef]
25. Li, S.; Zhao, T.; Alam, M.S.; Cheng, Z.; Wang, J.-Q. Probabilistic seismic vulnerability and loss assessment of a seismic resistance bridge system with post-tensioning precast segmental ultra-high performance concrete bridge columns. *Eng. Struct.* **2020**, *225*, 111321. [CrossRef]
26. Yang, C.; Zhang, L.; Zhang, Z.; Cao, X.; Khan, I.; Deng, K.; Xu, T. Effective stress-strain relationship for grouted sleeve connection: Modeling and experimental verification. *Eng. Struct.* **2020**, *210*, 110300. [CrossRef]
27. Liu, Y.; Li, X.; Zheng, X.; Song, Z. Experimental study on seismic response of precast bridge piers with double-grouted sleeve connections. *Eng. Struct.* **2020**, *221*, 111023. [CrossRef]
28. Zhanghua, X.; Jiping, G.; Youqin, L.; Faqiang, Q. Shake table study on precast segmental concrete double-column piers. *Earthq. Eng. Eng. Vib.* **2020**, *19*, 705–723. [CrossRef]
29. Li, S.; He, S.; Jin, Y. Scour Depth Determination of Bridge Piers Based on Time-Varying Modal Parameters: Application to Hangzhou Bay Bridge. *J. Bridge Eng.* **2017**, *22*, 04017107. [CrossRef]
30. Lu, W.J.; Li, B.; Hou, J.F.; Xu, X.W.; Zou, H.F.; Zhang, L.M. Drivability of Large Diameter Steel Cylinders During Hammer-Group Vibratory Installation for the Hong Kong–Zhuhai–Macao Bridge. *Engineering* **2023**, *20*, 180–191. [CrossRef]
31. Tamai, S.; Yagata, Y.; Hosoya, T. New technologies in fabrication of steel bridges in Japan. *J. Constr. Steel Res.* **2002**, *58*, 151–192. [CrossRef]
32. Liu, H.; He, X.; Gao, L.; Li, F. Numerical Simulation of the Wind Field of Assembled Steel Truss Bridge. *Appl. Mech. Mater.* **2013**, *340*, 841–847. [CrossRef]
33. Martínez-Muñoz, D.; García, J.; Martí, J.V.; Yepes, V. Discrete swarm intelligence optimization algorithms applied to steel–concrete composite bridges. *Eng. Struct.* **2022**, *266*, 114607. [CrossRef]
34. Campagne, F.V.L. Efficiently Including Reclaimed Steel Elements in a Truss Bridge Design by Performing a Stock-Constrained Shape and Topology Optimization. Master's Thesis, Delft University of Technology, Delft, The Netherlands, 2 December 2022.
35. Aydın, Z. Size, layout and tendon profile optimization of prestressed steel trusses using Jaya algorithm. *Structures* **2022**, *40*, 284–294. [CrossRef]
36. Hui, J.-Z.; Lei, J.-Y.; Wang, Y.-F.; Zhang, F.-Q.; Ding, K.; Zhang, J.-L. Multiattribute Decision-making Evaluation Method for Highway Steel Bridge Factory Manufacturing. *China J. Highw. Transp.* **2021**, *34*, 99–108.
37. Zhang, C.; Liu, Y.; Liu, J.; Yuan, Z.; Zhang, G.; Ma, Z. Validation of long-term temperature simulations in a steel-concrete composite girder. *Structures* **2020**, *27*, 1962–1976. [CrossRef]
38. Li, C.; Zhou, Z.; Wang, H.; Gao, Y.; Fan, L. Numerical Investigation on Novel Shear Connectors in Prefabricated Composite Beams. *Appl. Sci.* **2023**, *13*, 6130. [CrossRef]
39. Martínez-Muñoz, D.; García, J.; Martí, J.V.; Yepes, V. Deep learning classifier for life cycle optimization of steel–concrete composite bridges. *Structures* **2023**, *57*, 105347. [CrossRef]
40. Wang, X.; Mao, X.; Frangopol, D.M.; Dong, Y.; Wang, H.; Tao, P.; Qi, Z.; Tang, S. Full-scale experimental and numerical investigation on the ductility, plastic redistribution, and redundancy of deteriorated concrete bridges. *Eng. Struct.* **2021**, *234*, 111930. [CrossRef]
41. Standardization, C.A.F.E.C. *Technical Specification for Application of Self-Compacting Concrete*; China Planning Press: Beijing, China, 2021; pp. 1–67.
42. Fu, J.-D.; Wan, S.; Yang, Y.; Su, Q.; Han, W.-W.; Zhu, Y.-B. Accelerated corrosion behavior of weathering steel Q345qDNH for bridge in industrial atmosphere. *Constr. Build. Mater.* **2021**, *306*, 124864. [CrossRef]
43. Ji, W.; Li, X.; He, J.; Zhang, X.; Li, J. Research status of monitoring, detection, and intelligent identification of weathering steel bridges. *J. Constr. Steel Res.* **2024**, *220*, 108814. [CrossRef]
44. Guo, X.; Zhu, J.; Kang, J.; Duan, M.; Wang, Y. Rust layer adhesion capability and corrosion behavior of weathering steel under tension during initial stages of simulated marine atmospheric corrosion. *Constr. Build. Mater.* **2020**, *234*, 117393. [CrossRef]
45. Administration for Market Regulation of Sichuan province C. *Technical Specification for Variable Stiffness Bearings of Highway Beam Bridge*; China Communication Press: Beijing, China, 2019; pp. 1–38.

buildings

Review

Research Progress on Shear Characteristics and Rapid Post-Disaster Construction of Narrow-Width Steel Box–UHPC Composite Beams

Yunteng Chen [1,2], Jiawei Xu [3,*], Peilong Yuan [3,4], Qiang Wang [4], Guanhua Cui [3] and Xulin Su [3]

[1] School of Civil Engineering, Shaoxing University, Shaoxing 312000, China
[2] Shaoxing Communications Investment Group Co., Ltd., Shaoxing 312000, China
[3] School of Highway, Chang'an University, Xi'an 710064, China
[4] CCCC Second Highway Engineering Co., Ltd., Xi'an 710075, China
[*] Correspondence: jiaweixu@chd.edu.cn

Abstract: The narrow-width steel box girder is an important type of steel–concrete composite bridge structure, which is usually composed of reinforced concrete wing plates, narrow steel boxes partially injected with concrete, and shear connectors that promote shear force transfer. The utilization of narrow-width steel box girders, augmented by partially filled concrete, embodies the synthesis of steel and concrete elements, fostering structural efficiency. Moreover, its attributes, including reduced structural weight, diminished vertical profile, enhanced load-bearing capacity, and augmented stiffness, have prompted its gradual integration into bridge engineering applications. In this study, the calculated values of shear strength under three current design codes were reviewed, and the shear failure phenomena and its determinants of narrow-width steel box–ultra-high-performance concrete (UHPC) composite beams under negative bending moment conditions were investigated, which were mainly determined by shear span ratio, concrete wing plate, UHPC steel fiber content, UHPC plate thickness, and transverse partition inside the box. Concurrently, this paper evaluates two innovative structural designs, including a double-narrow steel box girder and a three-narrow steel box girder. In addition, strategies to reduce crack formation under the negative bending moment of long-span continuous narrow and wide box girder abutments are discussed, and we show that this measure can effectively control the formation of cracks to support the negative bending moment zone. At the same time, the scope of the application of a narrow-width steel box girder composite bridge is reviewed, and the conclusion is that a narrow-width steel box girder is mainly used in small-radius flat-curved bridges or widened-ramp bridges with a span of 30 m or more in interworking areas and in the main line with a 60–100 m span in mountainous or urban areas. Finally, the research direction of the shear resistance of the UHPC–narrow steel box girder under negative bending moments is proposed.

Keywords: narrow steel box girder; negative moment zone; shear capacity; rapid construction

1. Introduction

As economic development advances, the demand for infrastructure construction in China is experiencing steady growth. With the continuous development of bridge construction, the number of bridges in mountainous areas is increasing day by day [1], and the rapid development of bridges after a disaster has become an important research topic. Taking the mountainous area of Sichuan, China as an example, the region is prone to frequent geological disasters, such as debris flow and earthquakes [2]. The necessity of the rapid reconstruction of bridges after a disaster is reflected in the following aspects: providing temporary shelter and infrastructure, restoring social order and economic activities, reducing the long-term impact of disasters, and improving the efficiency of emergency rescue are important links in post-disaster reconstruction. Concurrently, the advancement of highway engineering is paralleled by ongoing progress and development in the construction techniques and theoretical frameworks of bridges and tunnels [3–26].

Composite system that integrates the strengths of both steel and concrete structures [27]. Steel–concrete composite structures offer the advantage of reducing the amount of steel required for construction compared to conventional steel structures [28]. They enhance structural stiffness, durability, stability, and fire and shock resistance [29]. This approach utilizes steel beam-laminated concrete bridge panels to maximize the exceptional tensile properties of steel and the favorable compressive properties of concrete. Amidst the ongoing evolution of various steel–concrete composite beams, a novel structure has arisen: narrow steel box–concrete composite beams, depicted in Figure 1. Its main features are that the main bearing steel beam is designed into a narrow box shape, the width of the box does not exceed 1/2 of its height, and the box is connected by a beam and then superimposed to a concrete bridge panel. Figure 1 briefly illustrates the difference between the narrow steel box girder structure and the ordinary steel box structure. In the follow-up study on the influence of the width/height ratio of the box girder on the performance of the composite beam bridge, it is necessary to reduce the thickness and ensure that the filling amount of concrete is equal.

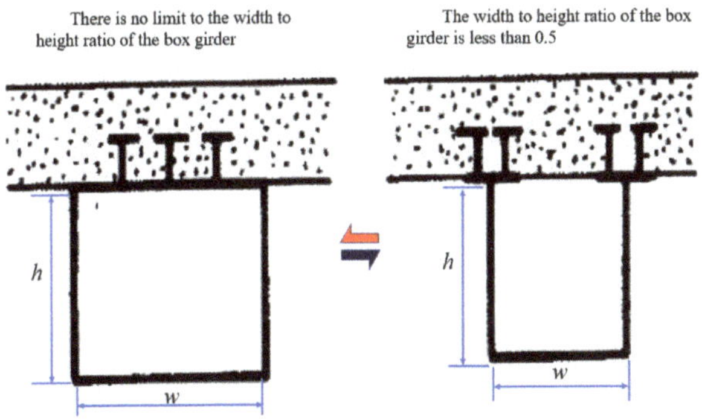

Figure 1. Comparison between concrete composite beams and narrow steel box beams.

Compared to other bridge types, narrow steel box–concrete composite bridges offer advantages such as reduced construction costs, low construction complexity, and shorter construction durations, as shown in Figure 2. This type of bridge is particularly well-suited for environmental conditions, such as mountainous terrains, where traditional construction methods may be challenging. Additionally, it holds the potential for widespread adoption and applications in the reconstruction of highways that have been severely impacted by earthquake disasters. Steel–concrete composite beam bridges offer unique advantages in mountain road construction, particularly in the aftermath of disrupted traffic conditions following an earthquake: (1) Compared to concrete structure bridges, the weight of the bridge span is significantly reduced. The steel beams utilize a separate body combination method, resulting in lightweight unit components that facilitate factory prefabrication and transportation in mountainous regions. (2) The bridge can be rapidly erected and constructed without the need for large lifting equipment. During construction, workers do not have to work in the steep gullies and riverbeds that are common in mountainous regions, leading to significant savings in support materials and labor [30]. The prefabrication of steel box girders and bridge panels can proceed concurrently with the construction of the substructure without being constrained by time or location. (3) Once the bridge pier is constructed, the installation and construction of the bridge span can proceed as usual, even during periods of flooding [31]. (4) The steel–concrete composite beam bridge makes full use of the advantages of steel and concrete, and through reasonable structural design, it can minimize the use of materials while ensuring the strength and bearing capacity, thus reducing the material cost. Compared to the traditional pure steel bridge or pure concrete bridge, the construction period of the steel–concrete composite beam bridge is relatively short. Due to the use of prefabricated components and the process of on-site assembly, the construction period can be greatly shortened, the usage time of labor and equipment can be reduced, and the construction cost can be reduced. Thanks to the aforementioned advantages of narrow-width steel box girders, they have increasingly been utilized in the rapid reconstruction efforts following earthquakes and in construction projects undertaken in challenging geographical environments, particularly in mountainous areas. The spacing between the two main I-beams of the Hospital Bridge [32] completed in France in 1990 is 12.6 m, indicating that the spacing between steel boxes can further expand the application range of the double-narrow steel box. The simple configuration of double main beams is clearly desired in both the design and construction phases. Structural analysis revealed that the structural system of a double main beam bridge, featuring a steel main beam roof with large-section stiffness and a concrete bridge panel, forms a composite structure.

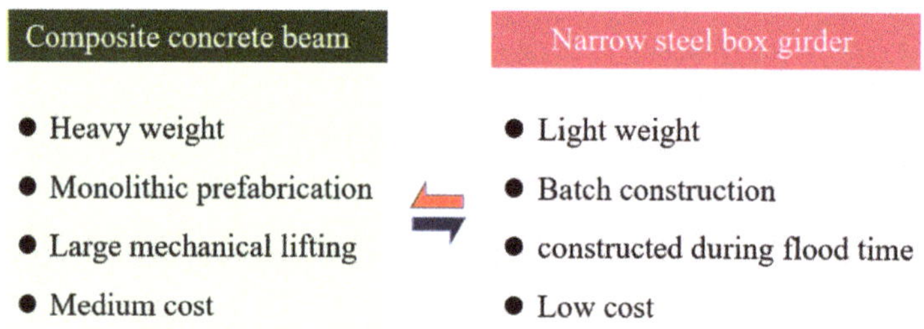

Figure 2. Advantages of the post-disaster reconstruction of the narrow-width steel box girder.

The economic span-to-height ratio of steel–concrete composite beams typically ranges from 1/20 to 1/18 [33]. This value can be further reduced with the introduction of external prestressing. The thickness of the steel box web, top, and bottom plates should be determined by considering factors such as steel strength, box height, and the presence of stiffening ribs in a comprehensive manner. Without stiffeners, the thickness of the steel plate should generally be at least 1/150 of the beam height. Longitudinal and transverse stiffeners can be added to the steel box as needed.

The introduction of narrow-width steel box girders offers a new structural option for concrete composite beam bridge systems. However, there is ongoing development in the design improvement of narrow-width steel box girders. Within the composite continuous beam system, the negative bending moment tends to concentrate close to the cross-section of the central support [34] and the conventional steel–concrete composite section of the concrete wing exhibits signs of separation and fracturing [35]. It fails to effectively fulfill its role in withstanding external forces, with the steel beam predominantly situated within the compression zone of the section, leading to overall stability concerns. Consequently, a densely packed narrow steel box–concrete composite continuous beam, illustrated in Figure 3, is proposed. This configuration typically comprises reinforced concrete wing plates, narrow steel boxes partially filled with concrete, and shear connectors facilitating shear force transmission. The densely packed narrow steel box–concrete composite continuous beam constitutes the interconnection of the concrete wing plate and the concrete-filled steel box beam via shear connectors [36]; given that the steel box adjacent to the side support, middle support, and negative moment region experiences significant shear forces, it may be infused with concrete to augment the coaction between the steel box and enhance the flexural and shear strength within this segment. The concrete wing slab in the negative moment zone of the densely packed narrow steel box and the concrete composite continuous beam are subjected to tensile forces. In the case of the NC (ordinary concrete) wing slab in the negative moment zone, it is suboptimal to postpone the initiation of initial cracking [37], while the concrete wing slab of the composite beam retains an ideal reserve of strength and stiffness after cracking; it is imperative to mitigate the accelerated corrosion of steel bars following cracking [38]. Minimizing crack propagation in the negative moment zone during normal usage is imperative. Moreover, the swift advancement of research on ultra-high-performance concrete (UHPC) has captivated the interest of numerous scholars owing to its exceptional mechanical properties. Mo [37] pioneered the utilization of UHPC wing plates, replacing conventional NC (ordinary concrete) wing plates, in the negative moment region of composite beams. This innovative approach effectively addressed the issue of cracking observed in narrow steel box girder wing plate concrete.

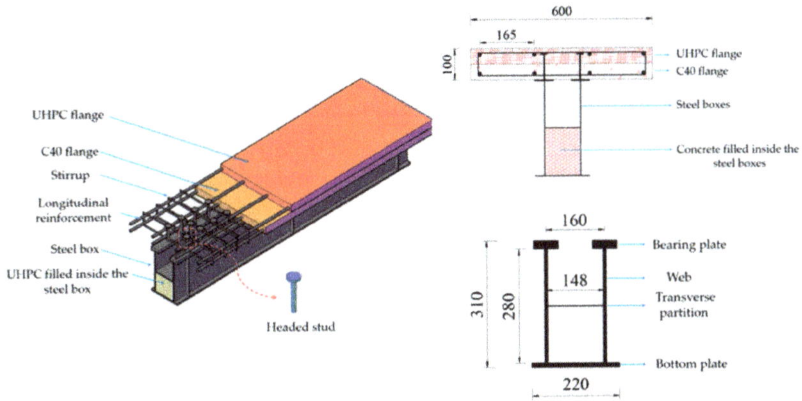

Figure 3. UHPC wing plane–narrow steel box continuous beam structure [35].

In recent years, research on the performance of ultra-high-performance concrete (UHPC) in composite beam structures has significantly advanced. Tadesse G. Wakjira et al. [39] employed the Conditional Form Generation Adjunct Network (CTGAN) and the Optuna-enhanced machine learning (ML) model to predict the stress–strain response of unconfined UHPC, accurately forecasting peak and ultimate axial stress–strain responses under different steel constraints. Roya Solhmirzaei et al. [40] proposed a data-driven ML framework to predict UHPC beam failure patterns and shear resistance, using algorithms like the support vector machine (SVM), artificial neural network (ANN), K-nearest neighbor (k-NN), and genetic programming (GP) to identify key parameters. Wakjira et al. [41] also developed a stress–strain constitutive model for UHPC using mixed ML techniques, which accurately predicts the confined UHPC stress–strain response.

The advancement of ML technology has greatly contributed to UHPC constitutive model research, aiding the design and application of narrow steel box–UHPC composite beam structures, especially in earthquake-prone mountainous areas. Despite UHPC's high performance, its initial cost can be prohibitive.

However, UHPC production can be costly due to high-cost materials like silica fume and superplasticizers, along with a substantial amount of cement. Using supplementary cementitious materials such as slag and fly ash, which are readily available, it is possible to reduce UHPC material costs. Notably, fly ash availability in North America has declined due to stricter environmental regulations on coal combustion for energy production. Furthermore, UHPC production has significant environmental impacts, necessitating a comprehensive evaluation of its overall environmental footprint. Therefore, optimizing UHPC mixtures for cost and sustainability is crucial. Wakjira et al. [42] created a reliable uniaxial compressive strength prediction model by combining tree and boost integration ML models, considering 19 objective functions, including cost and various environmental impact categories. This study offers valuable tools for designers to select optimal UHPC mixes for specific engineering needs. Reducing UHPC costs, considering sustainability indices, and optimizing mixtures under varying design requirements remain important research topics.

Numerous studies have been conducted to investigate the mechanical properties and influential parameters of steel–concrete composite structures [38–45]. Ensuring the optimal mechanical performance and durability of composite beams while operating under cracked conditions has emerged as a key research focus for numerous scholars. The predominant focus of the current research lies on steel beams featuring T-shaped or I-shaped sections [36,43,45]. Furthermore, limited research has been conducted on the interface of narrow steel box beams. In this study, we review the research progress and influencing factors related to the shear resistance in the negative moment zone of narrow-width steel box–UHPC composite beams. Additionally, we discuss the applicability and accuracy of various shear capacity codes and review the research progress on rapid construction methods following seismic events. Lastly, we present conclusions and suggestions for further research on the shear performance and rapid construction of UHPC–narrow-width steel box girders.

2. Overview of Shear Resistance Tests of Narrow-Width Steel Box–UHPC Composite Beams

2.1. Preparations for Experiments

The narrow-width steel box–UHPC composite beam primarily consists of a concrete wing plate and a steel box beam, connected through shear connectors and filled with ultra-high-performance concrete (UHPC) [46]. The production process of the narrow-width steel box–UHPC composite beam (NSBCB) structure entails several stages, including steel box fabrication, concrete filling within the box, steel mesh binding, wing mold preparation, and wing plate casting [47], as shown in Figure 4.

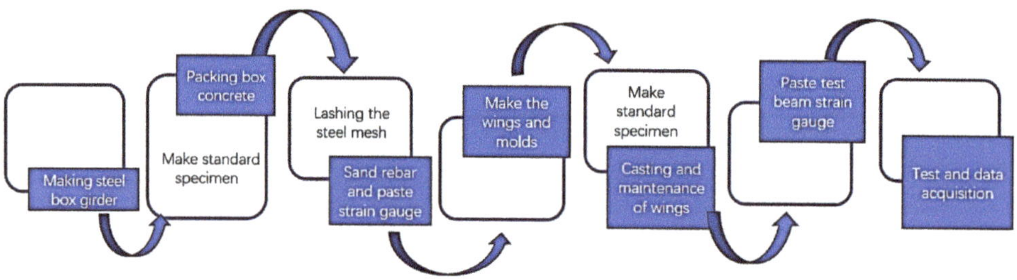

Figure 4. Specimen production process [47].

During the fabrication and welding of the steel box, flexible shear connectors, typically bolts, are used, and welding is initiated with a welding gun attached to the steel beam flange. High-strength steel plates are used to weld the steel box, and stiffeners are welded to the support area at the loading point. Liu et al. [48] found that infusing core concrete in the lower section of the negative moment zone creates a synergistic interaction between the materials, reducing compressive stress concentration and preventing inward buckling of the steel box girder.

After filling the box with concrete, a vibrator ensures uniform vibration of the core concrete. The steel box is then positioned horizontally for curing. To ensure the strength of the wing panels [49], material mixtures were strictly controlled to use the same batch of concrete. The concrete was quickly transported to the pouring platform and uniformly poured from the middle to both ends of the span. Delays in concrete preparation can cause bonding issues, reducing wing slab strength [50]. After 28 days of curing at room temperature, all specimens underwent tests for compressive strength, flexural strength, and splitting strength, as shown in Figure 5.

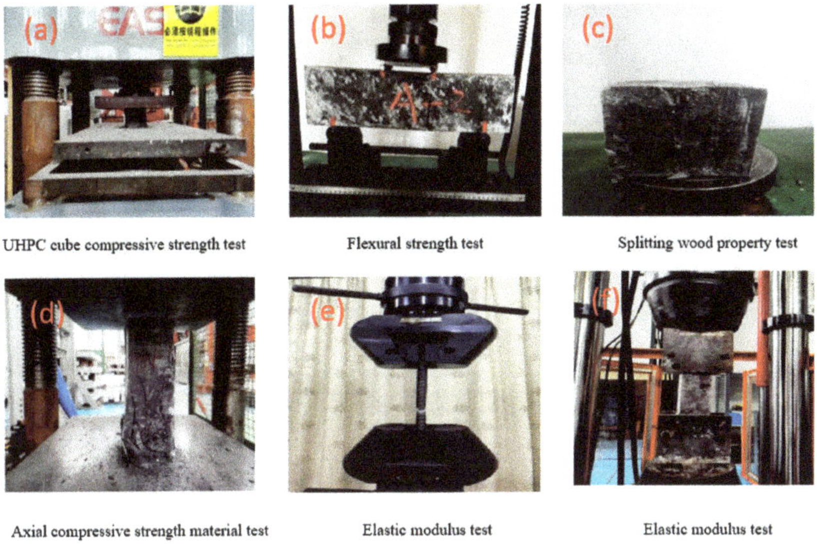

Figure 5. Material property test [50].

2.2. Experimental Tests

Prior to conducting the shear test on the narrow-width steel box–UHPC composite beam, it is essential to ascertain the strength and type of concrete employed. Concrete strength can be assessed through the examination of prepared specimens. According to He [51], augmenting the concrete strength can substantially enhance the stiffness, strength capacity, and bending moment capacity of narrow-width steel box–UHPC composite beams. The test primarily encompasses evaluations of interface slip between the wing plate and the steel beam, mid-span deflection, wing plate strain, steel beam strain, and wing plate crack propagation, among others [52].

A slip meter measures slip at the wing plate and steel box beam interface. Strain gauges measure strain distribution at various locations on the test beam. The locations include the mid-span steel box girder's web, the steel beam's web along its height, the main rib of the wing plate, the top surface of the wing span, and the mid-side of the span. A displacement meter is installed at the loading point and support position to generate the load–deflection curve [53]. Strain data are collected using a strain-collecting instrument, while displacement measurements are obtained manually. The crack distribution of the wing plate is monitored with a crack observation instrument during the test loading. Two common loading methods are used: three-point bending and four-point bending as shown in Figure 6. The four-point bending test (4PBT) is commonly used to assess shear behavior in the positive moment region, while the three-point bending method is typically used for shear tests on continuous beam bridges in practical engineering. However, some studies use the four-point bending method to comprehensively analyze the combined action of composite beam bridges [54].

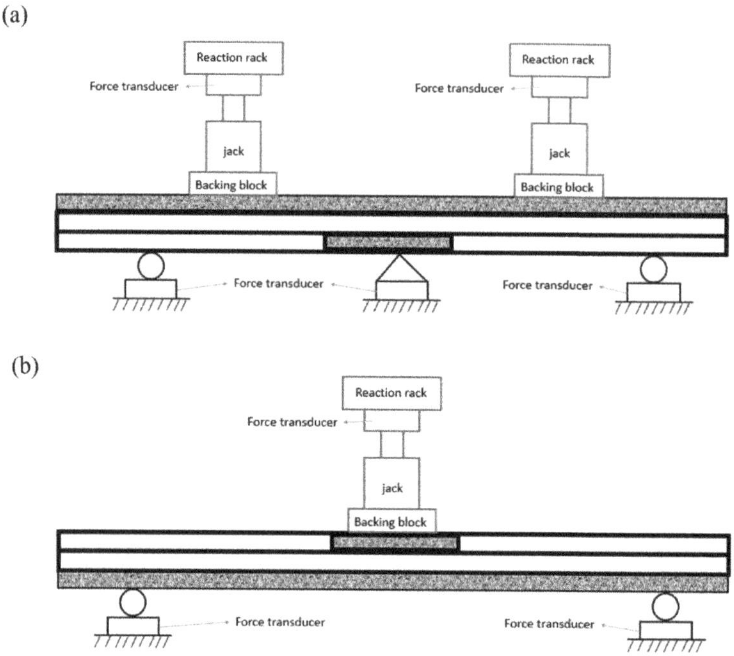

Figure 6. Loading mode and loading device (**a**) three-point bending (**b**) four-point bending [36].

2.3. Failure Modes

Steel–UHPC beams can fail in different ways based on loads, materials, and geometry. Maximum shear capacity is achieved when the wing plate and steel box fail simultaneously. Practical applications may experience overall failure due to single or multiple local failures. The literature review indicates that under negative bending moments, the failure of narrow-width steel box girder–UHPC composite beams is characterized by buckling deformation, cracks on the upper wing plate, and localized concrete crushing. The typical failure modes of narrow-width steel box girder–UHPC composite beams are summarized in Table 1. Mo et al. [37] reported that flexural shear failure occurred mainly due to shear action, and the mid-span of the wing plate was cracked by bending action. Due to the action of pegging, the lateral cracks on the wing plate developed slowly and appeared as steps between the UHPC–NC layers. The cracks in the upper part of the wing plate developed approximately equidistant from the stud spacing, which was due to the stress concentration caused by the local strengthening of the stud in the inner section of the wing plate. Shear oblique cracks appeared on the side of the wing plate at the support, accompanied by the slip between the UHPC–NC wings. The steel beam buckled at the loading point and the middle partition. Liu et al. [34] reported that the main fracture across the flange was observed on the fragile surface located at the midpoint of the test beam, and the final failure mechanism observed in the six test beams was remarkably similar. When the steel wire in UHPC was extracted, the delamination between the flange and the steel bar was observed at the crack location. Due to the relative displacement, longitudinal cracks and delamination appeared between the UHPC layer and the C40 layer of the flange, and diagonal cracks appeared on the side of the flange. In general, the test beams exhibited typical tensile fracture characteristics, such as local buckling and observable shear deformation in the lower part of the steel box girder. Xiong et al. [47] reported that filling concrete inside the steel box of UHPC composite beams can significantly improve the shear resistance of the specimens. When the steel box was not filled with concrete, the bending deformation of the box beams resulted in brittle failure of the test beams, while the other test beams had similar mechanical characteristics, resulting in bending and shear failure mainly caused by shear action; the wing plates were cracked by bending action in the middle span. According to the literature review, the main failure of the UHPC composite beam under the condition of negative moment zone is buckling deformation of the box girder, surface crack of the wing plate, and local concrete crushing, as shown in Figure 7.

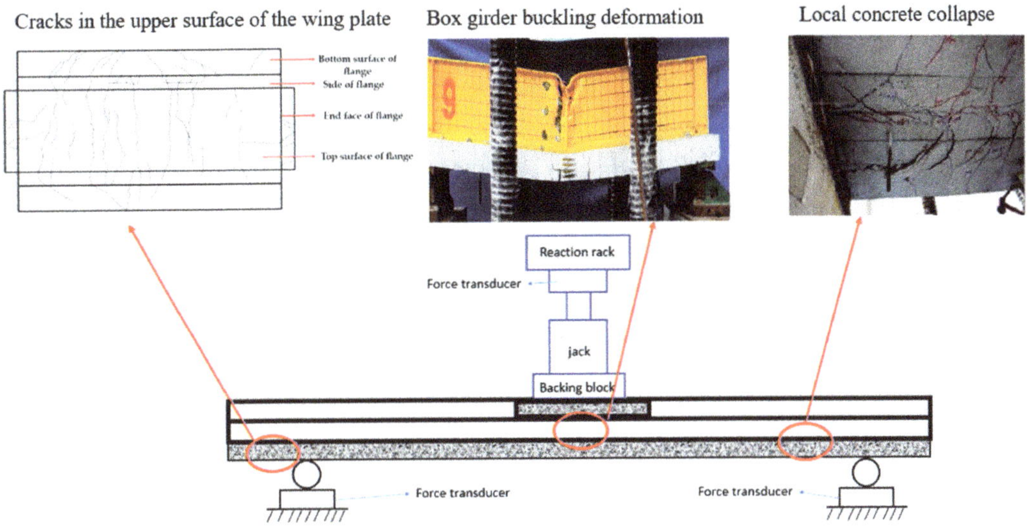

Figure 7. Typical failure mode [47].

Simultaneously, through a literature review of load–span deflection [27,34,36,37,43,47–49,51–57], we can roughly categorize the deflection curve into three stages.

(1) Elastic stage (V < Vy)

A linear relationship exists between the stage load and deflection, and the steel beam remains unyielding. However, under the negative bending moment, a few cracks emerged, and the composite beam wing showed some plastic deformation.

(2) Elastoplastic stage (Vy < V < VU)

As the load increases, the crack width of the primary crack in the middle span of the wing plate accelerates, leading to partial yielding of the steel bar and internal force redistribution within the section. Consequently, the proportion of plastic deformation in the beam increases, resulting in decreased stiffness, amplified deflection of the wing plate, and nonlinear deflection progression with increasing load. This signifies the onset of the failure stage.

(3) Plastic stage

As the bridge structure is further loaded, it enters the complete shaping stage. The full shaping stage is usually short as the deflection increases. Once in the full shaping stage, the bridge structure quickly reaches its ultimate load, and the load decreases as the deflection continues to increase beyond the ultimate load.

(4) Failure stage

Upon reaching the ultimate load, the test beam transitions into the failure stage. As the deflection of the test beam increases, the load diminishes to some extent, accompanied by the widening of the main crack. Subsequently, the steel box web plate exhibits baroclinic buckling, while the bottom plate of the steel beam bulges at the point of the concentrated force application. This leads to a sharp increase in deflection and a rapid reduction in the bearing capacity due to shear failure. Figure 8 illustrates the load–mid-span deflection curves of six specimen beams under various parameter conditions as reported by Liu et al. It is observed that upon entering the plastic stage, the stiffness of the UHPC composite beam notably decreases, with cracks appearing on the upper surface of the wing plate upon further loading (Figure 9). After entering the failure stage, the steel beam experiences partial buckling deformation (Figure 9), eventually rendering it incapable of fulfilling its intended function.

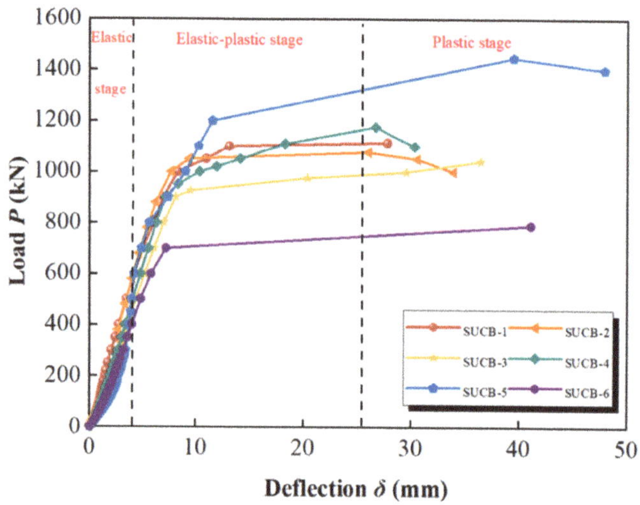

Figure 8. Load–mid-span deflection curve [34].

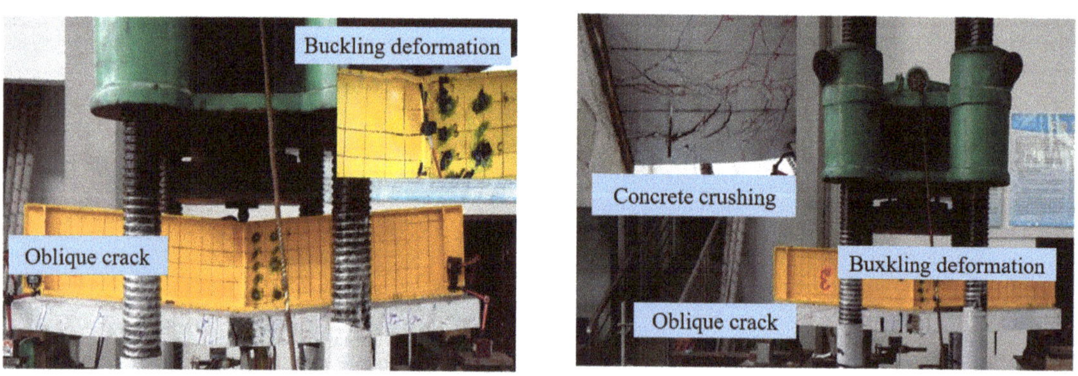

Figure 9. Failure characteristic [47].

Table 1. Typical shear failure mode of the UHPC composite beam.

References	Box Girder Concrete Filling Height/Box Girder Height	Box Girder Concrete Strength (MPa)	UHPC Wing/Wing Thickness	Loading Protocol	Comments	Typical Failure Mode
Xiong et al. [47]	0.5	38.54	0 0.5 1	Force control is used throughout. The formal loading starts at 0. Before the concrete wing cracks, the load of each stage is 10 kN after the cracking, and the load is adjusted to 50 kN. After the load is loaded, the load step is adjusted to 100 kN per stage when the load-displacement curve of the composite beam is no longer linear.	The failure characteristics include the diagonal buckling of the steel beam web and the loss of bearing capacity of the UHPC wing plate due to the pulling out of the steel fiber.	
Wang et al. [55]	0.5	45	0	The loading method of each stage is hierarchical, and the corresponding loading size of each stage is selected according to the observation requirements.	The failure stage of the test beam can be divided into the elastic stage, shaping stage, and failure stage.	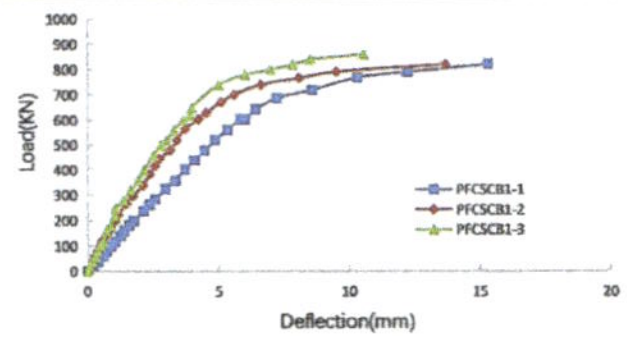
Liu et al. [34]	0.5	42	0.5	In the elastic stage, the force-controlled loading technique is used to increase the load by 10 kN per layer. After the transition to the elastoplastic stage, the load increment of each layer is increased by 50 kN.	The deflection at the middle point of the steel box girder increases the fastest, and the bottom plate and web plate of the steel box girder produce buckling deformation.	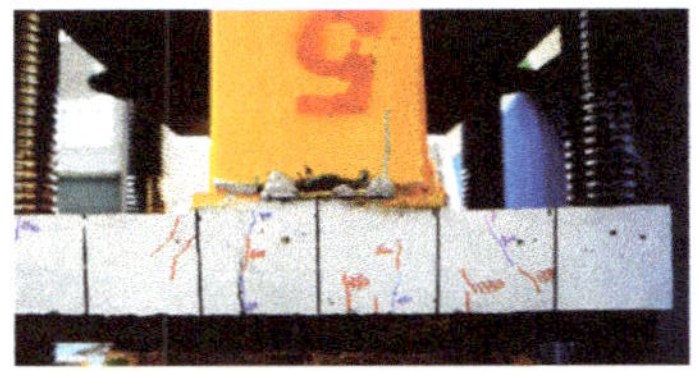

Table 1. *Cont.*

References	Box Girder Concrete Filling Height/Box Girder Height	Box Girder Concrete Strength (MPa)	UHPC Wing/Wing Thickness	Loading Protocol	Comments	Typical Failure Mode
Zhang et al. [56]	1	58	0	Force control	The load drops sharply after Vu, indicating brittle failure of the tested beam. During the whole loading process, the mid-span top concrete of all the beams was not crushed.	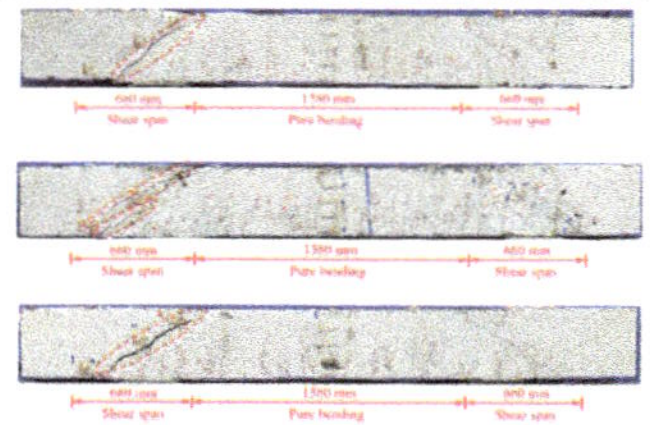
Mo et al. [37]	0.5	38.5	0 0.5 1	Loading protocol	Comments	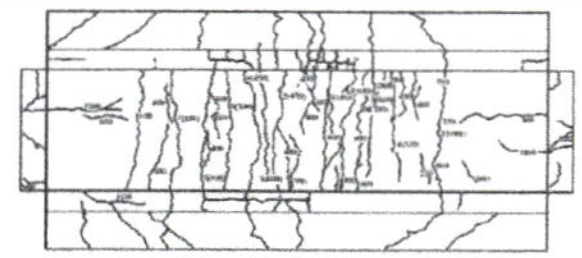

3. Review of the Shear Code Guidelines and Analysis of the Influencing Factors in the Negative Moment Zone of the Narrow-Width Steel Box–UHPC Composite Beam

3.1. Specification Guidelines for Shear Resistance in the Negative Moment Zone

To confirm the applicability of different design codes to the evaluation of the shear resistance of steel box–UHPC composite beams, we reviewed and analyzed the calculated values of three codes that are mainly used in the world today. These specifications will be used later in the shear behavior analysis.

(1) American AASHTO specification [58]

The American AASHTO bridge design code stipulates that only the contribution of composite steel beams is considered in the shear design, and the shear capacity of concrete bridge panels is ignored. Under the action of the negative bending moment, the calculation formula for composite beams with stiffened webs is as follows.

$$V_{cr} = V_p\left[C + \frac{0.07(1-C)}{\sqrt{1+(d_0+D)^2}}\right]$$

$$\frac{2D_t}{b_{fc}t_{fc}+b_n t_{ft}} \le 2.5n$$

$$V_p = 0.58\times 10^{-3}F_{yw}D_{tw},$$

$$C=\begin{cases}1.0, & \frac{D}{t_w}\le 1.12\sqrt{\frac{E_K}{F_{yw}}}\\ \frac{1.12}{D/t_w}\sqrt{\frac{E_k}{F_{yw}}}, 1.12\sqrt{\frac{E_K}{F_{yw}}}\le\frac{D}{t_w}\le 1.4\frac{E_K}{F_{yw}} & \\ \frac{1.57E_K}{\left(\frac{D}{t_w}\right)^2F_{yw}}, D/t_w & >1.4\frac{E_K}{F_{yw}}\end{cases}$$

where V_{cr} is the buckling capacity of the steel beam web, C is the ratio of the shear buckling capacity to the plastic shear capacity of the steel beam web, V_p is the plastic shear strength of the steel beam web, F_{yw} is the buckling strength of the steel beam web, D is the height of the steel beam web, t_w is the thickness of steel beam web, b_f is the height of the steel beam compression roof, t_{fc} is the thickness of the compression roof of the steel beam, tft is the thickness of the tensile roof of the steel beam, d0 is stiffener spacing for the web, and k is the shear buckling coefficient, which is set to 5 when there is no stiffener.

(2) European specification EC4 [59]

The formula for calculating the shear strength of EC4 steel-mixed composite beams considers the effect of buckling on shear performance by introducing a partial coefficient, i.e., γ_{Rd}, for local web plate buckling, with a value of 1.1. The vertical shear capacity of the steel-mixed composite beams is then determined by the formula V_b, R_d.

$$V_{b,Rd} = \frac{\tau_{ba}A_v}{\gamma_{Rd}}$$

Among them, the shear strength of the steel beam, i.e., τ_{ba}, will change with the change in the web length ratio, i.e., λ_w, after buckling.

$$\tau_{ba}=\begin{cases}\frac{0.9f_y}{\overline{\lambda_w}\sqrt{3}}, \overline{\lambda_\omega}\ge 1.2\\ \left[1-0.625\left(\overline{\lambda_w}-0.8\right)\right]\frac{f_y}{\sqrt{3}}, 0.8<\overline{\lambda_w}<1.2\\ \frac{f_y}{\sqrt{3}}, \overline{\lambda_w}<0.8\end{cases}$$

(3) Chinese Code for the Design of Steel–Concrete Composite Bridges [60]

When calculating the vertical shear capacity of composite beams, the current industry standard GB 50917-2013 [60] stipulates that only the contribution of the steel girder web to composite beams is considered, and the ultimate vertical shear capacity is calculated using the plastic limit design method.

$$\gamma_0 V \leq h_w t_w f_{vd} \times 10^{-3}$$

where γ_0 is the structural importance coefficient, V is the design value of the shear force of the composite beam, h_w is the effective height of the steel beam web, tw is the thickness of the steel beam web, and f_{vd} is the design value of the shear strength of the steel beam web.

3.2. Examination of the Experimental Results of the Narrow-Width Steel Box–UHPC Composite Beam

European and American codes for composite beams account for buckling and plastic deformation of steel beam webs, while the Chinese code only considers the plastic shear contribution and ignores local buckling effects. Consequently, the Chinese code adopts a conservative approach. A comparison shows that U.S. norms yield the highest shear capacities, while China's are the lowest, primarily due to conservative steel strength values.

The Table 2 shows that only considering the contribution of the web to the shear strength will make the calculated value of the shear strength much smaller. Among them, the calculated value provided by GB 50917-2013 is the most conservative, and the average value of the calculated value to the test value (v_c/v_t) is about 0.65, while the value (v_c/v_t) obtained by the American AASHTO specification is the largest. However, it is not difficult to observe that regardless of the above three specifications, the predicted shear strength obtained is less than 70% of the actual shear strength, which will lead to a great waste of design strength in the actual design process. At the same time, to obtain the shear strength to meet the design requirements, the box height of the composite beam will be increased, and the proportion of the bridge's self-weight load will be greatly increased, which is unfavorable to the bearing capacity of the composite beam. Therefore, it is necessary to study the calculation theory of shear strength, which is closer to practice.

Table 2. Comparison of the test values to those calculated by national norms.

ID	Test Value of Shear Capacity	American Calculated Value	Ratio	European Calculation	Ratio	Chinese Calculated Value	Ratio
1 [47]	625	312	0.4992	308	0.4928	299	0.4784
	600	312	0.52	308	0.513333	299	0.498333
	550	312	0.567273	308	0.56	299	0.543636
	650	312	0.48	308	0.473846	299	0.46
	700	312	0.445714	308	0.44	299	0.427143
	395	312	0.789873	308	0.779747	299	0.756962
2 [48]	315	312	0.990476	308	0.977778	299	0.949206
	341	312	0.914956	308	0.903226	299	0.876833
	376	312	0.829787	308	0.819149	299	0.795213
	304	312	1.026316	308	1.013158	299	0.983553
	393	312	0.793893	308	0.783715	299	0.760814
	393	312	0.793893	308	0.783715	299	0.760814
3 [54]	557	312	0.560144	308	0.552962	299	0.536804
	538.5	312	0.579387	308	0.571959	299	0.555246
	520	312	0.6	308	0.592308	299	0.575
	588.5	312	0.530161	308	0.523364	299	0.508071
	723.5	312	0.431237	308	0.425708	299	0.413269
	395	312	0.789873	308	0.779747	299	0.756962
Mean value			0.674566		0.665918		0.646459

A review of UHPC–NC composite beams and experimental studies reveals the crucial role of the wing plate in shear strength. Therefore, the shear capacity of each part should be considered and summed to obtain the total shear capacity. As Pu Zhang [56] reported, the shear carrying capacity of the UHPC–NC beam was divided into three parts, i.e., V_{NC} provided by the NC part, VUHPC provided by the UHPC formwork, and Vsg provided by the steel grid. Luaay Hussein [61] reported that the shear resistance of a UHPFRC-NSC/HSC beam is thus equivalent to the expected shear resistance of an NSC/HSC beam without shear reinforcement, i.e., V_c, plus the additional shear resistance provided by the fibers, i.e., V_f, due to the improved post-cracking resistance of the UHPFRC layer. The shear strength of UHPFRC-NSC/HSC composite beams can be calculated from $V_u = V_c + V_f$. Using the principle of partial superposition proposed by Shuai Liu [34], we formulated an equation for the numerical assessment of the influence of concrete filling on shear capacity. Men Pengfei [62] reported that it is necessary to comprehensively consider the influence of the section form of the wings and the tensile strength of concrete on the shear capacity of composite beams and proposed a calculation method for the shear capacity of the wings considering the buckling form of the web of steel beams:

$$V_{fuc} = \alpha_1 f_{tk} b_c h_c + \alpha_2 \rho f_{tk} b_c h_c = (\alpha_1 + \alpha_2 \rho) f_{tk} b_c h_c$$

Leveraging the principle of superposition calculation of the shear capacity for each component, numerous scholars have conducted additional investigations into the shear force calculation formula for narrow-width steel box girders. Nonetheless, a comprehensive and universally applicable calculation criterion remains elusive for narrow-width steel box girders. Hence, further review and study of the calculation formula for the shear capacity of narrow-width steel box girders are imperative.

3.3. Analysis of the Influencing Factors on the Shear Strength of the Narrow-Width Steel Box–UHPC Composite Beam

Similar to the traditional steel–concrete composite beam bridge structure, the factors affecting the shear resistance of narrow-width steel box–UHPC composite beams can be divided into geometric factors and material factors. Among them, a considerable number of the factors are the same as the narrow steel box girder and the concrete composite beam bridge (as shown in Figure 10), such as the shear span ratio, the strength of concrete and steel materials, and so on. Therefore, in this section, we focus on the unique factors affecting the shear strength of narrow-width steel box–UHPC composite beams (shown by the star in Figure 10).

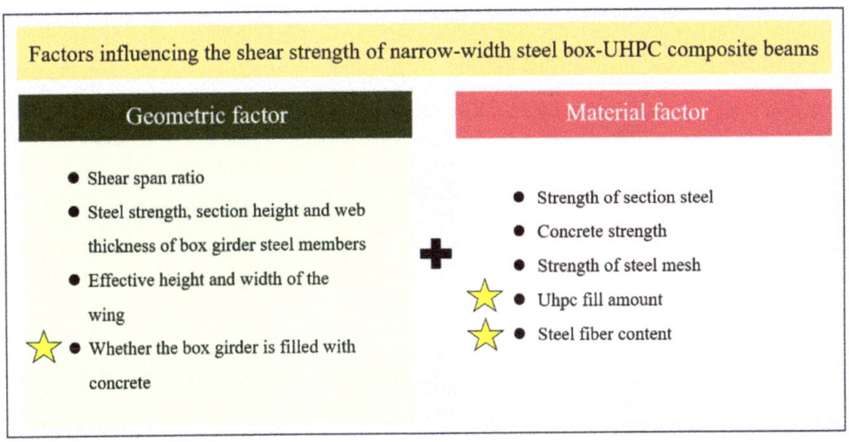

Figure 10. Influencing factors of shear resistance.

(1) Shear span ratio

The generalized shear span ratio of composite beam concrete panels closely resembles that of ordinary reinforced concrete beams. Research findings [46] indicate that for shear span ratios below 4, shear strength exhibits an inverse correlation with the shear span ratio.

(2) Steel box girder

The steel section of the composite beam significantly influences its shear strength. With all other factors held constant, increases in steel strength, section height, and web thickness correspondingly enhance the shear strength of the composite beam. In practical scenarios where the overall shear strength of the composite beam falls short of standards, adjustments typically involve augmenting the web thickness and elevating the section height of the steel beam.

(3) Concrete wing slab

The strength of concrete material also affects the shear strength of composite beams, albeit to a lesser extent than steel beams. Enhancing the effective height and width of the wings can marginally enhance shear strength. Additionally, incorporating steel mesh within the concrete slab augments the shear strength and stiffness of the composite beam. While the primary compressive stress in the concrete wing slab is borne by the steel mesh under positive bending moments, its impact on shear strength is minimal. Conversely, under negative bending moments, the concrete wing slab experiences predominantly tensile stresses, posing unfavorable conditions. However, the steel mesh effectively mitigates concrete slab cracking, thereby enhancing the cross-section stiffness and overall shear-bearing capacity of the composite beam [34]. When cracks were initiated within the flange, the steel fibers embedded in the UHPC material with diverse orientations exhibited a 'bridging effect', effectively mitigating the propagation of cracks.

In addition to the above factors, UHPC steel fiber content, UHPC board thickness, and filled concrete have a great influence on the narrow steel box–UHPC composite beam in the negative bending moment area. Through the review of the literature [36,46,54,55,59,61–65], we compiled a database of the test results, as shown in Table 2, and collected the detailed shear test data of narrow-width steel box–UHPC composite beams from 11 typical articles, as shown in Tables 1 and 2. Through a review of the above literature, we obtained the key influencing factors of narrow-width steel box–UHPC composite beams.

(1) UHPC steel fiber content

Introducing steel fibers effectively curtails wing plate cracking, leading to the development of numerous narrow and dense cracks upon failure. This demonstrates that the inclusion of steel fibers in UHPC enhances the shear strength of composite beams to a certain extent while substantially enhancing their ductility.

When the steel fiber content of UHPC changes [29], the steel fiber content increases from 0 to 2%, and the ultimate loads of the composite beams are 1300 and 1250 kN, respectively, with a difference of only 4%, as shown in Figure 11a. The steel fiber content has little effect on the bearing capacity of the composite beams. However, during the crack development of the specimen, it can be observed that the crack development speed of the test beam decreases after the addition of steel fibers. When the maximum crack width is 0.2 mm, the loads of the specimens SUCB-1 (steel fiber content of 2%) and SUCB-2 (without steel fiber addition) are 500 and 330 kN, respectively. Adding steel fibers can effectively improve the cracking resistance of composite beams.

(2) UHPC board thickness

As the composite beam is subjected to shear action mainly under the action of the negative bending moment, the concrete wing slab of the composite beam will suffer shear failure. According to the test results, when the other parameters remain unchanged and the thickness of the UHPC wing slab changes, compared to SUCB-3, the bearing capacity of SUCB-1 and SUCB-4 increased by 18.2% and 27.3% respectively, and the deformation capacity increased by 38.02% and 41.2% respectively, as shown in Figure 11b. The increase in the thickness of UHPC fins can effectively improve the mechanical properties of composite beam fins and significantly enhance the shear-bearing capacity and deformation resistance of composite beams [29].

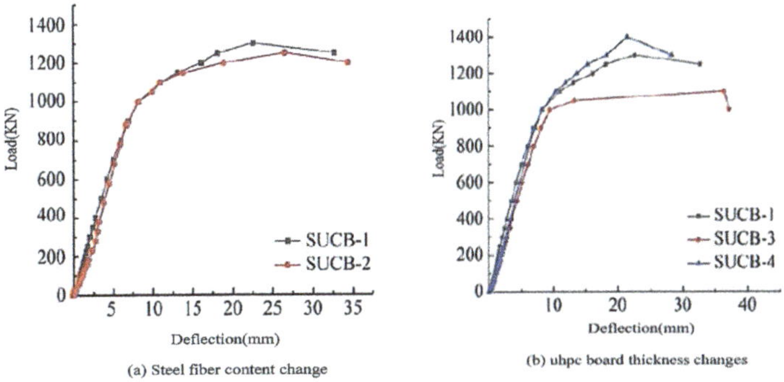

Figure 11. Influence of UHPC steel fiber content and UHPC plate thickness on shear strength [29].

(3) Filling concrete

Filling the narrow steel box with concrete improves the rigidity and stability of the composite beam and makes up for the lack of overall stability of the composite beam caused by reducing the effective width of the steel box. The load–deflection curve presents an upward trend with the increase in the UHPC height. By comparing the stiffness of composite beams at the initial loading stage, it is found that the stiffness of composite beams without UHPC filling inside steel beams is lower than that of composite beams with UHPC filling inside steel beams, as shown in Figure 12. In addition, the deformation ability of composite beams with half-filled UHPC narrow steel boxes is much greater than that of composite beams with full and unfilled narrow steel boxes. Filling UHPC inside steel boxes can not only improve the shear strength of composite beams but also enhance the deformation ability of composite beams. However, overall, half-filled UHPC narrow-width steel box composite beams are the best [43].

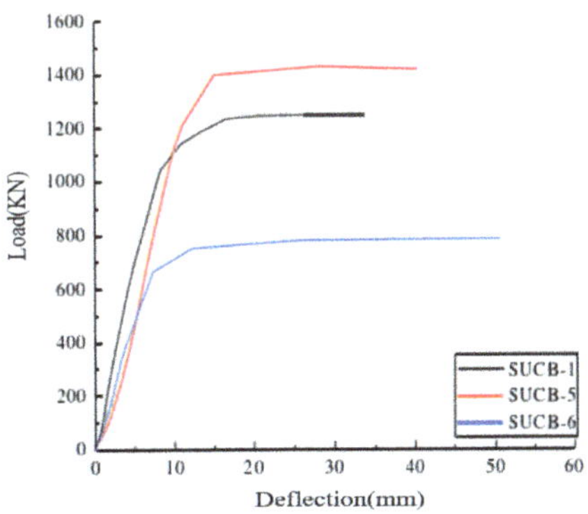

Figure 12. Effect of filling concrete on shear strength.

(4) Transverse partition in the box

The function of the transverse partition in the box is to prevent excessive deformation of the section, improve the torsion resistance of the section, and transfer the additional stress of the crossbeam between the boxes. According to the design guidelines of Japanese

Highway Steel Bridges [63], the spacing of the partitions in the bridge box is not more than 6 m. When the solid belly diaphragm is set, according to previous engineering experience and calculations, the stiffness of the diaphragm is relatively large, and the stress is generally not controlled by the design; so, the plate thickness of the cross-span diaphragm can be about 12 mm, and the fulcrum diaphragm generally needs to be thickened, which needs to be determined according to the checking calculation.

(5) Crossbeam between boxes

The function of the tinderbox beam is to maintain the structure shape during erection, reduce the relative deflection difference between the main beams, transfer the transverse load, reduce the stress of the single main beam through the transverse load distribution, and improve the torsional stiffness of the whole bridge. Because the steel–concrete composite beam is equipped with a strong concrete roof, the transverse load transfer and transverse load distribution effect of the beam are greatly reduced after erection.

(6) Shear connector

The shear connector plays a key role in combining the steel main beam and the bridge panel to work together [36]. According to existing engineering practice, stud connectors are flexible connectors with superior performance. Under the influence of concrete shrinkage, the longitudinal shear force on the joint surface of the composite bridge will increase. In addition, near the middle fulcrum, due to the large vertical shear force, the shear force of the joint surface in the local area is also large. Therefore, the beam end and the middle fulcrum should be strengthened against shear connectors. The strengthening range of the beam end should be less than 1/10 of the distance between the main beams and the span of the main beams.

In addition to confirming the factors affecting the shear strength of narrow-width steel box girders through indoor loading tests (as shown in Table 3), with the development of computer technology, using machine learning and probability statistics methods to predict the shear strength and seismic effect has gradually become an important research direction.

Tadesse G. Wakjira [64] adopted the interpretative machine learning (ML) method to obtain a prediction model of four damage states across ultra-high-performance concrete bridge columns based on the limit state of the drift ratio. The results showed that the axial load ratio and the aspect ratio were the main factors determining the drift ratio, and the higher the reinforcement grade, the smaller the vulnerability of ultra-high-performance concrete columns. However, the more serious the degree of damage is, the less influence the reinforcement grade has on the reinforcement effect.

Tadele [65] used the classification and regression tree (CART) and vulnerability analysis methods to reach a conclusion on the change of mechanical properties of reinforced concrete and its impact on the seismic design of reinforced concrete bridge columns. The results show that 0.12% of the tensile test data are lower than the minimum ASTM standard requirements, and the change in longitudinal steel yield strength significantly affects the seismic response of circular bridge columns.

Tadesse G. Wakjira [66] used a machine learning (ML) method to obtain a prediction model for the shear resistance of concrete beams reinforced by inorganic composite materials. The results showed that the Extreme Gradient enhancement (XGBoost) model was superior to the existing model in terms of accuracy, safety, and economy, and the resistance reduction factor was identified through reliability analysis. The target reliability indices were 3.5 and 4.0, respectively.

To date, the shear strength prediction based on experimental analysis, statistical analysis, and machine learning methods has been relatively comprehensive [67], but with the continuous development and improvement of narrow steel box girder structures, whether the relevant laws are applicable to new structures still needs to be reviewed and confirmed.

Table 3. Test of the influencing factors on the shear resistance of composite beams.

References	Box Girder Concrete Filling Height/Box Girder Height	Box Girder Concrete Strength (MPa)	UHPC Wing/Wing Thickness	Loading Protocol	Comments
Zhu et al. [68]	0	n/a	1	The formal loading starts at 0. Before the loading force reaches 500 kN, force-controlled loading is adopted, with 25 kN per stage.	The tribute of UHPC fins against the shear capacity is about 50%, and the bridge function of UHPC fin–steel fiber is good.
Chen et al. [69]	1	65	1	The formal loading starts at 0. Before the loading force reaches 500 kN, force-controlled loading is adopted, with 25 kN per stage. After the loading force reaches 500 kN, displacement-controlled loading is adopted.	The tribute of UHPC fins against the shear capacity is about 50%, and the bridge function of UHPC fin–steel fiber is good. Effectively reduce wing cracking
Xue et al. [70]	1	44.3	0	Through force-controlled loading, the preload is 20, 40, and 60 kN.	Two continuous composite beam specimens were studied, and the results show that the double combination can inhibit the crack development of the concrete wing slab during the loading stage.
Liu et al. [71]	1		0 0.1	Monotonic static loading	The shear-bearing capacity of the steel–composite beam under negative bending moments is borne by the steel beam and concrete wing slab.
Al-Osta [72]	1	54	n/a	The loading process is divided into three stages: preloading, force-controlled loading, and displacement-controlled loading.	A formula for calculating the shear strength of UHPC–NC (Normal Concrete) composite beams was obtained. The method can effectively reflect the influence of various parameters such as the UHPC layer, reinforcement of the UHPC layer, and the size effect on the shear capacity of UHPC–NC composite beams.
Hussein [61]	1	52	n/a	Force control	Both the bonding UHPC prefabricated slab and the cast-in-place gouge UHPC layer can effectively improve the bearing capacity of the composite structure, but the cast-in-place gouge casting method is superior to the bonding of the prefabricated slab.

4. Progress of the Rapid Construction Technology of the Narrow-Width Steel Box–UHPC Composite Beam after a Disaster

The narrow steel box composite beam enhances flange plate thickness by reducing the box chamber width, thus minimizing the number of longitudinal stiffeners in the top and bottom plates and eliminating the need for transverse stiffener plates. This design approach streamlines bridge components by increasing the panel span and eschewing small longitudinal girders in favor of durable steel–concrete or prestressed concrete bridge panels [73,74]. Reduced components, particularly within the box, facilitate manufacturing, while the lightweight steel main beam enables easy transportation and installation. This configuration ensures optimal force distribution, strong curve adaptability, and alignment with the green road ethos of factorization, standardization, and mechanization. Consequently, the narrow-width steel box composite beam has gained traction internationally, notably in Japan. However, its adoption remains limited in other regions, where it is still in an exploratory phase. Thus, this chapter examines recent advancements through two exemplary cases and explores progress in construction technology for double-narrow and three-narrow steel box girders, while also contemplating prospects for innovative narrow steel box girder composite structures.

4.1. Double-Narrow Steel Box Girder

Take the 37 m + 3 × 50 m + 37 m continuous double-narrow steel box–concrete girder bridge in a mountain area of Sichuan Province, China as an example. There are many mountains in this area [75], and the geological environment is complex and prone to strong geological activities, such as earthquakes, landslides, collapse, etc. Traditional large bridge prefabrication equipment is difficult to deploy quickly. According to the site environment and the needs of bridge design and use, the project uses the double-narrow steel box girder combined with the jacking method for rapid construction. The jacking method [76–78] was developed in the construction of prestressed concrete beam bridges, characterized by simple equipment and fast construction speed, and is very suitable for the rapid reconstruction of bridges after a disaster. The fixed thrust method combined with narrow-width steel box–UHPC composite beams can be used to reconstruct bridges quickly in a poor construction environment after an earthquake. As shown in Figure 13, when using the jacking method to carry out the rapid construction of narrow steel box beams, it can be pushed from one side of a river or valley to the other side as long as the transportation conditions on one side are met [76,77,79]. To reduce the reaction of the fulcrum of the pier on the general section of the steel box, the steel box and the bridge panel of the double-narrow steel box–concrete composite bridge are pushed successively.

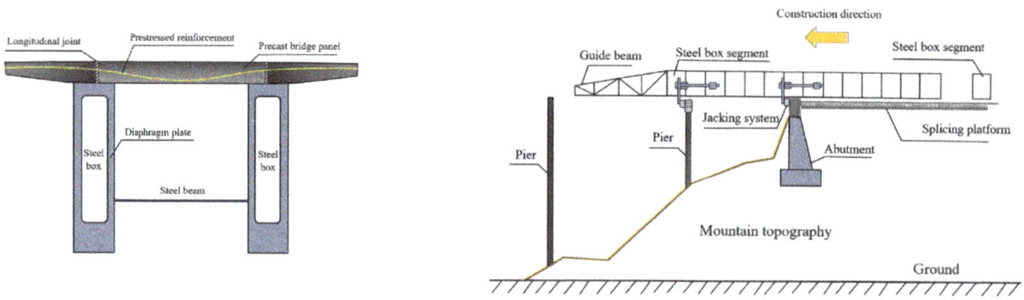

Figure 13. Rapid construction of a double-narrow steel box girder by the pushing method.

Under the specified viewing conditions, a platform is erected behind the abutment, or a steel box girder is positioned at the initial span. The multi-section steel box segments are interconnected and welded to create a unified steel box structure. Subsequently, sliding facilities for the steel box girder are installed. A pair of small-tonnage horizontal jacks are placed on each pier, with the pushing force being distributed across multiple points on the pier. Before the commencement of construction, careful examination of the pushing method's construction process is conducted, with particular attention paid to the localized stress on the steel box segment through the pier top. If deemed necessary, adjustments to the box's stiffener design can be made accordingly. A leading beam is installed during the pushing operation, with its length being 0.7 times that of the main span. Refer to the accompanying Figure 13 for visual representation. The analysis indicates that no intermediate temporary pier is required for spans of 50 m using this jacking scheme. Once the steel box is in position, prefabricated bridge panels are assembled on the splicing platform. Transverse prestress from the tensioned plates is applied to the box, after which the splicing platform is cleared for the installation of the next bridge panel section. This process is then repeated, as depicted in Figure 13. Similar procedures can be adopted for bridges to minimize the pushing distance. Connectors are installed at designated holes in the bridge panel after the full beam has been pushed into place. Welded bolts of perforated steel plate connectors may be welded or bolted as required.

The jacking method for double-narrow steel box–concrete composite beam bridges overcomes site, machinery, and concrete waiting constraints, significantly improving construction progress. It shortened the construction period by about 30%, crucial for rapid post-disaster reconstruction.

4.2. Three-Narrow Steel Box Girder

With advancements in design technology and concepts, narrow-width steel box girders have found application in the construction of long-span continuous beam bridges. Furthermore, the interface design of narrow-width steel box girders has evolved to introduce a novel form known as the three-narrow steel box girder. Illustrated by the case of a continuous three-narrow steel box composite girder bridge on an expressway in Zhejiang Province, China, spanning (50 + 85 + 50) m, the project mandates a short construction duration and uninterrupted traffic flow during execution. Hence, the construction plan involves the deployment of (50 + 85 + 50) m narrow-width steel box composite beams crossing the existing roadway in a left and right cross-hole layout, augmented by large-section lifting + bracket assembly. Narrow-width steel box composite beams can be flexibly arranged to accommodate various bridge widths. However, to fully exploit their structural advantages, main beam spacing should ideally be maximized. Employing steel composite bridge panels aids in reducing the weight of the bridge panel. Typically, the span of steel composite bridge panels does not exceed 6 m, with the overhang length limited to 0.4 times the span of the bridge panel between the main beams. Additionally, to meet the requirement of an 85 m span, the project introduces an innovative three-narrow steel box girder structure.

The web spacing of the steel box beam determines its width and distinguishes it from conventional steel box composite beams. For narrow steel box girders, the minimum web spacing should be at least 1.2 m to meet operational requirements. However, wider boxes require more steel. Due to the large span of the bridge and complexities in support stiffening and beam-to-roof positioning, the web spacing for the narrow steel box girder is set at 1.4 m, as shown in Figure 14.

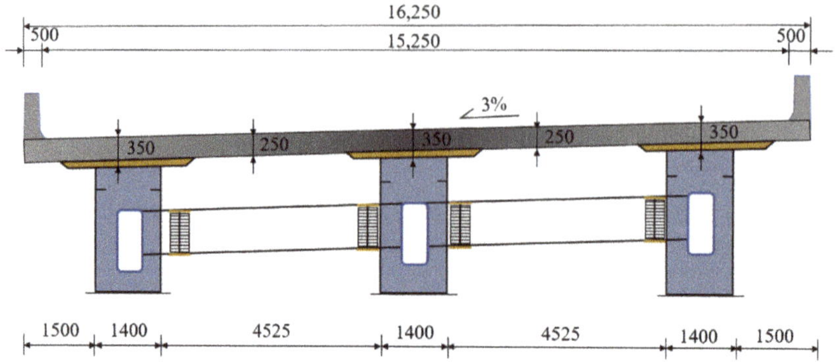

Figure 14. Three-narrow steel box girder section form.

4.3. Anti-Crack Control Measures of the Negative Bending Moment of the Fulcrum in the Narrow-Width Steel Box Continuous Beam with a Long Span

The bridge panel in the negative bending moment region at the fulcrum of a long-span continuous steel–concrete composite beam is susceptible to cracking. Common anti-cracking methods comprise prestressing, fulcrum forced displacement, ultra-high-performance concrete, weight compression, flexible connection, and construction procedure adjustment methods [75,80–84]. The prestressing method influences the internal forces [82,85] of the steel beam and entails a complex process. In practical applications, the flexible connection assembly method, represented by anti-pull and non-shear force nails, is typically employed in medium- and small-span composite beams measuring less than 60 m. Regarding the application of prestress and forced [86] displacement, the prestress effect diminishes [87] over time.

The compression method for large-span continuous steel–concrete composite beams is both costly and complex to implement. The forced displacement method necessitates substantial displacements of lifting and lowering beams. Utilizing ultra-high-performance concrete incurs higher costs. Based on calculations and practical experience with bridge construction, the method of adjusting the construction sequence is implemented for large-span continuous steel–concrete composite beams. Initially, concrete is poured in the positive bending moment zone of the bridge panel, followed by the pouring of concrete in the negative bending moment zone (within 0.15 times the span range around the central fulcrum). This approach does not significantly increase costs or complexity and facilitates convenient construction. This approach effectively controls the stress levels and crack widths in the bridge panel's negative bending moment region at the central fulcrum, meeting usage requirements comprehensively.

Section 4.2 uses the three-narrow steel box girder project in Zhejiang, China as an example. Prestressed UHPC wing slabs are used to control shear cracking in the negative bending moment area of the middle support, effectively reducing cracking and meeting stress and displacement requirements, as shown in Figure 15.

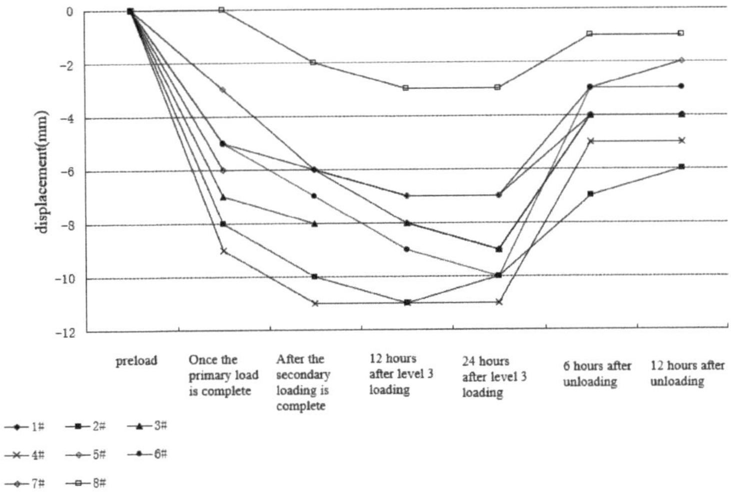

Figure 15. Test result of the precompression test in the negative bending moment area of the center fulcrum.

4.4. Narrow-Width Steel Box Girder Application Scenarios

The narrow steel box composite beam is a type of steel box composite beam. Considering the economic span of 45–75 m and the typical span of 60–100 m, it is important to maintain rationality in terms of processing, transportation, and structural forces. When the span exceeds 100 m, the height of the steel main beam needs to be above 4.2 m and divided during transportation. This compromises the advantages of convenient processing, transportation, and connection offered by narrow steel box girders, contradicting the design concept. Therefore, it is recommended that the maximum common span for narrow steel box composite girder bridges does not exceed 100 m.

(1) Small-radius flat-curved bridge or widened-ramp bridge in an interworking area with a span of 30 m or more

Curved bridges in the expressway interconnection face certain challenges, including cracking in reinforced concrete cast-in-place box girders and adaptability issues with prestressed concrete cast-in-place box girders. Furthermore, these curved bridges are often located beyond the height change point, resulting in possible lateral tilt changes for each span. Dealing with such situations using small box beams, T-beams, or short T-beams requires frequent adjustments to the steel mold and internal reinforcement arrangement of the prefabricated main beam, adding complexity to the construction process.

The narrow steel box girder adopts a closed steel box section, which has good torsional resistance and is more suitable for curved and inclined bridges. Compared to steel plate beams, it can adapt to the situation of curved bridges better. In addition, the narrow steel box composite beam box chamber is narrow, and with the bridge panel structure of a larger span, the transverse section layout is flexible, and it can also be applied to the greatly widened bridge.

According to the research of Ye Jianlong et al. [88], when span $L \leq 25$ m, the small-radius flat-curved bridge can be better solved by the assembled structure of superimposed T-beams. In addition, when the span is less than 30 m, the construction of a narrow-width steel box composite beam is unreasonable and uneconomical. Based on this, it can be concluded that the narrow-width steel box composite girder bridge can conveniently solve the dilemma of the small-radius flat-curved bridge or the widened-ramp bridge in the interworking area with a span of 30 m or more, which has little choice of structure and strong applicability and competitiveness in this scenario.

(2) The main-line 60 ~ 100 m span of a mountain or an urban bridge

When the span of the main bridge is 60 m or less, more economical steel composite beams can be used for a more convenient span [89]. When the span is large, the continuous steel box composite beam structure is a more reasonable choice. The following table is a comparison of the material indicators of the narrow steel box composite beam bridge in the Kezhu Expressway project and the conventional slot steel composite beam bridge in other projects this year. It can be seen that compared to the conventional slot steel composite beam structure, the steel amount per unit area of the main beam, the steel amount of the whole bridge, and the painting area of the main beam of the narrow steel box composite beam within the span of 60~100 m have been improved by different degrees, saving about 15%, 4.3%, and 21% respectively, as shown in Table 4.

Table 4. Comparison of key indices between narrow steel box composite continuous beams and conventional steel-mixed composite continuous beams.

Item	① Conventional 85 m Main-Span Steel–Concrete Composite Beam	② Kezhu High-Speed 85 m Main-Span Narrow Steel Box Girder	③ Conventional 75 m Main-Span Steel–Concrete Composite Beam	④ Kezhu High-Speed 75 m Main-Span Narrow Steel Box Girder	②/①	④/③
Steel beam (kg/m^2)	457	397	411	339	0.869	0.825
Steel beam + bridge panel steel (kg/m^2)	584	562	521	496	0.962	0.952
coating (m^2/m^2)	9.28	7.28	9.33	7.32	0.784	0.785

5. Discussion and Design Suggestions

Considering only the contribution of the web to the shear strength, the calculated value of the shear strength will be much smaller. Whether it is the American AASHTO, the European code EC4, or the Chinese code GB 50917-2013, the calculated value of the shear strength is too small, which will greatly waste the design performance of the narrow steel box girder.

Based on the principle of superposition calculations of the shear capacity of each part, different scholars have carried out further research on the calculation formula of the shear force of narrow-width steel box girders, but there is still a complete and applicable calculation criterion for narrow-width steel box girders. Therefore, it is necessary to further review and study the calculation formula of the shear capacity of narrow-width steel box girders. At the same time, the main factors affecting the shear performance of the narrow steel box girder are UHPC steel fiber content, UHPC plate thickness, filling concrete, transverse partition in the box, interbeam, shear connector, etc. When designing a narrow-width steel box girder, it is necessary to design and verify the parameters of each influencing factor according to the field requirements. Finally, the rapid post-disaster construction technology of narrow-width steel box–UHPC composite beam is increasingly being developed. The new sectional structure and construction method can greatly improve the adaptability of narrow-width steel box girders under various engineering requirements and can also generate more new technologies and instruments to promote the development of the rapid post-disaster reconstruction technology of bridges.

6. Conclusions and Perspective

In this paper, the development of narrow steel box–UHPC composite beam structures, the influencing factors of the shear strength, and the predicted value of the shear strength under different specifications are reviewed, and the influencing factors of the shear strength and the calculation method of the shear strength limit based on the superposition principle of the shear strength are summarized. The development of the rapid post-disaster construction technology of narrow-width steel box–UHPC composite beams in recent years is reviewed, and the new composite structure and post-disaster construction method of the narrow-width steel box are evaluated. Based on the above discussions, the following conclusions and recommendations can be derived:

1. Due to technological advancements in the use of high-strength materials and the increasing number of modern structures using concrete of high strength, some restrictions on narrow-width steel box–UHPC composite beams in the design specification need to be updated. In this paper, the Chinese code GB 50917-2013, the European code EC4, and the American code AASHTO are reviewed. The results show that the calculated values of the shear strength obtained by the above three codes are too conservative, and the predicted values of the shear strength obtained are less than 70% of the actual shear strength, resulting in a waste of shear strength. The calculation method of shear force based on the principle of superimposed shear capacity is not complete and lacks relatively complete and applicable calculation criteria.
2. According to the literature review, the main failures of UHPC composite beams under the condition of the negative moment zone are buckling deformation of the box girder, cracks on the upper surface of wing plates, and local concrete crushing. The main factors affecting the shear strength of the composite beam structure are shear span ratio, concrete wing plate, UHPC steel fiber content, UHPC plate thickness, and transverse partition in the box. The formula for calculating the shear capacity of a narrow-width steel box girder based on the superposition principle of shear capacity lacks a uniform conclusion regarding its applicability and needs further study.
3. The new double-narrow steel box girder structure and three-narrow steel box girder structure have a good application effect in the reconstruction of bridges after a disaster. For the long-span continuous beam bridge, the narrow steel box composite structure needs to use corresponding control measures to reduce the development of cracks at the middle support. At the same time, the scope of the application of the narrow-width steel box girder composite bridge is reviewed, and the conclusion is that the narrow-width steel box girder is mainly used in small-radius flat-curved bridges or widened-ramp bridges in interworking areas with a span of 30 m or more and in the main-line 60–100 m span in mountainous or urban areas.
4. The steel–concrete composite structure, with advantages in all aspects, has a wide range of applications in housing construction and large road and bridge construction. The construction method combined with narrow-width steel box–UHPC composite beams can be used to quickly rebuild bridges under poor construction environments after an earthquake, which is of great significance for post-disaster reconstruction projects.

Author Contributions: Conceptualization, Y.C.; Formal analysis and Writing—original draft, J.X.; Investigation, P.Y. and Q.W. and G.C.; Formal analysis, X.S. All authors have read and agreed to the published version of the manuscript.

Funding: The authors would like to acknowledge the financial support from the Innovation Capability Support Program of Shaanxi (Program No. 2023-CX-TD-35) and the Key Research and Development Program of Shaanxi (Program No. 2023KXJ-159).

Data Availability Statement: Not applicable.

Conflicts of Interest: Author Yunteng Chen was employed by the company Shaoxing Communications Investment Group Co., Ltd. Authors Peilong Yuan and Qiang Wang were employed by the company CCCC Second Highway Engineering Co., Ltd. The remaining authors declare that the research was conducted in the absence of any commercial or financial relationships that could be construed as a potential conflict of interest.

References

1. Meng, B.; Xiong, Y.P.; Zhong, W.H.; Duan, S.C.; Li, H. Progressive collapse behavior of composite substructure with large rectangular beam-web openings. *Eng. Struct.* **2023**, *295*, 116861. [CrossRef]
2. Gerasimidis, S.; Ellingwood, B. Twenty Years of Advances in Disproportionate Collapse Research and Best Practices since 9/11/2001. *J. Struct. Eng.* **2023**, *149*, 2. [CrossRef]
3. Fan, H.B.; Liu, T.; Zhang, S.Y.; He, H.J.; Zhu, Z.G.; Zhu, Y.Q.; Gao, X.Q. Effects of Jet-Grouting Piles on Loess Tunnel Foundation with Centrifugal Model Tests. *Int. J. Geomech.* **2023**, *23*, 3. [CrossRef]
4. Fang, Q.; Liu, X.; Zeng, K.H.; Zhang, X.D.; Zhou, M.Z.; Du, J.M. Centrifuge modeling of tunnelling below existing twin tunnels with different types of support. *Undergr. Space* **2022**, *7*, 1125. [CrossRef]
5. Jin, Z.X.; Zhang, C.P.; Li, W.; Tu, S.Q.; Wang, L.B.; Wang, S.X. Stability analysis for excavation in frictional soils based on upper bound method. *Comput. Geotech.* **2024**, *165*, 105916. [CrossRef]
6. Qiu, J.L.; Liu, Y.H.; Qian, X.Y.; Ma, C.; Liu, J.; Liu, T.; Han, H.X.; Ma, P.G. Guarantee rate statistics and product-moment correlation analysis of the optimal deformation allowance for loess tunnel in China. *Environ. Earth Sci.* **2024**, *in press*.
7. Li, Z.Q.; Lai, J.X.; Ren, Z.D.; Shi, Y.F.; Kong, X.G. Failure mechanical behaviors and prevention methods of shaft lining in China. *Eng. Fail. Anal.* **2023**, *143*, 106904. [CrossRef]
8. Qiu, J.L.; Liu, D.D.; Zhao, K.; Lai, J.X.; Wang, X.L.; Wang, Z.C.; Liu, T. Influence spatial behavior of surface cracks and prospects for prevention methods in shallow loess tunnels in China. *Tunn. Undergr. Space Technol.* **2024**, *143*, 105453. [CrossRef]
9. Tu, S.Q.; Li, W.; Zhang, C.P.; Chen, W. Effect of inclined layered soils on face stability in shield tunneling based on limit analysis. *Tunn. Undergr. Space Technol.* **2023**, *131*, 104773. [CrossRef]
10. Tu, S.Q.; Li, W.; Zhang, C.P.; Wang, L.B.; Jin, Z.X.; Wang, S.X. Seepage effect on progressive failure of shield tunnel face in granular soils by coupled continuum-discrete method. *Comput. Geotech.* **2024**, *166*, 106009. [CrossRef]
11. Wang, Z.C.; Cai, Y.C.; Xie, Y.L.; Zhang, M.Z.; Lai, J.X.; Qiu, J.L.; Liu, T. Laboratory study on mechanical behavior of hollow p-type steel-concrete composite support in loess tunnel. *Tunn. Undergr. Space Technol.* **2023**, *141*, 105280. [CrossRef]
12. Wu, K.; Zheng, X.M.; Zhao, N.N.; Shao, Z.S. Effect of compressible layer on time-dependent behavior of soft-rock large deformation tunnels revealed by mathematical analytical method. *Appl. Math. Model.* **2024**, *126*, 457. [CrossRef]
13. Xu, S.S.; Lai, J.X.; O'kelly, B.C.; Zhao, B.D. 3D deformation and strain fields in drying kaolinite obtained from tracking internal bubbles using X-ray CT and ANN. *Acta Geotech.* **2024**, *19*, 99. [CrossRef]
14. Xu, S.S.; Nowamooz, H.; Lai, J.X.; Liu, H.T. Mechanism, influencing factors and research methods for soil desiccation cracking: A review. *Eur. J. Environ. Civ. Eng.* **2023**, *27*, 3091. [CrossRef]
15. Yuan, P.L.; Zhang, B.D.; Yang, Y.T.; Jiang, T.F.; Li, J.; Qiu, J.L.; He, H.J. Application of polymer cement repair mortar in underground engineering: A review. *Case Stud. Constr. Mater.* **2023**, *19*, e02555. [CrossRef]
16. Zan, W.B.; Liu, L.J.; Lai, J.X.; Wang, E.B.; Zhou, Y.P.; Yang, Q. Deformation failure characteristics of weathered phyllite tunnel and variable-stiffness support countermeasures: A case study. *Eng. Fail. Anal.* **2023**, *153*, 107553. [CrossRef]
17. Liu, Y.H.; Qian, X.Y.; Qiu, J.L.; Chen, Y.H.; Liu, T.; Zhao, K.; Sun, H.; Liu, D.D. Disaster Analysis and Protection Measures of Coal and Gas Outburst for Coal Measures Strata Gas Tunnel in China. *KSCE J. Civ. Eng.* 2024; *in press*.
18. Zhao, Z.Z.; Sun, H.; Zhang, N.X.; Xing, T.H.; Cui, G.H.; Lai, J.X.; Liu, T.; Bai, Y.B.; He, H.J. Application of unmanned aerial vehicle tilt photography technology in geological hazard investigation in China. *Nat. Hazards* **2024**. [CrossRef]
19. Bekic, D.; Kerin, I.; Cahill, P.; Michalis, P.; Lapthorne, J.; Solman, H.; Gilja, G.; Potocki, K.; Pakrashi, V.; McKeogh, E. In BRIDGE SMS—INNOVATIVE SOLUTION FOR MANAGEMENT OF BRIDGES OVER WATER. In Proceedings of the 5th International Conference on Road and Rail Infrastructure (CETRA), Zadar, Croatia, 17–19 May 2018; pp. 57–63.
20. Niemierko, A. Modern bridge bearings and expansion joints for road bridges, 6th Transport Research Arena (TRA), Warsaw, Poland, April 18-21, 2016. *Transp. Res. Procedia* **2016**, *14*, 4040–4049. [CrossRef]
21. Tang, K.; Liu, D.; Xie, S.; Qiu, J.; Lai, J.; Liu, T.; Fang, Y. Analysis of loess water migration regularity and failure response of tunnel structure under rainfall environment. *Bull. Eng. Geol. Environ.* **2024**, *83*, 251. [CrossRef]
22. Zhang, Z.; Song, Z.; Lai, J.; Qiu, J.; Cheng, Y.; Zhang, J. Critical slowing down precursor information for the acoustic emission response characteristics of defective tuffs. *Theor. Appl. Fract. Mech.* **2024**, *129*, 104220. [CrossRef]
23. Li, Z.H.; Ma, E.L.; Lai, J.X.; Su, X.L. Tunnel deformation prediction during construction: An explainable hybrid model considering temporal and static factors. *Comput. Struct.* **2024**, *294*, 107276. [CrossRef]
24. Zan, W.B.; Lai, J.X.; Zhang, W.J.; Yang, Q.; Qin, Y.W.; Su, X.L. Experimental and applied research on similar materials to granular mixtures for the solid-liquid coupling model test of an underwater tunnel. *Constr. Build. Mater.* **2024**, *416*, 135170. [CrossRef]
25. Qin, Y.W.; Chen, Y.H.; Lai, J.X. Failures in Loess Slope-Tunnel System: An overview of trigging sources, acting mechanism and mitigation strategies. *Eng. Fail. Anal.* **2024**, *158*, 107996. [CrossRef]
26. Li, Z.; Lai, J.; Wang, M.; Qiu, J.; Zan, W. Collapse mechanism and treatments of a deep tunnel in the weathered granite fault zone. *Tunn. Undergr. Space Technol.* **2024**, *152*, 105891. [CrossRef]
27. Mao, M.; Yin, C.L.; Shen, S.Y.; Wan, Y. Experimental and Numerical Study on Flexural Behaviors of Steel-UHPC Composite Beams under Hogging Moment. *Ksce J. Civ. Eng.* **2024**, *28*, 2344–2354. [CrossRef]
28. Ataei, A.; Zeynalian, M.; Yazdi, Y. Cyclic behaviour of bolted shear connectors in steel-concrete composite beams. *Eng. Struct.* **2019**, *198*, 109455. [CrossRef]

29. Lu, X.Z.; Zhang, L.; Lin, K.Q.; Li, Y. Improvement to composite frame systems for seismic and progressive collapse resistance. *Eng. Struct.* **2019**, *186*, 227. [CrossRef]
30. Romero, M.L.; Espinós, A. Advances in Steel-Concrete Composite Structures (ASCCS 2018). *Structures* **2019**, *21*, 1. [CrossRef]
31. Zhang, Z.J.; Liu, Y.J.; Liu, J.; Xin, G.F.; Long, G.X.; Zhang, T.T. Experimental study and analysis for the long-term behavior of the steel-concrete composite girder bridge. *Structures* **2023**, *51*, 1305. [CrossRef]
32. Saiidi, M.; Douglas, B.; Feng, S. Prestress Force Efect on Vibration Frequency of Concrete Bridges. *J. Struct. Eng.* **1994**, *12*, 2233. [CrossRef]
33. Mcginnis, M.J.; Pessiki, S.; Turker, H. Application of Three-dimensional Digital Image Correlation to the Core—Drilling Method. *Soc. Exp. Mech.* **2005**, *4*, 359. [CrossRef]
34. Liu, S.; Wang, B.H.; Zheng, Y.; Mo, S.X.; Zhang, Q.Y.; Zhou, P.X. Experimental and Computational Research on the Shear Performance of Partially Filled Narrow-Width Steel Box-UHPC-Combined Girders under Negative Moment Action. *Buildings* **2023**, *13*, 2749. [CrossRef]
35. Wang, Y.C. Performance of steel-concrete composite structures in fire. *Prog. Struct. Eng. Mater.* **2005**, *7*, 86. [CrossRef]
36. Benedetty, C.A.; Santos, V.B.D.; Krahl, P.A.; Rossi, A.; de Andrade Silva, F.; Cardoso, D.C.T.; Martins, C.H. Flexural and shear behavior of steel-UHPC composite beams: A review. *Eng. Struct.* **2023**, *293*, 116649. [CrossRef]
37. Mo, S.; Xiong, H.; Zheng, Y.; Chai, L. Experimental study on shear strength of partially filled narrow steel box-UHPC composite beam. *J. Henan Polytech. Univ. (Nat. Sci.)* **2023**, *43*, 157–164.
38. Chen, H.; Nie, X.; Gan, S.; Zhao, Y.; Qiu, H. Interfacial imperfection detection for steel-concrete composite structures using NDT techniques: A state-of-the-art review. *Eng. Struct.* **2021**, *245*, 112778. [CrossRef]
39. Wakjira, T.G.; Alam, M.S. Peak and ultimate stress-strain model of confined ultra-high-performance concrete (UHPC) using hybrid machine learning model with conditional tabular generative adversarial network. *Appl. Soft Comput.* **2024**, *154*, 111353. [CrossRef]
40. Solhmirzaei, R.; Salehi, H.; Kodur, V.; Naser, M.Z. Machine learning framework for predicting failure mode and shear capacity of ultra high performance concrete beams. *Eng. Struct.* **2020**, *224*, 111221. [CrossRef]
41. Wakjira, T.G.; Abushanab, A.; Alam, M.S. Hybrid machine learning model and predictive equations for compressive stress-strain constitutive modelling of confined ultra-high-performance concrete (UHPC) with normal-strength steel and high-strength steel spirals. *Eng. Struct.* **2024**, *304*, 117633. [CrossRef]
42. Wakjira, T.G.; Kutty, A.A.; Alam, M.S. A novel framework for developing environmentally sustainable and cost-effective ultra-high-performance concrete (UHPC) using advanced machine learning and multi-objective optimization techniques. *Constr. Build. Mater.* **2024**, *416*, 135114. [CrossRef]
43. Chung, L.; Lim, J.-J.; Hwang, H.-J.; Eom, T.-S. Review of Design Flexural Strengths of Steel-Concrete Composite Beams for Building Structures. *Int. J. Concr. Struct. Mater.* **2016**, *10*, S109. [CrossRef]
44. Liu, J.; Liu, J.; Han, F.; Zheng, X. Research progress and application of coarse aggregate ultra high performance concrete used for lightweight steel-concrete composite structure. *J. Build. Struct.* **2022**, *43*, 36.
45. Nie, J.; Wang, J.; Gou, S.; Zhu, Y.; Fan, J. Technological development and engineering applications of novel steel-concrete composite structures. *Front. Struct. Civ. Eng.* **2019**, *13*, 1–14. [CrossRef]
46. Ranzi, G.; Leoni, G.; Zandonini, R. State of the art on the time-dependent behaviour of composite steel-concrete structures. *J. Constr. Steel Res.* **2013**, *80*, 252. [CrossRef]
47. Hao, X. Study of Shear Resistance of Partially Filled Narrow Steel Box-UHPC Composite Beam. Master's Thesis, Guilin University of Technology, Guilin, China, 2023.
48. Liu, X. Research on Bending Behavior of Narrow-Width Steel Box-UHPC Composite Beams Under Negative Moment. Master's Thesis, Guilin University of Technology, Guilin, China, 2023.
49. Luo, M.; Lin, P.; Yang, Z. Flexural Capacity of UHPC Reinforced Ordinary Concrete Composite Beam. *China Railw. Sci.* **2021**, *21*, 110722.
50. Yoo, S.W.; Choo, J.F. Evaluation of the flexural behavior of composite beam with inverted-T steel girder and steel fiber reinforced ultra high performance concrete slab. *Eng. Struct.* **2016**, *118*, 1–15. [CrossRef]
51. He, J.; Chao, L. Numerical analysis on shear resistance of ultra-high performance concrete-normal strength concrete composite beam. *Struct. Concr.* **2021**, *22*, 1128. [CrossRef]
52. Zhao, Q.; Xiao, F.; Zhang, H.; Fang, X.M. Behavior and reasonable design of steel-UHPC composite beams under negative moment. *J. Constr. Steel Res.* **2024**, *212*, 108268. [CrossRef]
53. Lu, W.L.; Peng, W.Q.; Zhu, L.; Ma, B.; Li, F.L. Study on mechanical behavior of steel-UHPC-NC composite beams under negative bending moment. *Case Stud. Constr. Mater.* **2022**, *17*, e01593. [CrossRef]
54. Yang, H. Finite Element Analysis of UHPC Flange-Narrow Width Steel Box Continuous Composite Beam's Double Combination Effects. Master's Thesis, Guilin University of Technology, Guilin, China, 2021.
55. Wang, Y. Experimental Study on Shear Strength of Concrete-Partial-Filled Steel Box Composite Continuous Beam under Negative Moment. Master's Thesis, Guilin University of Technology, Guilin, China, 2018.
56. Zhang, P.; Xu, F.; Liu, Y.; Sheikh, S.A. Shear behaviour of composite beams with permanent UHPC formwork and high-strength steel rebar. *Constr. Build. Mater.* **2022**, *352*, 128951. [CrossRef]

57. Chen, B.C.; Liu, A.R.; Zhang, J.P.; Zhang, F.; Bradford, M.A. Behavior of T-shaped embedded-nut bolted shear connectors in prefabricated steel-concrete composite beams. *Eng. Struct.* **2022**, *272*, 114983. [CrossRef]
58. American Association of State Highway and Transportation Officials. *AASHTO LRFD Bridge Design Specifications*; American Association of State Highway and Transportation Officials: Washington, DC, USA, 2017.
59. *EN 1994-2*; London Eurocode 4: Design of Composite Steel and Concrete Structures Part 2: General Rules and Rules for Bridges. British Standards Institution: London, UK, 2005.
60. *GB 50917-2013*; Code for Design of Steel-Concrete. Composite Bridges China Planning Publishing House: Beijing, China, 2013.
61. Hussein, L.; Amleh, L. Structural behavior of ultra-high performance fiber reinforced concrete-normal strength concrete or high strength concrete composite members. *Constr. Build. Mater.* **2015**, *93*, 1105. [CrossRef]
62. Men, P. Study on Shear Performance and Shear Capacity Calculation Method of Steel-Concrete Composite Girders in Negative Moment Regions. Ph.D. Thesis, Chongqing University, Chongqing, China, 2021.
63. Shozo, N. Study on Fatigue Design Load for Steel Bridge Piers and Steel Girder Bridges; Kaken. 2005. Available online: https://kaken.nii.ac.jp/en/grant/KAKENHI-PROJECT-17560429/ (accessed on 27 April 2024).
64. Wakjira, T.G.; Alam, M.S. Performance-based seismic design of Ultra-High-Performance Concrete UHPC) bridge columns with design example–Powered by explainable achine learning model. *Eng. Struct.* **2024**, *214*, 118346. [CrossRef]
65. Tadele, S.; Aldabagh, S.; Zhang, Q.; Alam, M.S. Variability of mechanical properties of steel rebars in North America and its effects on the seismic fragility of reinforced concrete columns. *Eng. Struct.* **2023**, *295*, 116805. [CrossRef]
66. Wakjira, T.G.; Ebead, U.; Alam, M.S. Machine learning-based shear capacity prediction and reliability analysis of shear-critical RC beams strengthened with inorganic composites. *Case Stud. Constr. Mater.* **2022**, *16*, e01008. [CrossRef]
67. Niu, F.; Liu, Y.; Xue, F.; Sun, H.; Liu, T.; He, H.; Kong, X.; Chen, Y.; Liao, H. Ultra-high performance concrete: A review of its material properties and usage in shield tunnel segment. *Case Stud. Constr. Mater.* 2024; *in press*.
68. Zhu, J.S.; Wang, Y.G.; Guo, X.Y. Experimental study on shear behaviors of steel-UHPC waffle plate composite beams. *China J. Highw. Transp.* **2020**, *33*, 11.
69. Chen, X.; Su, Q.; Chong, W. Experimental study on double composite action in the negative flexural region of two-span continuous composite box girder. *J. Constr. Steel Res.* **2011**, *67*, 10.
70. Xue, J.Y.; Cheng, G.; Zhao, H.T. Experimental study on shear performance of steel-concrete compositebeams under negative bending moment. *J. Build. Struct.* **2008**, *4* (Suppl. S1), 29.
71. Liu, C.; Ji, H. Shear resistance of ultra-high performance concrete-concrete composite beams. *J. Tongji Univ. Sci.* **2019**, *47*, 7.
72. Al-Osta, M.A.; Isa, M.N.; Baluch, M.H.; Rahman, M.K. Flexural behavior of reinforced concrete beams strengthened with ultra-high performance fiber reinforced concrete. *Constr. Build. Mater.* **2017**, *134*, 279. [CrossRef]
73. Chen, N.; Liu, L.; Deng, M.; Yang, C.; He, J. Tendency prediction and prevention strategy of geohazards triggered by Lushan Ms 7 earthquake, Sichuan Province, China. *J. Chengdu Univ. Technol. Sci. Technol. Ed.* **2013**, *40*, 371.
74. Hua, W.; Xin-Hua, Z. Earthquake-damages of reservoirs in Sichuan Province and rehabilitation countermeasures after disaster. *Water Resour. Hydropower Eng.* **2009**, *40*, 16.
75. Xu, X.; Zheng, F.; Guan, Y.; Yao, Y.; Qin, C. Spatiotemporal Characteristics of Earthquake Disasters and Their Damages within 2013 in China. *Res. Soil Water Conserv.* **2015**, *22*, 321.
76. Li, J.; Zeng, Y.; Chen, H.; Liu, X.; Song, X.; Ma, K. Study on mid-span jacking force for long span continuous rigid-frame bridges. *J. Railw. Sci. Eng.* **2015**, *12*, 335.
77. Wei, X.; Wei, G. Study on calculational method of ground settlement induced by level parallel pipe jacking. *Rock Soil Mech.* **2006**, *27*, 1129.
78. Li, G.; Feng, X. Study on Jacking Technology for Closure of Long-span Hybrid Girder Cable-stayed Bridge at Low Temperature. *J. Highw. Transp. Res. Dev.* **2019**, *36*, 69.
79. Wu, J.D.; Li, N.; Hu, A.J.; Zhang, P.; Liu, X.Q.; Wen, Y.T. Indirect Economic Loss Estimation of Wenchuan Earthquake to Sichuan Province in China Based on the ARIO Model. In Proceedings of the International Disaster and Risk Conference, Chengdu, China, 13–15 July 2009.
80. Gao, H.S.; Li, B.C.; Jian, J.S.; Yu, T.T.; Liu, H.B. Integral jacking of concrete continuous box beam bridge. *Structures* **2023**, *54*, 1026. [CrossRef]
81. Wang, P.; Liu, X. Primary Technical Investigation on Double Narrow Steel Box Girder of Bridge Rebuilt after Earthquake. *Technol. Highw. Transpor.* **2008**, *12*, 5.
82. Takayuki, K. Study on the Design Method of Prestressed Reinforced Concrete Beams Subjected to Torsion; Kaken. 1989. Available online: https://kaken.nii.ac.jp/en/grant/KAKENHI-PROJECT-01550380/ (accessed on 27 April 2024).
83. Jia, X.; Ji, L.; Cao, C.; Chen, X.; Yu, F. Influence of fulcrum and load positions on springboard mechanics. *J. Tsinghua Univ. (Sci. Technol.)* **2004**, *44*, 201.
84. Gao, S.; Wang, L. Research on Simply-Supported Continuous Bridge's Internal Force Optimized by Unequal Altitude Bearings. *Appl. Mech. Mater.* **2012**, *178*, 2353–2356. [CrossRef]
85. Lu, Z.-F.; Liu, M.-Y. Long-time Prestress Losses Analysis Method for Concrete Structure. *J. Wuhan Univ. Technol.* **2011**, *33*, 83.
86. Hitoshi, S. Nonlinear Hysteresis Behavior of Precast/Prestressed Concrete Member; Kaken. 1996. Available online: https://kaken.nii.ac.jp/en/grant/KAKENHI-PROJECT-08455247/ (accessed on 27 April 2024).

87. Rational Maintenance Method of Prestressed Concrete Bridge Considering Stress Corrosion Cracking of Prestressing Steel Wires; Kaken. 2022. Available online: https://kaken.nii.ac.jp/en/grant/KAKENHI-PROJECT-22K04259/ (accessed on 27 April 2024).
88. Ye, J.L.; Hu, S.X.; Yu, M.F.; Guo, B.Q. Research on New Prestressed Concrete Superposed Bridge. *Technol. Highw. Transp.* **2020**, *36*, 67–75+83.
89. Shen, D.W.; Ji, B.H.; Sun, Q.; Yang, X.H.; Zheng, H.; Rui, F. Analysis and estimation of the cost of the whole synchronous jacking project of bridge. *Constr. Econ.* **2012**, *5*, 32–34.

buildings

Article

Advanced Analysis of Structural Performance in Novel Steel-Plate Concrete Containment Structures

Guopeng Ren [1,*], Rong Pan [1], Feng Sun [1], Zhanfa Dong [2] and Tianyun Lan [2]

[1] Nuclear and Radiation Safety Center, Ministry of Ecology and Environment, Beijing 100082, China; panrong@chinansc.cn (R.P.); sunfeng@chinansc.cn (F.S.)
[2] China Nuclear Power Design Co., Ltd. (Shenzhen), Shenzhen 518031, China; dongzhanfa@cgnpc.com.cn (Z.D.); lantianyun@cgnpc.com.cn (T.L.)
* Correspondence: renguopeng@chinansc.cn

Abstract: This paper investigates the structural performance of novel steel-plate concrete containment structures, focusing on third-generation nuclear power plants. To address the challenges of increased complexities and costs associated with double-layer containment designs, this study explores the potential of steel-plate concrete structures to enhance safety, economic efficiency, and construction simplicity. The steel-plate concrete structure, characterized by its core concrete and dual steel plates, shows superior compressive strength, bending resistance, and elastoplasticity. Extensive numerical analyses, including finite element modeling and thermal-stress coupling, were conducted under various load conditions. Under structural integrity test conditions, the maximum radial displacement observed was 24.59 mm. Under design basis conditions, the maximum radial displacement was 47.61 mm; under severe accident conditions, it was 53.83 mm. The ultimate bearing capacity was 0.91 MPa, 2.17 times the design pressure. This study concludes that the steel-plate concrete containment structure maintains a high safety margin under all tested conditions, with stress and strain well within acceptable limits. It can effectively serve as a robust barrier against radioactive leakage and malicious impacts, providing a viable alternative to conventional containment designs.

Keywords: nuclear power plants; steel-plate concrete; containment structures; ultimate bearing capacity; structural performance; thermal-stress coupling; safety margin

1. Introduction

Currently, to enhance safety, most third-generation nuclear power technologies, such as those used in EPR and VVER reactors, have incorporated double-layer containment structures to prevent radioactive leakage in the event of commercial aircraft impacts. However, this design significantly increases the construction workload and complexity, leading to higher costs and extended project timelines [1,2]. The independently designed third-generation HPR1000 nuclear reactor by China faces the same issue, which impacts the economic viability of nuclear power. Ensuring safety and economic efficiency is crucial for the sustainability of nuclear energy. The new containment structure proposed in this study aims to address both needs by simplifying construction processes, shortening the construction period, reducing costs, and enhancing economic efficiency. The steel-plate concrete structure (SC structure) will be investigated as a new type of containment due to its superior performance, construction advantages, and contribution to enhancing containment leak tightness.

The SC structure consists of core concrete and steel plates on either side or on one side. A distinctive feature of this structure is that the core concrete lacks conventional stressed steel bars, tie bars, and stirrups. Instead, the steel plate and concrete are bonded by round head studs welded to the inner side of the steel plate, with the two layers of steel plates connected by shear connectors. This innovative design leverages the high compressive strength of concrete alongside the strong bending resistance and excellent elastoplasticity of the steel plate, significantly enhancing the overall bearing capacity, seismic resistance, and impact resistance of the structure [3,4]. As a result, SC structures have been widely adopted in infrastructure and major key projects. Furthermore, its construction characteristics facilitate modular construction in nuclear power projects, improve construction efficiency, shorten the construction period, and fully realize its advantages in ensuring nuclear power project safety.

Pan Rong (2014) comprehensively reviewed the development of steel-plate concrete in nuclear power engineering, covering its technical advancements, potential advantages, application experiences, lessons learned, ongoing research, and unresolved issues [5]. The content is highly comprehensive and broad. Since the 1980s, SC structures have been first applied in the immersed tube tunnel of the North Wales highway and have since been used in military defense, mine shafts, coastal caissons, TV towers, and high-rise buildings [6,7]. In 2002, Tokyo Electric Power Company first adopted SC structures in the auxiliary buildings of units 6 and 7 of the Kashiwazaki-Kariwa Nuclear Power Plant [8], and planned to use them in the reactor buildings of units 7 and 8 of the Fukushima Nuclear Power Plant. However, the plan was not implemented due to the Fukushima nuclear accident. Additionally, Westinghouse applied SC structures in the development of the AP1000 nuclear power plant, and Mitsubishi Heavy Industries used them in the development of the US-ABWR. China also adopted this structure in the cylindrical segment of the shielding building for the CAP1400 third-generation nuclear power technology demonstration project.

Over the past 30 years, Japanese researchers have conducted extensive experimental studies and numerical simulations on SC structures. Their research has focused on various aspects, including in-plane shear, out-of-plane shear and bending, axial compression, and the effects of temperature gradients under accidental loads. These comprehensive studies have contributed to the development and revision of Japan's relevant standards, resulting in the publication of JEAG 4618-2005 [9] and JEAC 4618-2009 [10]. Korea has developed the KEPIC-SNG-2010 guidelines for the design of SC structures related to the safety of nuclear facilities, drawing on Japanese research findings [11]. The American Institute of Steel Construction (AISC) has compiled AISC N690 [12] to guide the design of steel structures related to nuclear safety, including specific design requirements for SC composite walls. In China, based on the results of major national science and technology projects, the 'Technical Standard for SC Structures of Nuclear Power Plants' (GB/T 51340-2018) [13] has been compiled to guide the design, construction, inspection, and acceptance of SC structures and components in nuclear power plants. In recent years, many scholars have conducted a series of studies on the structural performance of SC structures. Zhang et al. (2019) conducted small-scale tests and analysis of corrugated-steel-plate–concrete composite members, introducing novel shear connectors for improved structural integrity [14]. Wang (2019) discussed the structural design and shock-resistant performance of steel beam-concrete column frame structures, highlighting the benefits of this composite structure in terms of mechanical performance, durability, and fire resistance [15]. Zhou et al. (2020) developed a biaxial steel plate concrete constitutive model for composite structures, validating its implementation in finite element analysis, which is essential for accurately predicting structural behavior under complex load conditions [16]. Wu et al. (2020) investigated the ultimate strength behavior of steel plate-concrete composite slabs through experimental and theoretical studies, focusing on the impacts of critical factors such as shear stud spacing and concrete slab thickness on load-carrying capacity [17]. Liu and Yang (2021) analyzed the fatigue performance of steel-plate-concrete composite slabs, focusing on the effects of shear ratio and steel plate layout under low-cycle fatigue loads [18]. Szewczyk and Szumigała (2021) presented optimal design methods for steel-concrete composite beams strengthened under load, focusing on maximizing bending capacity and minimizing cost [19]. Tusnin (2022) conducted numerical calculations on steel-concrete structures, emphasizing the importance of reliable shear force transfer between steel beams and concrete slabs for effective structural behavior [19]. Ma et al. (2022) studied the seismic behavior of coupled steel plate and reinforced concrete composite walls, noting that the coupling mechanism significantly enhanced the seismic performance of the structures [20]. Liu et al. (2023) examined the seismic design and performance of coupled steel plate and reinforced concrete composite walls, proposing an energy balance-based plastic design method for improving seismic performance [21]. Wang et al. (2023) explored the performance of steel-plate-reinforced concrete composite walls in tall structures, particularly in earthquake zones, highlighting the enhanced seismic performance and improved shear resistance due to the inclusion of steel plates [22].

Research on the structural performance of containment structures has primarily focused on prestressed concrete containment and steel containment. Since the late 1960s, a series of experimental studies have been conducted worldwide to evaluate the bearing capacity of containment structures, with initial model scales predominantly around 1:10, aimed at exploring internal pressure limits during leakage failure scenarios. Typical experiments include India's 1:12 scale model (Rao et al., 1975), which leaked due to excessive cracking [23], and Poland's 1:10 scale model (Donten, K et al., 1980), which demonstrated that cracks around the equipment gate can lead to leakage [24]. Although the British 1:10 scale model achieved an ultimate bearing pressure of 2.4 times the design pressure, it tilted due to bottom plate warping, resulting in broken concrete [25]. In the United States, the leakage rate of a 1:4 scale model (Hessheimer et al., 2003) increased significantly at 3.3 times the design pressure, accompanied by notable local concrete cracks [26]. Further experiments on structural failure morphology showed that the model ruptured at a midsection position of nearly 6 degrees at 3.6 times the design pressure, leading to complete structural destruction. The 1:3.2 scale model test of a prestressed concrete containment (Zhao et al., 2024) was conducted in Langfang City, Hebei Province, China [27]. This test validated the structural performance of the containment vessel under prolonged high temperature and high pressure, following a preset temperature-pressure curve. In studying the bearing capacity of containment structures, Zhao et al. (2013) used ABAQUS software to construct a three-dimensional finite element model of the CPR1000 reactor containment, analyzing its response under internal pressure loads ranging from 0 to 3 times the design pressure [28]. Their research proposed evaluation criteria for functional and structural failure, indicating that the containment would fail at 1.83 times the design internal pressure and would be structurally destroyed at 2.16 times. Chen-nin (2014) explored the ultimate compressive bearing capacity of nuclear-reinforced concrete (RC) containment structures, highlighting the role of prestress and steel liners in determining the containment's failure modes under internal pressure [29]. Ren et al. (2017) conducted an in-depth analysis and summary of the effective prestressing effect under the design basis internal pressure [30]. Chakraborty et al. (2017) assessed the ultimate load capacity of concrete containment structures against structural collapse, using non-linear FE analysis with ABAQUS and ANSYS to understand containment performance under severe accident conditions [31]. Tong et al. (2018) investigated the ultimate pressure-bearing capacity of prestressed concrete containment vessels (PCCVs), providing technical support for Severe Accident Management Guidelines (SAMG) development by analyzing the containment's behavior under increasing internal pressures [32]. In addition to internal pressure, the influence of temperature load is also a focus of research. The ISP 48 Benchmark research project (2005) on containment integrity, initiated by the Organisation for Economic Co-operation and Development-Nuclear Energy Agency (OECD-NEA), has attracted the participation of 11 organizations from 9 countries [33]. It was found that after considering the temperature load, there is no consistent conclusion on the change in ultimate bearing pressure of the containment, but the influence on the failure pressure is small (within 10%), while structural deformation increases significantly, and damage occurs mostly near the penetration. Huang et al. (2017) studied the long-term performance of a CANDU containment structure considering the internal temperature profile, support conditions, and temporary construction openings. The research indicated that time-dependent effects, including temperature loads, significantly influenced the ductile behavior of the structure under internal pressure [34]. Desai and Mahida (2020) investigated the structural response of a nuclear containment structure subjected to internal overpressure and high-temperature loading. The study used non-linear analysis with finite element modeling to assess the impact of temperature and pressure beyond design conditions, revealing that safety failure occurred at 1.2 times the design basis and structural failure at 1.22 times [35].

In order to fully consider the sealing performance of SC containment and the need to prevent malicious impacts from commercial aircraft, this study adopts a double SC containment structure, chosen for its excellent sealing capabilities and strong impact resistance. Using a typical section of the containment structure as a specific case, the thicknesses of the concrete and double steel plates are initially determined through thin film stress theory. Building on this foundation, the structural performance response of the SC containment is further analyzed under structural integrity test conditions, design basis conditions, and severe accident conditions. Additionally, OECD-NEA analysis is employed to calculate and examine the time history curves of pressure and temperature in the containment following a severe accident, thereby evaluating the ultimate bearing capacity of the SC containment. Finally, based on existing research theories and practical experience, specific criteria for assessing the failure of the SC containment structure are established. In summary, the research process is outlined in Figure 1, presenting the key steps in evaluating the structural performance of SC containment structures.

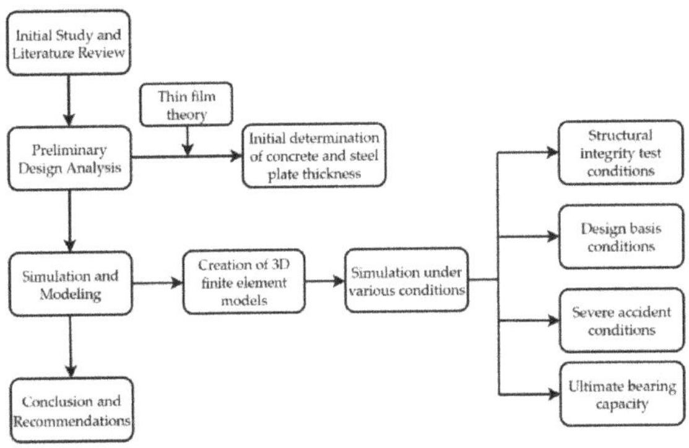

Figure 1. Research process for evaluating SC containment structures.

2. The Determination of SC Containment Scheme

2.1. The Main Types of Containment at Present

Currently, based on the structural system, containment can be classified into four types: reinforced concrete structure, prestressed concrete structure, steel structure, and SC. Based on the structural configuration, it can be divided into two types: single-containment and double-layer containment.

2.1.1. CPR1000 [28]

As the third safety barrier of a nuclear power plant, the CPR1000 containment employs a single-layer prestressed reinforced concrete structure with a design pressure of 0.42 MPa (relative pressure) and a design temperature of 145 °C. The containment vessel is cylindrical with an oval dome at the top. The outer dimensions are as follows: the outer diameter of the cylinder is 38.8 m, the wall thickness of the cylinder is 0.9 m, and the inner diameter is 37 m. The highest point elevation of the dome is at +7.20 m, the dome area covers elevations ranging from +44.83 to +57.20 m, with a height of 12.37 m, and the thickness of the dome concrete is 0.8 m as shown in Figure 2a. The inner surface of the containment vessel is covered with a 6 mm thick carbon steel liner to ensure sealing. Within the cylinder of the containment, 223 and 144 prestressed steel bundles are arranged in the circumferential and vertical directions, respectively, to enhance structural strength.

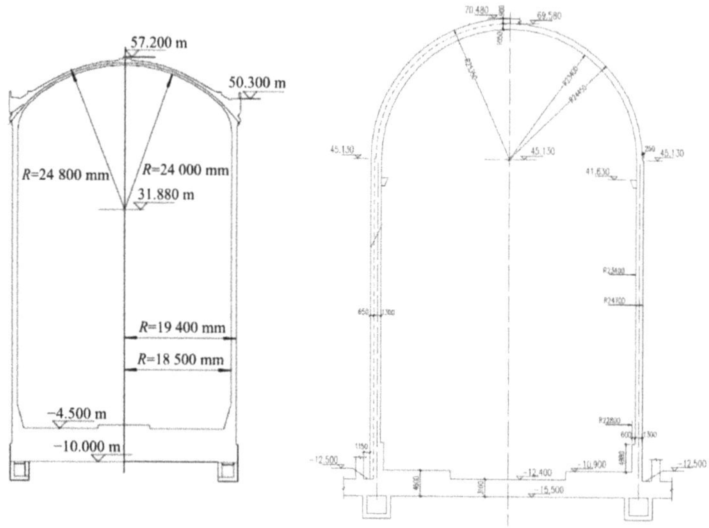

(a) (b)

Figure 2. (**a**) The cross-sectional structure of the inner containment of the CPR1000 reactor; (**b**) the cross-sectional structure of the inner containment of the HPR1000 reactor.

2.1.2. HPR1000 [36]

The HRP1000 containment is designed as a double-layered structure. The dome of the inner containment is hemispherical and directly connected to the cylindrical wall. Both the cylinder wall and the dome are prestressed reinforced concrete structures, and the inner side is fully covered by a sealed steel lining. The wall thickness of the inner shell is 1.3 m (with some local areas thickened), and the thickness of the dome is 1.05 m, as shown in Figure 2b. The steel lining is made of Q265 HR carbon steel, which is firmly anchored to the inner concrete wall and bottom plate by anchoring nails and angle steel. Besides the thickening treatment of local parts, the lining thickness is uniformly 6 mm. The shape of the outer containment is similar to that of the inner containment, but the thickness of the cylinder wall is 1.50 m (with the exposed area thickened to 1.80 m), and the thickness of the dome is also 1.80 m.

2.1.3. EPR [37]

The EPR reactor building utilizes a double-layer containment, with the inner layer being a prestressed concrete structure equipped with a sealed steel lining. The annular space between the inner and outer layers is 1.8 m wide, and negative pressure is maintained. The internal containment volume is 8000 m^3, the inner diameter is 46.8 m, the thickness is 1.30 m, and the height is 66.5 m, as shown in Figure 3a. C60/75 high-strength concrete is used, and the design parameters are based on ETC-C. The prestressed steel bundles include 119 horizontal, 47 vertical, and 104 gamma strands. The steel lining is 6 mm thick and the steel grade is P265GH.

2.1.4. VVER [38]

The VVER inner containment is a prestressed reinforced concrete structure composed of a cylindrical part and a dome, as shown in Figure 3b. The inner surface is lined with 6 mm of carbon steel to ensure effective sealing. The containment is installed on a tensioned corridor plate with an elevation of −1.250 m, and the elevation of the inner surface of the dome top is +66.600 m. The inner diameter of the cylinder is 44.0 m, and the height is 45.850 m. The thickness of the prestressed containment is determined through calculation.

The thickness of the cylinder is 1200 mm, and the thickness of the dome is 1100 mm. The concrete grade used is B60.

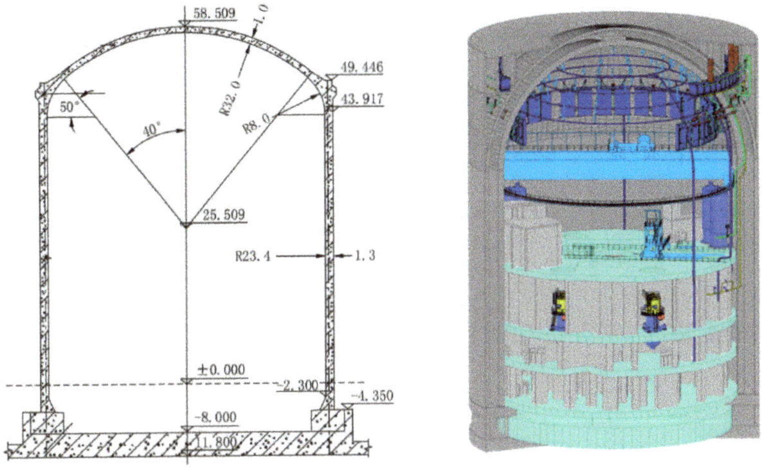

(**a**) (**b**)

Figure 3. (**a**) The cross-sectional structure of the inner containment of the EPR reactor; (**b**) schematic of the VVER reactor containment structure.

2.1.5. CAP1400 [39]

The CAP1400 shielded building is a composite structure surrounded by steel containment that serves multiple functions, such as radiation shielding, projectile protection, passive cooling, and resistance to extreme weather and earthquakes. Its main body is designed as a cylindrical concrete structure with an internal diameter of 46.0 m and a wall thickness of 1100 mm as shown in Figure 4a. The exposed part adopts the SC structure with a concrete strength of C55, steel plate Q345, and a thickness of 20 mm, which is classified as a seismic class I structure. The inner diameter and height of the containment are 43 m and 73.6 m, respectively. The design pressure is 0.443 MPa, the thickness of the containment is 52 mm, and the top elevation is 60.767 m, all of which are completely surrounded by the shielding workshop.

2.1.6. BWRX-300 [40]

The BWRX-300 consists of the steel-plate composite containment vessel (SCCV) and Reactor Building (RB) as shown in Figure 4b. The SCCV features high strength and compressive capacity, primarily designed to contain the internal pressure of the nuclear reactor and prevent the leakage of radioactive materials, ensuring integrity under extreme conditions such as earthquakes and impacts. The SC modules combine steel plates and concrete infill, enhancing the rigidity and durability of the structure through the composite action of materials. Diaphragm plates are used to separate different structural units, providing additional stability and support. The functions of the SC structure include seismic resistance, fire resistance, and construction efficiency. The design meets the requirements of the U.S. Nuclear Regulatory Commission (NRC) and the Canadian Nuclear Safety Commission (CNSC), and its performance under extreme conditions has been experimentally verified, ensuring the safety and reliability of the BWRX-300 reactor.

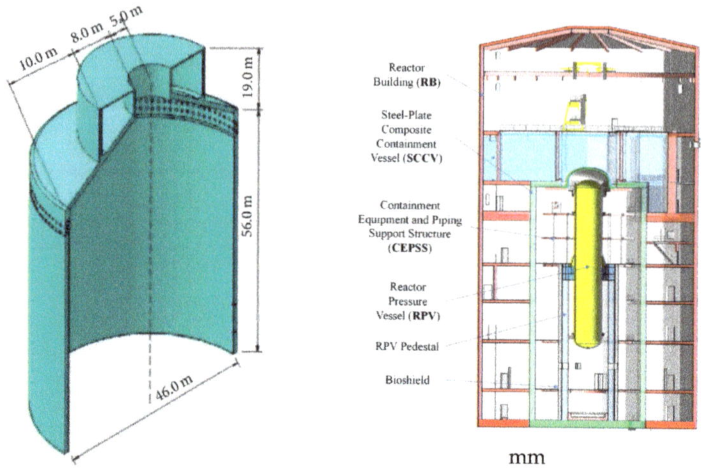

(**a**) (**b**)

Figure 4. (**a**) The cross-sectional structure of the inner containment of the CAP1400 reactor; (**b**) section view of the integrated reactor building of the BWRX-300.

2.2. Thin Film Stress Theory

The containment cylinder of a pressurized water reactor (PWR) is typically a cylindrical structure with two types of domes: flat shell and hemispherical. The entire containment structure is axisymmetric. Its thickness is relatively small compared to its radius. The design adheres to membrane stress theory, and the internal forces can be calculated using non-torque theory. For the containment cylinder, we have the following:

$$\sigma_m = \frac{PD}{4S}\ \sigma_\theta = \frac{PD}{2S} \tag{1}$$

In the formula, σ_m and σ_θ represent the vertical and circumferential stresses of the containment, respectively; P is the internal pressure load; D is the diameter of the inner wall of the cylinder; S is the thickness of the cylinder.

For the containment dome, we have the following:

$$\sigma_m = \sigma_\theta = \frac{PD}{4S} \tag{2}$$

For steel containment vessels, considering that thin-walled vessels are made of ductile materials, the third strength theory can be used for design, as follows:

$$\sigma_1 - \sigma_3 = \frac{PD}{2\sigma} - 0 \leq [\sigma] \tag{3}$$

Thus, the wall thickness design formula is obtained, as follows:

$$\sigma \geq \frac{PD}{2[\sigma]} + C \tag{4}$$

where σ_1 and σ_3 represent the first principal stress and the third principal stress, respectively; C represents the additional wall thickness accounting for processing and corrosion effects, as stipulated in the relevant design specifications.

2.3. Scheme Determination

The SC containment vessel used in this study utilizes the same type of steel plate as the AP1000 steel containment vessel, both of which are SA-738 Gr. B steel. Furthermore, C60 concrete is used to ensure the safety and stability of the structure. Referring to the preliminary design of the AP1000 steel containment, this paper adopts $P = 0.42$ MPa, $D = 45$ m, $[\sigma] = 184.3$ MPa, and substitutes these values into Formula (4), resulting in: $\sigma \geq 51$ mm. In this study, the inner steel plate is specified to be 30 mm thick and the outer steel plate is 25 mm thick. The concrete thickness of the dome and the cylinder is 1.1 m.

3. Structural Configuration

3.1. Steel-Plate Containment Model

The SC containment is composed of a raft foundation, containment cylinder wall, dome, and ASP tank. It is composed of C60 grade concrete and SA-738 Gr. B steel plate. Except for the outer and bottom of the raft foundation, the containment cylinder wall, dome, and ASP tank are all constructed with 20 mm thick steel plates. The cross-sectional structure of the containment is shown in Figure 5. The elevation of the cylinder wall is from −7.800 m to +45.4 m, the inner diameter is 45.000 m, the thickness is 1.100 m, and there are 20 mm thick steel plates inside and outside. The dome elevation is from +45.4 m to +60.00 m, and the thickness is also 1.100 m. The elevation of the raft foundation is from −11.800 m (without cushion) to −7.800 m, and the thickness is about 4 m. The elevation of the ASP water tank is from +38.1 m to +50.5 m, and the thickness of the water tank wall is 0.6 m. The inner steel plate is continuously arranged vertically, covering the entire containment dome, cylinder wall, bottom plate, and the inner surface of the ASP tank. The inner steel plate provides a leak-proof boundary. The bottom steel plate is anchored to the top of the raft foundation. The design of the steel plate ensures leak prevention under normal operating conditions, containment pressure test stages, and accident conditions. The outer steel plate is arranged on the outer surface of the dome, cylinder wall, and ASP tank, protecting against malicious impacts and tornado projectiles. At the same time, the internal and outer steel plates can be used as construction templates for the containment wall.

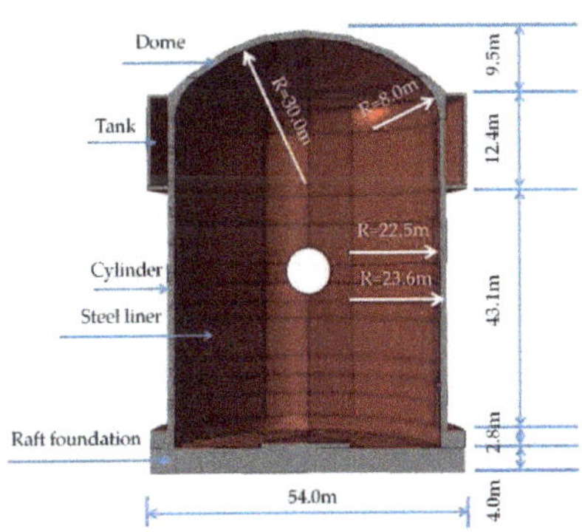

Figure 5. The cross-sectional structure of the containment.

The anchorage system of the steel plate is embedded in the concrete structure and welded to the steel plate simultaneously. It is composed of right-angled intersecting grids. Each grid is covered with 19 mm diameter and 152 mm length anchors, spaced 150 mm apart. Additionally, between the internal and outer steel plates, HRB335 grade ribbed steel with a diameter of 20 mm and a spacing of approximately 450 mm provides out-of-plane shear resistance. The anchoring system connects the steel plate to the concrete to ensure the stability of the steel plate during the construction and operation stages. The anchorage system ensures the deformation coordination of concrete and steel plate and limits the deformation of the steel plate caused by different thicknesses, temperatures, and elastic-plastic states between two adjacent grids.

There are three large gate penetrations on the containment cylinder: the equipment gate, with an axis angle of 12°, a center elevation of +20.80 m, and a diameter of 8.36 m; the personnel gate, with an axis angle of 83.5°, a center elevation of +2.30 m, and a diameter of 3.35 m; and the emergency personnel gate, with an axis angle of 50°, a center elevation of +18.60 m, and a diameter of 3.35 m.

3.2. Material Properties

3.2.1. Concrete

The mechanical properties of concrete at room temperature are calculated according to the Chinese code "Code for Design of Concrete Structures" (GB50010-2010) [41]. Under severe accident conditions, the maximum temperature considered is 200 °C, which is a conservative estimate based on typical accident scenarios. To account for the impact of high temperatures on the mechanical properties of materials such as concrete and steel, a reduction factor corresponding to this 200 °C threshold is applied to the concrete values. This approach ensures that the structural integrity is maintained even under elevated temperature conditions.

The relationship between the concrete strength reduction factor S_{RC} and temperature T is given by the following formula [33]:

$$S_{RC} = e^{-(T/632)^{1.8}} \tag{5}$$

The reduction coefficients for the concrete elastic modulus M_{RC} and concrete strength S_{RC} are given by the following formula [33]:

$$M_{RC} = \sqrt{S_{RC}} \tag{6}$$

Since the maximum temperature under severe accident conditions will not exceed 200 °C, this paper conservatively adopts the reduction coefficient value at 200 °C. Therefore, the mechanical properties of concrete at room temperature, considering the temperature effect, are shown in Table 1. The concrete strength grade typically used in the SC Containment Structures studied in this paper is C60, which aligns with the concrete strength grade used in the containment structures of China's HPR1000 reactor.

Table 1. The material mechanical properties of concrete.

Material Parameter	Room Temperature	High Temperature
Elastic modulus (MPa)	3.60×10^4	3.38×10^4
Compressive strength (MPa)	38.50	33.88
Tensile strength (MPa)	2.85	2.51
Density (kg/m^3)	2400	
Poisson ratio	0.2	

Note: The curing age for the mechanical properties listed in this table is 28 days.

The thermal characteristic parameters of concrete are determined according to the European standard (EN1994-1-2) "Design of composite steel and concrete structures: Structural fire Design" [42].

For thermal conductivity λ, we have the following:

$$\lambda = 2 - 0.2451(T/100) + 0.0107(T/100)^2 \text{ unit: W/m·°C} \tag{7}$$

For specific heat capacity c, we have the following:

$$c = \begin{cases} 900 & 20\ °\text{C} \le T \le 100\ °\text{C} \\ 900 + (T - 100) & 100\ °\text{C} < T \le 200\ °\text{C} \end{cases} \text{ unit: J/Kg·°C} \tag{8}$$

For the coefficient of thermal expansion, we have the following:

$$\alpha = -1.8 \times 10^{-4}/T + 9 \times 10^{-6} + 2.3 \times 10^{-11}T^2 \quad (9)$$

The thermal material characteristic values of concrete are shown in Table 2.

Table 2. The material thermal properties of concrete.

Temperature, °C	Thermal Conductivity, W/m·°C	Specific Heat Capacity, J/Kg·°C	Coefficient of Thermal Expansion
20	1.95	900	9.20×10^{-9}
40	1.90	900	4.54×10^{-6}
60	1.86	900	6.08×10^{-6}
80	1.81	900	6.90×10^{-6}
100	1.77	900	7.43×10^{-6}
120	1.72	920	7.83×10^{-6}
140	1.68	940	8.17×10^{-6}
160	1.64	960	8.46×10^{-6}
180	1.59	980	8.75×10^{-6}
200	1.55	1000	9.02×10^{-6}

This study employs design standards from multiple countries, primarily due to their conservative nature, which meets the safety requirements of engineering design. For example, the use of the Chinese code 'Code for Design of Concrete Structures' (GB50010-2010) is justified as the concrete materials in this project are sourced from China. The reduction factors for concrete strength and modulus at elevated temperatures are derived from OECD-NEA research, and the thermodynamic parameters are based on the EN1994-1-2 standard. These standards provide conservative values, which are advantageous for the safety and reliability of engineering design. The same approach was taken in selecting standards for steel materials.

The thermal properties calculations are derived from standard formulas applicable to typical concrete and steel materials used in nuclear containment structures. While these equations are broadly applicable, we acknowledge that variations in material composition may lead to differences in thermal behavior, which is a limitation of this study.

3.2.2. Steel

The mechanical properties of the inner and outer steel plates, anchors, and tendons of the SC containment at room temperature are based on ASME Volume II [43] and the Chinese code "Code for Design of Concrete Structures" (GB50010-2010) [41], respectively. The reduction relationship of high temperature on mechanical properties is based on the Chinese code "Code for fire safety of steel structures in buildings" (GB51249-2017) [44], with the elastic modulus given by the following formula:

$$E_{sT} = \chi_{sT} E_s \quad (10)$$

$$\chi_{sT} = \frac{7T - 4780}{6T - 4760} \quad (11)$$

where E_{sT} is the elastic modulus of steel at high temperature (N/mm^2); E_s is the elastic modulus of steel at room temperature (N/mm^2).

For steel at temperatures below 300 °C, the yield strength is considered to be the same as at room temperature [44]. Therefore, the mechanical properties of steel at room temperature, considering the temperature effect, are shown in Table 3.

Table 3. The material mechanical properties of steel.

Temperature, °C	Steel Plate		Anchors and Tendons	
	Yield Strength (MPa)	Elastic Modulus (MPa)	Yield Strength (MPa)	Elastic Modulus (MPa)
20	415	2.06×10^5	335	2.00×10^5
40		2.05×10^5		1.99×10^5
60		2.04×10^5		1.98×10^5
80		2.03×10^5		1.97×10^5
100		2.02×10^5		1.96×10^5
120		2.01×10^5		1.95×10^5
140		2.00×10^5		1.94×10^5
160		1.98×10^5		1.93×10^5
180		1.97×10^5		1.91×10^5
200		1.96×10^5		1.90×10^5

The European standard (EN1994-1-2) "Design of composite steel and concrete structures: Structural fire Design" provides the following relationship for the thermal characteristics of steel [42].

For thermal conductivity λ, we have the following:

$$\lambda = 54 - 3.33 \times 10^{-2} T \ 20\ ^\circ\text{C} \leq T \leq 800\ ^\circ\text{C unit}: \ \text{W/m}\cdot^\circ\text{C} \tag{12}$$

For specific heat capacity c, we have the following:

$$c = 425 + 7.73 \times 10^{-1} T - 1.69 \times 10^{-3} T^2 + 2.22 \times 10^{-6} T^3 \ 20\ ^\circ\text{C} \leq T \leq 600\ ^\circ\text{C unit}: \ \text{J/Kg}\cdot^\circ\text{C} \tag{13}$$

For the coefficient of thermal expansion, we have the following:

$$\alpha = -2.416 \times 10^{-4} T^{-1} + 1.2 \times 10^{-5} + 4 \times 10^{-9} T \ 20\ ^\circ\text{C} \leq T \leq 750\ ^\circ\text{C} \tag{14}$$

Therefore, the values of steel properties at room temperature, considering the temperature effect, are shown in Table 4.

Table 4. The material thermal properties of steel.

Temperature, °C	Thermal Conductivity, W/m·°C	Specific Heat Capacity, J/Kg·°C	Coefficient of Thermal Expansion
20	53.33	439.80	1.36×10^{-21}
40	52.67	453.36	6.12×10^{-6}
60	52.00	465.78	8.21×10^{-6}
80	51.34	477.16	9.30×10^{-6}
100	50.67	487.62	9.98×10^{-6}
120	50.00	497.26	1.05×10^{-5}
140	49.34	506.19	1.08×10^{-5}
160	48.67	514.51	1.11×10^{-5}
180	48.01	522.33	1.14×10^{-5}
200	47.34	529.76	1.16×10^{-5}

3.3. *FE Model of the SC Containment*

The concrete part of the SC containment is simulated by the three-dimensional solid element C3D8R and the temperature-displacement coupling element C3D8RT in the ABAQUS (version 6.13) finite element software, which is divided into four layers along the thickness direction. We fully consider the factors such as accuracy and convergence, and initially divide the concrete unit into about 800 mm. After the division, there are 96,812 concrete elements. The concrete FE model is shown in Figure 6a.

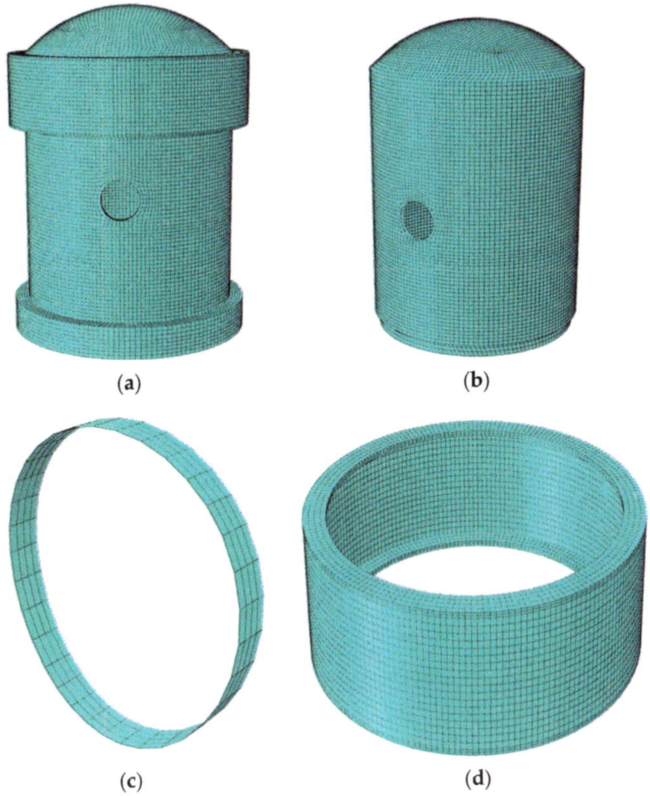

Figure 6. (a) The FE model of the concrete; (b) The FE model of the steel plate; (c) the FE model of the equipment penetration; (d) the FE model of the ASP tank.

The double steel plate is simulated by shell element S4R and temperature-displacement coupling element S4RT. The double steel plates and concrete are set in contact, and there are 46,334 shell elements in total. The FE models of steel plate and equipment penetration parts are shown in Figure 6b,c. The anchors and fibers are bonded to the steel plate and embedded into the concrete.

The role of the advanced secondary passive residual heat removal system (ASP) water tank is to export the decay heat of the core through ASP in the event of a loss of secondary side cooling function to prevent the core from melting. The water in the ASP tank is conservatively considered to be full of water, and the nonstructural mass is used in ABAQUS to consider the quality of the water. The FE model of the ASP tank is shown in Figure 6d.

The damage plasticity model is adopted to define the uniaxial stress–strain curves of concrete [45]. The concrete plastic damage model simulates concrete failure by considering two primary failure mechanisms: cracking due to tension and crushing due to compression. The model uses two hardening variables to describe the evolution of the yield surface—compressive equivalent plastic strain and tensile equivalent plastic strain. Key features of the model include the following: (1) The use of a damage variable to simulate the reduction in stiffness as concrete undergoes unloading, achieved by reducing the elastic stiffness matrix. (2) Incorporation of non-associated hardening in the constitutive relationship to better capture the plastic behavior of concrete under compression. (3) Options to control the characteristics of concrete before and after crack closure, enabling more realistic simulation of mechanical behavior under cyclic loading. This model effectively simulates uniaxial, cyclic, and dynamic loading conditions and exhibits good convergence, making it suitable for various loading scenarios in concrete behavior simulation. The stress–strain relationship can be represented by the following equations:

$$\sigma = (1 - d) \bullet E \bullet \varepsilon \quad (15)$$

$$\varepsilon = \varepsilon_e + \varepsilon_p \quad (16)$$

where σ is the internal force per unit area within a material that arises from externally applied forces. d is a parameter representing the degradation of material stiffness. The value of d ranges from 0 to 1, where $d = 0$ indicates no damage (the material retains full stiffness) and $d = 1$ indicates complete damage (the material has fully softened with zero stiffness). E is a measure of the material's stiffness during the elastic phase, defined as the ratio of stress to strain within the elastic limit. ε is the total deformation of the material under applied stress, encompassing both elastic and plastic components. ε_e is the recoverable deformation within the elastic range of the material. When the external load is removed, the material returns to its original shape. ε_p is the permanent, non-recoverable deformation that occurs once the material has yielded and undergone plastic deformation. The constitutive model of the steel components of the containment is elastoplastic material.

A mesh sensitivity analysis was conducted to evaluate the impact of mesh size on the simulation results. Three mesh sizes were considered: 1000 mm, 800 mm, and 500 mm. Under the structural integrity test conditions at a pressure of 0.42 MPa, key parameters such as maximum radial displacement, circumferential stress in the concrete, and the maximum Mises stress and strain in the inner and outer steel plates were calculated and compared. The analysis indicates that while the accuracy of the simulation improves as the mesh size decreases from 1000 mm to 500 mm, the differences between the results for the 800 mm and 500 mm meshes are minimal. As a result, the 800 mm mesh size was selected for subsequent simulations, providing an optimal balance between accuracy and computational efficiency. The results of this analysis are presented in Table 5, demonstrating that the 800 mm mesh size is a suitable choice for this study.

Table 5. Results of mesh sensitivity analysis under structural integrity test conditions.

Mesh Size, mm	Maximum Radial Displacement (mm)	Concrete Circumferential Stress (MPa)	Maximum Mises Stress (MPa)—Inner Steel Plate	Maximum Mises Stress (MPa)—Outer Steel Plate	Maximum Strain—Inner Steel Plate	Maximum Strain—Outer Steel Plate
1000	22.5	2.712	250.5	210.6	1.231×10^{-3}	1.002×10^{-3}
800	21.3	2.651	240.2	204.3	1.162×10^{-3}	9.736×10^{-3}
500	21.0	2.635	239.1	203.8	1.150×10^{-3}	9.654×10^{-3}

The results of this analysis are presented in Table 5, demonstrating that the 800 mm mesh size is a suitable choice for this study.

In this study, the modeling procedure involves several key steps. (1) We first distinguish between the sequential coupling and full coupling methods. The sequential coupling method begins by calculating the model's temperature field, which is then imported into the mechanical model, where material parameters are defined, boundary conditions are set, and internal pressure is applied for subsequent calculations. The full coupling method, on the other hand, simultaneously inputs thermal and mechanical material parameters, using a coupled temperature-displacement analysis step to solve the interactions between components while applying both temperature and internal pressure loads. (2) The concrete, inner and outer steel plates, water tank, anchors, and tendons are modeled based on the containment's sectional plan as shown in Figure 5. (3) Material properties are defined, with the mechanical and thermal properties of concrete provided in Tables 1 and 2, and those of the steel plates, anchors, and tendons in Tables 3 and 4. The mass of water in the tank, totaling 6922 tons, is applied using the *NONSTRUCTURAL MASS option in ABAQUS. (4) The components are then assembled according to their positions. (5) For the sequential coupling method, a temperature analysis step is used first, followed by a static analysis for the mechanical field, whereas the full coupling method directly sets a coupled temperature-displacement analysis step. (6) Tangential interactions between the steel plates and concrete are modeled using Coulomb friction, with normal interactions defined as hard contact. The anchors and tendons are tied to the steel plates and embedded within the concrete, and the water tank is also tied to the containment cylinder. (7) The gravity, temperature, and internal pressure loads are applied, with the temperature and pressure time-history curves from Figure 28 used for the ultimate bearing capacity analysis. (8) Mesh is generated for all components, and a mesh sensitivity analysis is performed to determine the appropriate mesh size. (9) Finally, a detailed analysis of each loading condition is conducted using the determined mesh size, and (10) the response and ultimate bearing capacity of the containment are evaluated based on the defined failure criteria.

3.4. Load Conditions and Analysis Methods

3.4.1. Load Conditions

This paper considers the structural performance response under structural integrity test conditions (self-weight, internal pressures of 0.42 MPa and 0.483 MPa), design basis conditions (self-weight, internal pressure of 0.42 MPa, and temperature of 145 °C), and severe accident conditions. The ultimate bearing capacity of the SC containment is calculated and analyzed using the time history curve of pressure and temperature in the containment after a severe accident, as provided by OECD-NEA.

The temperature and pressure of the containment of a nuclear power plant under two potential severe accident conditions are selected: (1) temperature of 150 °C, pressure of 0.55 MPa, with the corresponding containment heat removal system (EHR) effective; (2) temperature of 154 °C, pressure of 0.5 MPa, with EHR failure, but with the containment filtration and exhaust system (EUF) effective.

3.4.2. Analysis Methods

In finite element analysis, thermal stress analysis is classified into sequential coupling and full coupling. Sequential coupling analysis first addresses the heat conduction problem and then imports the temperature field as a predefined condition into the stress analysis. full coupled analysis simultaneously solves the stress/displacement and temperature fields, making it suitable for cases where thermal and mechanical solutions strongly interact. In ABAQUS/Standard, full coupled analysis employs the backward difference scheme for temperature integration and the Newton method to solve nonlinear coupling equations.

In the thermal coupling analysis, this study utilizes both sequential coupling and full coupling methods for comparative analysis. In sequential coupling analysis, heat transfer analysis is performed first, and the resulting temperature field is introduced into the finite element model to be analyzed alongside the internal pressure load condition. In full coupling analysis, the temperature boundary condition, gravity load, and internal pressure load are applied simultaneously. Additionally, for conservative consideration, the atmospheric temperature of the containment is directly applied to its inner surface, and the heat transfer coefficient of the external environment required for thermal analysis is determined by referencing the French standard RCC-CW [46]. The initial temperature condition of the outer surface of the containment is assumed to be the same as the ambient temperature, set at 20 °C.

3.4.3. Validation of Modeling and Analysis Methods

The validity of the modeling and analysis methods used in this study is primarily verified based on the prestressed concrete containment 1:3 scale model test benchmark project conducted by the French EDF Company [30]. A photograph of the containment mock-up and a general view of the containment are shown in Figure 7. The containment model and the arrangement of the prestressing tendons are illustrated in Figure 8. The modeling and analysis techniques employed in the benchmark project closely align with those used in this paper. The locations of sensors on the containment mock-up are presented in Figure 9. The strain, displacement, and concrete cracking patterns of the containment model under design basis pressure were compared with the results from the pressure testing. The strains and radial displacements at various locations under pressurization test conditions are compared with the experimental measurements in Tables 6 and 7. The actual cracking observed during the pressurization test is shown in Figure 10. Figure 10a presents the cracking observed in the dome area, while Figure 10b illustrates the cracking pattern at the bottom of the perimeter wall near the base slab. The predicted cracking pattern from the numerical analysis is illustrated in Figure 11. The comparison of strain and displacement under pressurization test conditions with experimental results showed good agreement, and the predicted locations and distribution patterns of concrete cracks were also consistent with the experimental findings. This comparison effectively validates the modeling and analysis methods applied in this study.

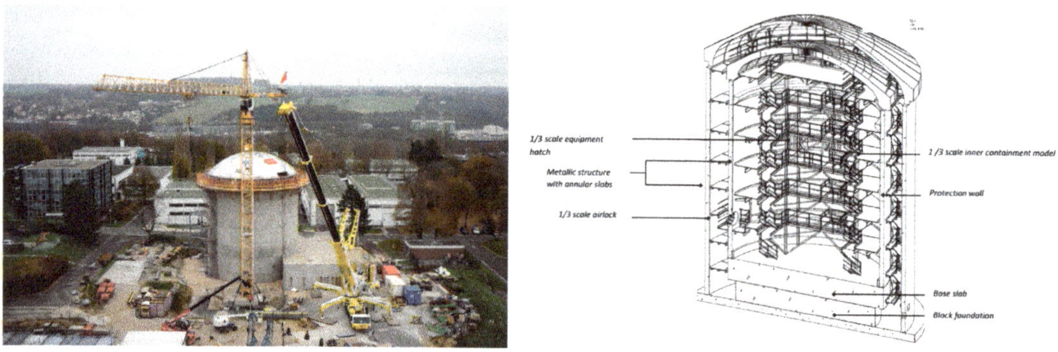

(a) (b)

Figure 7. (**a**) The photograph of the containment mock-up; (**b**) the general view of the containment.

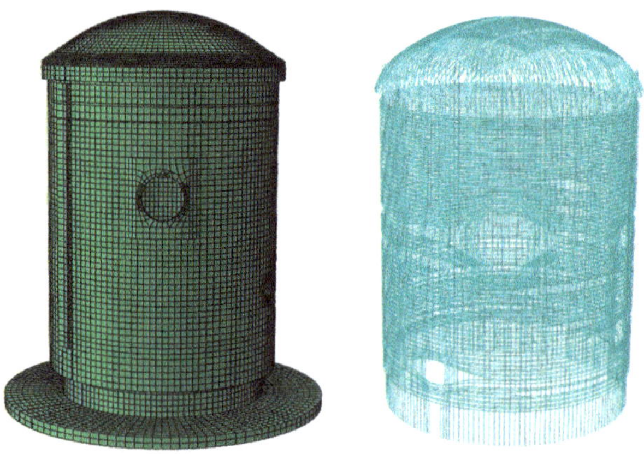

(a) (b)

Figure 8. (**a**) The containment model; (**b**) the arrangement of prestressing tendons.

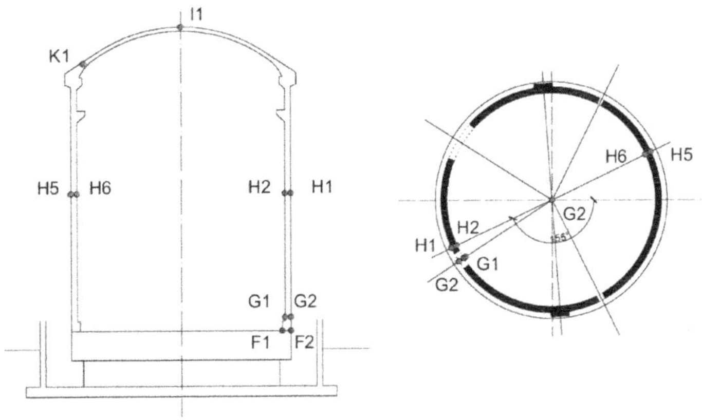

Figure 9. The locations of sensors on the containment mock-up.

Table 6. Comparison of strains at various locations under pressurization test conditions with experimental measurements.

Sensor	Direction	Deformation History in μm/m			
		End of Prestressing		At 0.52 MPa (abs.)	
		Test Result	Numerical Analysis	Test Result	Numerical Analysis
H1	T	−487.77	−405.45	−315.95	−244.45
	V	−271.83	−218.11	−230.71	−132.85
H2	V	−470.94	−347.14	−425.40	−262.70
H5	T	−551.17	−445.97	−369.32	−243.08
	V	−286.46	−165.31	−245.53	−139.27
H6	T	−574.84	−489.54	−391.03	−282.19
F1	T	−10.73	−6.36	−17.67	−33.41
F2	T	−39.76	−24.42	−41.08	−35.71
G1	V	−338.02	−369.54	−262.33	−163.53

Note: T—tangential; V—vertical.

Table 7. Comparison of displacement at various locations under pressurization test conditions with experimental measurements.

Level, m	Position, °	The Measured Radial Displacement during Pressurization Test, mm	The Calculated Radial Displacement during Pressurization Test, mm
4.0	60	1.6	1.44
	160	1.8	1.35
	260	1.4	1.23
	360	1.6	1.40
9.0	60	1.4	1.36
	160	1.5	1.38
	260	1.6	1.13
	360	1.4	1.42

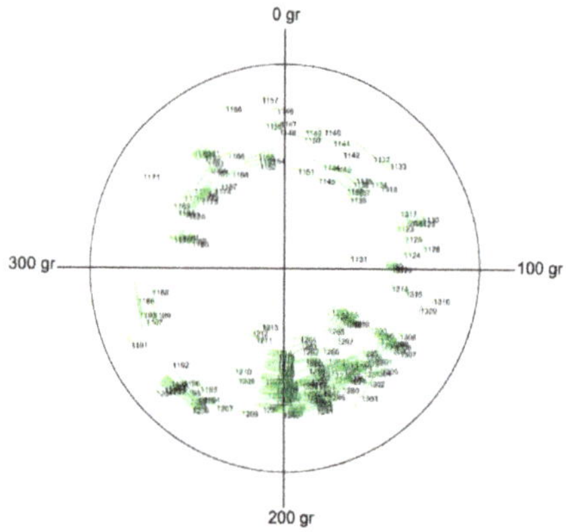

(a)

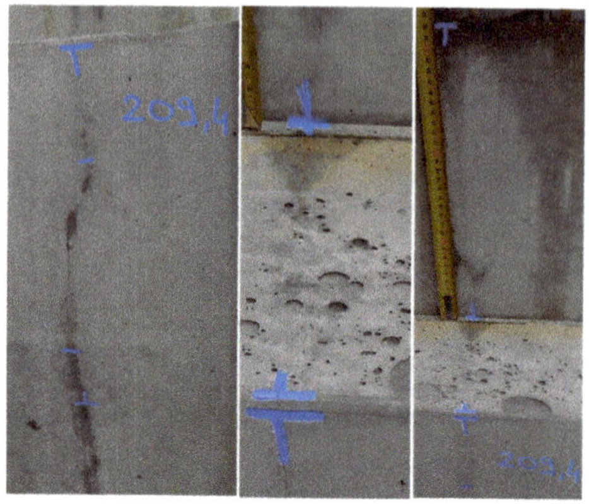

(b)

Figure 10. (**a**) The actual cracking observed in the dome area during the pressurization test; (**b**) the cracking pattern at the bottom of the perimeter wall near the base slab.

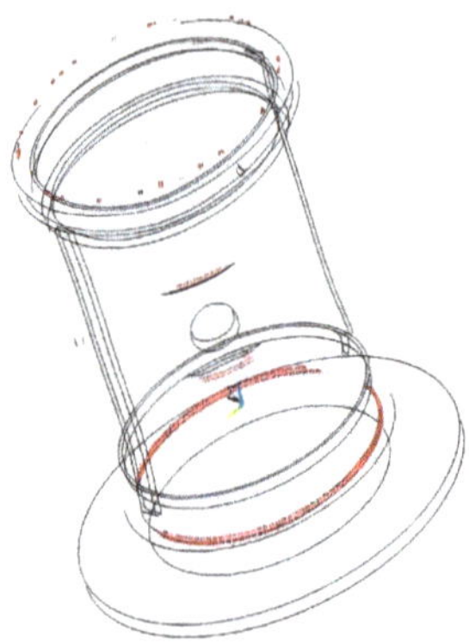

Figure 11. The predicted cracking pattern from numerical analysis.

4. Results

4.1. Structural Integrity Test Conditions

4.1.1. Deformation Analysis

The deformation displacement of the SC containment under pressure conditions of 0.42 MPa and 0.483 Mpa is shown in Figure 12. From Figure 12, it is evident that the maximum deformation area of the SC containment under pressure conditions is located in the 45° area on both sides of the equipment gate opening, which is essentially symmetrically distributed. The maximum radial deformation at 0.42 Mpa is 21.3 mm, and at 0.483 Mpa, it is 24.59 mm. There is a tendency for inward compression on the upper and lower sides of the equipment gate opening, resulting in more complex deformation. This region also experiences significant stress.

4.1.2. Stress and Strain Analysis

(1) Concrete.

The circumferential stress of the concrete part of the SC containment under pressure conditions of 0.42 Mpa and 0.483 Mpa is shown in Figure 13. It is evident from Figure 13 that under these pressure conditions, the stress distribution of the concrete standard section of the steel plate containment is relatively uniform, without any sudden changes. The circumferential stress at the interface between the raft foundation and the cylinder, as well as the lower part of the water tank, is relatively large. The maximum circumferential stress of the concrete is 2.651 Mpa under 0.42 Mpa pressure and 2.654 Mpa under 0.483 Mpa pressure. The circumferential stress in the cylinder and dome area generally ranges from 0.441 Mpa to 1.704 Mpa, which does not exceed the tensile strength of the concrete.

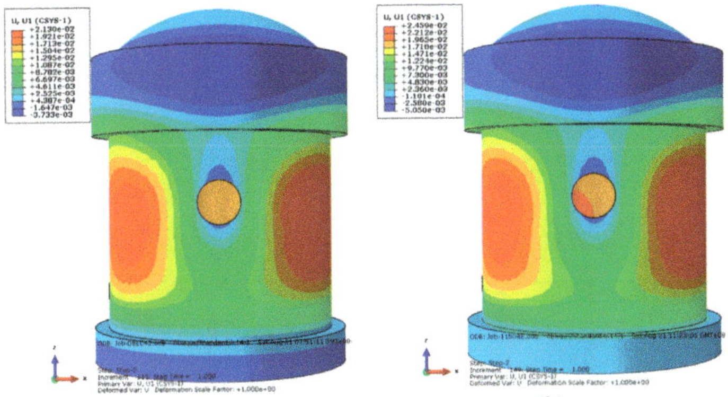

(**a**) (**b**)

Figure 12. (**a**) The overall deformation under a pressure of 0.42 Mpa; (**b**) the overall deformation under a pressure of 0.483 Mpa.

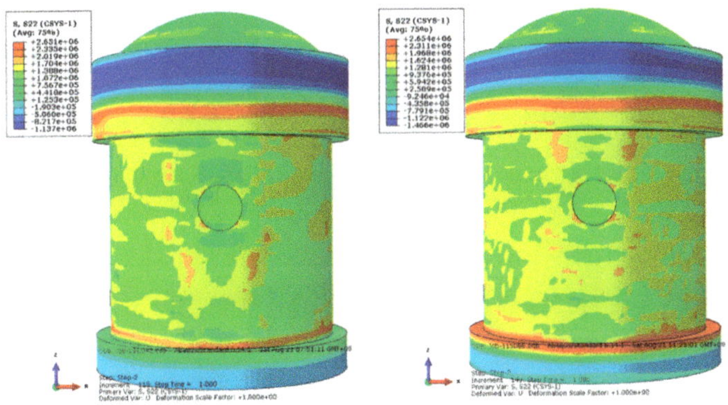

(**a**) (**b**)

Figure 13. (**a**) The circumferential stress of the concrete under a pressure of 0.42 Mpa; (**b**) the circumferential stress of the concrete under a pressure of 0.483 Mpa.

(2) Internal and outer steel plate.

The Mises stress, circumferential stress, and strain of the inner steel plate of the SC containment under pressure conditions of 0.42 Mpa and 0.483 Mpa are shown in Figure 14. From Figure 14, it is evident that under these pressure conditions, the stress distribution of the standard section of the steel plate on the inner side of the containment is relatively uniform, with no abrupt changes, and the maximum stress and strain occur on the upper and lower sides of the equipment gate opening. The maximum Mises stress of the inner steel plate at 0.42 Mpa is 240.2 Mpa, and at 0.483 Mpa, it is 277.6 Mpa, which does not exceed the yield strength of the steel plate. From Figure 14c,d, it is evident that the circumferential stress in the standard area of the inner steel plate ranges from approximately 122 Mpa to 210.8 Mpa. It is evident from Figure 14e,f that the maximum strain of the inner steel plate under the pressure of 0.42 Mpa is 1.162×10^{-3}, and under 0.483 Mpa, it is 1.352×10^{-3}.

Both values are less than the limit of 2×10^{-3} film strain under the service load of the lining plate, which plays a sealing role in the containment according to ASME Volume III.

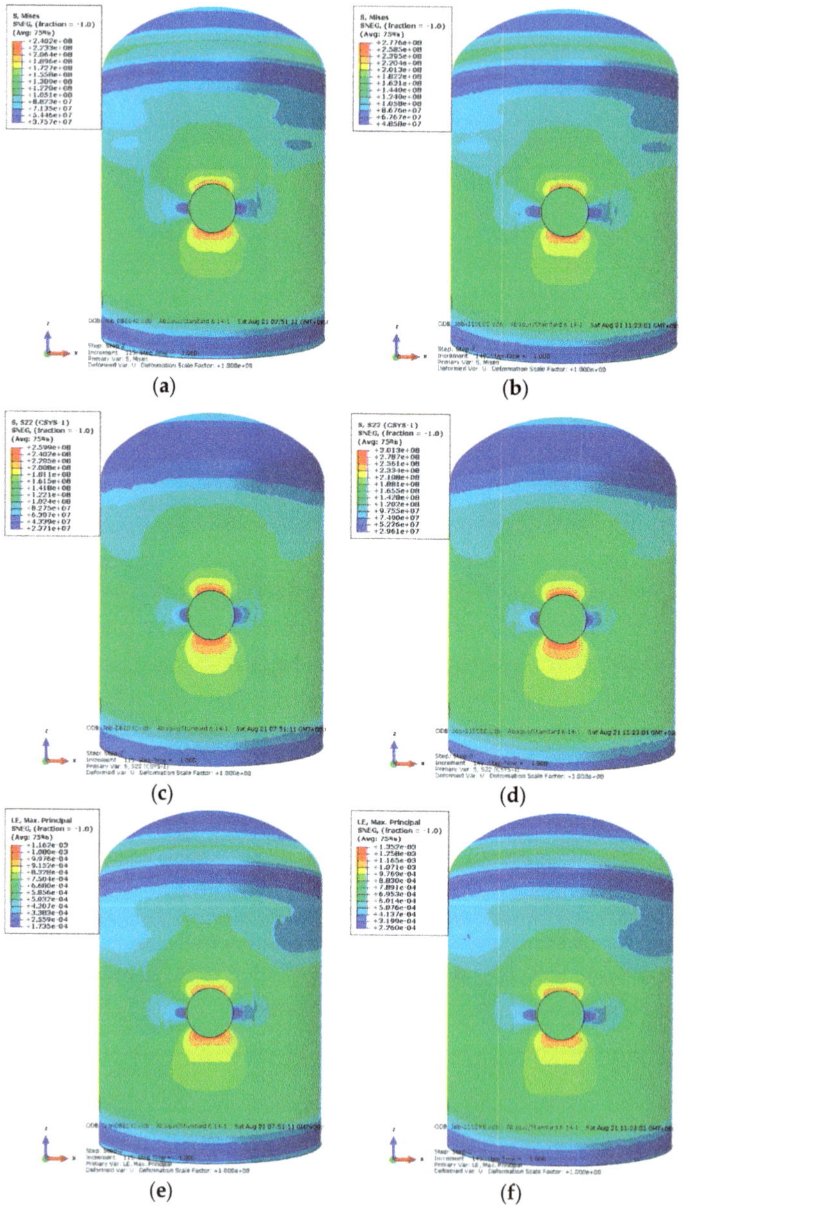

Figure 14. (**a**) The Mises stress of the inner steel plate under a pressure of 0.42 Mpa; (**b**) the Mises stress of the inner steel plate under a pressure of 0.483 Mpa; (**c**) the circumferential stress of the inner steel plate under a pressure of 0.42 Mpa; (**d**) the circumferential stress of the inner steel plate under a pressure of 0.483 Mpa; (**e**) the circumferential strain of the inner steel plate under a pressure of 0.42 Mpa; (**f**) the circumferential strain of the inner steel plate under a pressure of 0.483 Mpa.

The Mises stress, circumferential stress, and strain of the outer steel plate of the sc containment under pressure conditions of 0.42 Mpa and 0.483 Mpa are shown in Figure 15. it is evident from Figure 15 that under these pressure conditions, the stress distribution of the standard section of the outer steel plate of the containment is relatively uniform, with no abrupt changes. the stress and strain in the 45° downward area on both sides of the equipment gate are relatively large. the maximum mises stress of the outer steel plate at 0.42 Mpa is 204.3 Mpa, and at 0.483 Mpa, it is 237.4 Mpa, which is close to the stress of the inner steel plate and does not exceed the yield strength of the steel plate. it is evident from Figure 15c,d that the circumferential stress in the standard area of the outer steel plate ranges from approximately 93.65 Mpa to 192.9 Mpa. it is evident from Figure 15e,f that the maximum strain of the inner steel plate under the pressure of 0.42 Mpa is 9.736×10^{-4}, and under 0.483 Mpa, it is 1.144×10^{-3}. both values are less than the limit of 2×10^{-3} of the membrane strain under the use load of the lining plate, which plays a sealing role in the containment according to ASME Volume iii.

Therefore, from the above calculations and analysis, it is evident that the sealing of the sc containment can be ensured by the inner and outer steel plates under the given pressure conditions. the integrity of the containment can be ensured by the combined strength of the inner and outer steel plates and the concrete.

(3) Anchors and tendons.

The Mises stress of the anchors and tendons under pressure conditions of 0.42 Mpa and 0.483 Mpa are shown in Figure 16. The maximum Mises stress values of the anchors and tendons at 0.42 Mpa and 0.48 Mpa were 210.3 Mpa and 243.4 Mpa, respectively, which did not exceed their yield stress of 335 Mpa.

4.2. Design Basis Conditions

4.2.1. Deformation Analysis

The temperature distribution, radial displacement, and vertical displacement of the SC containment under the design basis condition (design basis pressure 0.42 Mpa, containment temperature 145 °C) are shown in Figures 17–19. Since the atmospheric temperature of the containment under accident conditions is 145 °C, convective heat transfer and thermal radiation occur between the atmosphere of the containment and the steel liner. In this study, a temperature of 145 °C is directly applied to the inner steel plate. It is evident from Figure 17 that the temperature is distributed in a gradient along the thickness of the containment wall, and the temperature distribution obtained by the sequential coupling method is consistent with that of the fully coupled method, demonstrating the reliability of both methods. From Figures 18 and 19, it is evident that the maximum radial displacement of the SC containment under the design basis condition appears in the water tank area, which also has a high temperature, and the maximum vertical displacement appears in the dome area. The displacement distribution obtained by the sequential coupling method is essentially the same as that obtained by the full coupling method. The maximum radial displacement and vertical displacement values obtained by the sequential coupling method are 47.61 mm and 90.85 mm, respectively. The maximum radial displacement and vertical displacement values obtained by the full coupling method are 46.76 mm and 90.40 mm, respectively. The results obtained by the sequential coupling method are slightly larger than those obtained by the full coupling method. This difference arises because the sequential coupling method is a linear superposition of temperature deformation and mechanical deformation, whereas the full coupling method considers the interaction between the two, resulting in slightly smaller values. The radial displacement value of the SC containment, considering the 145 °C, is twice as large as that without the temperature consideration. This finding is consistent with the OECD-NEA's containment benchmark research conclusion that the deformation of the containment structure increases significantly after considering the temperature load, especially after accounting for material performance degradation.

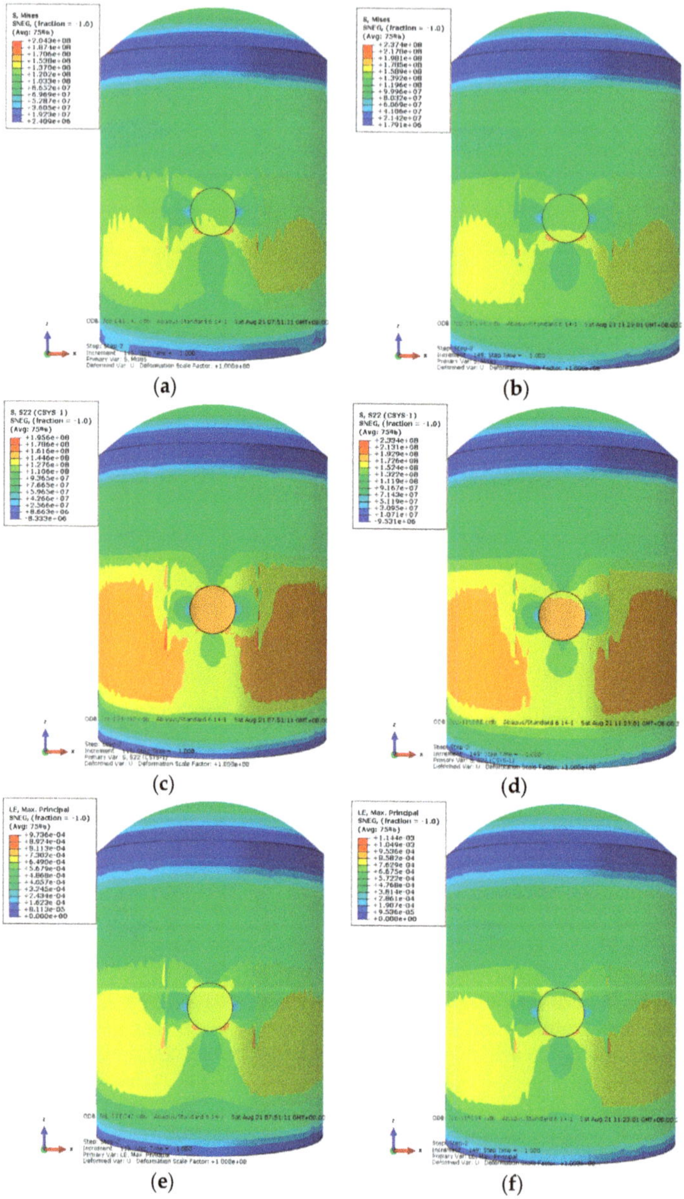

Figure 15. (**a**) The mises stress of the outer steel plate under a pressure of 0.42 Mpa; (**b**) the Mises stress of the outer steel plate under a pressure of 0.483 Mpa; (**c**) the circumferential stress of the outer steel plate under a pressure of 0.42 Mpa; (**d**) the circumferential stress of the outer steel plate under a pressure of 0.483 Mpa; (**e**) the circumferential strain of the outer steel plate under a pressure of 0.42 Mpa; (**f**) the circumferential strain of the outer steel plate under a pressure of 0.483 Mpa.

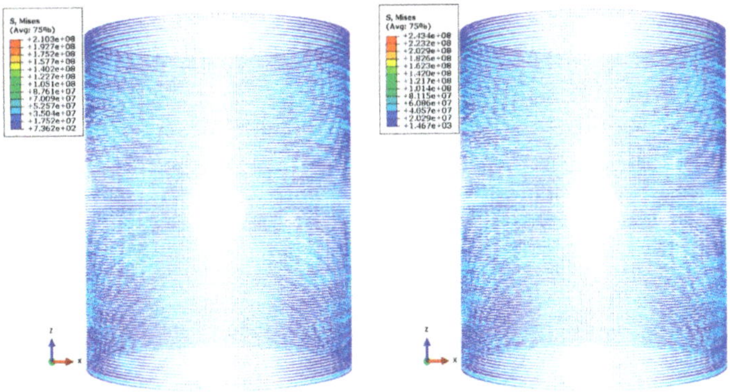

(**a**) (**b**)

Figure 16. (**a**) The Mises stress of the anchors and tendons under a pressure of 0.42 Mpa; (**b**) the Mises stress of the anchors and tendons under a pressure of 0.483 Mpa.

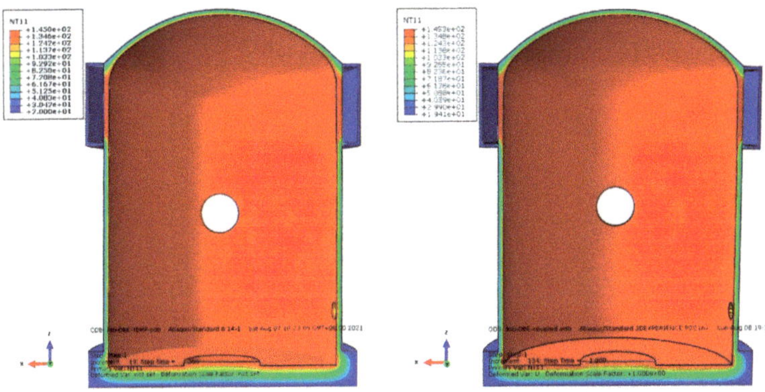

(**a**) (**b**)

Figure 17. (**a**) The temperature distribution map of the SC containment under the design basis condition is obtained using the sequential coupling method; (**b**) the temperature distribution map of the SC containment under the design basis condition is obtained using the full coupling method.

4.2.2. Stress and Strain Analysis

(1) Concrete.

The circumferential stress of the concrete part of the SC containment under the design basis condition is shown in Figure 20. It is evident from Figure 20 that under the design basis condition, the stress distribution of the concrete standard section of the steel plate containment is relatively uniform, with no abrupt changes. The tensile damage of the concrete occurs due to the combined action of membrane stress and bending stress under the design basis condition. At this time, the sealing and integrity of the SC containment are ensured by the inner and outer steel plates.

(2) Internal and outer steel plate.

The Mises stress, circumferential stress, and strain of the inner steel plate of the SC containment under the design basis condition are shown in Figure 21. It is evident from Figure 21 that under the design basis condition, the stress distribution of the standard section of the steel plate inside the steel containment is relatively uniform, with no abrupt changes. The stress and strain on the upper and lower sides of the equipment gate opening are higher. The maximum Mises stress of the inner steel plate under the sequential coupling method and the full coupling method is 270.7 Mpa and 257.3 Mpa, respectively, which does not exceed the yield strength of the steel plate. From Figure 21c,d, it is evident that the circumferential stress in the standard area of the inner steel plate is under pressure in some areas due to the combined action of bending stress and internal pressure. At this time, the internal contraction effect of the bending stress on the inner steel plate is greater than the external expansion effect caused by internal pressure. It is evident from Figure 21e,f that the maximum strain of the inner steel plate under the design basis condition is 2.293×10^{-3} and 2.271×10^{-3}, respectively, under the sequential coupling method and the full coupling method, which is less than the limit of 4×10^{-3} for the combined film and bending strain under the use load of the lining plate with a sealing effect, as specified in ASME Volume III.

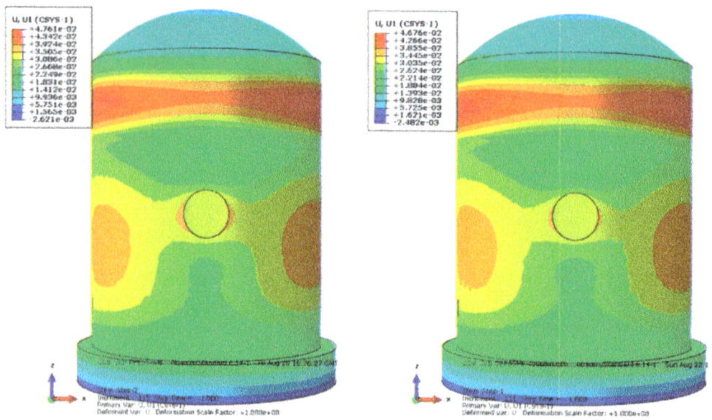

(**a**) (**b**)

Figure 18. (**a**) The radial displacement of the SC containment under design basis conditions based on the sequential coupling method; (**b**) the radial displacement of the SC containment under design basis conditions based on the full coupling method.

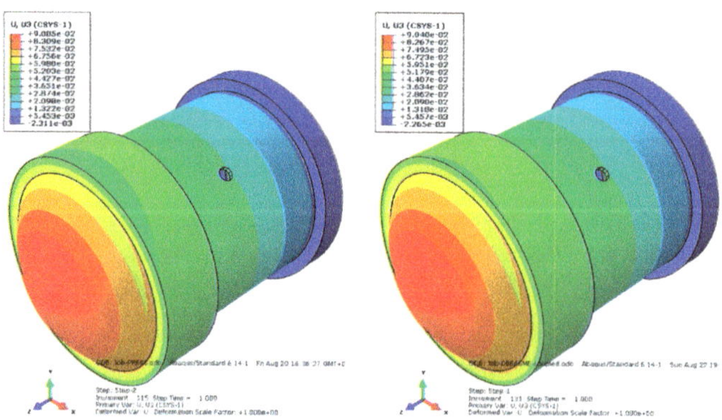

(**a**) (**b**)

Figure 19. (**a**) The vertical displacement of the SC containment under the design basis condition is calculated using the sequential coupling method; (**b**) the vertical displacement of the SC containment under the design basis condition is calculated using the full coupling method.

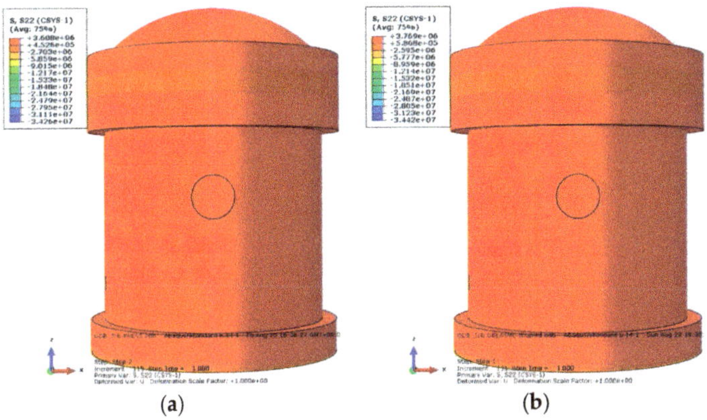

Figure 20. (**a**) The circumferential stress of the SC containment under design basis conditions based on the sequential coupling method; (**b**) the circumferential stress of the SC containment under design basis conditions based on the full coupling method.

The Mises stress, circumferential stress, strain, and equivalent plastic strain of the outer steel plate of the SC containment under the design basis condition are shown in Figure 22. It is evident from Figure 22 that the stress and strain of the outer steel plate are greater than those of the inner steel plate. Under the design basis condition, the Mises stress of the outer steel plate of the SC containment, under the combined action of bending stress and membrane stress, does not reach its yield strength. The maximum circumferential stress values of the outer steel plate under the sequential coupling method and the full coupling method are 422.2 Mpa and 398.1 Mpa, respectively, appearing in the upper and lower areas of the equipment gate opening. The circumferential stress in the standard area ranges between 309.9 Mpa and 384.7 Mpa, which does not reach its yield strength. It is evident from Figure 22e,f that the maximum strain values of the outer steel plate under the design basis conditions under the sequential coupling method and the full coupled method are 2.467×10^{-3} and 2.310×10^{-3}, respectively, which are less than the limit of 4×10^{-3} for the combined film and bending strain under the use load of the lining plate with a sealing effect, as specified in ASME Volume III. From Figure 22g,h, it is evident that the outer steel plate has not yet exhibited plastic strain under these working conditions.

Therefore, it is evident from the above calculations and analysis that the stress and strain of the outer steel plate of the SC containment are significantly greater than those of the inner steel plate under the design basis condition, and the sealing and integrity of the containment are mainly ensured by the inner and outer steel plates.

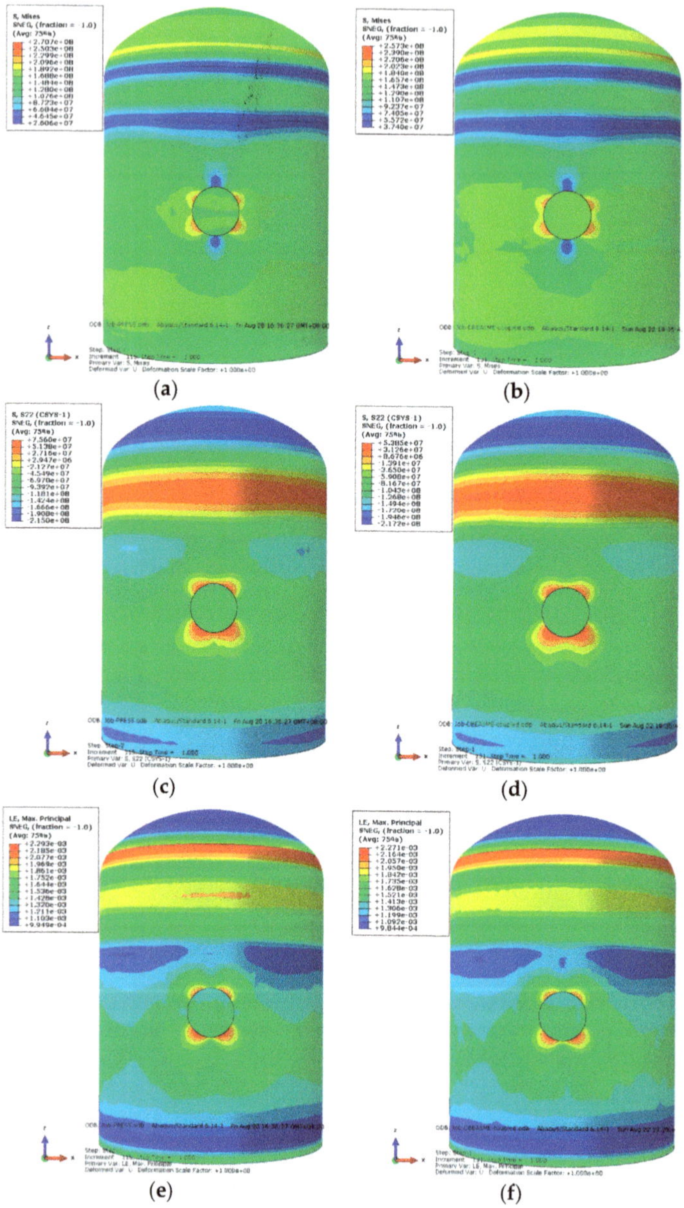

Figure 21. (**a**) The Mises stress of the inner steel plate under the design basis condition based on the sequential coupling method; (**b**) the Mises stress of the inner steel plate under the design basis condition based on the full coupling method; (**c**) the circumferential stress of the inner steel plate under the design basis condition based on the sequential coupling method; (**d**) the circumferential stress of the inner steel plate under the design basis condition based on the full coupling method; (**e**) the strain of the inner steel plate under the design basis condition based on the sequential coupling method; (**f**) the strain of the inner steel plate under the design basis condition based on the full coupling method.

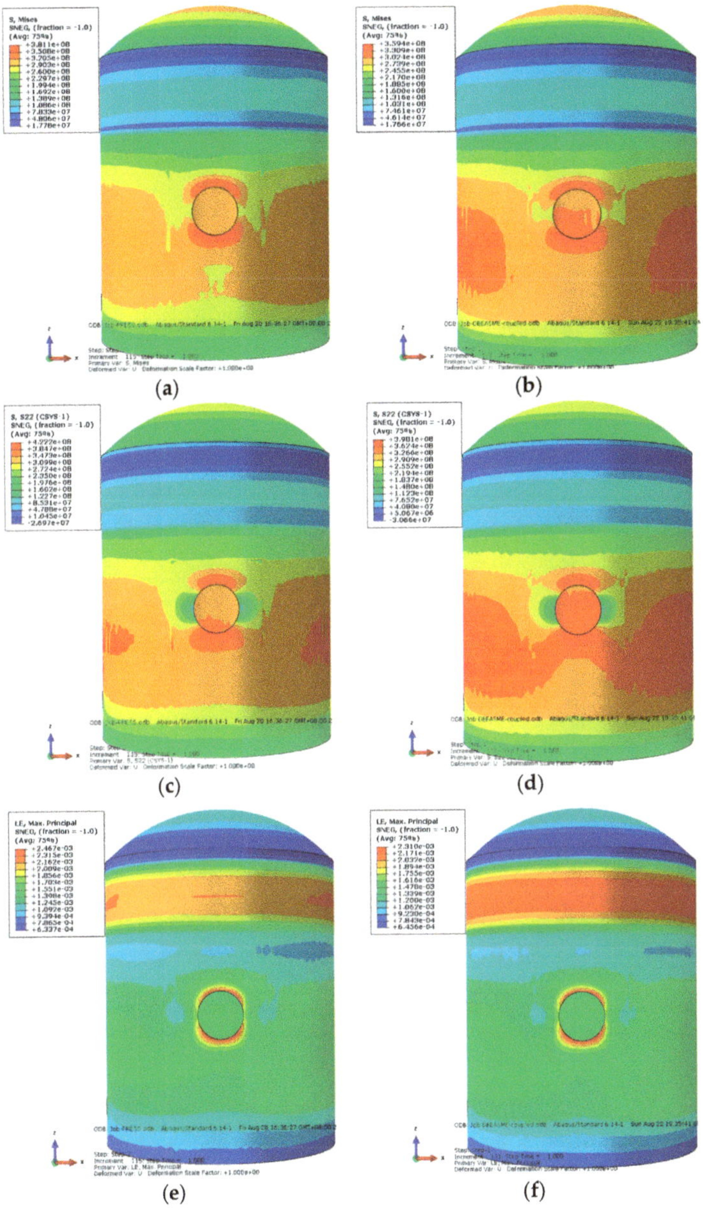

Figure 22. *Cont.*

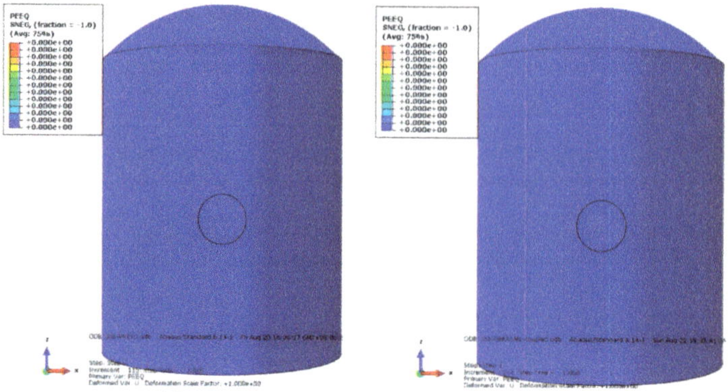

(g) (h)

Figure 22. (**a**) The Mises stress of the outer steel plate under the design basis condition based on the sequential coupling method; (**b**) the Mises stress of the outer steel plate under the design basis condition based on the full coupling method; (**c**) the circumferential stress of the outer steel plate under the design basis condition based on the sequential coupling method; (**d**) the circumferential stress of the outer steel plate under the design basis condition based on the full coupling method; (**e**) the strain of the outer steel plate under the design basis condition based on the sequential coupling method; (**f**) the strain of the outer steel plate under the design basis condition based on the full coupling method; (**g**) the effective plastic strain of the outer steel plate under the design basis condition based on the sequential coupling method; (**h**) the effective plastic strain of the outer steel plate under the design basis condition based on the full coupling method.

4.3. Severe Accident Conditions

4.3.1. Deformation Analysis

The radial displacement of the SC containment under two severe accident conditions (temperature of 150 °C, pressure of 0.55 Mpa, and temperature of 154 °C, pressure of 0.5 Mpa) is shown in Figure 23. From Figure 23, it is evident that the maximum radial displacement of the SC containment under severe accident conditions appears in the water tank area, which is consistent with the displacement distribution area of the design basis condition. The displacement obtained by the sequential coupling method and the full coupling method is essentially the same. The radial deformation value of the containment at 150 °C and 0.55 Mpa is 53.83 mm, which is greater than the deformation value of 53.46 mm at 154 °C and 0.50 Mpa. Therefore, the following analysis mainly focuses on the calculation results for a temperature of 150 °C and pressure of 0.55 Mpa.

4.3.2. Stress and Strain Analysis

(1) Concrete.

The circumferential stress of the concrete part of the SC containment under severe accident conditions is shown in Figure 24. It is evident from the diagram that under severe accident conditions, the stress distribution of the standard section of steel containment concrete is relatively uniform, with no abrupt changes. Tensile damage occurs in the concrete due to the combined action of membrane stress and bending stress under severe accident conditions. At this time, the sealing and integrity of the SC containment are ensured by the inner and outer steel plates.

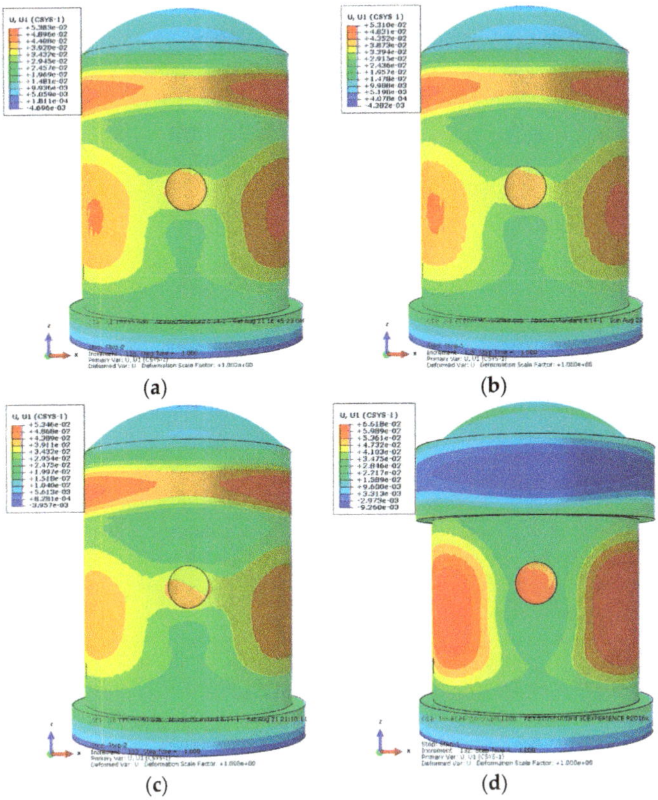

Figure 23. (**a**) The radial displacement of the SC containment under severe accident conditions based on the sequential coupling method (temperature of 150 °C, pressure of 0.55 Mpa); (**b**) the radial displacement of the SC containment under severe accident conditions based on the full coupling method (temperature of 150 °C, pressure of 0.55 Mpa); (**c**) the radial displacement of the SC containment under severe accident conditions based on the sequential coupling method (temperature of 154 °C, pressure of 0.50 Mpa); (**d**) the radial displacement of the SC containment under severe accident conditions based on the full coupling method (temperature of 154 °C, pressure of 0.50 Mpa).

(2) Internal and outer steel plate.

The Mises stress, circumferential stress, and strain of the inner steel plate of the SC containment under severe accident conditions are shown in Figure 25. It is evident from Figure 25 that under severe accident conditions, the stress distribution of the standard section of the steel plate inside the steel containment is relatively uniform, with no abrupt changes. The stress and strain on the upper and lower sides of the equipment gate opening are higher, which is consistent with the stress and strain distribution under the design basis condition. The maximum circumferential stress of the inner steel plate under the sequential coupling method and the complete coupling method is 143.7 Mpa and 118.7 Mpa, respectively, which does not exceed the yield strength of the steel plate. The maximum strain of the inner steel plate under severe accident conditions is 2.631×10^{-3} and 2.622×10^{-3}, respectively, under the sequential coupling method and the complete coupling method, which is less than the limit of 4×10^{-3} for the combined film and bending strain under the use load of the lining plate with a sealing effect, as specified in ASME Volume III, and less than the limit of 1×10^{-2} for the combined film and bending strain under the ultimate design load.

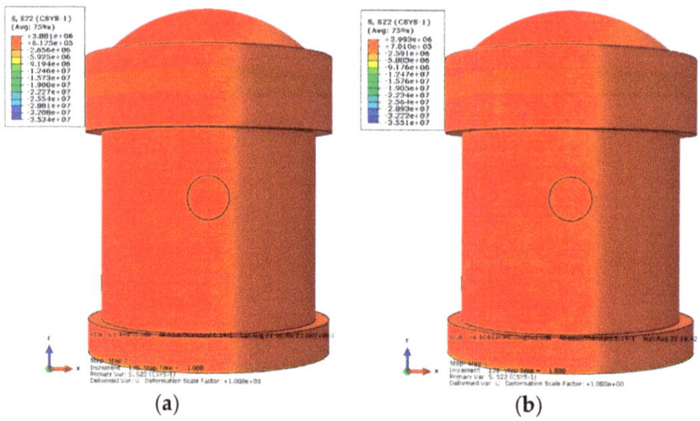

Figure 24. (**a**) The circumferential stress of the SC containment under severe accident conditions based on the sequential coupling method; (**b**) the circumferential stress of the SC containment under severe accident conditions based on the full coupling method.

The Mises stress, circumferential stress, strain, and equivalent plastic strain of the outer steel plate of the SC containment under severe accident conditions are shown in Figure 26. It is evident from Figure 26 that the stress and strain of the outer steel plate are greater than those of the inner steel plate. Under severe accident conditions, the maximum Mises stress and circumferential stress of the outer steel plate of the SC containment shell under the sequential coupling method and the complete coupling method appear on the upper and lower sides of the equipment gate opening. The maximum values are 464.3 Mpa and 451.2 Mpa, respectively, which exceed their yield strength but do not reach their tensile strength. The maximum strain values of the outer steel plate under the sequential coupling method and the complete coupling method are 2.815×10^{-3} and 2.601×10^{-3}, respectively, which are less than the limit of 1×10^{-2} for the combined film and bending strain under the ultimate design load of the lining plate with a sealing effect, as specified in ASME Volume III. It is evident from Figure 26g,h that under the sequential coupling method, the plastic deformation of the outer steel plate begins to appear near the gate opening of the equipment, and the equivalent plastic deformation value is about 3.273×10^{-4}.

Therefore, it is evident from the above calculations and analysis that under severe accident conditions, the stress and strain of the outer steel plate of the containment are significantly greater than those of the inner steel plate, and the sealing and integrity of the containment are mainly ensured by the inner and outer steel plates.

(3) Anchors and tendons.

The Mises stress and strain of anchors and tendons under severe accident conditions are shown in Figure 27. It is evident from the diagram that the Mises stress of anchors and tendons in some areas reaches their yield strength of 335 Mpa, which is mainly concentrated near the interface between the ASP tank and the cylinder, where the force is also more complex. The maximum strain value is 4.155×10^{-3}, which is less than the limit of 1×10^{-2} for the combined film and bending strain under the ultimate design load specified in ASME Volume III.

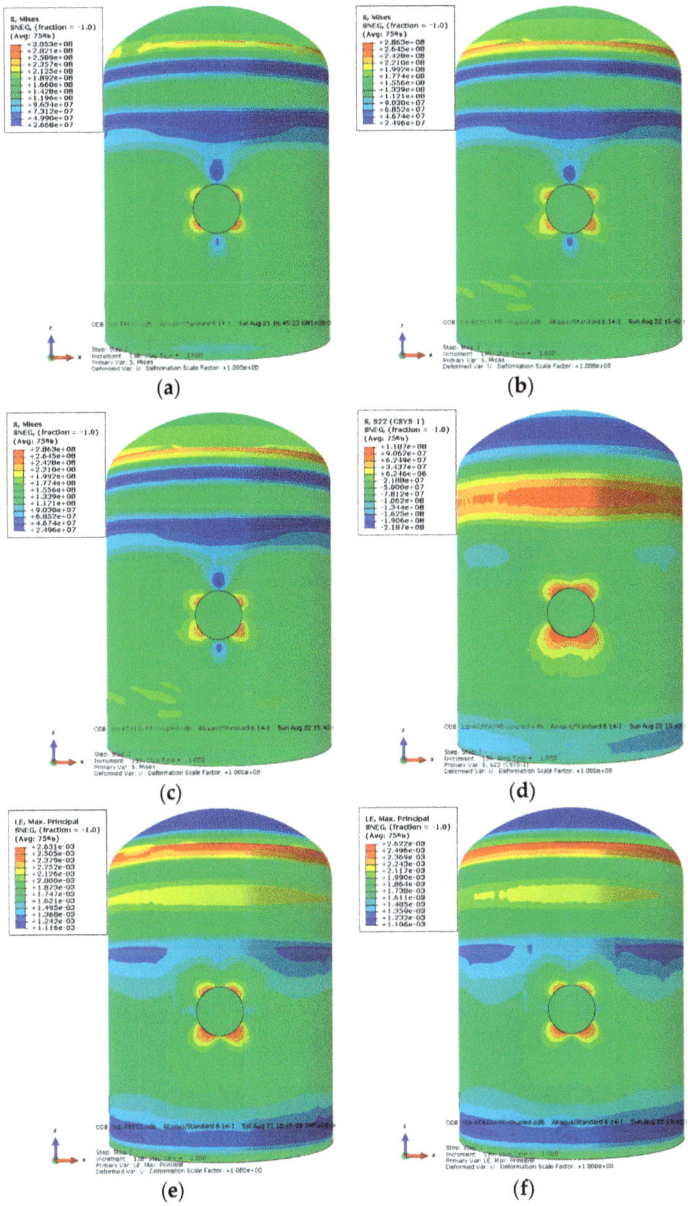

Figure 25. (**a**) The Mises stress of the inner steel plate under severe accident conditions based on the sequential coupling method; (**b**) the Mises stress of the inner steel plate under severe accident conditions based on the full coupling method; (**c**) the circumferential stress of the inner steel plate under severe accident conditions based on the sequential coupling method; (**d**) the circumferential stress of the inner steel plate under severe accident conditions based on the full coupling method; (**e**) the strain of the inner steel plate under severe accident conditions based on the sequential coupling method; (**f**) the strain of the inner steel plate under severe accident conditions based on the full coupling method.

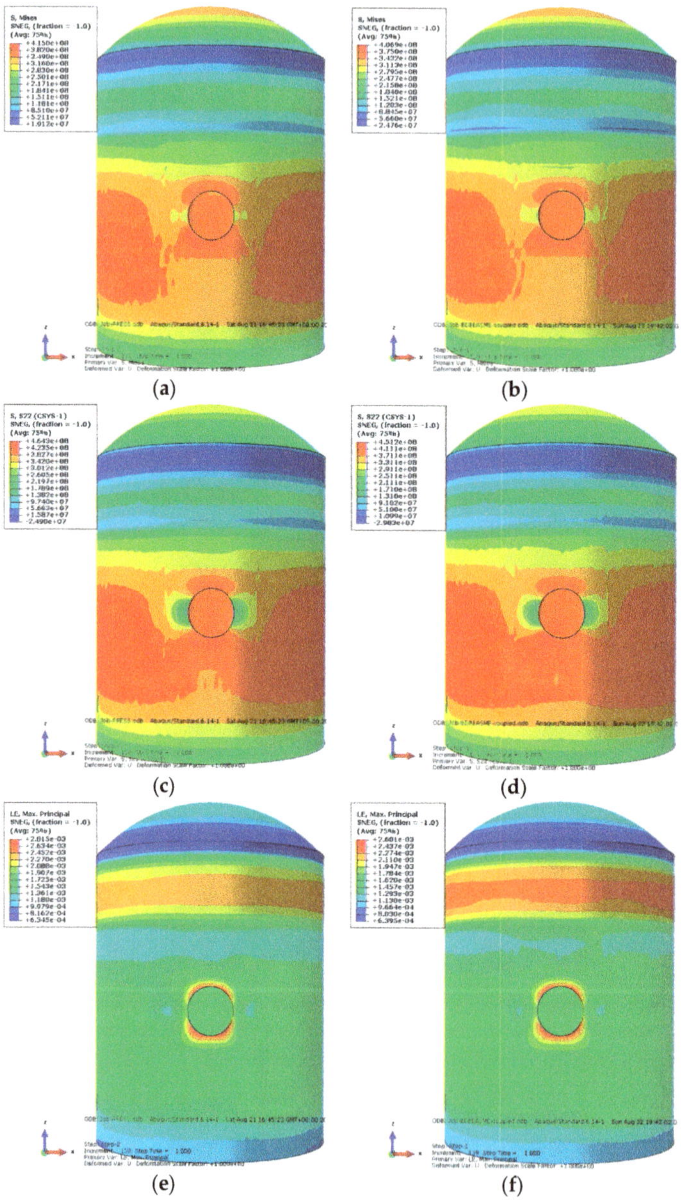

Figure 26. *Cont.*

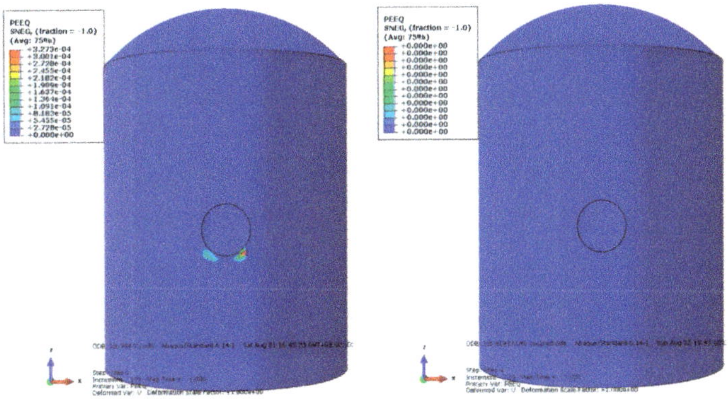

(g) (h)

Figure 26. (**a**) The Mises stress of the outer steel plate under severe accident conditions based on the sequential coupling method; (**b**) the Mises stress of the outer steel plate under severe accident conditions based on the full coupling method; (**c**) the circumferential stress of the outer steel plate under severe accident conditions based on the sequential coupling method; (**d**) the circumferential stress of the outer steel plate under severe accident conditions based on the full coupling method; (**e**) the strain of the outer steel plate under severe accident conditions based on the sequential coupling method; (**f**) the strain of the outer steel plate under severe accident conditions based on the full coupling method; (**g**) the effective plastic strain of the outer steel plate under severe accident conditions based on the sequential coupling method; (**h**) the effective plastic strain of the outer steel plate under severe accident conditions based on the full coupling method.

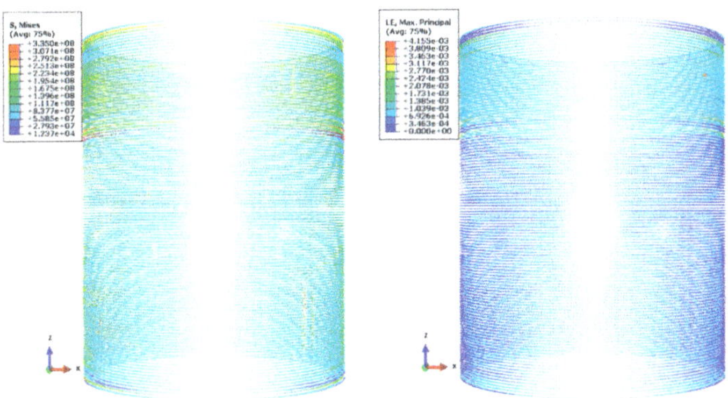

(a) (b)

Figure 27. (**a**) The Mises stress of the anchors and tendons under severe accident conditions based on the sequential coupling method; (**b**) the strain of the anchors and tendons under severe accident conditions based on the sequential coupling method.

4.4. Failure Criterion and Ultimate Bearing Capacity

4.4.1. Failure Criterion

The containment is the last barrier for nuclear power plants to prevent the release of nuclear fission products to the environment under accident conditions. The realization of this function depends on the sealing effect formed by the integrity of the steel plate in the containment. Therefore, the ultimate bearing capacity of the containment structure will depend on whether the containment can continue to withstand internal pressure and ensure the integrity of the steel plate.

In this paper, one of the four conditions is used as the criterion for the loss of integrity of the containment:

(1) The regional yield of the inner steel plate occurs, and the yield area reaches more than 10 m^2. This acceptance criterion is based on the engineering practice of some nuclear power plants.

(2) According to the research results of OECD-NEA on the international standard of containment integrity (ISP48) [33]: When the maximum equivalent plastic strain of the containment steel lining exceeds 0.30% ± 0.15%, the steel lining is likely to be regionally torn. Therefore, when the maximum equivalent plastic strain of the inner steel plate reaches 0.15%, the containment structure reaches the ultimate bearing capacity.

(3) According to the provisions of ASME III CC-3710 [47]: When the steel lining is under the ultimate design load, the combined film and bending strain of the steel lining should not be more than 1%. In this paper, the maximum strain value of the inner steel plate reaches 1% as the containment reaches the ultimate bearing capacity.

(4) According to the provisions of RG1.216 [48]: Except for the prestressed steel bundle material, the overall strain of other materials in the discontinuous region should not exceed 0.4%. Therefore, in this paper, the maximum strain value of the discontinuous region reaches 0.4% as the containment reaches the ultimate bearing capacity.

4.4.2. Ultimate Bearing Capacity

Due to the conservative nature of the sequential coupling method's calculation results compared to the full coupling method, this paper primarily presents the calculation results of the sequential coupling method. This paper utilizes OECD-NEA guidelines to analyze and examine the time history curves of pressure and temperature in the containment following a severe accident, as depicted in Figure 28, to evaluate the ultimate bearing capacity of the SC containment. Representative observation points near the standard section area and the equipment hatch section area are selected to observe the strain-pressure curve and the equivalent plastic strain-pressure curve, as shown in Figures 29 and 30. The diagram indicates that the SC containment undergoes elastic, local plastic, elastic-plastic, and large deformation states during heating and pressurization. According to the failure criterion, when the strain value of the standard area reaches approximately 0.4%, the internal pressure is about 1.19 Mpa. When the maximum equivalent plastic strain of the inner steel plate reaches 0.15%, the internal pressure is about 1.14 Mpa. When the maximum strain value of the inner steel plate reaches 1%, the internal pressure is about 1.25 Mpa. When the yield area of the inner steel plate reaches 10 m^2, the internal pressure is about 0.91 Mpa, as shown in Figure 31. Therefore, it can be concluded that when the internal pressure reaches 0.91 Mpa, the containment reaches its ultimate bearing capacity.

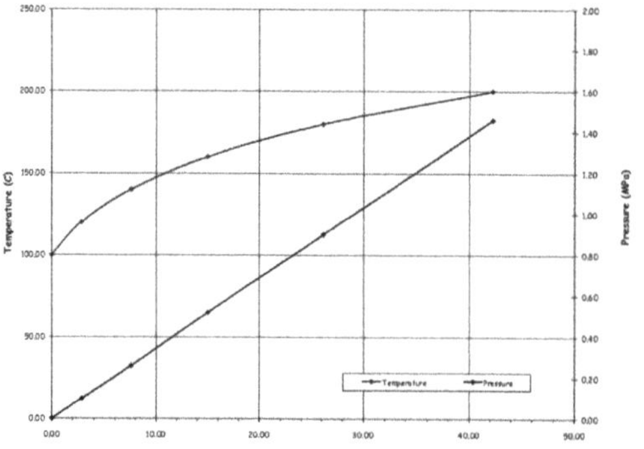

Figure 28. The time history curves of pressure and temperature in the containment following a severe accident according to OECD-NEA.

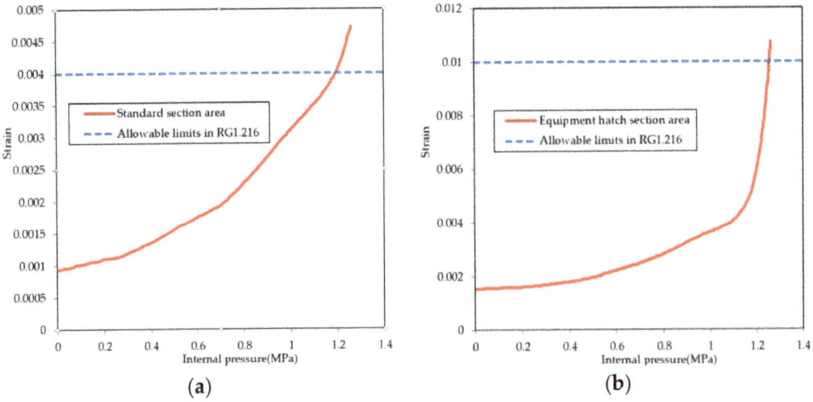

Figure 29. (**a**) The strain-pressure curve in the standard section area; (**b**) the strain-pressure curve in the equipment hatch section area.

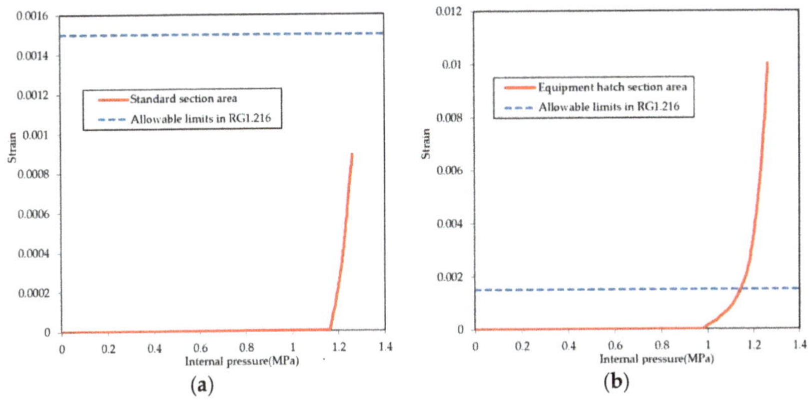

Figure 30. (**a**) The equivalent plastic strain-pressure curve in the standard section area; (**b**) the equivalent plastic strain-pressure curve in the equipment hatch section area.

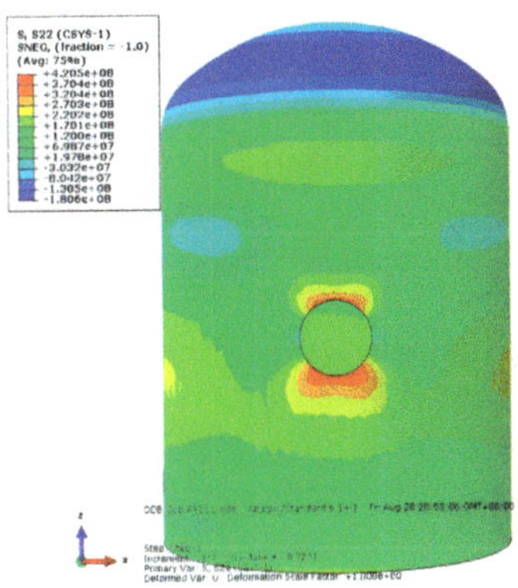

Figure 31. The circumferential stress is when the yield area of the inner steel plate reaches 10 m^2.

5. Conclusions and Recommendations

This study demonstrates that steel-plate concrete containment structures offer significant advantages in strength and safety margins for third-generation nuclear power plants. The findings support their use as a viable alternative to traditional designs, with the potential for further optimization and broader application in diverse reactor systems:

1. The steel-plate concrete containment structure is recommended to adopt a single-layer containment design, with the inner steel plate continuously covering the entire containment dome, cylinder wall, bottom plate, and the inner surface of the ASP tank. This steel plate serves as a pressure boundary to prevent leakage, while the outer steel plate protects against external impacts. The internal and outer steel plates can also serve as construction templates. The recommended thicknesses are 30 mm for the outer steel plate and 25 mm for the inner steel plate, with a concrete thickness of 1.1 m. Both steel plates should be made of SA-738 Gr. B steel, as used in the AP1000 containment. The anchorage system, consisting of intersecting grids, should be embedded within the concrete structure and welded to the steel plate, ensuring stability and resistance to out-of-plane shear forces.

2. The steel-plate concrete containment has a high safety margin under pressure conditions, with stress and strain levels well within safe limits. The maximum deformation occurs symmetrically on both sides of the equipment gate opening, confirming the containment's robustness.

3. Under design basis conditions, the containment maintains a high safety margin, with stress and strain levels far below critical thresholds. The maximum radial and vertical displacements occur in high-temperature areas, such as the water tank and dome, but remain within acceptable limits. The findings align with previous studies, highlighting the increased deformation under temperature effects.

4. Under severe accident conditions, the containment exhibits significant resilience, with stress and strain levels in the steel plates remaining below yield strengths. The sequential coupling method's conservative calculations confirm the containment's integrity, even under extreme conditions.

5. The ultimate bearing capacity of the containment is reached when the internal pressure exceeds 0.91 MPa, which is 2.17 times the design pressure. This threshold marks the containment's ability to withstand extreme internal pressures, ensuring its role as a final safety barrier.

Author Contributions: Conceptualization, G.R. and R.P.; methodology, G.R. and R.P.; software, G.R., Z.D. and T.L.; validation, F.S.; formal analysis, G.R. and R.P.; investigation, G.R. and F.S.; resources, G.R.; data curation, G.R.; writing—original draft preparation, G.R.; writing—review and editing, R.P., F.S. and Z.D.; visualization, G.R. and T.L.; project administration, R.P. and F.S.; funding acquisition, G.R. All authors have read and agreed to the published version of the manuscript.

Funding: This research received no external funding.

Data Availability Statement: The original contributions presented in this study are included in the article; further inquiries can be directed to the corresponding author.

Conflicts of Interest: Authors Zhanfa Dong and Tianyun Lan were employed by the company China Nuclear Power Design Co., Ltd. (Shenzhen). The remaining authors declare that the research was conducted in the absence of any commercial or financial relationships that could be construed as a potential conflict of interest.

References

1. Frano, R.L.; Forasassi, G. Dynamic Response of a Nuclear Power Plant Subjected to External Accident Event. In Proceedings of the 18th International Conference on Nuclear Engineering, Xi'an, China, 17–21 May 2010; pp. 607–615.
2. Ryzhov, S.B.; Mokhov, V.A.; Nikitenko, M.P.; Bessalov, G.G.; Podshibyakin, A.K.; Anufriev, D.A.; Gadó, J.a.; Rohde, U. VVER-Type Reactors of Russian Design. In *Handbook of Nuclear Engineering*; Cacuci, D.G., Ed.; Springer: Boston, MA, USA, 2010; pp. 2249–2320. [CrossRef]
3. Sener, K.C.; Varma, A.H.; Wang, S.; Bhardwaj, S.R.; Gallocher, S. Modular steel-plate composite (SC) wall steelbricks: Experimental and numerical evaluations. *Nucl. Eng. Des.* **2019**, *350*, 224–233. [CrossRef]

4. Zhu, X.; Li, J.; Lin, G.; Pan, R.; Li, L. Sensitivity Analysis of Steel-Plate Concrete Containment against a Large Commercial Aircraft. *Energies* **2021**, *14*, 2829. [CrossRef]
5. Pan, R.; Wu, J.; Zhang, X. Application and development of steel plate reinforced concrete structure in nuclear power engineering. *Ind. Constr.* **2014**, *44*, 1–6.
6. Ning, J.; Bao, F.; Fang, J. Experimental research on seismic behavior of low shear-span ratio composite shear wall with double steel plates and infill concrete. *J. Build. Struct.* **2011**, *32*, 74–81. [CrossRef]
7. Chen, J.; Jun, Z. Construction Technology of Composite Structure of Double Steel Plate and Shear Wall in Yancheng Broadcast Television Tower. *Constr. Technol.* **2011**, *40*, 17–20.
8. Tokyo Electric Power Company (TEPCO). *Improved Construction and Project Management*; TEPCO: Tokyo, Japan, 2001.
9. *JEAG 4618-2005*; Technical Guidelines for Aseismic Design of Steel Plate Reinforced Concrete Structures: Buildings and Structures. Japan Electric Association: Tokyo, Japan, 2005.
10. *JEAG 4618-2009*; Technical Code for Seismic Design of Steel Plate Reinforced Concrete Structures: Buildings and Structures. Japan Electric Association Nuclear Safety: Tokyo, Japan, 2009.
11. Korea Electric Power Industry Code-Structural Nuclear Grade. *Specification for Safety-Related Steel Plate Concrete Structures for Nuclear Facilities. Board of KEPIC Policy, Structural Committee*; Korea Electric Association: Seoul, Republic of Korea, 2010.
12. *ANSI/AISC N690s1-15*; Specification for Safety-Related Steel Structures for Nuclear Facilities. American Institute of Steel Construction: Chicago, IL, USA, 2015; pp. 1–15.
13. Ministry of Housing and Urban Development–Regulation Construction. *GB/T 51340-2018*; Technical Standard for Steel Plate Concrete Structures of Nuclear Power Plants. China Planning Press: Beijing, China, 2018.
14. Zhang, J.-L.; Liu, B.-D.; Zhang, P.-Y.; Wang, Z.-H. Small-scale test and analysis of corrugated-steel-plate–concrete composite member adopting novel shear connectors. *Eng. Struct.* **2019**, *184*, 369–383. [CrossRef]
15. Wang, Q. *The Structural Design and Shock-Resistant Performance of Steel Beam-Concrete Column Frame Structure*; Springer: Cham, Switzerland, 2020; pp. 1020–1026.
16. Zhou, T.; Wang, J.; Sawab, J.; Chen, H.; Mo, Y.L. Biaxial steel plated concrete constitutive models for composite structures: Implementation and validation. *J. Constr. Steel Res.* **2021**, *177*, 106452. [CrossRef]
17. Wu, L.; Wang, H.; Lin, Z. Ultimate strength behavior of steel plate-concrete composite slabs: An experimental and theoretical study. *Steel Compos. Struct.* **2020**, *37*, 741–759. [CrossRef]
18. Liu, R.; Yang, Y. Research on fatigue performance of steel-plate-concrete composite slab. *Thin-Walled Struct.* **2021**, *160*, 107339. [CrossRef]
19. Szewczyk, P.; Szumigała, M. Optimal Design of Steel–Concrete Composite Beams Strengthened Under Load. *Materials* **2021**, *14*, 4715. [CrossRef]
20. Ma, Z.; Wu, Y.; Zhang, J.; Zhang, M. Experimental Study on Seismic Behavior of Coupled Steel Plate and Reinforced Concrete Composite Wall. *Buildings* **2022**, *12*, 2036. [CrossRef]
21. Liu, A.; Wu, Y.; Wang, B.; Chen, X. Seismic Design and Performance Evaluation of Coupled Steel Plate and Reinforced Concrete Composite Walls. *Buildings* **2023**, *13*, 2242. [CrossRef]
22. Wang, B.; Chen, P.; Jiang, H.; Wang, J. Performance of steel-plate-reinforced concrete composite walls in tall structures. *Proc. Inst. Civ. Eng.—Civ. Eng.* **2024**, *177*, 118–125. [CrossRef]
23. Appa Rao, T.V.S.R. *Behavior of Concrete Nuclear Containment Structures up to Ultimate Failure with Special Reference to MAPP-1 Containment, Inelastic Behavior*; Structural Engineering Research Centre: Madras, India, 1975.
24. Donten, K.; Knauff, M.; Sadowski, A.; Scibak, W. *Tests on Model of Prestressed Reactor Containment*; Institute of Nuclear Research: Otwock, Poland, 1980; pp. 231–245.
25. *NUREG/CR-5825*; Post-Test Report on Testing and Analysis of the Sizewell-B 1:10-Scale Containment Model. ANATECH Research Corp.: San Diego, CA, USA, 1991.
26. *NUREG/CR-6810*; Overpressurization Test of a 1:4-Scale Prestressed Concrete Containment Vessel Model. Sandia National Laboratories: Albuquerque NM, USA, 2003.
27. Zhao, X.; Yao, D.; Gao, J.; Tan, J.; Hao, X. A simulation research on thermal-pressure coupling loading of a 1:3.2 containment model. *J. Nucl. Sci. Technol.* **2024**, *150*, 1–14. [CrossRef]
28. Zhao, C.; Li, Z.L.; Dong, Z. Ultimate bearing capacity analysis on containment structure of CPR1000. *Comput. Aided Eng.* **2013**, *22*, 393–398.
29. Song, C.; Hou, G.; Zhou, G. Ultimate Capacity and Influenced Factors Analysis of Nuclear RC Containment Subjected to Internal Pressure. *Chin. J. Nucl. Sci. Eng.* **2014**, *34*, 228–235.
30. Ren, G.; Pan, R.; Sun, F. Imulation of behaviors of a one-third scale containment model test. In Proceedings of the 25th International Conference on Nuclear Engineering, Shanghai, China, 15–19 April 2017.
31. Chakraborty, M.K.; Acharya, S.; Pisharady, A.S.; Roshan, A.D.; Bishnoi, L.R. Assessment of Ultimate Load Capacity of concrete containment structures against structural collapse. *Nucl. Eng. Des.* **2017**, *323*, 417–426. [CrossRef]
32. Tong, L.; Zhou, X.; Cao, X. Ultimate pressure bearing capacity analysis for the prestressed concrete containment. *Ann. Nucl. Energy* **2018**, *121*, 582–593. [CrossRef]
33. Nuclear Energy Agency/Committee on the Safety of Nuclear Installations. *International Standard Problem No. 48 Containment Capacity*; Nuclear Energy Agency/Committee on the Safety of Nuclear Installations: Paris, France, 2005.

34. Huang, X.; Kwon, O.-S.; Bentz, E.; Tcherner, J. Evaluation of CANDU NPP containment structure subjected to aging and internal pressure increase. *Nucl. Eng. Des.* **2017**, *314*, 82–92. [CrossRef]
35. Mahida, P.A.; Desai, D. Structural response evaluation of reinforced-concrete nuclear containment structure subjected to internal overpressure with high-temperature loading. *Int. J. Nucl. Energy Sci. Technol.* **2020**, *14*, 281–290. [CrossRef]
36. Yuqin, G. *The Study on the Ultimate Bearing Capability of Nuclear Power Plant's Containment with Temperature Effect*; Harbin Engineering University: Harbin, China, 2015.
37. Lin, S.; Hu, Z.; Han, J. Experimental Studies on Friction Loss of Prestress of Containment Vessel in an EPR Nuclear Power Plant. *Spec. Struct.* **2022**, *39*, 57–60.
38. Asmolov, V.G.; Gusev, I.N.; Kazanskiy, V.R.; Povarov, V.P.; Statsura, D.B. New generation first-of-the kind unit—VVER-1200 design features. *Nucl. Energy Technol.* **2017**, *3*, 260–269. [CrossRef]
39. Liu, J.; Han, P. Numerical Analyses of a Shield Building Subjected to a Large Commercial Aircraft Impact. *Shock. Vib.* **2018**, *2018*, 7854969. [CrossRef]
40. GE-Hitachi Nuclear Energy Americas, LLC. *BWRX-300 Steel-Plate Composite Containment Vessel (SCCV) and Reactor Building (RB) Structural Design–Non-Proprietary Information*; GE-Hitachi Nuclear Energy Americas, LLC: Wilmington, NC, USA, 2023.
41. *GB-50010-2010*; Specification for Design of Concrete Structures. China Construction Industry Press: Beijing, China, 2011.
42. *EN1994-1-2*; Eurocode 4-Design of Composite Steel and Concrete Structures-Part 1-2: General Rules—Structural Fire Design. British Standards Institution: London, UK, 2005.
43. The American Society of Mechanical Engineers. *ASME Boiler & Pressure Vessel Code II Part A*; The American Society of Mechanical Engineers: New York, NY, USA, 2007.
44. *GB 51249-2017*; Code for Fire Safety of Steel Structures in Buildings. China Planing Press: Beijing, China, 2017.
45. Jin, S.; Li, Z.; Lan, T.; Dong, Z.; Gong, J. Nonlinear Finite Element Analysis of Prestressed Concrete Containment Vessel under Severe Accident Loads. *KSCE J. Civ. Eng.* **2020**, *24*, 816–825. [CrossRef]
46. Federal Association of Fire Protection Engineers Design. *Rules for Design and Construction of PWR Nuclear Civil Works: RCC-CW. Paris*; French Association for Design: Paris, France, 2015; pp. 7–8.
47. American Society of Mechanical Engineers. *ASME Boiler & Pressure Vessel Code III Division 2-Code for Concrete Containments:ACI 359*; The American Society of Mechanical Engineers: New York, NY, USA, 2004.
48. Nuclear Regulatory Commission. *Rule 1.206, Revision. Containment Structural Integrity Evaluation for Internal Pressure Loadings above Design Basis Pressure*; Nuclear Regulatory Commission: Rockville, MD, USA, 2010.

buildings

MDPI

Article

Effect of Preload on Box-Section Steel Columns Filled with Concrete under Axial Load: A Numerical Study

Ahmed Mohamed Sayed [1,2]

[1] Department of Civil and Environmental Engineering, College of Engineering, Majmaah University, Al-Majmaah 11952, Saudi Arabia; a.sayed@mu.edu.sa or ahmed_m_sg@aun.edu.eg
[2] Department of Civil Engineering, College of Engineering, Assiut University, Assiut 71511, Egypt

Abstract: External loads applied to a box-section steel column before it is filled with concrete to increase its efficiency due to modifications in structural systems or design errors may reduce its ultimate capacity and change its structural behavior. To examine this effect, finite element modeling (FEM) has been used to simulate these columns under preloading at different ratios with many variables in the geometric dimensions of the columns. The FEM results have been investigated using 38 experimental specimens obtained from previous studies without preloading. The results demonstrated high accuracy in modeling these columns in structural behavior and ultimate load capacity. After verifying the results, 84 Concrete-Filled Steel Columns (CFSC) were modeled under different preload ratios. The results indicated that some variables have directly affected the value of the decrease in column capacity in terms of its height, wall thickness, yield stress, and preload ratios, while others were inversely proportional in terms of the cross-section dimensions and concrete strength. The preload effect ratio had two separate limits, where when it reached 70%, the maximum value of the decrease in column capacity was 10.90%. The value increased sharply reaching 19.90% when there was a preload equal to 80%. New equations have been proposed to predict the ultimate capacity of CFSC under preloading with suitable accuracy with a correlation coefficient of no less than 0.949.

Keywords: preloaded steel columns; concrete-filled box sections; finite element modeling; axial load capacity; structural behavior prediction

1. Introduction

Steel columns in structural systems may be defined as the backbone of steel buildings and great attention during design is given to them. Many shapes are used as steel columns, including I-sections, boxes (rectangles and squares), and built-up sections of angles or channels. These sections can be bonded to concrete by filling or encasing them to form a composite section. It might encounter certain problems that appear in the steel construction sites as a result of an increase in loads on the box columns as a result of a change in the structural system or an error in the design. It is categorically pivotal to build up these steel columns to resist any increase in the loads applied to them. Therefore, one of the strengthening methods is to use concrete to fill the box columns though they are under the influence of the already existing loads (preloading). Through literature review, it has been found out that many studies and even existing codes (ANSI/AISC 360-22 [1] and Eurocode-4 [2]) focused on this type of column labeled (composite section), but no preloading has been applied. That is, the column is just filled with concrete and then exposed to loads. This method of loading does not represent the application in reality if the columns are filled with concrete while they are under the influence of existing loads due to their presence in the current construction system.

Recently, many studies have paid attention to studying the behavior of Concrete-Filled Steel Columns (CFSC), whether it is a normal-strength [3–5] or high-strength concrete [6–8].

Among the normal-strength concrete studies was Schneider [9]. A number of 14 steel columns with different cross-sections consisting of 3 circular, 6 rectangular, and 5 square were tested under axial load. The concrete cylinder strength used ranged from 24 to 31 MPa and the yield stress ranged from 312 to 537 MPa. Different column wall thicknesses were used, which resulted in column width- or diameter-to-thickness ratios ranging from 17 to 50.8. The most important results are that the axial load behavior is greatly affected by the cross-section shape and the ratio of column dimensions to wall thickness. Uy [10] performed an experiment testing 19 square CFSC with concrete strength ranging from 28 to 32 MPa and a steel yield stress of 750 MPa. The test results indicated that the Eurocode-4 [2] approach was unsafe for the specimens exposed to pure axial load or bending. The reason was the Eurocode-4, which assumes concrete to be plastic (fully crushed) and steel to be elastic (fully yielded). Han [11] conducted experimental tests on 24 steel columns consisting of 20 rectangles and 4 squares tested under axial load. The concrete strength was 48.3 MPa and the steel yield stress ranged from 194 to 228 MPa. Han [11] concluded that the steel column behavior was influenced by the tube width ratio and material properties. Dondo [12] has conducted experimental studies on the behavior of 29 square steel tubes filled with concrete under compression load. The cross-section of the steel tubes ranges from 60 × 60 × 3.0 mm to 150 × 150 × 4.5 mm. The slenderness ratio ranges from 6.67 to 27.0, while the depth-to-wall column thickness ranges from 33.87 to 50.00. Specimens with high depth-to-wall thickness ratios (b/t) had more local buckling or bulging compared to those with small b/t ratios. All concrete-filled square columns experienced a decrease in their strength with an increase in the slenderness ratio. Further tests have been conducted by Shaker-Khalil and Molly [13] on concrete-filled rectangular hollow sections with 3 to 9 m heights with dimensions of 120 × 80 × 5 mm and 150 × 100 × 5 mm. The concrete strength ranged from 40.8 to 47.2 MPa while the steel yield stress ranged from 340 to 363 MPa. The conclusion showed that filling the section with concrete increased the capacity of the cross sections of columns 120 × 80 × 5 mm and 150 × 100 × 5 mm by 12% and 65%, respectively. Column failure occurred due to overall flexural buckling of the column with no sign of local buckling. Zhu et al. [14] have investigated the behavior of 30 square CFSC with concrete strength ranging from 24.47 to 26.58 MPa and yield stress ranging from 235 to 345 MPa. Five different cross-section designs, including unstiffened and stiffened cross-sections, have been identified. The results concluded that the inner stiffeners more significantly affected the failure mode, deformability, and overall strength of the composite columns. To study this type of composite columns with stiffened cross-section, Tao et al. [15–17] and Liu et al. [18,19] conducted experimental and numerical studies on thin-walled tube columns where the steel walls were stiffened to improve their structural behavior. Some other studies, including experimental tests [20–24] and numerical analysis [25–27], took the effect of concrete filling on double-skin steel tubular columns into consideration. Zhang et al. [28,29] studied the static and dynamic loads of steel columns with unstiffened and stiffened cross-sections with thicknesses of 1.25 to 1.48 mm. It was also observed that the ultimate capacity increased with the number of stiffeners, and the columns with more stiffeners had better ductility.

In recent years, much attention has been paid to the behavior of composite columns constructed from high-strength concrete and steel [30–34]. Based on the experimental test results of 41 circular steel columns with yield stress ranging from 410 to 435 MPa and concrete strength ranging from 58 to 96 MPa. Kilpatrick and Rangan [35] confirmed that the slenderness and load eccentricity significantly affected the strength of columns. Experimental testing of 22 specimens under axial compressive load was performed by Liu [36]. The test variables include a cross-section aspect ratio of 1.0 to 2.0, a steel-to-concrete volume ratio of 0.13 to 0.17, a steel yield stress of 495 MPa, and a concrete strength of 60 to 89 MPa. The confinement achieved by the steel section increases the strength of the concrete. Strength improvement is negatively affected by the section aspect ratio. Sakino et al. [37] axial loading tests have been performed on 48 square and 36 circular steel columns. The test parameters were material strengths with yield stress ranging from 262 to 853 MPa, concrete strength ranging from 25 to 91 MPa, and volumetric steel-to-concrete ratio from 0.09 to 0.26. The rest results showed that the strength improvement was attributed to the strain hardening of the steel section instead of the confinement effect. The concrete filling increased the strength of square columns more than the circular ones. In addition, some studies [10,38–40] have mainly focused on investigating the reliability of various existing codes, ANSI/AISC 360-22 [1] and Eurocode-4 [2], in predicting the ultimate capacity of CFSC.

For several reasons, many studies in the last three decades have relied directly on using a finite element modeling (FEM) system. It saved time and effort and reduced the costs of preparing experimental tests. There were some special tests such as preload or load cycle tests that were difficult to perform in the laboratory due to the unavailability of testing machines or preload systems. Added to this difficulty were the long duration of the tests and the lack of a safety factor while conducting these tests. Therefore, the researchers selected FEM programs with the highest accuracy in modeling structural elements to obtain the required reliability. FEM analysis was used to model steel tube columns whether unfilled [41–43] or filled with concrete [44–50] under the influence of axial load, or with the use of flexural load [51–53]. Wang et al. [43] conducted FEM analysis on 20 specimens of steel tube columns without concrete filling, and the results compared to experimental tests were very accurate. The average ultimate capacity ratio was 1.016 with a Coefficient of Variation, COV, equal to 5.66%. Dai and Lam [46] carried out FEM simulations on 21 specimens of concrete-filled steel tube columns with elliptical hollow sections. The results of comparing the FEM analysis with experimental tests were an excellent match in both structural behavior (load-axial shortening curve), failure mode, and ultimate load capacity. The average ultimate capacity ratio was 1.013 with COV equal to 2.75%. Tao et al. [49] have conducted a FEM analysis on 340 samples collected from previous studies. The results showed great accuracy using FEM analysis, as the average ultimate capacity reached 1.024, 1.009, and 1.037 for each of the 142 circular columns, 154 square columns, and 44 rectangular columns with a standard deviation equal to 0.065, 0.078, and 0.079, respectively. One of the applications of FEM analysis was the preload modeling. Sayed [54] has conducted a FEM study on reinforced concrete columns under preloading, which were strengthened using steel jacketing. The average ultimate capacity ratio of columns with or without strengthening was 1.003, with COV = 3.16% and a correlation coefficient, r, of 0.987. These results demonstrated the accuracy and reliability of using the FEM simulation system in modeling these structural elements. From the literature review, it was found that it was difficult to conduct experimental tests of these columns under the influence of preload due to the long time of the test and the method of representing the loads required for this. It was also found that using the FEM system had great accuracy in modeling this type of structural elements in obtaining the behavior of these elements and their ultimate capacity.

In this study, the FEM analysis will be applied to simulate these concrete-filled columns with or without preloading. Many of the previously mentioned variables will be used specifically compared to those that are likely to have an impact on the ultimate capacity of the steel columns under preload. These variables will be modeled in terms of column height, column wall thickness, yield stress, cross-section dimensions, and concrete strength under the influence of preload ratios equal to 40%, 60%, 70%, and 80% of the ultimate capacity of the steel column without concrete filling. First, the results of the FEM analysis will be verified through comparison with specimens available from previous studies [12,14,36,38,39] without preloading. Also, the results will be compared with the prediction of existing codes, ANSI/AISC 360-22 [1] and Eurocode-4 [2], to obtain the accuracy of their use in such columns and the extent of safety in the design. Second, the effect of preload ratios will be obtained with all these variables on the ultimate axial load capacity of CFSC. Thirdly, the possibility of proposing a new model that can predict the ultimate axial load capacity of CFSC under preloading is tangible, taking into account all the variables under study.

2. Existing Models

Many existing models, including international codes [1,2] or scientific papers, have provided equations that could predict the ultimate capacity of these types of structural elements, especially CFSC. The model is available through the ANSI/AISC 360-22 [1] in the chapter on designing composite elements, as in Equation (1). This model can predict the plastic axial compressive strength of CFSC without consideration of the column length effect.

$$N_P = f_y A_s + K f\prime_c A_c \quad (1)$$

where f_y is steel column yield stress, A_s is the steel column cross-section, f'_c is the compressive strength of the concrete-filled steel column, A_c is the area of the concrete section, and K is a coefficient that depends on the cross-sectional shape of the steel column equal to 0.95 for columns with a round cross-section and 0.85 for other shapes. Some conditions have been set for using this model, as the concrete strength must not be less than 21 MPa and not more than 69 MPa. The yield stress of the steel column must not be more than 525 MPa. To take the column length into account, Equation (2) must be used to calculate the critical elastic buckling load.

$$N_e = \pi^2 (EI)_{eff} / L^2{}_c \quad (2)$$

where L_c is the effective column length, $(EI)_{eff}$ is the effective stiffness of composite section calculated by Equation (3), E_s and E_C are the modulus of elasticity of steel and concrete, respectively, I_s and I_c are the moment of inertia of steel and concrete column, respectively, and C_3 is the coefficient for effective rigidity.

$$(EI)_{eff} = E_s I_s + C_3 E_c I_c \quad (3)$$

$$C_3 = 0.45 + 3\left(\frac{A_s}{A_c}\right) \leq 0.9 \quad (4)$$

There are two column conditions based on the ratio between the width and the thickness of the steel column wall. The first is that the column is a compact section, when the ratio ($\lambda = b/t$) is less than Equation (5), in which the column nominal axial compressive strength $N_{no} = N_P$. The second case is that the column section is noncompact, when the ratio ($\lambda = b/t$) is less than Equation (6), so the N_{no} calculation is based on Equation (7).

$$\lambda_P = 2.26\sqrt{\frac{E_s}{f_y}} \quad (5)$$

$$\lambda_r = 3.00\sqrt{\frac{E_s}{f_y}} \quad (6)$$

$$N_{no} = N_P - \left(\left(\frac{N_P - N_y}{\lambda_r - \lambda_P}\right)(\lambda - \lambda_P)^2\right) \quad (7)$$

$$N_y = f_y A_s + 0.7 f\prime_c A_c \quad (8)$$

To calculate the ultimate axial load capacity of the CFSC, two cases where the ratio between the column's nominal axial compressive strength and the critical elastic buckling column load (N_{no}/N_e) are used, and their comparison is 2.25 where

$$N_n = N_{no}\left(0.658^{\left(\frac{N_{no}}{N_e}\right)}\right) \text{ When} \frac{N_{no}}{N_e} \leq 2.25 \quad (9)$$

$$N_n = 0.877 N_e \text{ When} \frac{N_{no}}{N_e} > 2.25 \quad (10)$$

To use the Eurocode-4 [2] in designing such a case, Equation (11) can be used. The axial compressive plastic resistance load of the composite column can be obtained as follows:

$$N_n = f_y A_s + 0.85 f\prime_c A_c \quad (11)$$

Equation (11) applies to concrete partially and encased steel sections. For CFSC, the coefficient of 0.85 is replaced with 1.0. Therefore, the Equation (11) is modified to

$$N_n = f_y A_s + f'_c A_c \quad (12)$$

3. Finite Element Model Study

3.1. Concrete Modeling and Properties

To perform the numerical analysis process, the ANSYS program was used. To model the concrete used to fill steel columns, the element SOLID65 is accordingly used [55]. This element contains eight points each containing three degrees of freedom (DOFs) in the X, Y, and Z directions. It can also provide high-accuracy modeling in deforming the plastic, breaking, and crushing in three directions. For the ANSYS program to employ the characteristics required for concrete modeling, some important characteristics must be defined. The stress–strain relationship for the concrete is used, as well as the shear transfers coefficient β, uniaxial tensile stress, modulus of elasticity, and Poisson's ratio. The modulus of elasticity can be calculated from Equation (13), and through it, the rest of the characteristics of the stress–strain relationship in the elastic and plastic stages can be obtained [49,56].

$$E_c = 4700\sqrt{f'_c} \quad (13)$$

$$f_{el} = 0.30\, f'_c \quad (14)$$

$$E_c = f_{el}/\varepsilon_{el} \quad (15)$$

$$\varepsilon_0 = \frac{2f'_c}{E_c} \quad (16)$$

$$f = \frac{E_c\varepsilon}{1 + (\varepsilon/\varepsilon_o)^2} \quad (17)$$

wherein

f_{el} = the stress at the elastic strain (ε_{el}) in the elastic range.
ε_o = the strain at the ultimate cylinder compressive strength.
f = the stress at any strain ε.

Regarding the shear transfer coefficient β [55], the values range from 0.0 for a smooth crack to 1.0 for a rough crack [57,58]. By reviewing previous studies [56,59], the best shear transfer coefficient was obtained, with the coefficient of open crack at, β_t = 0.2, and the coefficient of closed crack, β_c = 0.8. Also, the uniaxial tensile stress, f_r, can be calculated from Equation (18), and Poisson's ratio of 0.2 for concrete was applied [49]. Through these equations, the stress–strain relationship for different grades of concrete can be calculated, which will be used in the FEM analysis process, as listed in Table 1.

$$f_r = 0.623\sqrt{f'_c} \quad (18)$$

Table 1. The stress–strain relationships for concrete applied to the FEM analysis.

Concrete Grade 25		Concrete Grade 26.6		Concrete Grade 28.8		Concrete Grade 46.6		Concrete Grade 60	
Stress (MPa)	Strain	Stress (MPa)	Strain	Stress (MPa)	Strain	Stress (MPa)	Strain	Stress (MPa)	Strain
7.50	0.00032	7.98	0.00033	8.64	0.00034	13.98	0.00044	18.00	0.00049
19.25	0.0010	20.07	0.0010	21.16	0.0010	28.68	0.0010	33.34	0.0010
23.55	0.0015	24.78	0.0015	26.43	0.0015	38.00	0.0015	45.24	0.0015

Table 1. *Cont.*

	Concrete Grade 25		Concrete Grade 26.6		Concrete Grade 28.8		Concrete Grade 46.6		Concrete Grade 60	
	Stress (MPa)	Strain	Stress (MPa)	Strain	Stress (MPa)	Strain	Stress (MPa)	Strain	Stress (MPa)	Strain
	24.95	0.0020	26.49	0.0020	28.55	0.0020	43.53	0.0020	53.22	0.0020
	25.00	0.0021	26.60	0.0022	28.80	0.0023	46.08	0.0025	57.78	0.0025
	25.00	0.0025	26.60	0.0025	28.80	0.0025	46.60	0.0029	59.74	0.0030
	25.00	0.0030	26.60	0.0030	28.80	0.0030	46.60	0.0035	60.00	0.0033
									60.00	0.0035
E_c (GPa)	23.5		24.24		25.22		32.08		61.41	
f_r (MPa)	3.12		3.21		3.34		4.25		4.83	

3.2. Steel Column Modeling and Characteristics

To model the steel columns, element SOLID185 was used [55]. This element contains eight points with 3-DOFs in three directions: x, y, and z. There are many properties of this element that it is capable of plasticity, creep, hyperelasticity, large deflection, and larger capabilities of strain. This is also suitable for asymmetric meshes and the ability to simulate deformations of elastic–plastic materials. The properties of the steel used to model the steel columns are introduced from the relationship between stress and strain, as well as the modulus of elasticity, based on the values included in previous studies from experimental tests [12,14,36,38,39]. A Poisson's ratio of 0.3 for steel was used. Table 2 shows the stress–strain relationship available in previous studies and used in modeling steel in the FEM analysis process.

Table 2. The stress–strain relationships for steel applied to the FEM analysis.

	Young and Liu [39] (R1L)		Young and Liu [39] (R2L)		Young and Liu [39] (R3L)		Young and Liu [39] (R4L)		Young and Ellobody [38] (SHS2)	
	Stress (MPa)	Strain	Stress (MPa)	Strain	Stress (MPa)	Strain	Stress (MPa)	Strain	Stress (MPa)	Strain
	350	0.0018	424	0.0022	366	0.0019	443	0.0023	448	0.00224
	381	0.009	477	0.011	398	0.013	482	0.011	474	0.013
	412	0.05	512	0.058	447	0.056	508	0.052	504	0.047
	449	0.09	540	0.11	514	0.14	550	0.13	574	0.13
	536	0.21	604	0.23	577	0.26	608	0.27	637	0.22
	626	0.45	661	0.41	629	0.43	659	0.41	676	0.31
	646	0.63	675	0.55	647	0.53	676	0.51	695	0.42
	649	0.72	676	0.61	648	0.64	678	0.61	699	0.5
	625	0.74	634	0.64	622	0.66	642	0.63	641	0.51
	601	0.75	571	0.66	581	0.68	596	0.64	596	0.52
Es (GPa)	198		194		193		194		200	

The contact elements of TARGE170 and CONTA174 [55] were used to model the contact between the steel column wall and the filling concrete that functions as Coulomb's friction model. CONTA174 applies to 3D geometrical elements with an 8-node intended for flexible–flexible and rigid–flexible contact analysis, which can be applied to the contact between solid bodies or shells. The CONTA174 contact element is associated with the 3D target segment element TARGE170 located on the surface of 3D solid elements. It has the same geometric characteristics as the underlying elements. TARGE170 was applied with 3D surface-to-surface contact elements with the CONTA174. In the basic Coulomb friction model ($\tau_{\lim} = \mu P + b$), two surfaces in contact can carry shear stresses, where μ = isotropic friction coefficient, p = normal contact pressure, and b = contact cohesion. Through the CONTA174 element, an option can be provided to specify the maximum equivalent frictional stress τ_{max}. Therefore, sliding will occur if the magnitude of the equivalent frictional stress reaches this value, and when the equivalent shear stress is less than the limit of frictional stress, no motion occurs between the two surfaces, which is known as sticking.

3.3. Model Studies of Structure

Using the ANSYS program, 122 specimens were modeled, as they contain many of the variables that were taken in this study. These specimens were divided into two main groups.

The first group numbers 38 specimens obtained through data available in previous studies from experimental tests [12,14,36,38,39], as listed in Table 3. Modeling of these steel columns was performed, whether they were filled with concrete (30 specimens) or not (8 specimens). These specimens contain many variables, including cross-section dimensions, height, concrete strength, yield stress, and wall thickness of the steel columns. All these specimens were tested to the failure point under axial load. This group will verify the accuracy of FEM analysis results in modeling this type of structural element under the influence of many variables.

The second group consists of 84 specimens that were selected based on some available data conducted as a literature review [12,14,36,38,39] and were modeled with dimensions similar to reality. These specimens were tested under three stages of axial load. The first stage is based on a hollow steel column without a concrete filling. The second stage uses CFSC without preloading. In the third stage, these columns were tested at four percentages of preload, where there was 40%, 60%, 70%, and 80% of the ultimate load of the steel column without concrete filling. Then, add concrete to the steel column after reaching this percentage of preload until the failure load. Table 4 contains all the variables in terms of dimensions and characteristics of the materials used in this study.

Table 3. Summary of columns on previous studies used to verify FEM results.

FE Model Based on	Column Specimen	Steel Column Dimensions					f_y (MPa)	Concrete f'_c (MPa)
		a (mm)	b (mm)	t (mm)	L (mm)	a/b		
Liu [36]	R1-1	120	120	4.00	360	1.0	495	60.0
	R2-1	150	100	4.00	450	1.5	495	60.0
	R3-2	180	90	4.00	540	2.0	495	60.0
	R4-2	130	130	4.00	390	1.0	495	60.0
	R5-1	160	110	4.00	480	1.5	495	60.0
	R6-1	190	100	4.00	570	1.9	495	60.0
	R7-1	106	106	4.00	320	1.0	495	89.0
	R8-1	130	90	4.00	390	1.4	495	89.0
	R9-2	160	80	4.00	480	2.0	495	89.0
	R10-1	140	140	4.00	420	1.0	495	89.0
	R11-1	160	125	4.00	480	1.3	495	89.0
Zhu et al. [14]	Pa-6-1	197	197	6.40	600	1.0	438	26.6
	Pa-10-1	201	201	10.10	600	1.0	382	26.6
Young and Ellobody [38]	SHS1C40	150.5	150.5	5.834	450	1.0	497	46.6
	SHS1C60	150.6	150.6	5.829	450	1.0	497	61.9
	SHS2C0	150.5	150.5	2.796	600	1.0	448	----
	SHS2C40	150.5	150.5	2.782	450	1.0	448	46.6
	SHS2C60	150.5	150.5	2.780	450	1.0	448	61.9
	RHS1C40	197.1	109.5	4.060	600	1.8	503	46.6
	RHS3C0	140.0	78.8	3.075	600	1.8	486	----
	RHS3C40	140.2	80.1	3.100	420	1.8	486	46.6
	RHS3C60	140.2	80.1	3.100	420	1.8	486	61.9

Table 3. *Cont.*

FE Model Based on	Column Specimen	Steel Column Dimensions					f_y (MPa)	Concrete f'_c (MPa)
		a (mm)	*b* (mm)	*t* (mm)	*L* (mm)	*a*/*b*		
Dundu [12]	S1-101/1	101.6	101.6	3.00	1000	1.0	484	25.0
	S1-101/1.5	101.6	101.6	3.00	1500	1.0	484	25.0
	S1-120/1	120	120	3.50	1500	1.0	401	25.0
	S1-120/2	120	120	3.50	2000	1.0	401	25.0
	S1-150/1.5	150	150	3.50	1500	1.0	405	25.0
	S1-150/2	150	150	3.50	2000	1.0	405	25.0
	S1-150/2.5	150	150	3.50	2500	1.0	405	25.0
	S3-120/2.7	120	120	3.00	2700	1.0	401	28.8
	S3-150-3/2.7	150	150	3.00	2700	1.0	429	28.8
	S3-150-4.5/2.7	150	150	4.50	2700	1.0	335	28.8
Young and Liu [39]	R1L1200	120.1	40	1.94	1199	3.0	350	----
	R1L2000	120.2	40	1.95	2000	3.0	350	----
	R2L1200	119.7	39.9	5.31	1200	3.0	424	----
	R3L1200	119.9	80.1	2.82	1200	1.5	366	----
	R3L2000	120	80	2.80	2000	1.5	366	----
	R4L2000	120.1	80.4	6.02	2000	1.5	443	----

Table 4. Summary of the columns created in this study by FEM analysis.

FEM Specimen	Steel Column Dimensions					f_y (MPa)	Concrete f'_c (MPa)
	b (mm)	*t* (mm)	*L* (mm)	*b*/*t*	Compact limit		
R1-1	120	4	360	30	46.1	495	60
R1-L = 1500	120	4	1500	30	46.1	495	60
C1	200	5	4000	40	46.1	495	60
C2	150	5	4000	30	46.1	495	60
C3	250	5	4000	50	46.1	495	60
C4	200	5	3000	40	46.1	495	60
C5	200	5	5000	40	46.1	495	60
C6	200	8	4000	25	46.1	495	60
C7	200	10	4000	20	46.1	495	60
C8	200	12	4000	16.7	46.1	495	60
C9	200	5	4000	40	46.1	495	21
C10	200	5	4000	40	46.1	495	40
C11	200	5	4000	40	55.3	334.8	40
C12	200	5	4000	40	50.6	405	40

3.4. Choosing the Optimal Mesh Convergence

The size of the elements used in FEM analysis plays a major role in the accuracy of the results and in the time taken to perform the analysis. Therefore, this section aims to determine the appropriate element mesh size in FEM analysis to obtain acceptable accuracy in the results with the least possible number of elements. To obtain this result, the density of the mesh was changed by changing the size of the elements from 2.5 mm to 20 mm, while monitoring the effect of this on the ultimate load capacity. The CFSC (S3-150-4.5/2.7) [12] was chosen as the largest specimen dimension for this operation. The relationship between the number of elements and the corresponding ultimate axial load capacity resulting from the FEM analysis is plotted, as shown in Figure 1. It is clear from Figure 1 that after the number of elements equals 389,680, the effect on the ultimate load capacity is not noticeable. From this result, the mesh dimensions corresponding to this number of elements can be adopted in conducting FEM analysis tests on the remaining specimens. Therefore, the FEM analysis model was divided into small elements with maximum typical dimensions of 5 mm in width, length, and height, respectively.

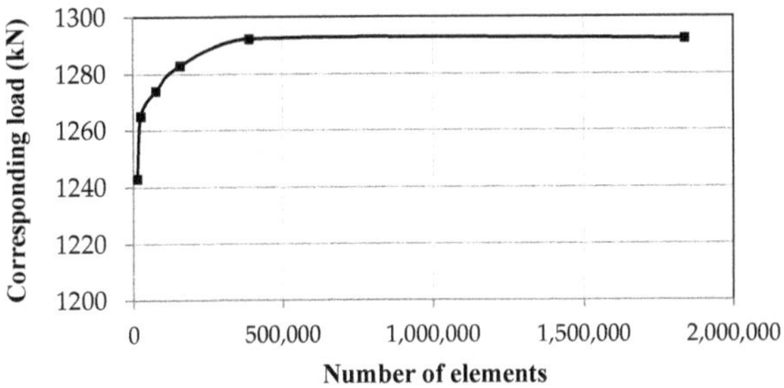

Figure 1. Evaluating the mesh sensitivity on the ultimate load capacity.

4. Verify the FEM Results and Existing Codes with Experimental Tests

4.1. Verify the FEM Results

The ultimate strength of structural elements is often the most important thing that can be obtained from testing and analyzing specimens. To verify the reliability and validity of the FEM analysis results, the ultimate axial load capacity results were compared with the experimental test results, as shown in Table 5, for CFSC while Table 6 for steel columns is not filled (with concrete).

Table 5. Comparison of FEM results of CFSC and existing codes with experimental tests.

FE Model Based on	Column Specimen	N_{Exp} (kN)	FEM Results		ANSI/AISC 360-22		Eurocode 4	
			N_{FEM} (kN)	N_{FEM}/N_{Exp}	N_{AISC} (kN)	N_{AISC}/N_{Exp}	N_{EC4} (kN)	N_{EC4}/N_{Exp}
Liu [36]	R1-1	1701.0	1685.7	0.991	1555.9	0.915	1671.4	0.983
	R2-2	1778.0	1764.5	0.992	1618.9	0.911	1742.2	0.980
	R3-2	1795.0	1836.3	1.023	1746.2	0.973	1883.8	1.049
	R4-2	2018.0	2028.1	1.005	1754.1	0.869	1891.0	0.937
	R5-1	1982.0	1942.4	0.980	1822.1	0.919	1967.8	0.993
	R6-1	2049.0	2006.0	0.979	1959.7	0.956	2121.4	1.035
	R7-1	1749.0	1728.0	0.988	1531.4	0.876	1662.6	0.951
	R8-1	1752.0	1764.3	1.007	1590.3	0.908	1729.9	0.987
	R9-1	1878.0	1896.8	1.010	1734.4	0.924	1892.7	1.008
	R10-1	2752.0	2713.5	0.986	2390.5	0.869	2627.9	0.955
	R11-1	2580.0	2518.1	0.976	2434.5	0.944	2679.7	1.039
Zhu et al. [14]	Pa-6-1	2730.0	2710.9	0.993	2669.5	0.978	2780.8	1.019
	Pa-10-1	3980.0	3852.6	0.968	3432.9	0.863	3539.8	0.889
Young and Ellobody [38]	SHS1C40	2768.1	2690.6	0.972	2424.1	0.876	2562.8	0.926
	SHS1C60	2972.0	3037.4	1.022	2675.9	0.900	2859.8	0.962
	SHS2C40	1381.5	1376.0	0.996	1233.3	0.893	1381.9	1.000
	SHS2C60	1620.0	1660.5	1.025	1505.3	0.929	1702.7	1.051
	RHS1C40	1627.2	1685.9	1.036	1672.7	1.028	1813.8	1.115
	RHS3C40	1048.7	1021.4	0.974	951.9	0.908	1024.9	0.977
	RHS3C60	1096.9	1135.3	1.035	1079.2	0.984	1175.8	1.072
Dundu [12]	S1-101/1	718.9	745.5	1.037	751.7	1.046	797.1	1.109
	S1-101/1.5	691.2	720.3	1.042	738.1	1.068	797.1	1.153
	S1-120/1	997.2	968.3	0.971	903.5	0.906	970.1	0.973
	S1-120/2	899.7	933.8	1.038	889.2	0.988	970.1	1.078
	S1-150/1.5	1330.1	1316.8	0.990	1245.0	0.936	1338.6	1.006
	S1-150/2	1307.6	1312.8	1.004	1232.0	0.942	1338.6	1.024
	S1-150/2.5	1198.3	1224.6	1.004	1215.4	1.014	1338.6	1.117
	S3-120/2.7	825.2	812.0	0.984	821.4	0.995	937.0	1.136
	S3-150-3/2.7	1087.1	1095.8	1.008	1085.6	0.999	1187.9	1.093
	S3-150-4.5/2.7	1300.4	1292.3	0.994	1342.6	1.032	1449.4	1.115
Average ratio				1.022		0.945		1.024
COV				2.25%		5.74%		6.81%
r				0.999		0.998		0.998

Table 6. Comparison of FEM results for unfilled concrete steel columns and existing codes with experimental tests.

FE Model Based on	Column Specimen	N_{Exp} (kN)	FEM Results		ANSI/AISC 360-22		Eurocode 4	
			N_{FEM} (kN)	N_{FEM}/N_{Exp}	N_{AISC} (kN)	N_{AISC}/N_{Exp}	N_{EC4} (kN)	N_{EC4}/N_{Exp}
Young and Ellobody [38]	SHS2C0	408.6	409.8	1.003	405.2	0.992	405.8	0.993
	RHS3C0	558.2	563.2	1.009	552.9	0.991	557.0	0.998
Young and Liu [39]	R1L1200	167.0	169.0	0.988	189.6	1.135	154.4	0.925
	R1L2000	141.3	147.0	1.040	162.9	1.153	139.4	0.986
	R2L1200	717.4	730.3	1.018	567.6	0.791	646.6	0.901
	R3L1200	398.3	390.7	0.981	381.1	0.957	391.3	0.982
	R3L2000	394.0	382.2	0.970	361.4	0.917	389.1	0.987
	R4L2000	970.4	998.5	1.029	873.6	0.900	960.9	0.990
Average ratio				1.005		0.980		0.970
COV				2.41%		12.00%		3.62%
r				0.999		0.996		0.999

Table 5 shows the values of the ultimate axial load for both FEM analysis and experimental tests and the ratio between them, N_{FEM}/N_{Exp}, and the relationship is drawn in Figure 2. From Figure 2 and Table 5 for CFSC, the average ratio of N_{FEM}/N_{Exp} equals 1.002, with COV = 2.25% and r equal to 0.999. Also, according to Table 6 and Figure 2, for steel columns not filled with concrete, the average ratio of N_{FEM}/N_{Exp} equals 1.005, with COV equal to 2.41% and r equal to 0.999. From a statistical perspective, whether steel columns are filled with concrete or not, the reliability and validity of the results measured from the FEM analysis are an excellent match. This high accuracy and agreement in the results are consistent with the results of previous studies [43–54] that used FEM analysis in modeling such structural elements.

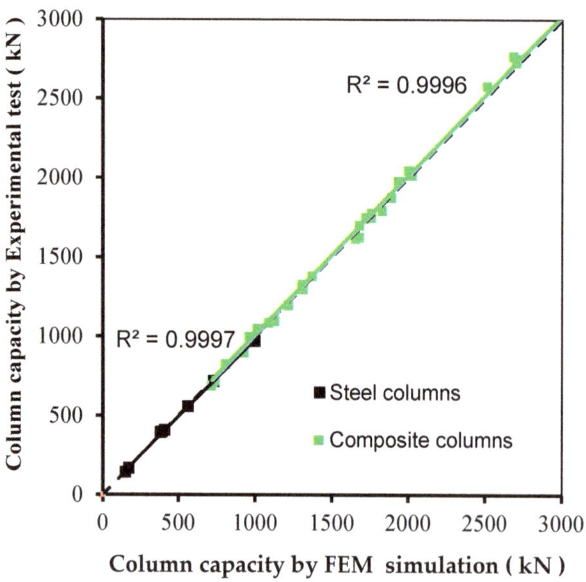

Figure 2. A comparison of the FEM results with the experimental test for axial load capacity.

Through the load and axial shortening curves measured from the experimental tests and resulting from the FEM analysis of these specimens, the relationship between the load and axial shortening was drawn, as shown in Figures 3 and 4. These specimens represent many considered variables. Where the compressive strength of concrete ranges from 26.6

to 89 MPa, the yield stress of steel ranges from 382 to 495 MPa, and the dimensions of columns range in ratios from 1.0 to 2.0; whereas, the wall thickness of columns ranges from 4.0 to 10.1 mm, as well as the height of the columns, which ranges from 360 to 600 mm. Figure 3 represents the relationship between load and axial shortening resulting from the FEM analysis, which has a very high accuracy when compared with the same experimental test results for these specimens (R1-1, R2-2, R3-2, R7-1, R8-1, and R9-1) provided in Liu [36]. Figure 4 represents the load-strain relationship obtained from FEM analysis, which is in excellent agreement with the same experimental test specimens Pa-6-1 and Pa-10-1 provided in Zhu et al. [14]. From Figures 3 and 4, the accuracy resulting from comparing experimental results and FEM analysis of the behavior of the columns in applied load and axial shortening curves is an excellent match.

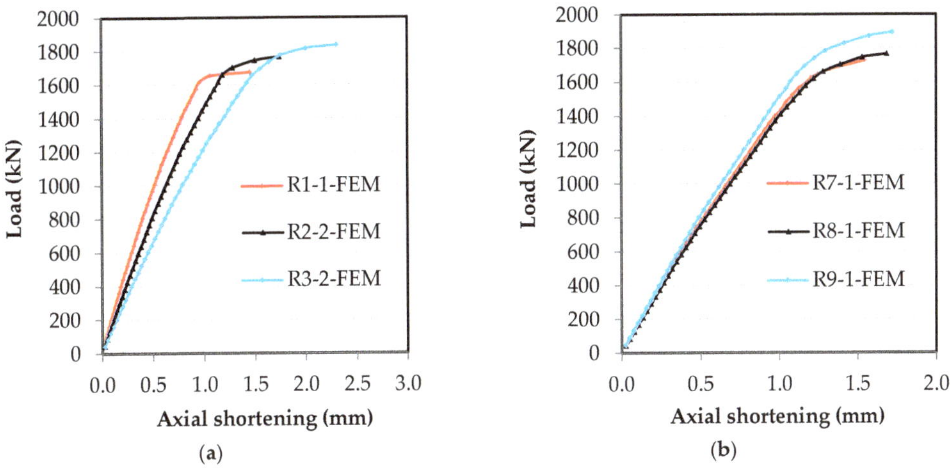

Figure 3. Load and axial shortening curves generated by FEM results.

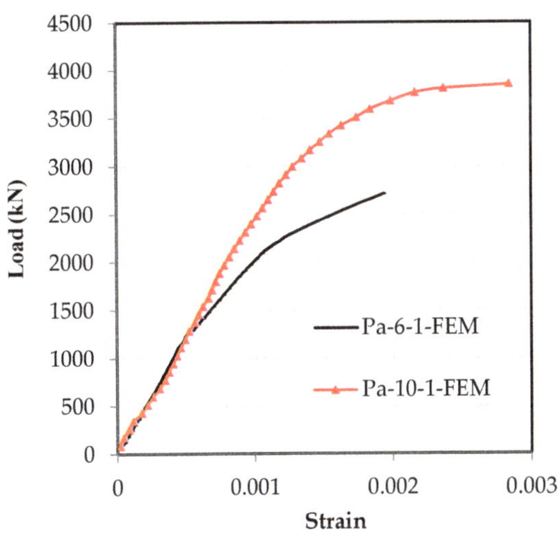

Figure 4. Load and strain curves generated by FEM results.

Here are the deformations and locations of the failure modes available through previous studies [14,36] from experimental tests and the same failure modes that can be obtained from FEM analysis. The degree of similarity between experimental tests and FEM analysis is an excellent match, as shown in Figure 5. A similar match of the deformation shape in Figure 5a was found with the same specimen Pa-6-1 available in Zhu et al. [14], and also in Figure 5b with the same specimen R4-2 available in Liu [36]. It clearly shows that the failure location most observed in these columns filled with concrete is the occurrence of local buckling, which is in the upper third of the column at the location of the applied load.

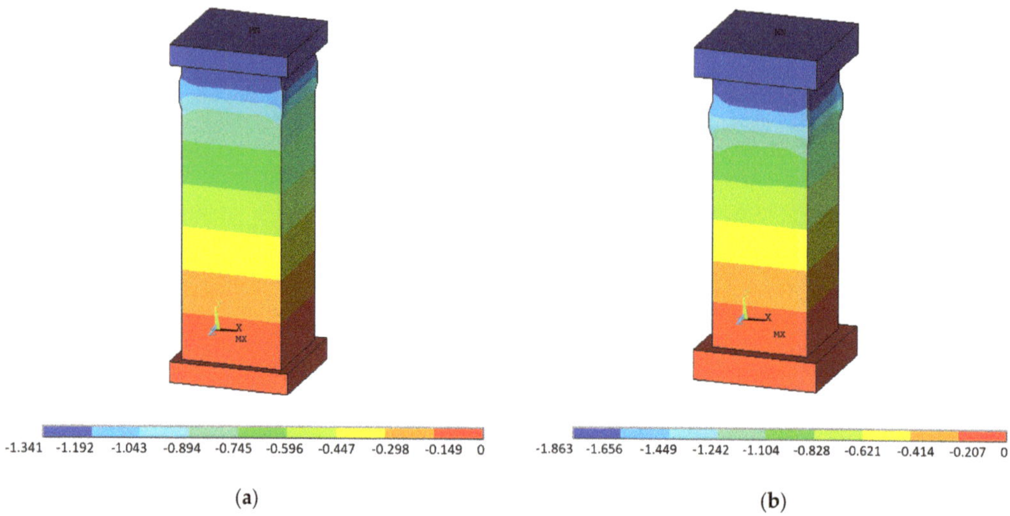

Figure 5. Deformations (axial shortening (mm)) and failure locations were obtained from FEM results (**a**) Specimen Pa-6-1 Zhu et al. [14] and (**b**) Specimen R4-2 Liu [36].

4.2. *Verify the Existing Code Results*

The most two important codes were selected from the existing codes, ANSI/AISC 360-22 [1] and Eurocode-4 [2], on which most models in studies or other codes are based, namely, ANSI/AISC 360-22 and Eurocode 4. The ultimate axial load of these steel columns was calculated through the equations provided in these codes, which appear in this study as Equations (1)–(12). These values were listed in Tables 3 and 4, and a relationship was drawn between the calculated load and the experimental test results, as shown in Figure 6 for the ANSI/AISC 360-22, and Figure 7 for the Eurocode 4.

The tables and figures show that according to the ANSI/AISC 360-22, the ultimate capacity of columns filled with concrete can be predicted with an average ratio $N_{\mathrm{AISC}}/N_{\mathrm{Exp}}$ equal to 0.945, with COV equal to 6.08% and r equal to 0.998. As for columns not filled with concrete, the average ratio $N_{\mathrm{AISC}}/N_{\mathrm{Exp}}$ is equal to 0.980, with COV equal to 12.25%, and r equal to 0.996. From a statistical perspective, in most cases, the values calculated from the ANSI/AISC 360-22 are less than the ultimate capacity of steel columns, whether filled with concrete or not. This means that it is on the safe side of the design.

As for the Eurocode-4, the ultimate capacity of columns filled with concrete can be predicted with an average ratio $N_{\mathrm{EC4}}/N_{\mathrm{Exp}}$ equal to 1.024, with COV equal to 6.65% and r equal to 0.997. As for columns not filled with concrete, the average ratio $N_{\mathrm{EC4}}/N_{\mathrm{Exp}}$ is equal to 0.970, with COV equal to 3.73% and r equal to 0.996. It is clear from these results that Eurocode 4 predicts values greater than the ultimate capacity of CFSC, i.e., on the unsafe side of the design. However, it predicts values less than the ultimate capacity of steel columns not filled with concrete, i.e., on the safe side of the design. All these results relate to steel columns without preload.

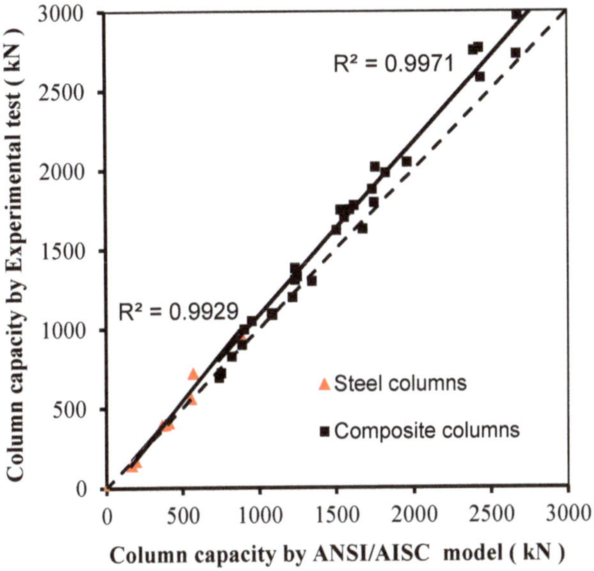

Figure 6. A comparison of the ANSI/AISC 360-22 results with the experimental test.

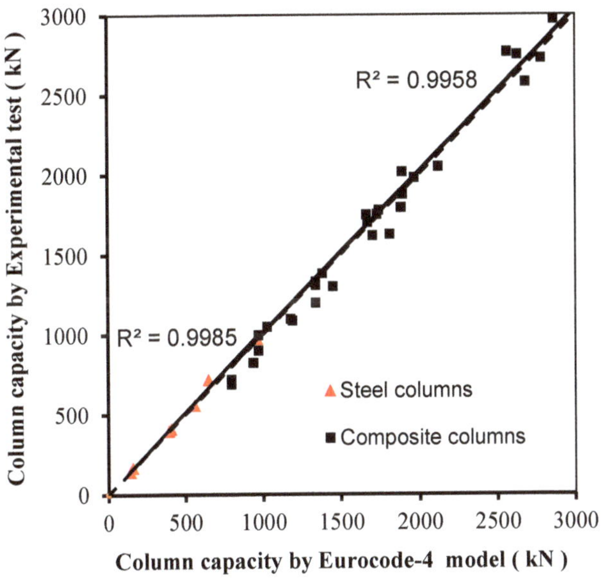

Figure 7. A comparison of the Eurocode-4 results with the experimental test.

5. Results and Discussion

To evaluate the effect of preload on the ultimate axial load capacity of CFSC, FEM analysis was performed on 84 specimens. Preloading was performed on the columns at ratios equivalent to 40%, 60%, 70%, and 80% of the ultimate load capacity of the steel columns not filled with concrete. In other words, these preload ratios were applied to the columns with several variables, cross-section dimensions, column height, concrete strength, yield stress, and wall thickness of steel columns. Meanwhile, existing codes, ANSI/AISC

360-22 and Eurocode-4, were used to compare with these results from preload, as shown in Table 7.

Table 7. Summary of FEM results for steel columns with and without concrete filling under preloading.

FEM Specimen	% Preloadingof Steel	FEM Results				ANSI/AISC 360-22		Eurocode 4	
		N_{FEM} (kN)	N_{PL}/N_{Uf}	N_C (kN)	N_{CPL}/N_C	N_{AISC} (kN)	N_{AISC}/N_{FEM}	N_{EC4} (kN)	N_{EC4}/N_{FEM}
R1-1-Without preloading	0	1685.7	1.000	623.8	1.000	1555.9	0.923	1671.4	0.991
R1-1-P40%	40	1685.7	1.000	623.8	1.000	1555.9	0.923	1671.4	0.991
R1-1-P60%	60	1685.7	1.000	623.8	1.000	1555.9	0.923	1671.4	0.991
R1-1-P70%	70	1655.5	0.982	593.6	0.952	1555.9	0.940	1671.4	1.010
R1-1-P80%	80	1619.2	0.961	557.3	0.893	1555.9	0.961	1671.4	1.032
R1-1-Only steel	---	1061.9	---	---	---	917.4	0.864	922.8	0.869
R1-L = 1500-Without preloading	0	1614.2	1.000	610.2	1.000	1514.7	0.938	1671.4	1.035
R1-L = 1500-P40%	40	1614.2	1.000	610.2	1.000	1514.7	0.938	1671.4	1.035
R1-L = 1500-P60%	60	1614.2	1.000	610.2	1.000	1514.7	0.938	1671.4	1.035
R1-L = 1500-P70%	70	1564.0	0.969	560.0	0.918	1514.7	0.968	1671.4	1.069
R1-L = 1500-P80%	80	1471.8	0.912	467.8	0.767	1514.7	1.029	1671.4	1.136
R1-L = 1500-Only steel	---	1004.0	---	---	---	895.6	0.892	922.8	0.919
C1-Without preloading	0	3971.5	1.000	2084.3	1.000	3491.5	0.879	4096.5	1.031
C1-P40%	40	3971.5	1.000	2084.3	1.000	3491.5	0.879	4096.5	1.031
C1-P60%	60	3898.1	0.982	2010.9	0.965	3491.5	0.896	4096.5	1.051
C1-P70%	70	3787.6	0.954	1900.4	0.912	3491.5	0.922	4096.5	1.082
C1-P80%	80	3367.7	0.848	1480.5	0.710	3491.5	1.037	4096.5	1.216
C1-Only steel	---	1887.2	---	---	---	1810.2	0.959	1762.6	0.934
C2-Without preloading	0	2428.5	1.000	1103.5	1.000	2138.8	0.881	2611.5	1.075
C2-P40%	40	2394.3	0.986	1069.2	0.969	2138.8	0.893	2611.5	1.091
C2-P60%	60	2332.9	0.961	1007.8	0.913	2138.8	0.917	2611.5	1.119
C2-P70%	70	2283.5	0.940	958.4	0.869	2138.8	0.937	2611.5	1.144
C2-P80%	80	2038.0	0.840	712.9	0.646	2138.8	1.049	2611.5	1.281
C2-Only steel	---	1325.1	---	---	---	1277.9	0.964	1310.5	0.989
C3-Without preloading	0	5826.4	1.000	3335.4	1.000	5060.5	0.869	5881.5	1.009
C3-P40%	40	5826.4	1.000	3335.4	1.000	5060.5	0.869	5881.5	1.009
C3-P60%	60	5826.4	1.000	3335.4	1.000	5060.5	0.869	5881.5	1.009
C3-P70%	70	5648.7	0.970	3157.7	0.947	5060.5	0.896	5881.5	1.041
C3-P80%	80	5131.8	0.881	2640.8	0.792	5060.5	0.986	5881.5	1.146
C3-Only steel	---	2491.0	---	---	---	2328.6	0.935	2425.5	0.974
C4-Without preloading	0	4041.1	1.000	2002.9	1.000	3611.4	0.894	4096.5	1.014
C4-P40%	40	4041.1	1.000	2002.9	1.000	3611.4	0.894	4096.5	1.014
C4-P60%	60	4041.1	1.000	2002.9	1.000	3611.4	0.894	4096.5	1.014
C4-P70%	70	3945.1	0.976	1906.9	0.952	3611.4	0.915	4096.5	1.038
C4-P80%	80	3635.0	0.900	1597.8	0.798	3611.4	0.993	4096.5	1.127
C4-Only steel	---	2038.2	---	---	---	1861.9	0.913	1762.6	0.865
C5-Without preloading	0	3735.5	1.000	1943.8	1.000	3343.1	0.895	4096.5	1.097
C5-P40%	40	3666.1	0.981	1874.4	0.964	3343.1	0.912	4096.5	1.117
C5-P60%	60	3615.2	0.968	1823.5	0.938	3343.1	0.925	4096.5	1.133
C5-P70%	70	3504.8	0.938	1713.1	0.881	3343.1	0.954	4096.5	1.169
C5-P80%	80	3036.1	0.813	1244.4	0.640	3343.1	1.101	4096.5	1.349
C5-Only steel	---	1791.7	---	---	---	1745.9	0.974	1762.6	0.984
C6-Without preloading	0	4891.8	1.000	1892.2	1.000	4431.6	0.906	5072.6	1.037
C6-P40%	40	4891.8	1.000	1892.2	1.000	4431.6	0.906	5072.6	1.037
C6-P60%	60	4742.0	0.969	1742.4	0.921	4431.6	0.935	5072.6	1.070
C6-P70%	70	4596.1	0.940	1596.5	0.844	4431.6	0.964	5072.6	1.104
C6-P80%	80	4122.8	0.843	1123.2	0.594	4431.6	1.075	5072.6	1.230
C6-Only steel	---	2999.6	---	---	---	2846.6	0.949	3005.6	1.002
C7-Without preloading	0	5745.3	1.000	1957.2	1.000	5029.0	0.875	5706.0	0.993
C7-P40%	40	5673.4	0.987	1885.4	0.963	5029.0	0.886	5706.0	1.006
C7-P60%	60	5459.4	0.950	1671.4	0.854	5029.0	0.921	5706.0	1.045
C7-P70%	70	5297.1	0.922	1509.0	0.771	5029.0	0.949	5706.0	1.077
C7-P80%	80	4733.9	0.824	945.8	0.483	5029.0	1.062	5706.0	1.205
C7-Only steel	---	3788.1	---	---	---	3516.4	0.928	3762.0	0.993
C8-Without preloading	0	6497.5	1.000	2022.3	1.000	5611.4	0.864	6325.4	0.974
C8-P40%	40	6306.2	0.971	1831.0	0.905	5611.4	0.890	6325.4	1.003
C8-P60%	60	6107.2	0.940	1632.0	0.807	5611.4	0.919	6325.4	1.036
C8-P70%	70	5914.9	0.910	1439.7	0.712	5611.4	0.949	6325.4	1.069
C8-P80%	80	5288.4	0.814	813.2	0.402	5611.4	1.061	6325.4	1.196
C8-Only steel	---	4475.2	---	---	---	4169.4	0.932	4466.9	0.998
C9-Without preloading	0	2551.5	1.000	664.3	1.000	2417.1	0.947	2688.6	1.054
C9-P40%	40	2451.8	0.961	564.6	0.850	2417.1	0.986	2688.6	1.097
C9-P60%	60	2355.0	0.923	467.8	0.703	2417.1	1.026	2688.6	1.142
C9-P70%	70	2281.9	0.894	394.7	0.594	2417.1	1.059	2688.6	1.178
C9-P80%	80	2042.6	0.801	155.4	0.234	2417.1	1.183	2688.6	1.316
C9-Only steel	---	1887.2	---	---	---	1810.2	0.959	1762.6	0.934

Table 7. *Cont.*

FEM Specimen	% Preloadingof Steel	FEM Results				ANSI/AISC 360-22		Eurocode 4	
		N_{FEM} (kN)	N_{PL}/N_{Uf}	N_C (kN)	N_{CPL}/N_C	N_{AISC} (kN)	N_{AISC}/N_{FEM}	N_{EC4} (kN)	N_{EC4}/N_{FEM}
C10-Without preloading	0	3255.8	1.000	1368.6	1.000	2942.8	0.904	3374.5	1.036
C10-P40%	40	3188.9	0.979	1301.7	0.951	2942.8	0.923	3374.5	1.058
C10-P60%	60	3157.8	0.970	1270.6	0.928	2942.8	0.932	3374.5	1.069
C10-P70%	70	3050.9	0.937	1163.6	0.850	2942.8	0.965	3374.5	1.106
C10-P80%	80	2702.0	0.830	814.8	0.595	2942.8	1.089	3374.5	1.249
C10-Only steel	---	1887.2	---	---	---	1810.2	0.959	1762.6	0.934
C11-Without preloading	0	2548.1	1.000	1270.9	1.000	2403.1	0.943	2749.7	1.079
C11-P40%	40	2548.1	1.000	1270.9	1.000	2403.1	0.943	2749.7	1.079
C11-P60%	60	2507.3	0.984	1230.1	0.968	2403.1	0.958	2749.7	1.097
C11-P70%	70	2448.8	0.961	1171.6	0.922	2403.1	0.981	2749.7	1.123
C11-P80%	80	2192.8	0.861	915.5	0.720	2403.1	1.096	2749.7	1.254
C11-Only steel	---	1277.2	---	---	---	1248.7	0.978	1272.6	0.996
C12-Without preloading	0	2886.5	1.000	1336.8	1.000	2649.2	0.918	3023.5	1.047
C12-P40%	40	2886.5	1.000	1336.8	1.000	2649.2	0.918	3023.5	1.047
C12-P60%	60	2827.2	0.979	1277.9	0.956	2649.2	0.937	3023.5	1.069
C12-P70%	70	2742.5	0.950	1193.2	0.893	2649.2	0.966	3023.5	1.102
C12-P80%	80	2450.4	0.849	901.1	0.674	2649.2	1.081	3023.5	1.234
C12-Only steel	---	1549.3	---	---	---	1497.6	0.967	1491.4	0.963

5.1. The Behavior of Applied Load and Axial Shortening

The behavior of steel columns is directly explained through the relationship between load and axial shortening. In general, the ultimate load increases when concrete is used to fill steel columns. This increase tends to subside when there is preload on steel columns not filled with concrete. However, the decrease in the ultimate capacity of the columns does not occur at all with preload ratios or the same value, as the variables mentioned above affect the value of this effect. For example, when using preload ratios of 40%, 60%, 70%, and 80% on a column C1 with a height of 3000 mm, the load decrease was 1.00, 0.982, 0.954, and 0.848, respectively, of the ultimate capacity of the column filled with concrete without preload, as shown in Figure 8a. The value of this decrease in load entirely changes when using column C5 with a height of 5000 mm. The value of load decreases at the same preload ratios and becomes 0.982, 0.968, 0.938, and 0.813, respectively, as shown in Figure 8b.

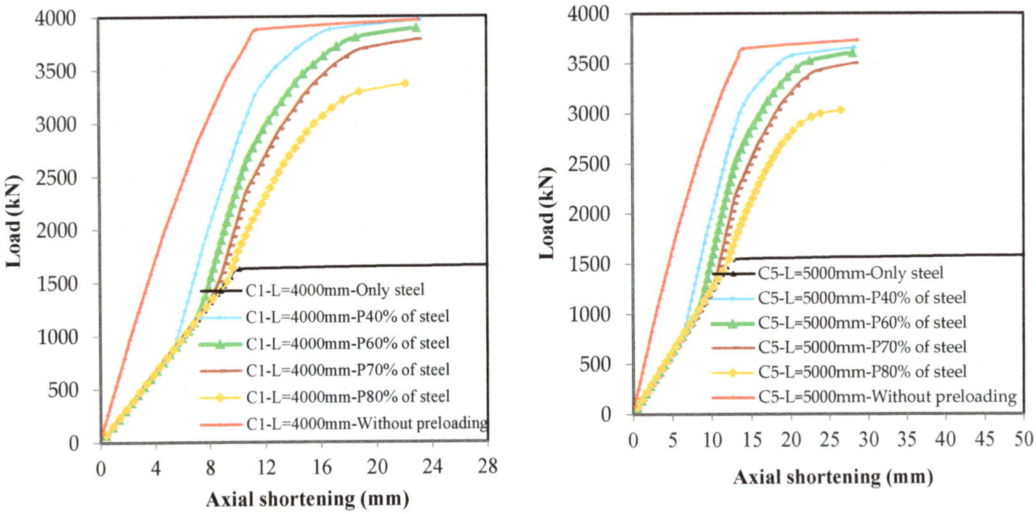

(a) (b)

Figure 8. Load and axial shortening curves obtained from FEM simulation with different preload ratios. (**a**) Column C1 with L = 4000 mm; (**b**) Column C5 with L = 5000 mm.

5.2. The Effect of Variables and the Preload Ratio on the Efficiency of Columns

To obtain the effect of different variables and preload ratio on the efficiency of CFSC, Figures 9–13 were drawn. Figures 9–13 show the relationship between the ultimate capacity of preloaded columns to the ultimate capacity of columns without preload and the ratios of the effect of preload for each variable separately.

5.2.1. Effect of Column Cross-Section Dimensions

Figure 9 shows this relationship regarding changing the dimensions of the cross-section of the steel column, as was changed from 150 to 250 mm. It precisely shows that the effect of the cross-section dimensions is inversely proportional to the effect of the preload ratio. As the dimensions of the column's cross-section increase, the effect of the preload ratio on the ultimate capacity of the column decreases. In addition, not all preload ratios have the same effect when the column's cross-section varies. With preload ratios of 40%, 60%, 70%, and 80% and a column with dimensions of 150 mm, the ratios (N_{PL}/N_{uf}) were 0.986, 0.961, 0.940, and 0.840, respectively, and through changing the column dimensions to 250 mm, they were 1.00, 1.00, 0.970, and 0.881, respectively. Through this, as the cross-section of the steel column increases, the efficiency and effect of the concrete fill increases.

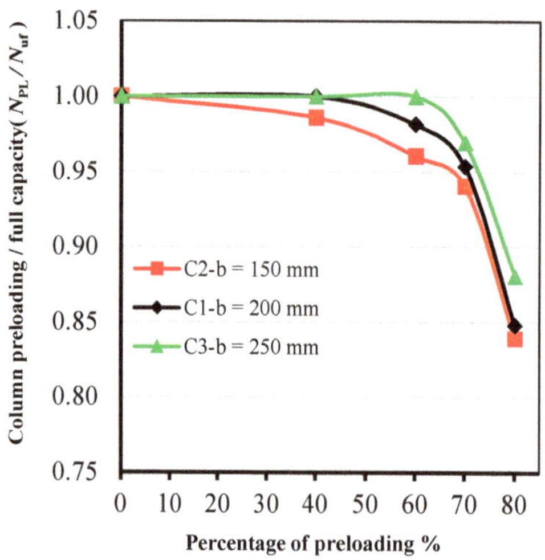

Figure 9. The relationship between the ratio of the capacity with to without preload for the effect of cross-section dimensions.

5.2.2. Effect of Column Height

Figure 10 shows this relationship regarding changing the steel column height, as it was changed from 3000 to 5000 mm. It clearly shows that the effect of column height is directly proportional to the effect of the preload ratio. As the height of the column increases, the impact of the preload ratio on the ultimate capacity of the column increases. With preload ratios of 40%, 60%, 70%, and 80% and with a column height of 3000 mm, the ratios N_{PL}/N_{uf} were 1.00, 1.00, 0.976, and 0.900, respectively, and by changing the column height to 5000 mm, they were 0.981, 0.968, 0.938, and 0.813, respectively. From this, as the height of the steel column increases, the efficiency and effect of the concrete fill decreases.

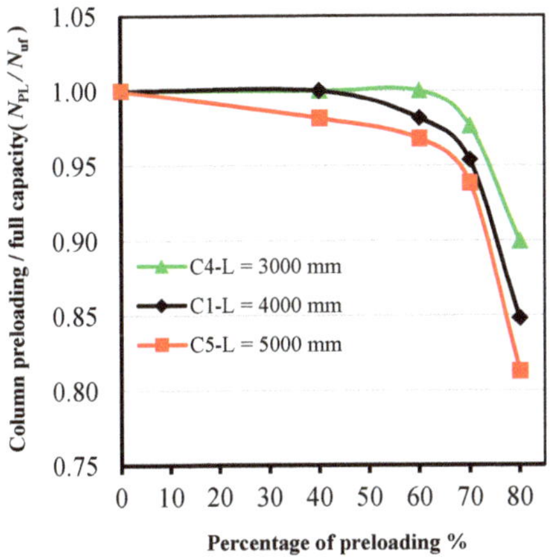

Figure 10. The relationship between the ratio of the capacity with to without preload for the effect of column height.

5.2.3. Effect of Column Wall Thickness

Figure 11 shows this relationship regarding changing the steel column wall thickness, as it was changed from 5.0 to 12.0 mm. The effect of the column wall thickness is directly proportional to the effect of the preload ratio. As the wall thickness of the column increases, the impact of the preload ratio on the ultimate capacity increases. With preload ratios of 40%, 60%, 70%, and 80% and with a column wall thickness of 5.0 mm, the ratios N_{PL}/N_{uf} were 1.00, 0.982, 0.954, and 0.848, respectively, and by changing the column wall thickness to 12.0 mm, they were 0.971, 0.940, 0.910, and 0.814, respectively. Therefore, as the wall thickness of the steel column increases, the efficiency and effect of the concrete fill decreases.

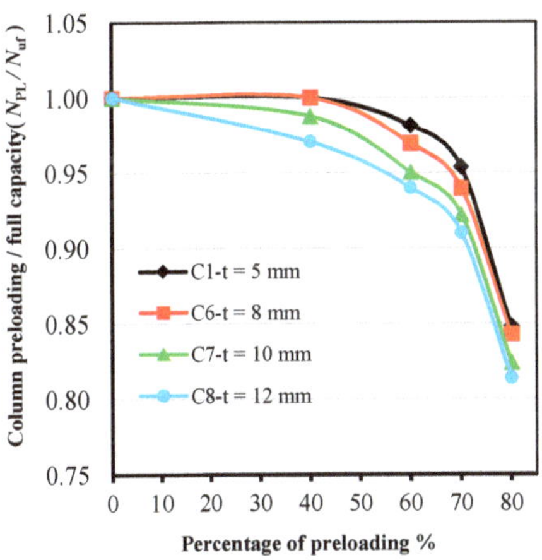

Figure 11. The relationship between the ratio of the capacity with to without preload for the effect of column wall thickness.

5.2.4. Effect of Concrete Strength

Figure 12 shows this relationship regarding changing the concrete compressive strength, as it was changed from 21 to 60 MPa. Concrete compressive strength 21 MPa was chosen as it is the lowest value that can be used according to the ANSI/AISC 360-22. The effect of the concrete strength is inversely proportional to the effect of the preload ratio. As the concrete strength increases, the impact of the preload ratio on the ultimate capacity decreases. With preload ratios of 40%, 60%, 70%, and 80% and with concrete compressive strength of 21 MPa, the ratios N_{PL}/N_{uf} were 0.961, 0.923, 0.894, and 0.801, respectively, and by changing it to 60 MPa, they were 1.00, 0.982, 0.954, and 0.848, respectively. Accordingly, as the concrete compressive strength increases, the efficiency and effect of the concrete fill increases.

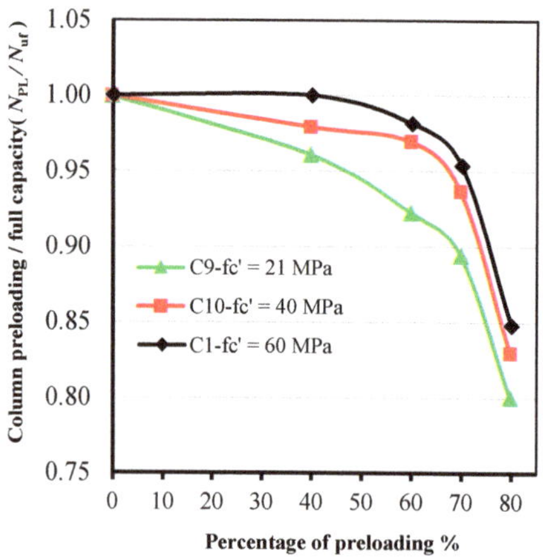

Figure 12. The relationship between the ratio of the capacity with and without preload for the effect of concrete strength.

5.2.5. Effect of Steel Column Yield Stress

Figure 13 shows this relationship regarding changing the steel column yield stress, as it was changed from 334.8 to 495 MPa. The effect of the steel column yield stress is directly proportional to the effect of the preload ratio. As the steel column yield stress increases, the impact of the preload ratio on the ultimate capacity increases. With preload ratios of 40%, 60%, 70%, and 80% and with a yield stress of 334.8 MPa, the ratios N_{PL}/N_{uf} were 0.979, 0.970, 0.937, and 0.830, respectively, and by changing it to 495 MPa, they were 1.00, 0.984, 0.961, and 0.861, respectively. Thus, as the steel column yield stress increases, the efficiency and effect of the concrete fill decreases.

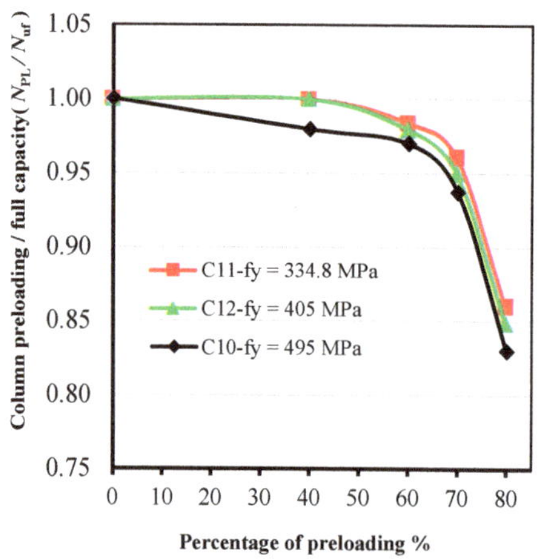

Figure 13. The relationship between the ratio of the capacity with to without preload for the effect of steel yield stress.

6. Prediction of Ultimate Axial Load Capacity under Preload

The second group of FEM analysis, which contains 84 specimens, was used to determine the effect of preload on the ultimate capacity of CFSC. By considering all of the variables taken into account, all of the results are listed in Table 7, which also contains the values calculated from existing codes (ANSI/AISC 360-22, and European-4).

First stage: Through the results of steel columns that were modeled with or without concrete fill and without preload, these results were used to verify the reliability of the existing codes, ANSI/AISC 360-22 [1] and Eurocode-4 [2], as shown in Figure 14. Through Figure 14a, where the ratio between the values calculated from the ANSI/AISC 360-22 and the results of FEM analysis, all specimens, whether columns filled or not filled with concrete, fall on the safe side of the design, where the values are less than the ultimate capacity of the columns. The average ratio N_{AISC}/N_{FEM} was equal to 0.918 with a COV of 4.05%. In Eurocode-4, the calculated values for columns filled with concrete mostly fall in the unsafe zone, i.e., the calculated values are greater than the ultimate capacity of the column, as shown in Figure 14b. The average ratio N_{EC4}/N_{FEM} was equal to 1.030 with COV of 3.71%, while in steel columns not filled with concrete, the calculated values fall on the safe side of the design. The average ratio N_{EC4}/N_{FEM} was equal to 0.947 with a COV of 5.38%. This result is completely consistent with the same result when comparing these codes with the experimental test results available from previous studies.

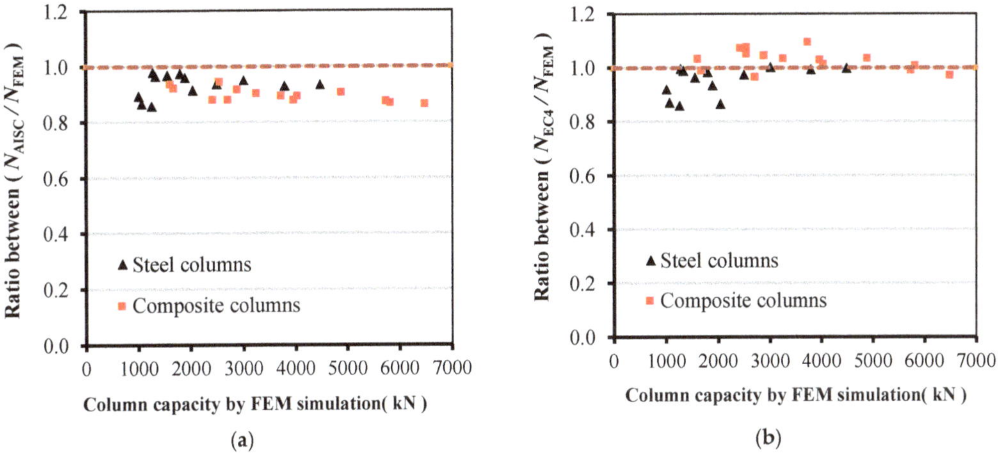

Figure 14. A comparison of the existing codes with the FEM results: (**a**) ANSI/AISC 360-22 and (**b**) Eurocode-4.

Second stage: Through Table 7, the ultimate capacity of the columns was obtained under the influence of preload (N_{PL}). Hence, the ratio (N_{PL}/N_{uf}) of decrease in this capacity was obtained relative to the CFSC without preload. It is also possible to obtain the contribution of concrete (N_{C}.) in increasing the load of CFSC with or without preloading. This is determined by the difference between the ultimate capacity of a CFSC and an unfilled steel column. Also, from Figures 9–13, the effect of preload is almost linear in reducing the capacity of the column up to a percentage equal to 70% of the load of the steel column that is not filled with concrete. Then, this effect suddenly decreases with a preload percentage equal to 80%. This is due to the fact that the preload percentages of up to 70% of the steel column that are not filled with concrete fall within the elastic region of the stress–strain relationship for the steel used. When the preload percentage reaches 80%, it falls within the elastic–plastic region, where the strains increase and the contribution of steel strength to increasing the column capacity decreases. Therefore, the effect of preload ratios will be divided into two groups. The first group preloads at rates from 40% to 70%.

The second group is at a preload equal to 80%. Through Table 7, the percentage decrease in the contribution of concrete with preload was obtained relative to the contribution of concrete for columns without preload. Through these ratios, the effect of preload on the contribution of concrete is greater than the effect on the total capacity of CFSC. The largest percentage decrease in ultimate capacity was recorded at 80% preload for specimen C9 at a rate of 0.801, while for the same specimen, the value of the decrease in concrete contribution was equal to 0.234.

From Figures 9–13, factor K was determined to take the effect of all variables into account. This factor was basically imposed on the effect of each variable on the value of a decrease in the ultimate capacity of the column. It was found out that the effect of each of the variables, such as column height, column wall thickness, yield stress, and preload ratios, R_{PL}, is directly proportional to the value of the decrease in column capacity. The variables such as the cross-section dimensions and the concrete strength are inversely proportional. Therefore, all these variables are collected in one factor, K, as in Equation (19).

$$K = \frac{L \times t \times f_y \times R_{PL}}{b \times f'_c} \quad (19)$$

The relationship between the ratios of decrease in concrete contribution resulting from preloading and the factor K was drawn as shown in Figure 15. So, this figure clearly shows that as the value of the factor K increases, the value of the concrete contribution is reduced. Through the polynomial relationship between these ratios, an obtained equation can predict the ultimate capacity of columns filled with concrete under the influence of preload on the steel column. Equation (20) is used with preload ratios not exceeding 70% of the capacity of the steel column not filled with concrete, with r equal to 0.966. Whereas, Equation (21) is used with preload ratios equal to 80% with r equal to 0.992.

$$\frac{(N_{PL.FEM} - N_{S.FEM})}{N_{C.FEM}} = -0.0014K^2 - 0.0022K + 1.00 \quad (20)$$

$$\frac{(N_{PL.FEM} - N_{S.FEM})}{N_{C.FEM}} = -0.0006K^2 - 0.0236K + 0.9001 \quad (21)$$

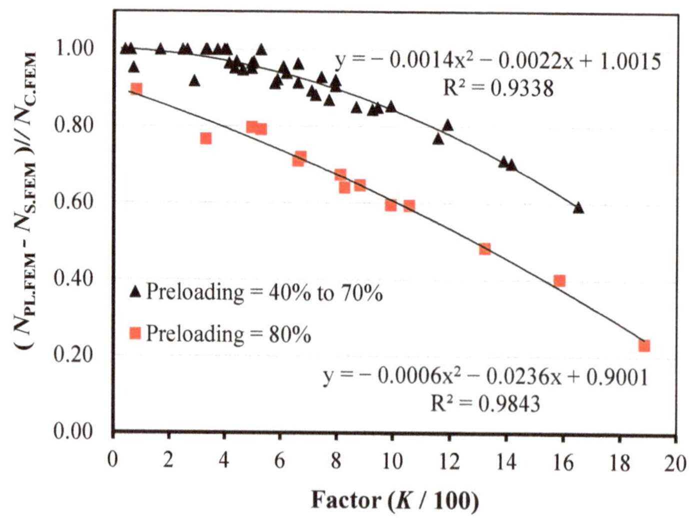

Figure 15. The relationship between the concrete contribution ratio and the factor k obtained from FEM analysis.

Through previous relationships, Equations (20) and (21), it is extremely difficult to perform an FEM analysis when there is a need to calculate the application of these equations. Therefore, one of the existing codes was used, and the ANSI/AISC 360-22 was selected to assist in the application. Its selection was based on previous analysis results and comparison with specimens, whether experimental or FEM results. It turns out that in all cases, these predictions fall on the safe side of the design. However, all these applications were without any preloading on steel columns. Through the results of the FEM analysis, the deficiency percentages were obtained for each preload case. The ANSI/AISC 360-22 model was used to calculate the concrete contribution according to Equation (22).

$$N_{c.AISC} = 0.85f'_{c}A_{c} \tag{22}$$

The ultimate capacity of the steel column with or without concrete filling and without preload was calculated according to Section 2. Then, the deficiency ratios resulting from the FEM analysis were used to find the ultimate remaining capacity for these specimens. The relationship between the ratio of concrete contribution of the specimens under preloading relative to the contribution of concrete without preloading was drawn with factor *K*, as shown in Figure 16. By applying the same polynomial relationship, Equations (23) and (24) were obtained to be used to determine the ultimate capacity of columns under preload.

$$\frac{(N_{PL} - N_{S.AISC})}{N_{C.AISC}} = -0.0002K^2 - 0.0235K + 1.00 \tag{23}$$

$$\frac{(N_{PL} - N_{S.AISC})}{N_{C.AISC}} = -0.0002K^2 - 0.0235K + 1.00 \tag{24}$$

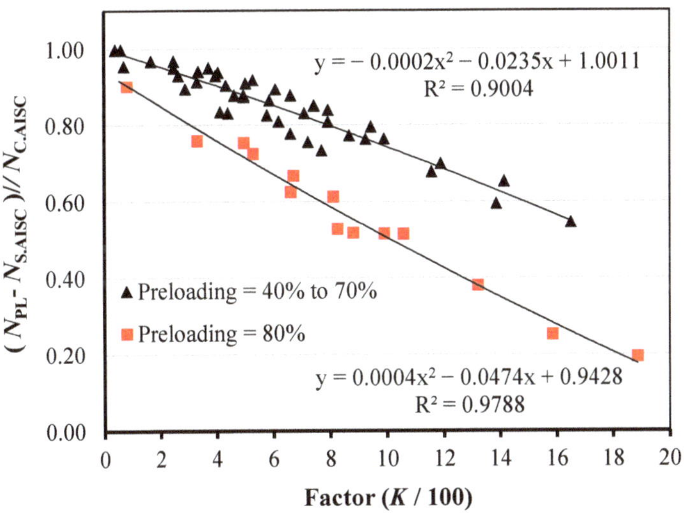

Figure 16. The relationship between the concrete contribution ratio and the factor k obtained from ANSI/AISC 360-22 [1].

Equation (23) is used with preload ratios not exceeding 70% of the capacity of the steel column not filled with concrete, with *r* equal to 0.949. Whereas, equation 24 is used with preload ratios equal to 80% with *r* equal to 0.989. Through the value of the correlation coefficient, *r*, whether accompanying the FEM analysis equations or accompanying the results of the ANSI/AISC 360-22, these new models can predict the ultimate capacity of CFSC under preload as a percentage of the capacity of the steel column without concrete with high accuracy.

7. Conclusions

There is a high possibility of using concrete to fill an existing steel column exposed to external loads to increase the efficiency of resisting external loads resulting from structural modifications or design errors. When such columns are difficult to implement in experimental tests, FEM analysis is performed to model these types of columns under preload. First, the FEM results were verified by comparing them with specimens available in previous studies of columns without preload. The results were very accurate in simulating these columns, as the average capacity ratio reached 1.002, COV was 2.25%, and r equal to 0.999. Secondly, FEM analysis was performed on the steel columns with several variables under different preload ratios. Through the proposed new models, the ultimate axial load capacity of CFSC under preload can be predicted. In light of this research, the conclusions drawn are presented below:

- For steel columns without concrete filling, prediction values calculated from existing codes, ANSI/AISC 360-22 and Eurocode-4, were always less than the actual ultimate capacity, average ratio is 0.951, whether resulting from experimental or FEM tests. This is a result of the existing codes relying only on the elastic region of the steel, while the behavior of the columns uses the actual properties of the available steel, as it enters the elastic–plastic region;
- Through the behavior of columns under preload, some variables are directly proportional to the value of the decrease in column capacity, in terms of column height, column wall thickness, yield stress, and preload ratios, while others are inversely proportional, in terms of cross-section dimensions and concrete strength;
- With some limits in the values of the geometric properties of steel columns and with preload ratios up to 60%, they do not affect the ultimate capacity of CFSC;
- In the case of a preload ratio not exceeding 70%, the value of the decrease in column capacity is approximately linear, and in small values, where the largest decrease ratio was equal to 0.891;
- With a preload ratio equal to 80%, there is a significant decrease in the percentage of remaining capacity of the columns reaching 0.801. This obviously affects the contribution of concrete, as the decreased value reaches 0.234;
- The proposed new models, Equations (23) and (24), using the ANSI/AISC 360-22 models, can predict the ultimate capacity of the axial load of CFSC under preloading. These models have appropriate accuracy, with r of 0.949 and 0.989 for preload ratios not exceeding 70% and equal to 80%, respectively.

Future research could focus on studies that can address the drawbacks of this new model. One of the drawbacks of this new model is the lack of diversity in the cross-sectional shape, such as the fact that rectangular, circular, double-skin, and sections of steel columns stiffened internally or externally remain to be studied. The reliability and accuracy of the new model have not been verified, so experimental tests on these columns under preload should be carried out and the results compared with FEM analysis. This model also relates to axial load only, so other types of preloading, such as eccentric and cyclic loads, can be used.

Funding: This research received no external funding.

Data Availability Statement: Data will be made available on request.

Acknowledgments: The author would like to thank the Deanship of Scientific Research at Majmaah University for supporting this work under project no. R-2024-1296.

Conflicts of Interest: The authors declare no conflicts of interest.

References

1. *ANSI/AISC 360-22*; Specification for Structural Steel Buildings. American Institute of Steel Construction: Chicago, IL, USA, 2022.
2. *EN 1994-1-1: 2004*; Design of Composite Steel and Concrete Structures, Part 1.1 General Rules and Rules for Buildings. European Committee for Standardization: Brussels, Belgium, 2004.

3. Liu, J.; Gao, P.; Lin, X.; Wang, X.; Zhou, X.; Chen, Y.F. Experimental assessment on the size effects of circular concrete-filled steel tubular columns under axial compression. *Eng. Struct.* **2023**, *275 Pt A*, 115247. [CrossRef]
4. Lei, M.; Li, Y.-L.; Li, Y.-Q.; Zhang, J.; Liu, D. Behavior of concrete-filled T-shaped steel tubular beam-columns under biaxial compressive loads. *Eng. Struct.* **2023**, *277*, 115409. [CrossRef]
5. Wang, J.; Yang, Z.-M.; Zheng, X.-L.; Ding, Y. Axial compression behavior of square section concrete-filled steel tubes reinforced with internal latticed steel angles. *J. Constr. Steel Res.* **2024**, *213*, 108414. [CrossRef]
6. Yang, Z.Y.; Liu, M.; Sun, L.; Shen, M.S. Size effect of axial compressive bearing capacity of high-strength concrete filled high-strength square steel tubular short columns. *Structures* **2024**, *60*, 105978. [CrossRef]
7. Wang, Y.-B.; Song, C.; Xiong, M.-X.; Liew, J.Y.R. Mechanical behavior and design applicability of ultra-high-performance concrete-filled circular steel tubular short columns under axial compression. *Eng. Struct.* **2024**, *299*, 117112. [CrossRef]
8. Hu, H.-S.; Yang, Z.-J.; Xu, L.; Zhang, Y.-X.; Gao, Y.-C. Axial compressive behavior of square concrete-filled steel tube columns with high-strength steel fiber-reinforced concrete. *Eng. Struct.* **2023**, *285*, 116047. [CrossRef]
9. Schneider, S.P. Axially loaded concrete-filled steel tubes. *J. Struct. Eng.* **1998**, *124*, 1125–1138. [CrossRef]
10. Uy, B. Strength of short concrete filled high strength steel box columns. *J. Construct. Steel Res.* **2001**, *57*, 113–134. [CrossRef]
11. Han, L.-H. Tests on stub columns of concrete-filled RHS sections. *J. Construct. Steel Res.* **2002**, *58*, 353–372. [CrossRef]
12. Dundu, M. Column buckling tests of hot-rolled concrete filled square hollow sections of mild to high strength steel. *Eng. Struct.* **2016**, *127*, 73–85. [CrossRef]
13. Shakir-Khalil, H.; Mouli, M. Further tests on concrete-filled rectangular hollow-section columns. *J. Struct. Eng.* **1990**, *68*, 405–413.
14. Zhu, A.; Zhang, X.; Zhu, H.; Zhu, J.; Lu, Y. Experimental study of concrete filled cold-formed steel tubular stub columns. *J. Constr. Steel Res.* **2017**, *134*, 17–27. [CrossRef]
15. Tao, Z.; Han, L.H.; Wang, Z.B. Experimental behaviour of stiffened concrete-filled thin-walled hollow steel structural (HSS) stub columns. *J. Constr. Steel Res.* **2005**, *61*, 962–983. [CrossRef]
16. Tao, Z.; Han, L.H.; Wang, D.Y. Strength and ductility of stiffened thin-walled hollow steel structural stub columns filled with concrete. *Thin Walled Struct.* **2008**, *46*, 1113–1128. [CrossRef]
17. Tao, Z.; Uy, B.; Han, L.H.; Wang, Z.B. Analysis and design of concrete-filled stiffened thin-walled steel tubular columns under axial compression. *Thin Walled Struct.* **2009**, *47*, 1544–1556. [CrossRef]
18. Liu, Y.J.; Zhang, J.G.; Xu, K.L.; Yao, W. Experimental research on stub columns of concrete filled steel box with longitudinal stiffener under axial load. *J. Build. Struct.* **2011**, *32*, 159–165.
19. Liu, Y.J.; Cheng, G.; Zhang, N.; Zhang, J. Experimental research on concrete-filled square steel tubular columns stiffened with PBL. *J. Build. Struct.* **2014**, *35*, 39–46.
20. Zhu, W.; Wang, J.; Liu, Y.; Sun, L. Experimental study on axial compressive behavior of concrete-filled double-skin thin-walled steel box columns with and without PBL. *Structures* **2023**, *47*, 1927–1939. [CrossRef]
21. Yuan, W.-b.; Yang, J.-j. Experimental and numerical studies of short concrete-filled double skin composite tube columns under axially compressive loads. *J. Constr. Steel Res.* **2013**, *80*, 23–31. [CrossRef]
22. Chen, J.; Ni, Y.-Y.; Jin, W.-l. Column tests of dodecagonal section double skin concrete-filled steel tubes. *Thin Walled Struct.* **2015**, *88*, 28–40. [CrossRef]
23. Liang, W.; Dong, J.; Wang, Q. Mechanical behaviour of concrete-filled double-skin steel tube (CFDST) with stiffeners under axial and eccentric loading. *Thin Walled Struct.* **2019**, *138*, 215–230. [CrossRef]
24. Qiao, Q.; Ding, R.; Cao, W. Experimental investigation on axial local compressive behavior of square concrete-filled double tubular stub columns. *Structures* **2023**, *55*, 2157–2174. [CrossRef]
25. Hassanein, M.F.; Elchalakani, M.; Karrech, A.; Patel, V.I.; Yang, B. Behaviour of concrete-filled double-skin short columns under compression through finite element modelling: SHS outer and SHS inner tubes. *Structures* **2018**, *14*, 358–375. [CrossRef]
26. Ding, F.-x.; Wang, W.-j.; Lu, D.-r.; Liu, X.-m. Study on the behavior of concrete-filled square double-skin steel tubular stub columns under axial loading. *Structures* **2020**, *23*, 665–676. [CrossRef]
27. Ayough, P.; Sulong, N.H.R.; Ibrahim, Z.; Hsiao, P.-C. Nonlinear analysis of square concrete-filled double-skin steel tubular columns under axial compression. *Eng. Struct.* **2020**, *216*, 110678. [CrossRef]
28. Zhang, Y.C.; Chen, Y. Experimental study and finite element analysis of square stub columns with straight ribs of concrete-filled thin-walled steel tube. *J. Build. Struct.* **2006**, *27*, 16–22.
29. Zhang, Y.C.; Xu, C.; Lu, X.Z. Experimental study of hysteretic behaviour for concrete filled square thin-walled steel tubular columns. *J. Constr. Steel Res.* **2007**, *63*, 317–325. [CrossRef]
30. Varma, A.H.; Ricles, J.M.; Sause, R.; Lu, L.-W. Experimental behaviour of high strength square concrete-filled steel tube beam-columns. *J. Struct. Eng.* **2002**, *128*, 309–318. [CrossRef]
31. Liu, D.; Gho, W.-M.; Yuan, J. Ultimate capacity of high-strength rectangular concrete-filled steel hollow section stub columns. *J. Constr. Steel Res.* **2003**, *59*, 1499–1515. [CrossRef]
32. Gho, W.-M.; Liu, D. Flexural behaviour of high-strength rectangular concrete-filled steel hollow sections. *J. Construct. Steel Res.* **2004**, *60*, 1681–1696. [CrossRef]
33. Li, G.-C.; Chen, B.-W.; Zhu, B.-W.; Yang, Z.-J.; Ge, H.-B.; Liu, Y.-P. Axially loaded square concrete-filled steel tubular long columns made of high-strength materials: Experimental investigation, analytical behavior and design. *J. Build. Eng.* **2022**, *58*, 104994. [CrossRef]

34. Guo, L.H.; Wang, Y.Y.; Zhang, S.M. Experimental study of concrete-filled rectangular HSS columns subjected to biaxial bending. *Adv. Struct. Eng.* **2012**, *15*, 1329–1344. [CrossRef]
35. Kilpatrick, A.E.; Rangan, B.V. Tests on high-strength concrete-filled steel tubular columns. *ACI Struct. J.* **1999**, *96*, 268–275.
36. Liu, D. Tests on high-strength rectangular concrete-filled steel hollow section stub columns. *J. Construct. Steel Res.* **2005**, *61*, 902–911. [CrossRef]
37. Sakino, K.; Nakahara, H.; Morino, S.; Nishiyama, I. Behaviour of centrally loaded concrete-filled steel-tube short columns. *J. Struct. Eng.* **2004**, *130*, 180–188. [CrossRef]
38. Young, B.; Ellobody, E. Experimental investigation of concrete-filled cold-formed high strength stainless steel tube columns. *J. Constr. Steel Res.* **2006**, *62*, 484–492. [CrossRef]
39. Young, B.; Liu, Y. Experimental investigation of cold-formed stainless steel columns. *J. Struct. Eng.* **2003**, *129*, 169–176. [CrossRef]
40. Zeghichea, J.; Chaoui, K. An experimental behaviour of concrete-filled steel tubular columns. *J. Construct. Steel Res.* **2005**, *61*, 53–66. [CrossRef]
41. Pavlovčič, L.; Froschmeier, B.; Kuhlmann, U.; Beg, D. Finite element simulation of slender thin-walled box columns by implementing real initial conditions. *Adv. Eng. Softw.* **2012**, *44*, 63–74. [CrossRef]
42. Yang, D.; Hancock, G.J. Numerical simulation of high-strength steel box-shaped columns failing in local and overall buckling modes. *J. Struct. Eng.* **2006**, *132*, 541–549. [CrossRef]
43. Wang, Y.-B.; Li, G.-Q.; Chen, S.-W.; Sun, F.-F. Experimental and numerical study on the behavior of axially compressed high strength steel box-columns. *Eng. Struct.* **2014**, *58*, 79–91. [CrossRef]
44. Khan, M.; Uy, B.; Tao, Z.; Mashiri, F. Concentrically loaded slender square hollow and composite columns incorporating high strength properties. *Eng. Struct.* **2017**, *131*, 69–89. [CrossRef]
45. Thai, H.-T.; Uy, B.; Khan, M.; Tao, Z.; Mashiri, F. Numerical modelling of concrete-filled steel box columns incorporating high strength materials. *J. Construct. Steel Res.* **2014**, *102*, 256–265. [CrossRef]
46. Dai, X.; Lam, D. Numerical modelling of the axial compressive behaviour of short concrete-filled elliptical steel columns. *J. Construct. Steel Res.* **2010**, *66*, 931–942. [CrossRef]
47. Ellobody, E.; Young, B. Nonlinear analysis of concrete-filled steel SHS and RHS columns. *Thin Walled Struct.* **2006**, *44*, 919–930. [CrossRef]
48. Kumar, S.; Gupta, P.K.; Iqbal, M.A. Experimental and numerical study on self-compacting alkali-activated slag concrete-filled steel tubes. *J. Constr. Steel Res.* **2024**, *214*, 108453. [CrossRef]
49. Tao, Z.; Wang, Z.-B.; Yu, Q. Finite element modelling of concrete-filled steel stub columns under axial compression. *J. Construct. Steel Res.* **2013**, *89*, 121–131. [CrossRef]
50. Ali, L.; Isleem, H.F.; Bahrami, A.; Jha, I.; Zou, G.; Kumar, R.; Sadeq, A.M.; Jahami, A. Integrated behavioural analysis of FRP-confined circular columns using FEM and machine learning. *Compos. Part. C Open Access* **2024**, *13*, 100444. [CrossRef]
51. Somodi, B.; Kövesdi, B. Flexural buckling resistance of welded HSS box section members. *Thin Walled Struct.* **2017**, *119*, 266–281. [CrossRef]
52. Zhao, T. Bearing capacity studies on square steel tube confined steel reinforced concrete column under eccentric load. *Adv. Civ. Eng.* **2020**, *2020*, 4212049. [CrossRef]
53. Lin, S.; Li, Z.; Zhao, Y.-G. Behavior of eccentrically loaded circular concrete-filled steel tube stub columns with localized corrosion. *Eng. Struct.* **2023**, *288*, 116227. [CrossRef]
54. Sayed, A.M.; Diab, H.M. Modeling of the axial load capacity of RC columns strengthened with steel jacketing under preloading based on FE simulation. *Model. Simul. Eng.* **2019**, *2019*, 8653247. [CrossRef]
55. ANSYS. *ANSYS Help User's Manual*; Version 22; Swanson Analysis Systems, Inc.: Houston, PA, USA, 2022.
56. Kachlakev, D.; Miller, T.; Yim, S. Finite Element Modeling of Reinforced Concrete Structures Strengthened with FRP Laminates. Oregon Department of Transportation: Washington, DC, USA, 2001.
57. Terec, L.; Bugnariu, T.; Păstrav, M. Non–linear analysis of reinforced concrete frames strengthened with in filled walls. *Rom. J. Mater.* **2010**, *40*, 214–221.
58. Kwan, A.K.H.; Dai, H.; Cheung, Y.K. Non-Linear seismic responce of reinforced concrete slit shear walls. *J. Sound. Vibr.* **1999**, *226*, 701–718. [CrossRef]
59. Sayed, A.M.; Xin, W.; Zhishen, W. Finite element modeling of the shear capacity of RC beams strengthened with FRP sheets by considering different failure modes. *Constr. Build. Mater.* **2014**, *59*, 169–179. [CrossRef]

buildings

MDPI

Article

A Framework for Evaluating the Reasonable Internal Force State of the Cable-Stayed Bridge Without Backstays

Tao Xu [1], Jiqian Ma [2], Guojie Wei [3], Boxu Gong [4] and Jiang Liu [4,*]

1 Xiamen Company of China First Highway Engineering Co., Ltd., Xiamen 361000, China; 18050669300@139.com
2 Hainan Jiaotou Huandao Tourism Highway Investment and Development Co., Ltd., Haikou 570203, China; 18389798508@163.com
3 Guangdong Province Nanyue Transportation Lianying Highway Administration, Qingyuan 513047, China; weiguojie19@163.com
4 School of Highway, Chang'an University, Xi'an 710064, China; gongboxumooc@163.com
* Correspondence: liu-jiang@chd.edu.cn

Abstract: The synchronous construction of the pylon and cables of a cable-stayed bridge without backstays has the characteristics of a short construction period and reduced support costs. However, it also increases the difficulty of construction control, making the reasonable completion state of the bridge more complex. To investigate the impact of various load parameters on the structural state of a cable-stayed bridge without backstays during the synchronous construction process, and to ensure a rational final bridge state, this study proposes an assessment framework for evaluating the internal forces of the bridge. The framework initially uses the response surface method to establish explicit equations relating the control indicators of the bridge's final state to various load parameters. Subsequently, through sensitivity analysis, the degree of influence of each load parameter on the structural response of the cable-stayed bridge without backstays is examined. The most sensitive factors are identified to create a bridge parameter influence library, which helps reduce computational costs. Based on this, a method for controlling construction errors and predicting cable forces is proposed. This method utilizes the pre-established bridge parameter influence library, combined with the internal force state of the bridge at the current construction stage, to accurately predict the tension force of the stay cables in the subsequent stage, thereby ensuring a rational final bridge state. The framework is ultimately validated through a case study of the Longgun River Bridge to assess its rationality and effectiveness.

Keywords: bridge engineering; cable-stayed bridge without backstays; steel–concrete composite pylon; bridge internal force state assessment; finite element modeling; reasonable finished bridge state

1. Introduction

The cable-stayed bridge without backstays balances the dead load and live load through the tilting of the pylon. Its innovative structure and elegant design often make it a landmark in the region once completed [1]. At the same time, in pursuit of innovation and aesthetics, the design and construction methods of cable-stayed bridges without backstays are continuously innovating and developing. Due to the ability of the synchronous construction method of pylon and cables to avoid the influence of geological conditions and improve construction efficiency, it has gradually become one of the main construction methods. However, its complex construction process significantly affects the internal force state of the bridge due to changes in structural parameters. If the construction errors during the process are not accurately managed, deviations will occur between the actual bridge state and the ideal design state, which will affect the safe operation of the bridge [2].

Compared to traditional cable-stayed bridges, the cable-stayed bridge without backstays exhibits significant differences in mechanical performance and force transmission paths. On the one hand, because there are no back cables, there are both vertical and horizontal forces, resulting in more complex stresses at the pylon root, as it must endure the combined effects of axial forces, shear forces, bending moments, and girder torsion [3,4]. On the other hand, the pylon is highly sensitive to live loads, which causes the front side of the pylon to be compressed and the back side to be pulled [5]. Under this condition, over time, various safety risks appear in the bridge, such as fatigue of the cables, concrete shrinkage and creep, and settlement and displacement of the supports [6–9]. Therefore, properly controlling the internal force state of the bridge during construction is essential for ensuring its long-term operation.

In the past few decades, research has focused on the analysis of parameter sensitivity of cable-stayed bridges without backstays. In 1992, Casas [10] first proposed the monitoring points for the cable-stayed bridge without backstays, noting that an error of 10% in the final weight of the pylon or deck with respect to design could lead to a variation of up to 70% in internal forces. After this, to ensure that the internal forces and deformations of the stay cable bridge structure remain within a safe range during construction, the impact of the variability of structural load parameters on the stress state of the cable-stayed bridge without backstays was explored. During the construction process, scholars analyzed various factors, including the weight of the main girder, tension in the cables, elasticity modulus of the cables, and temporary construction loads, providing numerous control recommendations [11–15]. The methods mentioned above mostly involve pre-construction assessments. However, during the bridge construction process, there are many influencing parameters, and the actual state of the bridge can differ significantly from the theoretical state. Pre-assessments cannot make real-time adjustments based on the construction conditions, which may result in a substantial difference between the final bridge state and the optimal bridge state. In addition, the aforementioned studies conducted sensitivity analyses based solely on single-parameter adjustment methods. This approach involves repeatedly modifying the finite element model to calculate changes in the structural response, which is cumbersome and does not fully reflect actual construction conditions, thus limiting its practical application.

Therefore, this paper presents a novel assessment framework for construction state control to ensure reasonable internal forces in cable-stayed bridges without backstays. This framework can quickly predict the parameters of a specific construction stage during the simultaneous construction of the python and cable of the cable-stayed bridge without backstays, without the need for additional finite element modeling by construction personnel. The assessment framework first uses multi-parameter analysis to identify the main sensitive parameters during construction, establishing these parameters as variables and creating a pre-generated library of bridge parameter influences. Next, based on the measured internal force state and displacement of the bridge at the current construction stage, corresponding constraints are set. Ultimately, the framework outputs construction control instructions that bring the bridge status closer to the expected reasonable completed state.

This paper proposes a novel framework for the assessment of construction state control and applies it to the Longgun River Bridge in Hainan Province, where it rapidly predicts and controls key parameters during the construction process. By using this framework, the rationality of the internal forces and alignment of the bridge is ensured, the need for secondary post-tensioning is avoided, and construction efficiency is significantly improved. The framework proves highly applicable to similar steel–concrete composite structure projects.

2. Bridge Internal Force Assessment Framework

This Section presents how the control assessment framework will be used by construction personnel to control the internal force management during the construction of a cable-stayed bridge without backstay cables, which is due to significant differences in the internal forces of completed bridges caused by various bridge types and construction methods. Therefore, in order to accurately grasp the internal force of the bridge, this method focuses specifically on the steel–concrete composite pylon structure of a cable-stayed bridge and the synchronous construction method of the pylon and cables. At the same time, due to structural differences, parameters such as the dead weight of the main girder, the dead weight of the bridge pylon, and installation errors have varying impacts on the structural state of the bridge during the construction process. Firstly, it is necessary to identify the main sensitive influence parameters during the construction process of the structure, such as stress limits at critical sections, displacement magnitudes, cable forces, and other restrictive constraints specified in the relevant codes and standards. Then, generate the impact of these main sensitive parameters on each construction stage of the bridge to create a parameter influence library specific to the bridge. An in-depth discussion on the establishment of the main sensitive parameters and the parameter influence library for the bridge is provided in later sections.

Figure 1 is a flowchart illustration of how the framework will be used to perform bridge internal force control. The user provides initialization parameters that describe the structural makeup and geometry of the bridge ("Structural" from Figure 1). Data for structural initialization parameters come from as-built design drawings of the bridge of interest. The user needs to determine the parameters in the bridge construction process, and use the internal force state and displacement of the key sections as the output variables to calculate the influence of each parameter on the structure, and form the bridge parameter influence library. Subsequently, construction personnel can match the corresponding construction stages based on the primary sensitivity parameters during the bridge construction process. By incorporating the casting errors and installation errors ("Known errors" from Figure 1) along with the monitored internal force state of the bridge ("Internal force state" from Figure 1), it is possible to seek the optimal construction control errors that ensure the internal force state of the bridge meets the required standards. During the synchronous construction of the pylon and cables for a cable-stayed bridge without backstays, the main girder is first constructed using full framing. Next, the steel box for the bridge pylon is hoisted and concrete is poured inside the pylon. Finally, the cables are tensioned. Therefore, during the construction process of the current stage, some errors (such as the bulk density of the main girder and that of the bridge pylon, the overall temperature, the lifting accuracy, etc.) can be regarded as a determined value. At this point, by combining the pre-tensioned cable force and performing iterative calculations on the structural forces, the ultimate optimal cable tension can be identified to meet the design requirements.

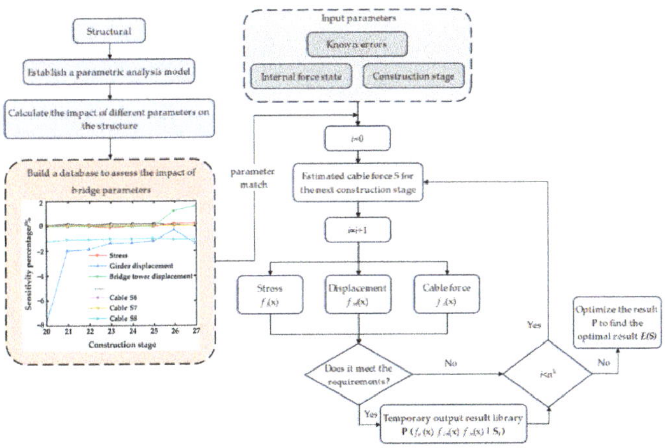

Figure 1. Flowchart illustrating how the framework conducts reasonable internal force assessment of cable-stayed bridges without backstays.

3. Methods of Assessment Framework

During the construction of bridges, various parameters affect the internal force state of the bridge structure differently. Therefore, it is essential to conduct a sensitivity analysis of the structural parameters to identify the critical sensitive parameters and improve analytical efficiency. Sensitivity analysis methods include the single-parameter analysis method and the response surface method, with their respective advantages and disadvantages shown in Table 1. By comparison, it can be concluded that the single-parameter analysis is suitable for simple systems or when the focus is on a specific parameter, whereas the response surface method is more appropriate for comprehensively evaluating the combined effects of multiple parameters. Hence, in this paper, the response surface method is used to systematically analyze the influence of each parameter on the internal force state of the bridge structure, so as to avoid the limitations of the single-parameter method.

Table 1. Comparison of parameter sensitivity analysis methods.

Parameter Analysis Methodology	Advantages	Disadvantages
Single-parameter methodology	Simple and easy to understand and calculate, helps identify key parameters.	Ignoring interactions between parameters may lead to misleading conclusions, fail to capture the behavior of complex systems, and result in substantial computational effort when the number of parameters is large.
Response surface methodology	Considers interactions between multiple parameters, suitable for complex systems, improves decision quality.	Dependence on response surface equations, especially when these equations are complex, can lead to high computational demands, increased analytical complexity.

3.1. *Response Surface Methodology*

3.1.1. Basic Principle

The key to the analysis of parameter sensitivity by the response surface method is to establish the relationship between the selected design parameters and the structural response control index. This method approximates the real structural response surface by fitting a quadratic surface based on the finite element analysis results of multiple parameters changing simultaneously. This method is widely applied and demonstrates high solution efficiency, effectively meeting the requirements of practical engineering problems [16,17].

Considering the numerous load parameters studied in this paper, which results in a large number of design variables in the equations, and to enhance computational efficiency, the response surface equation is formulated using a polynomial that includes only linear and quadratic terms of individual design variables, while excluding any quadratic interaction terms between variables, as shown in Equation (1). Therefore, when the number of design parameters is the same, the undetermined factors in the equation are significantly reduced, which in turn decreases the number of samples required to construct the structural response equation.

$$f(X) = m + \sum_{i=1}^{n} n_i X_i + \sum_{i=1}^{n} c_i {X_i}^2 \tag{1}$$

where $f(X)$ represents the objective function; m, n_i, and c_i are the unknown coefficients; and X_i denotes the design variables.

To further assess the accuracy of the response surface equation, the coefficient of determination (R^2) from regression analysis can be used for evaluation. This coefficient indicates how well the response surface approximates the true values, ranging from 0 to 1.

A value closer to 1 indicates that the response surface model more accurately reflects the actual situation.

$$R^2 = 1 - \frac{\sum\limits_{i=1}^{n} \left(f_i - \widehat{f}_i \right)^2}{\sum\limits_{i=1}^{n} \left(f_i - \overline{f}_i \right)^2} \tag{2}$$

where f_i, $\widehat{f}_i$, $\overline{f}_i$ represent the finite element analysis values for each response control index, the values fitted by the response surface model, and the mean of the finite element analysis values, respectively.

3.1.2. Sensitivity Factor

In previous studies of parameter sensitivity for cable-stayed bridges without backstays, sensitivity was often assessed by comparing structural responses resulting from changes in a single control parameter or by using the normalization method. The calculation results of these methods do not directly indicate the change range of the control target after a parameter change, lacking a clear standard for parameter sensitivity assessment. A more accurate method for assessing parameter sensitivity is to use the functional relationship between the control target and the design parameters established by the response surface method to perform the calculation. The structural response $f(X)$ is related to n structural parameters $X_1, X_2, \ldots, X_i, \ldots, X_n$ by the function relationship (1). The sensitivity S_{X_i} of the control objective $f(X)$ with respect to a specific structural parameter X_i is given by:

$$S_{X_i} = \frac{\partial F(X_1, X_2, \ldots, X_i, \ldots, X_n)}{\partial X_n} \tag{3}$$

The application of Equation (1) to determine parameter sensitivity requires the explicit function $F\,(X_1, X_2, \ldots, X_i, \ldots, X_n)$. However, in practical construction, the large number of parameters and continuous changes in the structural system make it difficult to construct the explicit function. In order to facilitate the calculation, based on perturbation theory, when the variation in the key parameter X_i is controlled to be Δx within a reasonable disturbance range, the variation in structural response is $\Delta f(X)$. From Equation (3), the parameter sensitivity S_{X_i} can be approximately expressed as:

$$S_{X_i} = \frac{|\Delta f(X)| / f(X)}{\partial X_n} \tag{4}$$

where S_{X_i} represents a set of dimensionless non-negative real numbers, reflecting the sensitivity of the control objective $f(X)$ to the structural parameter X in the bridge state. By computing the parameter sensitivity, a quantitative assessment of it can be achieved.

3.2. *Assessment Framework*

Using the proposed methods, the prediction of tensile forces for the next phase largely relies on the parameter influence library generated prior to deployment. However, modeling and analyzing all possible combinations of different bridge designs, loading conditions, and potential failure mechanisms is tricky. This Section will explain the creation method of the bridge parameter influence library and the internal force control method.

3.2.1. Construction of Parameter Influence Library

Due to the negligible difference between geometric non-linearity and elastic analysis results in the cable-stayed bridge without backstays [18], this study establishes the relationships between key parameters and bridge control parameters through elastic analysis during the creation of the construction error threshold control model, as shown in Equation (5). These relationships are then combined to form a parameter influence library, as indicated in Equation (6). The library describes the influence of various parameters on bridge stress, displacement, and cable force.

$$O_i = \begin{bmatrix} m_1(x_1) & m_1(x_1) + \frac{m_1(-x_1)-m_1(x_1)}{n} & \cdots & m_1(x_1) + (n-1)\frac{m_1(-x_1)-m_1(x_1)}{n} & m_1(-x_1) \\ \vdots & \vdots & \vdots & \vdots & \vdots \\ m_i(x_1) & m_i(x_1) + \frac{m_i(-x_1)-m_i(x_1)}{n} & \cdots & m_i(x_1) + (n-1)\frac{m_i(-x_1)-m_i(x_1)}{n} & m_i(-x_1) \end{bmatrix} \quad (5)$$

where O_i represents the influence of a specific parameter on an output variable at different construction stages and error levels; $m_i(x)$ indicates the influence of the parameter on the output variable at a value of x at different construction stages; and n is the discretization accuracy.

Then, the influence library of bridge parameters can be expressed as a set of Formula (6):

$$Y_{ij} = \begin{bmatrix} O_{11} & \cdots & O_{1j} \\ \vdots & \vdots & \\ O_{i1} & \cdots & O_{ij} \end{bmatrix} \quad (6)$$

where Y_{ij} represents the parameter influence library, while O_{ij} denotes the influence of parameter i on output variable j at different construction stages.

3.2.2. Prediction Method of Reasonable Internal Force State of Bridge

After establishing the bridge parameter influence library, the current state of the structure can be obtained based on monitoring results, and the structural state of the bridge can be predicted by integrating input variables. The influence of known errors on different construction stages is:

$$D_j(x) = D(Y_{ij} | i = a, j = b, m = c) \quad (7)$$

where $D_j(x)$ represents the impact of known errors on various parameters at different construction stages; a is the known load error; b is the parameter affected; and c is the error level.

Under the influence of different cable tensions, the parameter error expression that meets construction requirements is as follows:

$$f_i(x) = f([A_i(x) + \sum D_j(x) + \sum Y_j(x) \,|S] \leq v_i) \quad (8)$$

where S is the set of cables that need to be predicted; $f_i(x)$ represents the structural response of the output variables after the cable tension at different construction stages; $A_i(x)$ is the measured value of the corresponding parameter; $\sum Y_j(x) \mid S$ sums the influence from set S on the structural state at the same construction stage under varying parameters; and v_i represents the limit values corresponding to the output variables.

The resulting library that meets the requirements can be expressed as:

$$P = [f_j(x), f_m(x), f_n(x) \,|S] \quad (9)$$

where $f_j(x)$, $f_m(x)$, and $f_n(x)$ represent the changes in structural stress, displacement, and tensioned cables related to the cable S, respectively.

3.2.3. Optimistic Algorithm

The above algorithm can obtain more results. Considering the construction errors and the reliability of the structure, it is necessary to find the optimal solution with the minimum error. Thus, the final optimization result can be expressed as:

$$E(S) = P\left(\min\left(\sum_{j=1}^{a} (f_{ij}(x) - h_j)^2 + \sum_{m=1}^{b} (f_{im}(x) - h_m)^2 + \sum_{n=1}^{c} (f_{in}(x) - h_n)^2\right) |S_i\right) \quad (10)$$

where $E(S)$ is the optimal cable force result, which represents the minimum equation of the parameter results related to different cable forces; h_i is the theoretical value corresponding to the output results; a, b, and c are the structural stress, displacement, and the number of tensioned cable forces that need attention, respectively.

4. Case Study

4.1. Engineering Background

4.1.1. Overall Layout of the Bridge

The Longgun River Bridge in Hainan Province was selected as a case study to verify the functionality and accuracy of the framework. The bridge is a 30 + 80 m curved special-shaped pylon cable-stayed bridge without backstays. The main girder is constructed using prestressed concrete Π-shaped girders, and the material used is the C60 concrete. The construction method is synchronous construction of the cables and pylon. The cables are arranged in a spatial double-plane harp shape, with 8 pairs of 16 cables throughout the bridge, as shown in Figure 2.

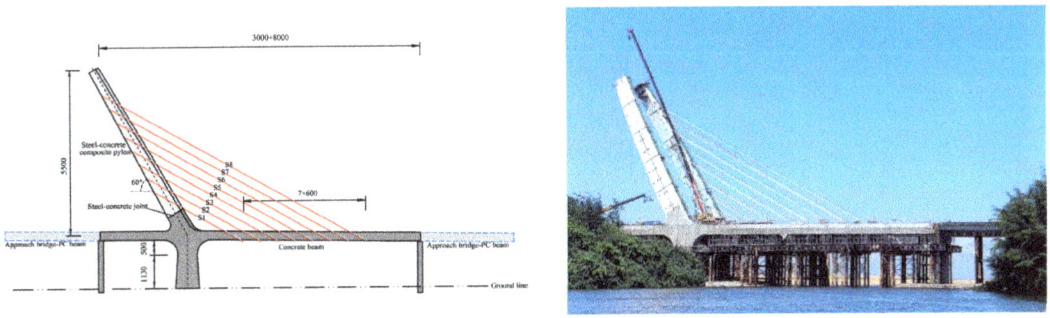

(a) (b)

Figure 2. Layout of Longgun River Bridge: (**a**) Schematic diagram of bridge structure (Unit: cm); (**b**) Construction process.

The pylon adopts a steel–concrete structure, with the structural characteristics illustrated in Figure 3. The lower pylon column and the range of 4.85 m above the bridge deck are concrete pylons, while the remaining upper pylon columns consist of single-cell steel boxes filled with iron sand heavy-weight concrete. The pylon is an 'arch' structure in the transverse direction. The pylon alignment is based on the vertical central axis of the main girder standard girder rib as the interface, and the vertical structure is from the bottom to the top of the pile cap, with a height of 17.7 m. The angle between the pylon and the side span is 60°, the vertical projection height is 55.0 m, and the oblique height is 63.5 m. Between the lower and upper pylon column is the pylon girder consolidation section, with an arc transition connecting the two components.

4.1.2. Construction Method

The bridge pylon of the Longgun River Bridge adopts the cantilever assembly construction method. Firstly, the full support is set up and the main girder is poured. Then, the cantilever construction is carried out by segmental cyclic hoisting, with the construction process illustrated in Figure 4a. The bridge pylon is divided into 7 sections, with 2 pairs of cables on each of the 2#~5# segments, and the cables are tensioned after lifting completion. Figure 4b shows the weight of each section of iron sand heavy-weight concrete (DL), the weight of each section of steel box (WS), and the l segment length (LS).

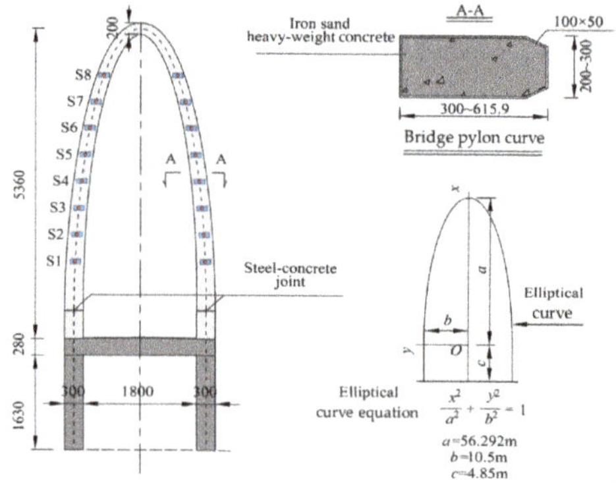

Figure 3. Bridge pylon parameters.

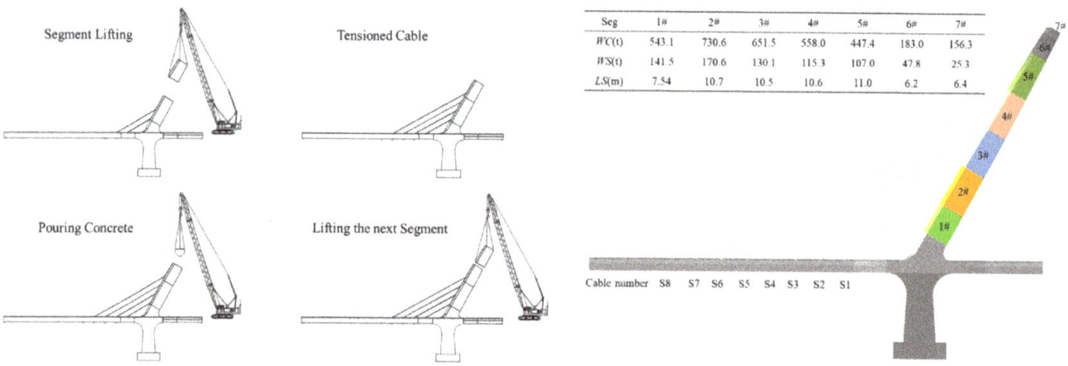

(a) (b)

Figure 4. Construction information: (**a**) construction process; (**b**) segment information.

4.1.3. Finite Element Model

The analysis model is established by ABAQUS 2019 software. The steel structure is simulated by the S3/S4R element, and the concrete structure is simulated by the DC3D8R/C3D6 element. The mesh division of the bridge pylon is illustrated in Figure 5a.

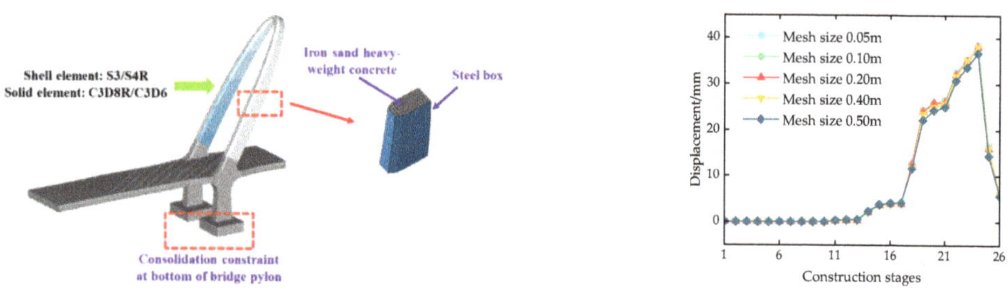

(a) (b)

Figure 5. Establishment of finite element model: (**a**) schematic diagram of bridge model; (**b**) influence of mesh size.

In the process of establishing a finite element model, the mesh size plays a crucial role in both the accuracy of the results and the computational efficiency. For this reason, the influence of mesh size on the structural calculation results was analyzed, with the mid-span vertical displacement of the main beam as the subject of the analysis. The results are shown in Figure 5b. When the mesh size is smaller than 0.4 m, the corresponding calculation results nearly coincide. However, when the mesh size is 0.5 m, a noticeable discrepancy appears in the displacement curve, with an error of approximately 2 mm. Considering the computational efficiency of the model, a mesh size of 0.4 m was chosen for the finite element model.

4.2. Establishment of Response Surface Model

During the construction process of cable-stayed bridges without backstays, parameter changes cause significant changes in the internal force state of the bridge structure. For the construction of the main girder, the deviation and the expansion of the formwork during the pouring process will cause a change in the weight of the main girder. For pylons, although the weight and geometry are relatively precise due to factory-prefabricated steel boxes and more controllable internal concrete volumes, meeting material density requirements for design-weighted iron sand concrete can be challenging. This may lead to changes in the weight of the pylon, affecting its geometry. Additionally, the precision of the steel box installation will also affect the alignment of the pylon. Site construction conditions significantly influence the tension in cables and other critical structural loads.

Combined with the relevant research, nine key structural parameters X_i were selected for comparative analysis [2,13,15,19]. These parameters are the weight of the main girder, the density of the pylon (the weight of the iron sand heavy-weight concrete), the elastic modulus of the pylon, the longitudinal installation accuracy of the steel box, the tension of cables, the overall temperature variation, the temperature difference in the stay and girder, the temperature gradient of the main girder, and the temperature difference on the exterior surface of the pylon. The specific flowchart is illustrated in Figure 6.

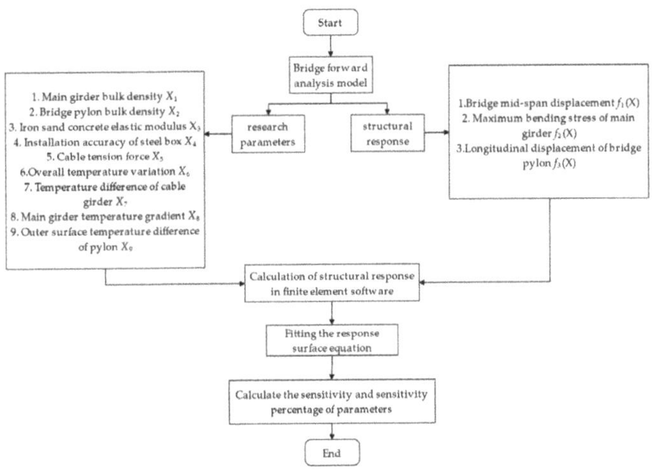

Figure 6. Flowchart illustrating how to analyze the sensitivity of bridge parameters.

In the design of iron sand heavy-weight concrete, weight is the primary consideration. However, when it is combined into a composite structure with the pylon, the modulus of elasticity of the composite structure becomes altered and is difficult to measure. Therefore, the elastic modulus of the pylon should also be considered. In this study, the elastic modulus of the iron sand heavy-weight concrete was 1% of C40 concrete, with variations accounting for ±20%. All other parameter values are detailed in Table 2. The sample for the data simulation experiments is presented in Table 3, comprising 23 distinct parameter sample groups. The functions $f_1(X)$~$f_3(X)$ serve as control indicators for the structural responses, and X_1~X_9 represent the load design parameters chosen in this study, and the meanings of each symbol are shown in Figure 6.

Table 2. Sensitivity parameters and selected ranges.

Control Parameters	X_1 (kN/m^3)	X_2 (kN/m^3)	X_3 ($\times 10^4$ MPa)	X_4 (mm)	X_5 (kN)	X_6 (°C)	X_7 (°C)	X_8 (°C)	X_9 (°C)
Theoretical value	25.0	38.0	0.0325	Design value	Design value	20	10	14	5
Variation range of error	±5% ±10%	−10% −15%	±20%	±10 ±20	±5%	12	±10 ±15	16, 20	6, 9

Table 3. Numerical simulation experiment samples.

Number	X_1	X_2	X_3	X_4	X_5	X_6	X_7	X_8	X_9	$f_1(X)$ (mm)	$f_2(X)$ (MPa)	$f_3(X)$ (mm)
1	1.1	1	1	0	1	20	10	14	5	−6.7	−11.0	−76.1
2	1.05	1	1	0	1	20	10	14	5	−11.2	−10.7	−69.2
3	0.95	1	1	0	1	20	10	14	5	14.5	−9.6	−85.3
4	0.9	1	1	0	1	20	10	14	5	28.6	−10.2	−92.1
5	1	0.9	1	0	1	20	10	14	5	−18.2	−10.5	−47.9
6	1	0.85	1	0	1	20	10	14	5	−27.5	−10.8	−31.3
7	1	1	1.2	0	1	20	10	14	5	−6.2	−9.7	−73.5
8	1	1	0.8	0	1	20	10	14	5	−4.4	−10.0	−73.6
9	1	1	1	20	1	20	10	14	5	3	−9.9	−79.5
10	1	1	1	10	1	20	10	14	5	2.9	−9.9	−79.7
11	1	1	1	−10	1	20	10	14	5	−4.6	−10.0	−73.6
12	1	1	1	−20	1	20	10	14	5	−4.8	−10.0	−73.7
13	1	1	1	0	1.05	20	10	14	5	2.6	−10.0	−75.9
14	1	1	1	0	0.95	20	10	14	5	−6.3	−10.1	−74.1
15	1	1	1	0	1	12	10	14	5	−13.8	−10.0	−65.7
16	1	1	1	0	1	20	15	14	5	−5.9	−10.1	−78.6
17	1	1	1	0	1	20	10	14	5	−4.5	−10.0	−73.5
18	1	1	1	0	1	20	−10	14	5	1.1	−9.7	−53.5
19	1	1	1	0	1	20	−15	14	5	2.6	−9.6	−48.4
20	1	1	1	0	1	20	10	16	5	−4.2	−10.0	−73.6
21	1	1	1	0	1	20	10	20	5	−3.8	−9.9	−73.8
22	1	1	1	0	1	20	10	14	6	−4.3	−10.0	−76.1
23	1	1	1	0	1	20	10	14	9	−3.8	−10.0	−83.6

Note: In the table, the research parameters X_1~X_3 and X_5 reference values 1 in the table correspond to the basic state values in Table 2, while the variation values correspond to the change error. The longitudinal displacement of the pylon indicated as "−" represents a direction away from the main span.

In establishing the response surface equation according to Equation (1), in this paper, nine design variables are used, necessitating the determination of nineteen unknown coefficients, as detailed in Equation (11). Consequently, nineteen independent equations are required to solve for these coefficients. After excluding the data samples numbered 2, 4, 12, and 19 from Table 3, the response surface equation was constructed, and the coefficient obtained coefficients are presented in Table 4.

$$f(X) = k_0 + k_1X_1 + k_2X_2 + k_3X_3 + k_4X_4 + k_5X_5 + k_6X_6 + k_7X_7 + k_8X_8 + k_9X_9 + \\ k_{10}X_1^2 + k_{11}X_2^2 + k_{12}X_3^2 + k_{13}X_4^2 + k_{14}X_5^2 + k_{15}X_6^2 + k_{16}X_7^2 + k_{17}X_8^2 + k_{18}X_9^2 \tag{11}$$

Table 4. Coefficients of the response surface equation.

Coefficient	$f_1(X)$	$f_2(X)$	$f_3(X)$
k_0	58.223	−1.754	−22.772
k_1	−4136.100	30.300	2860.950
k_2	502.167	14.167	−996.417
k_3	145.000	−5.250	−92.500
k_4	0.302	0.004	−0.242
k_5	−279.000	65.000	−342.000
k_6	332.279	−10.011	−129.958
k_7	−0.171	−0.014	−1.095
k_8	−6.170	−0.215	5.665
k_9	−7.921	−0.113	4.269
k_{10}	1948.667	−19.333	−1365.667
k_{11}	−180.667	−4.667	379.667
k_{12}	−74.750	3.000	46.375
k_{13}	0.000	0.000	0.000
k_{14}	184.000	−32.000	162.000
k_{15}	−11.833	0.358	4.612
k_{16}	−0.022	−0.001	0.018
k_{17}	0.174	0.007	−0.159
k_{18}	0.539	0.008	−0.451

To assess the accuracy of the obtained response surface equations, samples numbered 2, 4, 12, and 19, which were not used in the equation-solving process, were substituted into Equation (11) to compute the fitted values of the response surface models for each control indicator. These fitted values were then compared with the finite element analysis results to evaluate the precision of the equation fitting. Based on Equation (2), the R^2 values for the objective functions $f_1(X)$~$f_3(X)$ were 0.88, 0.91, and 0.87, respectively. These results indicate that the fitted response surface equations accurately reflect the relationship between the bridge state performance objectives and the various design load parameters, and can be used for further structural parameter sensitivity analysis.

4.3. Parameter Sensitivity Analysis

According to Equation (4), the sensitivity of X_1~X_9 parameter changes to the bridge state response control objectives such as bridge alignment, pylon stress, and deviation are calculated, and the sensitivity trends of structural parameters are quantitatively analyzed.

4.3.1. Mid-Span Displacement of Main Girder

The main girder alignment is an important index of the bridge state of the cable-stayed bridge. The vertical displacement of the main girder in the bridge state changes with the load parameters throughout the construction process. By analyzing the influence of various parameters on the main girder displacement, key influencing parameters are identified based on sensitivity analysis results, allowing for targeted control measures to ensure the proper alignment of the main girder.

The calculated sensitivities of various loading parameters on the mid-span deflection of the main girder are illustrated in Figure 7a. To further reflect the influence of each parameter, sensitivity percentages are introduced, as shown in Figure 7b. From the figures, it is evident that the most sensitive parameter affecting the mid-span deflection of the main girder is the cable tension, with a maximum sensitivity of 37.77 and a sensitivity percentage of 56.35%. Specifically, a 1% increase in cable tension results in a corresponding increase of 37.77% in the mid-span deflection of the main girder. The dead load of the main girder and pylon, the elastic modulus of the pylon, and the installation accuracy of the pylon are secondary influencing factors, with sensitivities of 6.55, 8.73, 3.41, and 4.25, and sensitivity percentages of 9.78%, 13.03%, 5.09%, and 6.35%, respectively. The effects of temperature variations on the main girder deflection are relatively minor, with sensitivities of the overall temperature variation, temperature difference of cable girder, main girder temperature gradient, and surface temperature difference of the pylon being 2.18, 1.55, 1.68, and 0.90, respectively.

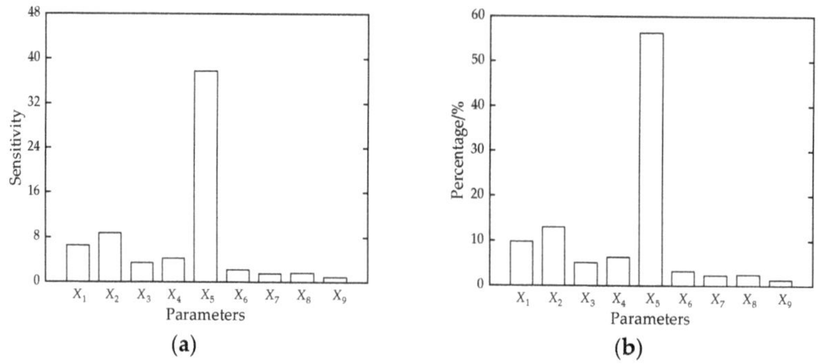

Figure 7. Sensitivity analysis of mid-span deflection to different parameters: (**a**) parameter sensitivity; (**b**) parameter sensitive percentage.

The main girder is a Π-shaped girder, characterized by relatively low overall structural stiffness. The weight of the main girder is balanced by that of the pylon. In addition to the weight of the main girder directly affecting mid-span displacement, changes in the stiffness and weight of the pylon can also make the pylon more likely to deviate under the load of cable forces. This deviation, in turn, induces variations in the displacement of the main girder.

Therefore, during the construction process of the cable-stayed bridge without back-stays, when the weight of the main girder, the weight of the pylon, the elastic modulus of the iron sand heavy-weight concrete, and the tension in the cables will significantly affect the mid-span deflection of the main girder. It is essential to ensure precise control over the tension, the weight of the iron sand heavy-weight concrete, and the volume of concrete poured for the main girder during the construction process. Additionally, the monitoring model should be accurately adjusted based on the actual elastic modulus of the pylon throughout construction to ensure a reasonable final bridge profile.

4.3.2. Maximum Bending Stress of Main Girder

The bending stress of the main girder reflects its loading conditions. If the bending stress exceeds the standard limit, the risk of crushing and cracking of the main girder may occur, which could affect the service life of the bridge. The sensitivity and sensitivity percentages of each load parameter to the maximum bending stress of the main girder are presented in Figure 8a,b. Changes in the weight of the main girder and pylon affect the maximum bending stress of the main girder, with sensitivities of 0.94 and 0.5, respectively, and sensitivity percentages of 39.25% and 21.16%. Additionally, the second significant parameter affecting the maximum bending stress of the main girder is the overall temperature variation, the temperature difference between the cables and girder, the temperature gradient of the main girder, and the surface temperature difference of the pylon, with sensitivities of 0.24, 0.21, 0.2, and 0.12, respectively.

4.3.3. Longitudinal Displacement of Bridge Pylon

The displacement of the top of the pylon is easily affected by the weight of the pylon, the weight of the main girder, the temperature load, and other factors. In severe cases, this can not only affect the alignment of the pylon but also fail to meet the requirements of the code, potentially jeopardizing the overall structural integrity. The sensitivity of each load parameter to the longitudinal displacement at the top of the pylon and the corresponding sensitivity percentages are illustrated in Figure 9a,b. This analysis indicates that the reduction in the unit weight of iron sand heavy-weight concrete results in a decrease in the pylon's weight, significantly affecting longitudinal displacement changes. The maximum sensitivity factor is 5.74, which means that a relative change of 1% in the unit weight of the iron sand heavy-weight concrete within the pylon results in a relative change of 5.74% in the longitudinal displacement of the pylon. Therefore, in the construction process, the mixing quality of the iron sand heavy-weight concrete of the pylon should be strictly controlled to ensure its weight effect. The impacts of overall temperature variation, the tension in the cables, and the temperature differences on the surface of the pylon are secondary, with sensitivities of 0.34, 0.13, and 0.14, respectively. The influence of other parameters is relatively minor, all remaining below 0.2.

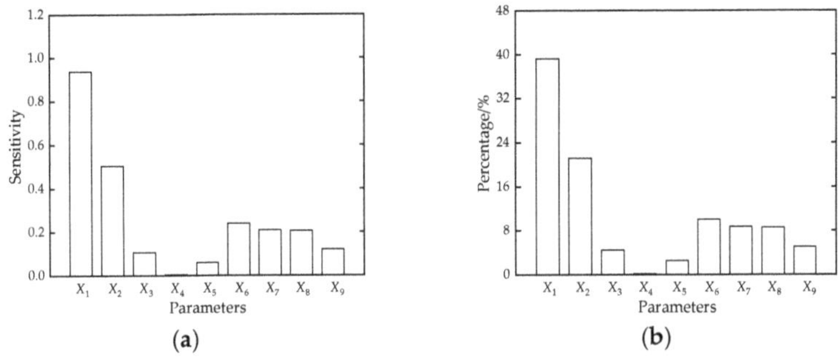

Figure 8. Sensitivity analysis of different parameters on the maximum bending stress of the main girder: (**a**) parameter sensitivity; (**b**) parameter sensitive percentage.

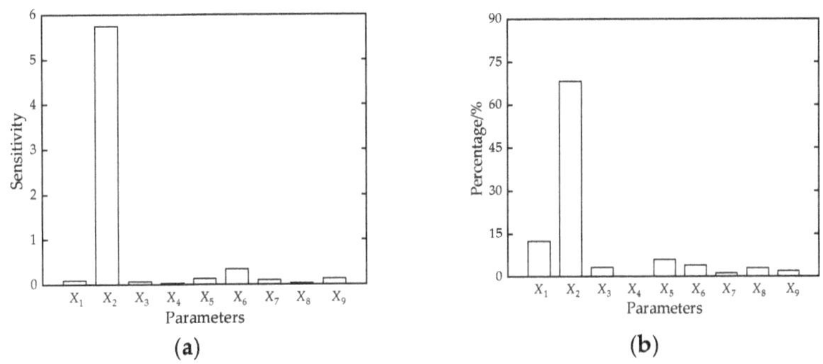

Figure 9. Sensitivity analysis of different parameters on the longitudinal displacement of pylon: (**a**) parameter sensitivity; (**b**) parameter sensitive percentage.

4.4. Internal Force Control of Bridge Construction

During the synchronous construction of pylon and cables in cable-stayed bridges without backstays, the parameters that most significantly affect the internal force and line shape of the bridge include the weight of the main girder and pylon, and the cable force. This paper establishes a parameter influence library for the bridge based on these three parameters. As the concrete volume and density during the pouring process are known, these can be considered as known errors. By considering the pouring error, stress, displacement, and cable force during the bridge construction process, the assessment framework is utilized to control the cable force at various stages of construction, thereby ensuring reasonable internal forces of the bridge. The measurement methods during the construction phase are illustrated in Figure 10, and the assessment framework is shown in Figure 1.

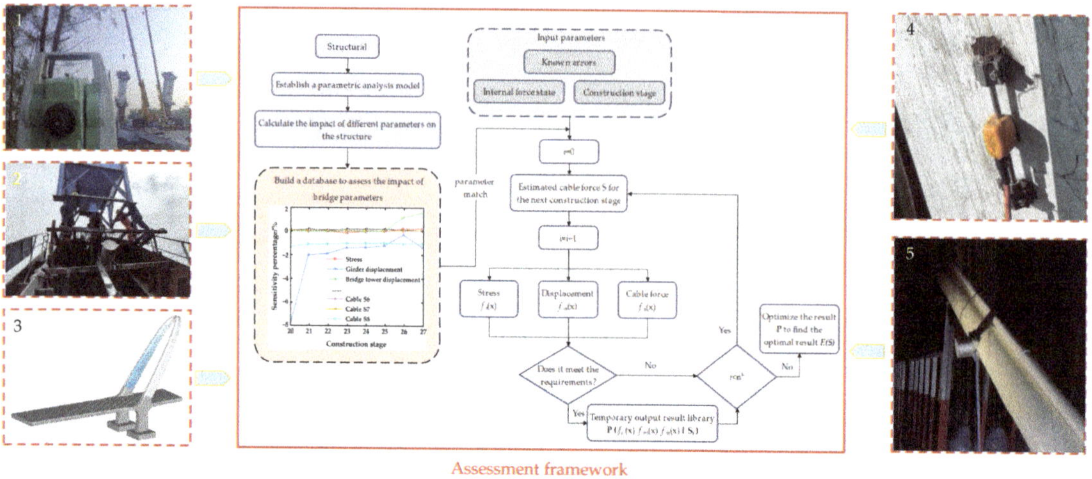

Figure 10. Data required for the assessment framework: (**1**) displacement measurement; (**2**) concrete volume measurement; (**3**) theoretical model; (**4**) stress measurement; (**5**) cable force measurement.

During the analysis, it was found that all sectional stresses exhibited considerable safety margins. Therefore, the stress conditions of the sections are not presented. After using the assessment framework, the predicted cable forces for cables S3 and S4 during tensioning are shown in Figure 11, where 'x' represents the downstream side and 's' represents the upstream side of the cables. Prior to the tensioning of S3 and S4, monitoring revealed that the overall cable forces of S1 and S2 were relatively high, with a maximum error of 3.3% on the upstream side of S1. Based on predictions and theoretical analysis, it is advisable to increase the cable forces of S3 and S4 to correct the tension error in S1 and S2. Given that the error magnitude of the upstream cables is significant, higher tension forces should be considered for the upstream cables S3 and S4. After applying the evaluation framework for prediction and combining it with measured data, it was found that the error in the cable forces of S1 and S2 decreased from a maximum of 3.3% to 1.3%, demonstrating the effectiveness of the assessment framework.

Figure 12 shows the cable force error and bridge displacement error in different construction stages. Because there are two sides of the pylon and cable in the construction stage, this paper only presents the maximum error values for the two sides of the pylons and the cables with the same sign. In the figure, construction stage 11 refers to the tension forces of cables S1 and S2 following the completion of the concrete pouring for segment 2#. Stages 12~20 are the cyclic tension, pouring, and hoisting of the S3~S8 cables and the corresponding steel boxes. Stages 21~23 are the pouring and hoisting of the remaining steel box segments, and 24 and 25 are the bridge deck pavement and the removal of the main girder bracket.

The analysis of the cable force shows that when there is an error in the cable force, it can be controlled by the cable force in the subsequent construction stage. Following the application of the assessment framework, the maximum error during the construction process is only 4.26% and the maximum error in the final cable tension is merely 2.7%, which meets the error range of ±5% required by Chinese code 'Technical Specification for Construction Monitoring and Control of Highway Bridges'. It is noted that cable tension errors can exhibit significant variations after the tensioning of different cables during construction, primarily due to the unpredictability of pylon displacement, which results in substantial discrepancies between theoretical and measured values. The maximum error between adjacent construction stages can reach 4.3%.

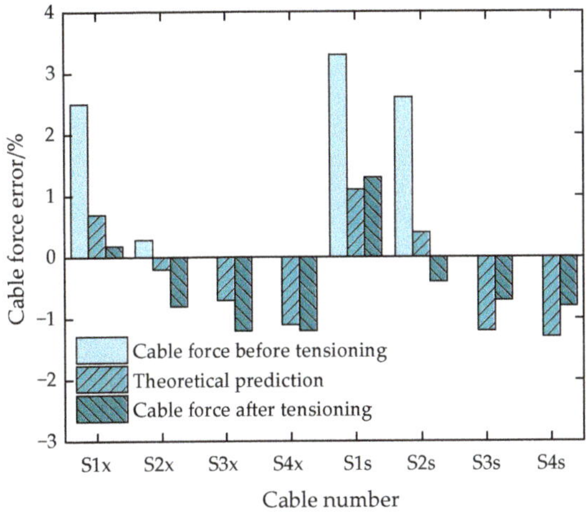

Figure 11. Changes in bridge cable forces.

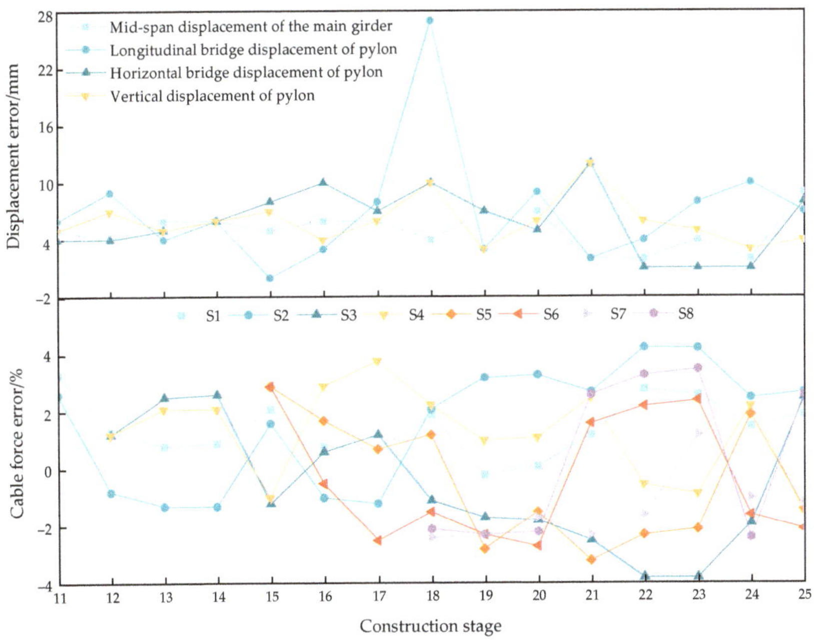

Figure 12. Measured values of bridge cable forces and displacement errors in different construction stages following the use of the assessment framework.

Analysis of the displacement of the bridge indicates that the mid-span displacement of the main girder and the displacement errors of the bridge pylon during various construction stages generally remain below 10 mm. The maximum error observed after the tensioning of cables S7 and S8 was 27 mm; however, after the casting of iron sand heavy-weight concrete in the subsequent construction stage, the maximum displacement error of the bridge pylon was reduced to 3 mm. The displacement has little influence on the construction control of the bridge, indicating that the evaluation framework meets the requirements for bridge construction control.

Based on the analysis of the measured results from the bridge, the proposed method effectively controls the cable force and structural displacement of the cable-stayed bridge without backstays during the synchronous construction method of the pylon and cables. Compared to other methods [20,21], the maximum error in cable force can be controlled within 4%, which represents an improvement of approximately 2% in error reduction. This method not only effectively manages the threshold of bridge construction errors, but also avoids the need for secondary cable adjustment after the bridge is completed, greatly improving construction efficiency and possessing considerable application value.

5. Conclusions

This paper presents a novel assessment framework for the control of construction states aimed at ensuring reasonable internal forces in cable-stayed bridges without backstays. The framework utilizes a prior generated library of bridge parameter influences and establishes corresponding constraints to evaluate reasonable internal forces of the bridge. When integrated with the measured internal force states and displacement conditions during construction phases, this framework allows for real-time assessment. However, the applicability of this framework is strictly limited to the range of pre-generated fault surface libraries. In other words, the framework can only be applied to the synchronous construction of pylons and cables in cable-stayed bridges without backstays. The main results of this study are as follows:

(1) To reduce computational costs, the response surface methodology is used to determine the primary sensitive parameters during the construction process. By utilizing the response surface methodology, an explicit relationship was established between the structural control objective response and the design parameters. This approach reduced the number of sensitive parameters in the construction process to three primary sensitivity parameters, which significantly enhanced the efficiency of structural analysis. The results of the response surface method are also useful for parameter sensitivity analysis, which can accurately evaluate the influence of different parameters on the structural response.

(2) The weight of the main girder and pylon, as well as the tension forces in the cables, significantly influence the internal forces and alignment of the bridge, which are sensitive parameters in the construction process. In contrast, the overall temperature variation of the bridge, the temperature difference between the cables and the girder, and the temperature gradient of the main girder are considered secondary sensitive parameters. The elastic modulus of the pylon, the installation accuracy of the steel box, and the temperature difference on the surface of the pylon have minimal impact on the internal force state of the bridge and can be considered non-sensitive parameters. The bridge parameter influence library for the synchronous construction of pylon and cables in cable-stayed bridges without backstays can be established by using the weight of the main girder and pylon, along with the tension forces in the cables.

(3) The finite element model of the Longgun River Bridge in Hainan Province is created to evaluate the proposed assessment framework. Results from the case study indicate that the proposed assessment method can accurately represent the structural responses of the full bridge model for the specific structure in question. Furthermore, it was found during the construction process that the proposed method can eliminate the need for secondary cable adjustments after bridge completion, significantly enhancing construction efficiency and demonstrating considerable practical value.

(4) For the concept verification proposed in this study, the scope of the framework remains narrow, focusing solely on the specific structure of the selected case study. The ongoing work aims to generate an accurate and efficient library of bridge parameter influences within the framework of defined structural responses. This library will be applicable to bridge structures with multiple parameters and outputs, with the aim of creating a general and comprehensive decision support tool for maintenance engineers.

Author Contributions: Conceptualization, T.X., J.M. and G.W.; methodology, T.X., J.M., G.W. and J.L.; software, G.W., J.L. and B.G.; validation, T.X. and J.M.; formal analysis, B.G. and J.L.; investigation, T.X.; supervision, J.L.; project administration, J.M.; writing—original draft preparation, T.X. and J.M.; writing—review and editing, B.G., T.X., J.M., G.W. and J.L. All authors have read and agreed to the published version of the manuscript.

Funding: This research was funded by the Key Research and Technologies for the Construction of the Main Pylon of Curved Special-shaped Pylon Cable-stayed Bridge without Backstays (Grant No. T-GL-XMGS(J)-HAN-HDLYGL-02-JS-034).

Data Availability Statement: The original contributions presented in the study are included in the article, further inquiries can be directed to the corresponding author.

Conflicts of Interest: Author Tao Xu was employed by the company Xiamen Company of China First Highway Engineering Co., Ltd. Author Jiqian Ma was employed by the company Hainan Jiaotou Huandao Tourism Highway Investment and Development Co., Ltd. The remaining authors declare that the research was conducted in the absence of any commercial or financial relationships that could be construed as a potential conflict of interest.

References

1. Shao, X.; Zhao, H.; Li, L.; Peng, W.; Liu, G.; Yan, B. Design and experimental study of a harp-shaped single span cable-stayed bridge. *J. Bridge Eng.* **2005**, *10*, 658–665. [CrossRef]
2. Li, Q.; Bu, Y.Z.; Zhang, Q.H. *Introduction and Application of Geometric Control in the Whole Construction Process of Long Span Cable Stayed Bridges*, 1st ed.; Southwest Jiaotong University Press: Chengdu, China, 2009; pp. 104–106.
3. Zhang, Y.; Fang, Z.; Jiang, R.; Xiang, Y.; Long, H.; Lu, J. Static performance of a long-span concrete cable-stayed bridge subjected to multiple-cable loss during construction. *J. Bridge Eng.* **2020**, *25*, 04020002. [CrossRef]
4. Wang, J.F.; Zhang, J.T.; Xu, R.Q.; Yang, Z.X. Evaluation of thermal effects on cable forces of a long-span prestressed concrete cable-stayed bridge. *J. Perform. Constr. Facil.* **2019**, *33*, 04019072. [CrossRef]
5. Zhang, K.; Qi, T.; Xue, X.; Zhu, Z.; Sun, Q. Study on the influence of cable/sling damage on the natural vibration characteristics of special-shaped cable-stayed arch bridge without back cable. *Stavebni Obz. Civil. Eng. J.* **2020**, *29*, 507–517. [CrossRef]
6. Qin, S.; Zhang, J.; Huang, C.; Gao, L.; Bao, Y. Fatigue performance evaluation of steel-UHPC composite orthotropic deck in a long-span cable-stayed bridge under in-service traffic. *Eng. Struct.* **2022**, *254*, 113875. [CrossRef]
7. Zhang, L.; Qiu, G.; Chen, Z. Structural health monitoring methods of cables in cable-stayed bridge: A review. *Measurement.* **2021**, *168*, 108343. [CrossRef]
8. Zhang, Z.H.; Li, W.B.; Ding, Z.X.; Wu, X.Y. An approach to the selection of target reliability index of cable-stayed bridge's main girder based on optimal structural parameter ratio from cost-benefit analysis. *Structures* **2020**, *28*, 2221–2231. [CrossRef]
9. Valiente, A.; Iordachescu, M.; Aznar, A.; Abreu, M. Damage tolerance of the prestressing steel in a bridge stay-cable after thirty years of service in a graded, extremely adverse environment. *Constr. Build. Mater.* **2021**, *285*, 122945. [CrossRef]
10. Casas, J.R.; Aparicio, A.C. Monitoring of the alamillo cable-stayed bridge during construction. *Exp. Mech.* **1998**, *38*, 24–28. [CrossRef]
11. Lei, B.; Lv, Y.Z.; Ren, W. Analysis on the impact of the backstays forms on the performance of inclined tower cable-stayed bridge. *Appl. Mech. Mater.* **2014**, *533*, 127–134. [CrossRef]
12. Liang, L.; Liang, B.; Wang, W.S. Research on seismic displacement response of cable stayed bridge without back stays. *Adv. Mater. Res.* **2014**, *919*, 1039–1042.
13. Zhu, Y.L. Research on construction technology of inclined pylon of single pylon cable-stayed bridge without backstays. *Appl. Mech. Mater.* **2011**, *90*, 1168–1172. [CrossRef]
14. Duan, H.R.; Liu, H.B.; Sun, Y.; Gao, H. Determination of reasonable internal force state for cable-stayed bridge without backstays. *J. Civ. Struct. Health* **2023**, *13*, 1243–1263. [CrossRef]
15. Zhang, Q.C.; Sun, Q.S. Analysis of parameter sensitivity in construction control of inclined pylon cable-stayed bridge without backstays. *Adv. Mater. Res.* **2011**, *255*, 851–855. [CrossRef]
16. Wong, F.S. Uncertainties in dynamic soil-structure interaction. *J. Eng. Mech.* **1984**, *110*, 308–324. [CrossRef]
17. Fang, S.E.; Perera, R. Damage identification by response surface based model updating using D-optimal design. *Mech. Syst. Signal Process.* **2011**, *25*, 717–733. [CrossRef]
18. Hu, J.R. *The Geometric Nonlinearity Analysis of a Single Pylon Cable-Stayed Bridge Without Backstays*; Hefei University of Technology: Hefei, China, 2009.
19. Gong, B.X.; Feng, L.J.; Liu, J.; Wang, Z.; Liu, Y.J. Finite-element analysis of temperature field and effect on steel-concrete composite pylon of cable-stayed bridge without backstays. *Buildings* **2024**, *14*, 1731. [CrossRef]

20. Ju, M.J.; Liu, S.M.; Liu, Y.J. Influence of Random Deviation of Cable Tension Force on Stress State of Composite Girder Cable-stayed Bridge without Back-stay. *J. Highw. Transp. Res. Dev.* **2016**, *33*, 60–66.
21. Lai, J.Z.; Zhang, Z.J. Study on dynamic optimization method of cable force of cable-stayed bridge considering construction process. *Eng. Constr.* **2022**, *54*, 42–48.

buildings

Article

Statistical Evaluation of Uniform Temperature and Thermal Gradients for Composite Girder of Tibet Region Using Meteorological Data

Yujuan Liu [1], Zhiyuan Ma [2,*] and Jiang Liu [2,*]

1 School of Transportation Engineering, Shanxi Vocational University of Engineering Science and Technology, Jinzhong 030606, China; liuyujuan@sxgkd.edu.cn
2 School of Highway, Chang'an University, Xi'an 710064, China
* Correspondence: mazhiyuan@chd.edu.cn (Z.M.); liu-jiang@chd.edu.cn (J.L.)

Abstract: To accurately assess the temperature action and effect of steel-concrete composite girder bridges in plateau and cold areas, this paper investigated nearly 50 years of historical meteorological data from 26 meteorological observation stations in the Tibet region of China. Based on the most unfavorable extreme meteorological data at each meteorological station, a finite element model was used to analyze the temperature field of the composite girder. The most unfavorable values of temperature were obtained. The regional differences in temperature actions at different meteorological stations were analyzed, and the isotherm maps of the extreme values of the uniform temperature and thermal gradients were further obtained based on spatial interpolation methods in the ArcGIS program. The study shows that the uniform temperature is significantly affected by the climatic environment, and the isotherm maps provide a visual representation of the geographic distribution pattern of temperature extremes. The maximum and minimum uniform temperatures in Tibet range from 18.28 °C to 42.27 °C and from −41.07 °C to 4.71 °C respectively. The maximum regional difference of positive and negative thermal gradient reaches 11.32 °C and 7.69 °C respectively. The temperature effects calculated using the most unfavorable values of the isotherm map are all more unfavorable than the specification calculations. In particular, the tensile stress of the concrete under the positive thermal gradient reaches 2.91 MPa, which exceeds the standard value of the tensile strength of concrete. This is a significant risk factor for cracking. The compressive stress of the steel girder under a negative thermal gradient reaches 19.35 MPa, which represents a 136% increase compared to the specified value. This increase elevates the risk of instability in the steel girder.

Keywords: bridge engineering; steel-concrete composite girders; thermal analysis; regional differences in temperature; historical meteorology; isotherm map

1. Introduction

Steel-concrete composite girders can fully exploit the advantages of two materials, steel and concrete. Due to the advantages of low weight, high load capacity, good dynamic performance, ease of construction and lower cost, steel-concrete composite girders have become a very competitive bridge type for small and medium-span bridges [1].

Bridge structures exposed to the environment are directly affected by solar radiation, air, temperature and wind speed. The temperature action has a significant impact on the safety and durability of bridges [1–3]. The composite girder consists of two types of materials with the thermal conductivity of steel being much higher than that of concrete, which makes the temperature effect more complicated than that of concrete bridges. Temperature actions in bridges can be divided into two categories, in which the uniform temperature action controls the longitudinal deformation of the main girder, which is a key factor in determining the length of the girder bridge splitting coupling as well as the selection of the bearing and expansion joints [4,5], and unreasonably uniform temperatures may result in increased manufacturing costs for bearings and expansion joints. The other category is the thermal gradient action, the thermal stress caused by the non-uniformity of the temperature distribution of the cross-section is an important cause of concrete cracking [6,7].

The temperature-action value obtained in a particular region is often difficult to apply to other regions due to significant regional differences in meteorological conditions. Therefore, it is reasonable to suggest that full consideration of the regional differences in meteorological factors for zoning values is essential. National codes have made provisions for regional differences in the uniform temperature extremes of concrete main girders. The AASHTO-LRFD specification employs two methodologies to account for the uniform temperature extremes of concrete main girders [8]. The first method delineates the United States into cold and mild regions, thereby establishing the uniform temperature extremes of concrete bridges in each region. The second method presents an isothermal map of the design temperature extremes of concrete bridges across the United States. The Eurocode [9] provides the uniform temperature extremes of concrete bridges based on the shade air temperature extremes, which are obtained through isothermal plots. The General Specifications for Design of Highway Bridges and Culverts divides China into three regions based on temperature extremes: extremely cold, cold, and warm. It stipulates the uniform temperature extremes for bridges in each region and refers to the British specification to give the formula for calculating the uniform temperature extremes based on the average daily temperature extremes of the local calendar year. National Railway Administration of the People's Republic of China Code for design on railway bridges and culverts (TB 10002-2017) [10] establishes the uniform temperature extremes based on the style and size of the structure and the local outside air temperature, among other factors. The outside air temperature is determined based on the national average air temperature charts for January and July. A review of the aforementioned codes reveals that, with the exception of the General Specifications for Design of Highway Bridges and Culverts, the remaining codes employ the isothermal map to calculate uniform temperature extremes. In comparison to the zonal value of temperature actions, the isothermal map represents a more precise methodology for accounting for the regional differences of temperature actions in the context of value determination. It can thus be concluded that the zoning method employed in the General Specifications for Design of Highway Bridges and Culverts may result in the determination of inaccurate values for the uniform temperature extremes.

The systematic investigation of thermal gradient research also shows obvious regional variations [11], paralleling those seen in uniform temperatures. Potgieter et al. [12] employed numerical simulation based on historical meteorological data from 26 meteorological stations in the United States to investigate the role of thermal gradients for concrete box girders in different regions of the United States. Accordingly, the AASHTO code subdivides the United States into four regions for the purpose of thermal gradient consideration, based on the level of solar radiation intensity. Mirambell [13] provided isotherm maps for the temperature difference taken of concrete girders in cross-sections in different regions of the Iberian Peninsula, taking into account various meteorological parameters in the summer season. In contrast, Liu Jiang et al. [14] divided China into four regions for thermal gradient taking by establishing a correlation between the temperature action of concrete box girders and geographic and meteorological parameters in terms of cities. The General Specifications for Design of Highway Bridges and Culverts stipulates that the thermal gradient action in different regions be taken as the same value, without considering the regional differences. With the development of bridge construction to the cold region, the codal specified values of uniform temperature and thermal gradient for the design of composite girders may lead to non-conservatism in design, which brings a a great risk for the safety and durability of bridges in extreme regions. The implementation of the 'One Belt, One Road' initiative and the 'Western Development Strategy' is facilitating the gradual extension of construction activities to the cold regions of the plateau. In the construction of the Qinghai-Tibet Plateau Golmud to Lhasa section of the highway, approximately 125 km of bridge will be constructed using composite girder bridges, marking the first instance of composite girder bridges being built at such an altitude. The construction of bridges in the plateau and cold regions is subject to the dual challenges of extreme solar radiation and extreme air temperature. The temperature-related effects are more complex and pronounced, with the detrimental impact on bridge safety being particularly pronounced in these harsh environments.

It is time-consuming and laborious to measure the meteorological data and the temperature field of the main girder cross-section in different areas on site, and it is impossible to measure the temperature data at all locations of the main girder cross-section and the temperature field of the main girder cross-section under the extreme meteorological conditions due to the limited number of measuring points and the length of monitoring time [15–18]. Therefore, in the absence of on-site measured meteorological information, the use of long-term historical meteorological data from meteorological stations to simulate the temperature field of the bridge cross-section under extreme meteorological conditions has become the key to accurately predicting the value of the bridge temperature action [19,20].

In this paper, 26 meteorological observation stations in Tibet, China, were investigated for nearly 50 years of historical meteorological data. Based on the most unfavorable meteorological data at each meteorological station, a finite element model was used to analyze the temperature field of the composite girder. The most unfavorable values of uniform temperature and positive and negative vertical thermal gradients were obtained. Then, the regional differences of the temperature actions at different meteorological stations are analyzed, and the Isotherm maps of the extreme values of the uniform temperature and thermal gradient actions are further obtained by using the spatial interpolation method, which improves the accuracy of the temperature actions. Related research can provide guidance for the refined design of composite girder bridges.

2. Numerical Simulation Method for Temperature Field

The temperature distribution of a bridge structure under insolation depends on both heat exchange with the outside world and heat conduction inside. The accuracy of numerical simulation results also depends on this [21,22].

Solving the temperature field of a bridge structure involves addressing a spatial heat conduction problem. The heat conduction within the bridge structure is governed by the partial differential equation for heat conduction proposed by Fourier (see Equation (1)).

$$\rho c\frac{\partial T}{\partial t}=\frac{\partial}{\partial x}\left(k\frac{\partial T}{\partial x}\right)+\frac{\partial}{\partial y}\left(k\frac{\partial T}{\partial y}\right)+\frac{\partial}{\partial z}\left(k\frac{\partial T}{\partial z}\right)+Q \tag{1}$$

where: t, x, y, z are the time and space coordinates; T is the temperature of the bridge; ρ, is the density, c, is specific heat capacity, k is thermal conductivity; Q is the heat generated by the structure per unit of volume of itself.

The transfer of heat between the bridge structure and the external environment under the influence of solar radiation is primarily attributable to three mechanisms: solar radiation, convection heat transfer, and radiation heat transfer. This phenomenon represents the boundary condition for the solution of the aforementioned partial differential equation of heat transfer. For further clarification, please refer to Figure 1. See Appendix A for detailed calculations.

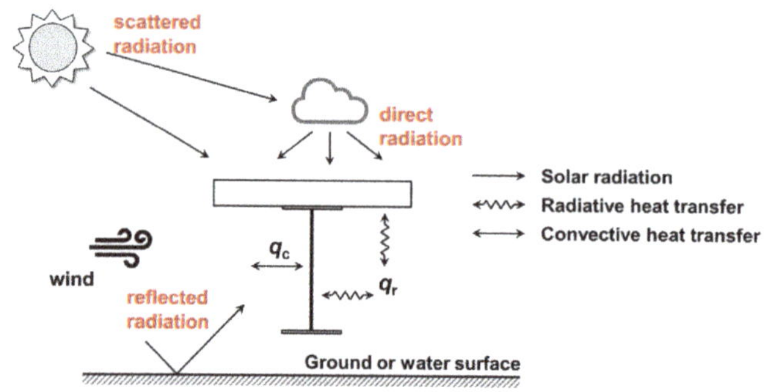

Figure 1. Schematic diagram of thermal loads and boundaries.

3. Temperature Actions on Composite Girder Bridges

3.1. Uniform Temperature

The temperature of a composite girder section can be categorized into four parts: uniform temperature, transverse linear temperature difference, vertical linear temperature difference, and residual temperature difference [15], where the uniform temperature causes longitudinal elongation and shortening of the bridge. Composite girder consists of two materials with very different thermophysical properties, steel and concrete, the temperature-induced longitudinal deformation of the steel girder and concrete deck slab must be coordinated, and its uniform temperature can be calculated by the following equation [15].

$$T_e = \frac{\sum E_s A_{si} T_{si} + \sum E_c A_{ci} T_{ci}}{\sum E_s A_{si} + \sum E_c A_{ci}} \tag{2}$$

where T_e is the uniform temperature of the composite girder section, E_s and E_c are the modulus of elasticity, and T_{si}, T_{ci}, A_{si}, and A_{ci} are the temperature value and cross-sectional area, respectively.

3.2. Thermal Gradient

The nonlinear temperature difference in the bridge cross-section due to daytime warming and nighttime cooling is usually defined as positive and negative thermal gradients, respectively.

The thermal gradient curves of bridges can generally be expressed in the form of multiple parabolas, exponential curves and multifolds, which can better reflect the effect of insolation on the warming of the top of bridges, in which the Chinese JTG D60-2015 [23], AASHTO and Eurocode-1 general model adopt the form of bifold line.

For the negative thermal gradient mode, China JTG D60-2015 and AASHTO take −0.5 and −0.3 of the respective positive thermal gradient, respectively. Only the general mode of Eurocode-1 considers that there exists an obvious nonlinear distribution of temperature on the steel girder. The measured and calculated studies have shown that the negative thermal gradient model of Eurocode-1 is more in line with the form of temperature distribution of the composite girder at night or cold wave cooling, in the daytime sunshine the concrete temperature is higher than that of the steel girder, at night or cold wave cooling, the top of the concrete deck plate cools down quickly while the interior cools down slowly, forming a negative temperature difference in the bridge deck plate, while the heat conductivity of the steel girder is about 50 times of that of the concrete deck plate. The temperature drops rapidly, but not as rapidly at the top where it is connected to the concrete; so a significant negative temperature difference is formed at the top of the steel girder.

For comprehensive consideration, the positive thermal gradient model in China JTG D60-2015 is selected for the subsequent analysis, which is more in line with the characteristics of China's climate environment. Eurocode-1 negative thermal gradient model is selected for negative thermal gradient analysis, which better reflects the temperature distribution characteristics of composite girders under cooling conditions (see Figure 2).

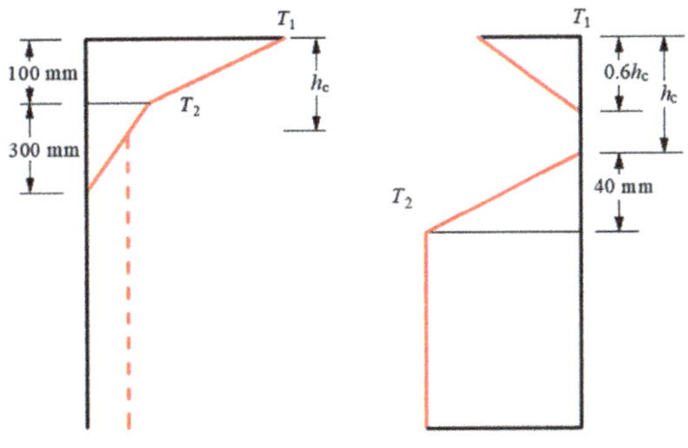

Figure 2. Thermal gradient model of composite girder bridge: (**a**) Positive thermal gradient. (**b**) Negative thermal gradient.

4. Temperature Action Extremes Based on Meteorological Data

4.1. Historical Meteorological Data Research

The temperature of the bridge structure is significantly affected by environmental factors; in order to investigate the differences in temperature actions in different regions, 26 meteorological stations were collected, and the distribution of those is shown in Figure 3. The meteorological data collected include daily maximum and minimum temperatures, daily average wind speed and total solar radiation. The meteorological data cover the period from 1 January 1953 to 31 December 2015. Meteorological data have been processed for outliers using the Hampel filter method [24].

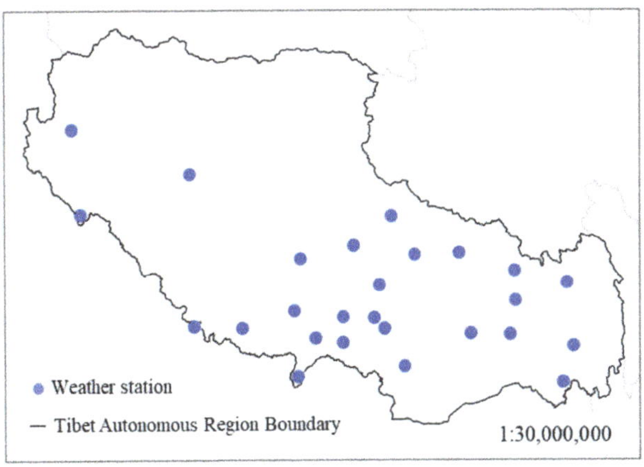

Figure 3. Distribution of meteorological stations in Tibet.

4.2. Extreme Meteorological Conditions

It is essential to ascertain the impact of extreme temperature on the bridge by analysing meteorological data representing extreme conditions. Consequently, an extreme value analysis was conducted using long-term data on air temperature, solar radiation, and wind speed, resulting in the identification of extreme meteorological values for each meteorological station over a 100-year period. These values exhibited a normal distribution with a 95% confidence level. The daily value data were decomposed into hourly segments using the methodology in the literature [25,26].

A sinusoidal function is used to simulate the daily ambient temperature profile. When the daily maximum temperature $T_{\max}$ and daily minimum temperature $T_{\min}$ are known, the temperature at any moment of the day can be obtained by the following formula [27,28].

$$T(t) = \frac{T_{\max} + T_{\min}}{2} + \frac{T_{\max} - T_{\min}}{2} \sin\left[(t - t_0)\frac{\pi}{12}\right] \tag{3}$$

where, $T(t)$ is the daily change of air temperature; $T_{\max}$ and $T_{\min}$ are the Maximum and minimum air temperature; t_0 indicates the time of occurrence of the maximum and minimum temperatures.

4.3. Extreme Value Calculation of Temperature Action

4.3.1. Computational Models

The quadrilateral heat-transfer element DC2D4 with 4-node was employed for the modelling of the cross-section. The dimensions of the element were selected to be approximately 3.0 × 2.5 cm and 3.0 × 1.0 cm for the concrete deck and steel beam, respectively, with a total of 848 elements and 1014 nodes over the cross-section. The interface between the concrete deck and the steel girder in the model was defined as a bond constraint, ensuring the continuity of temperature and heat flux at the interface, which is a reasonable approximation [29]. The finite element model is illustrated in Figure 4 and the thermal parameters of steel, concrete and asphalt in the model are shown in Table 1.

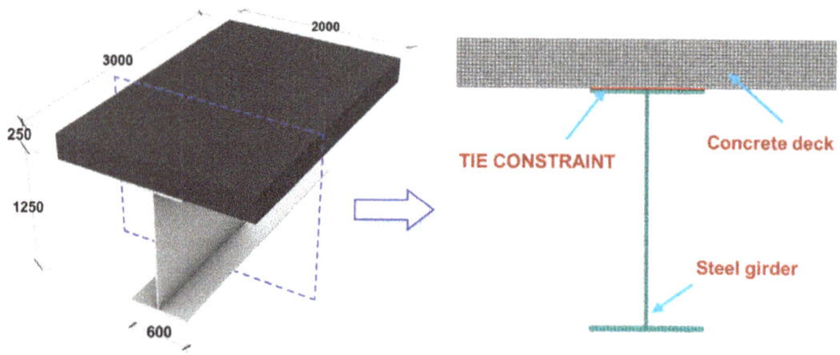

Figure 4. Schematic of the finite element model.

Table 1. Material properties used in FE analysis.

Characteristics	Steel	Concrete
Density $\rho/(\text{kg}\cdot\text{m}^{-3})$	7840	2150
Specific heat capacity $c/[\text{J}/(\text{kg}\cdot{}^\circ\text{C})]$	465	910
Thermal conductivity $\lambda/[\text{W}/(\text{m}\cdot{}^\circ\text{C})]$	55	2.8
Absorptivity ξ	0.5	0.45
Emittivity ε	0.8	0.85

4.3.2. Calculation Results

Based on the aforementioned extreme meteorological conditions setting method, the most unfavorable meteorological boundary conditions of 26 meteorological stations in Tibet were obtained. ABAQUS 6.14 software is used to establish the temperature field calculation model of the composite girder. Finally, each meteorological station has uniform temperature and thermal gradient extremes, see Table 2. To illustrate, the change pattern of the maximum and minimum uniform temperatures within a day at the Linzhi station is presented in Figure 5.

Table 2. Extreme value of temperature actions in Tibet.

Meteorological Station Information			Geographic Information			Uniform Temperature/°C		Positive Gradient/°C		Negative Gradient/°C	
Serial	Station Number	Name of Meteorological Station	Latitude	Longitude	Altitude/m	$T_{e,max}$	$T_{e,min}$	T_1^+	T_2^+	T_1^-	T_2^-
1	55228	Shiquan River	32.3	80.05	4278.6	33.00	−35.25	25.50	6.91	4.22	14.19
2	55248	Gerze	32.09	84.25	4414.9	32.69	−40.94	25.30	6.56	5.18	13.90
3	55279	Bangor	31.23	90.01	4700	29.53	−40.94	25.30	3.45	3.60	15.87
4	55294	Amdo	32.21	91.06	4800	26.51	−25.24	25.30	4.34	3.47	15.87
5	55299	Nagchu	31.29	92.04	4507	27.22	−26.66	25.99	5.18	4.63	13.19
6	55437	Burang	30.17	81.15	3900	28.26	−27.07	21.80	3.90	3.76	11.03
7	55472	Shenzha	30.57	88.38	4672	30.84	−24.39	21.80	7.37	3.35	11.03
8	55493	Dangxiong	30.29	91.06	4200	27.42	−24.51	22.85	5.09	3.90	13.16
9	55578	Shigatse	29.15	88.53	3836	31.27	−13.33	18.50	4.21	3.81	12.43
10	55585	Nimu	29.26	90.1	3809.4	31.45	−15.15	18.24	4.14	3.76	12.42
11	55591	Lhasa	29.4	91.08	3648.9	32.02	−14.03	18.38	4.09	3.36	11.08
12	55598	Tsedang	29.16	91.46	3560	32.52	−10.42	18.47	4.24	3.45	11.53
13	55655	Nyalam	28.11	85.58	3810	33.45	−11.74	18.47	4.42	3.16	9.66
14	55664	Tingri	28.38	87.05	4300	25.21	−15.35	21.80	5.02	4.03	15.99
15	55680	Gyangzê	28.55	89.36	4040	28.07	−20.30	21.10	4.82	3.94	12.33
16	55681	Nankazi	28.58	90.24	4431.7	30.59	−16.17	18.89	4.19	3.30	11.03
17	55696	Lhunzi	28.25	92.28	3860	22.63	−28.96	19.76	4.46	4.23	13.76
18	55773	Pari	27.44	89.05	4300	29.28	−14.42	19.76	5.07	3.82	13.76
19	56106	Suo County	31.53	93.47	4022.8	39.81	−5.20	19.09	4.19	3.65	11.74
20	56116	Tingqing	31.25	95.36	3873.1	29.49	−23.70	22.73	4.36	2.87	12.04
21	56137	Chamdo	31.09	97.1	3315	36.90	−11.92	24.50	5.42	3.98	13.57
22	56223	Lhokhorn	30.45	95.5	3640	26.08	−22.24	23.43	4.45	2.98	12.17
23	56227	Bomi	29.52	95.46	2736	32.34	−12.31	18.47	3.95	3.09	12.18
24	56312	Linzhi	29.4	94.2	2991.8	38.52	−0.35	17.33	4.01	3.40	10.86
25	56331	Zuogang	29.4	97.5	3780	32.87	−6.13	18.18	3.73	2.97	12.42
26	56434	Qasumi	28.39	97.28	2327.6	38.79	2.28	16.85	3.46	2.66	10.80

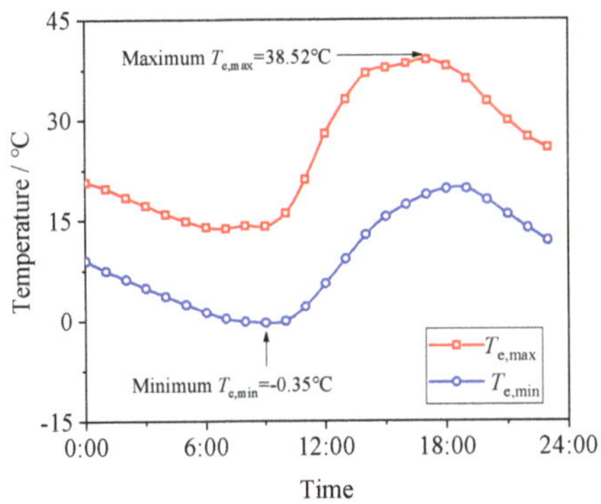

Figure 5. The fluctuations in uniform temperature throughout the diurnal cycle.

5. Temperature Actions Extremes Based on Geographic Variation

5.1. Regional Differences in Temperature Actions

The box plots of the results of uniform temperature and thermal gradient calculations for the composite girders are given in Figure 6, and the maximum and minimum values of each temperature action in the 26 meteorological stations are indicated. It can be clearly seen that the representative values of each temperature action have significant variability in each station. Among them, the values of $T_{e,max}$ and $T_{e,min}$ in the 26 stations range from 22.63 to 39.81 °C and −40.94 to −2.28 °C, respectively, and the maximum difference between stations reaches 17.18 °C and 43.22 °C, respectively; the values of T_1^+ and T_2^+ in the positive thermal gradient range from 16.85 to 25.99 °C and 3.45 to 7.37 °C, respectively, with the maximum difference reaching 9.14 °C. The values of T_1^+ and T_2^+ in the positive thermal gradient ranged from 16.85 to 25.99 °C and 3.45 to 7.37 °C, with the maximum difference of 9.14 °C and 3.92 °C, respectively; the values of T_1^- and T_2^- in the negative thermal gradient ranged from 2.66 to 5.18 °C and 9.66 to 15.99 °C, with the maximum difference of 2.52 °C and 6.33 °C, respectively; and the regional differences of $T_{e,min}$ was the most significant.

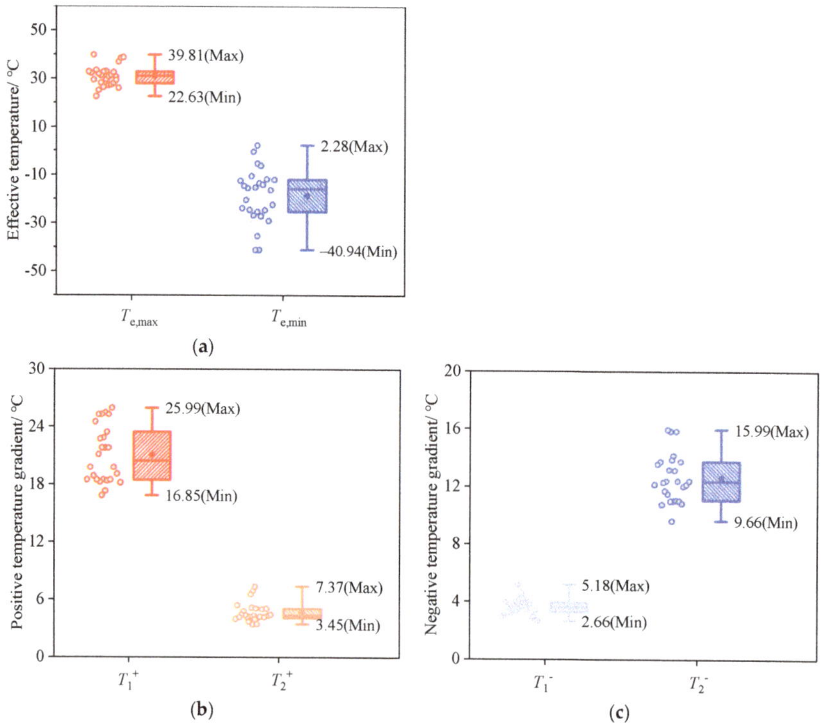

Figure 6. Calculation results of temperature action extremes of composite girders: (**a**) Uniform temperature. (**b**) Positive thermal gradient. (**c**) Negative thermal gradient.

5.2. Isotherm Maps of Extreme Temperatures Actions

On the basis of the long-term calculation of the temperature-action representative values of the 26 stations using finite elements, the spatial interpolation calculation is carried out by using ArcGIS 10.7 software, and the temperature action Isotherm maps are further drawn. To ensure that the data of the sample points remain unchanged after spatial interpolation and that the interpolation curve is smooth, spatial interpolation calculation using the spline function method averaging is performed.

5.2.1. Uniform Temperature

The Isotherm map of the maximum uniform temperature in Tibet is plotted in Figure 7a. For the uniform temperature T_e, the interpolated maximum uniform temperature $T_{e,max}$ takes values ranging from 18.28 °C to 42.27 °C, and the range of variation is slightly larger than that of the finite element calculation results, which also explains the reasonableness of the spatial interpolation method to some extent. Among them, the highest $T_{e,max}$ occurs at the junction of northern Tibet and Xinjiang Tarim Basin, and the $T_{e,max}$ is also larger at the junction of eastern Tibet and Sichuan and Yunnan. This is due to the fact that the uniform temperature is mainly affected by the annual average temperature, and these areas have lower altitudes and higher temperatures. The lowest $T_{e,max}$ occurs at the border of southwest Xinjiang and India, which is located on the Tibetan Plateau and has lower annual temperatures. Figure 7b shows the Isotherm map of the minimum uniform temperature, and it can be seen that the value of the minimum uniform temperature, $T_{e,min}$, ranges from −41.07 °C to 4.71 °C. The variation range of $T_{e,min}$ reaches 45.47 °C, which is significantly higher than that of the maximum uniform temperature, $T_{e,min}$ is also significantly affected by the air temperature, and the minimum value mainly occurs in the Tibetan Plateau region of northwestern Tibet, mainly due to the sharp decrease of the minimum temperature with the increase in altitude. This is mainly due to the sharp decrease in the minimum temperature with the increase in altitude.

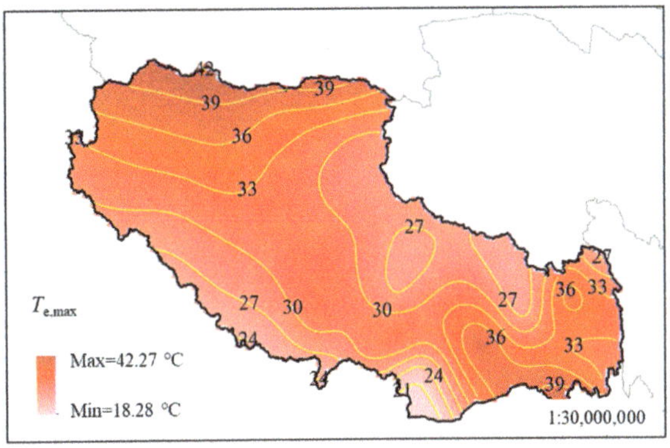

(a)

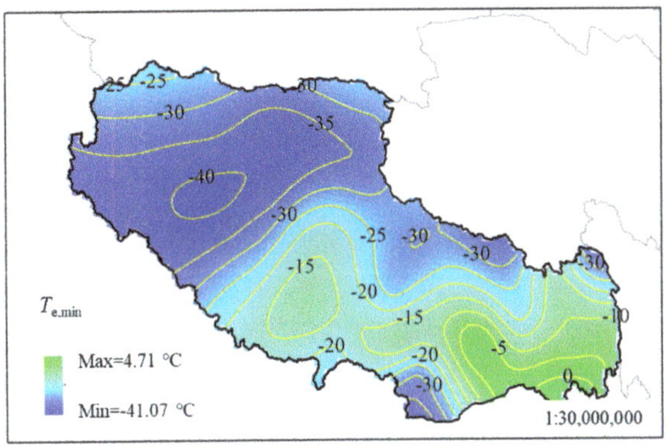

(b)

Figure 7. Uniform temperature extremum Isotherm maps: (**a**) Maximum uniform temperature. (**b**) Minimum uniform temperature.

Figure 8 shows the zoning map of the uniform temperature in the General Specifications for Design of Highway Bridges and Culverts, and the comparison with the General Specifications for Design of Highway Bridges and Culverts, it can be seen that in the General Specifications for Design of Highway Bridges and Culverts, the northern part of Tibet is in the cold region, the southern part is in the cold region, and the maximum uniform temperature code suggests that the values are all 39 °C, and comparing with the results of this paper, except for a few areas in the northern part of Tibet, the other areas are all less than the code suggests the value. This shows that the Chinese code for the maximum uniform temperature is safe enough, but the economy may be insufficient. The normative recommended values for the minimum uniform temperature in severe cold regions and cold regions are −32 °C and −15 °C, respectively, and the range of variation is much smaller than the calculated results of this paper, which are −41.07 °C to 4.71 °C. The minimum uniform temperature in cold regions is −41.07 °C to −4.71 °C. If the value is taken according to the specification, it is obviously too rough, and the minimum uniform temperature of the composite girder in most areas of the plateau region in northern Tibet is seriously underestimated, while the design of the composite girder bridge in the southern region of Tibet may be too conservative.

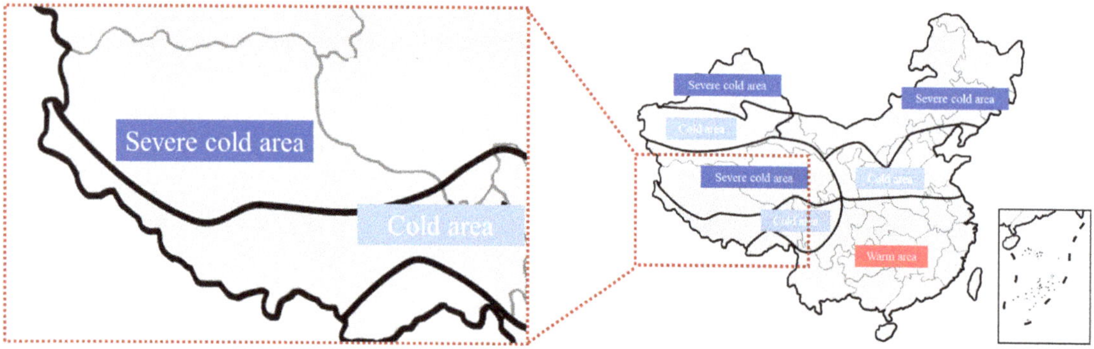

Figure 8. Zoning map in general specifications for design of highway bridges and culverts.

5.2.2. Positive Thermal Gradient

The national Isotherm maps of T_1^+ and T_2^+ in the positive thermal gradient of the composite girder bridge are plotted in Figure 9, in which the values of T_1^+ range from 15.63 °C to 26.95 °C, with a geographic variation of 11.32 °C. The T_1^+ is mainly caused by the solar radiation at the top of the deck slabs, which is stronger at higher elevations. Therefore, the maximum value occurs at the highest elevation in the southwestern and northeastern parts of the Tibetan Plateau. The temperature difference T_2^+ is accompanied by T_1^+, so the distribution pattern is similar to that of T_1^+. However, after the heat transfer process of the concrete of the bridge deck slab, the value decreases significantly; the value of T_2^+ ranges from 3.25 °C to 7.48 °C. The temperature difference T_2^+ is the same as that of T_1^+.

For the deck slab thickness of not more than 0.4 m of the composite girders, the Chinese highway specification of the top temperature difference T_1^+ takes the value of 25 °C, which basically can cover the unfavorable value in the calculation results of this paper. However, for most areas in Tibet, the specification is still conservative, and the value of T_2^+ is 6.7 °C, which is not more than 1 °C different from the average value of T_2^+ calculated in this paper.

5.2.3. Negative Thermal Gradient

The national Isotherm maps of T_1^- and T_2^- in the negative thermal gradient of the composite girder bridge are plotted in Figure 10, where the values of T_1^- are taken in the range of 2.66 °C to 5.58 °C, and the regional differences are not significant. This phenomenon can be attributed to the inadequate thermal conductivity of concrete, which results in a negligible temperature differential within the concrete deck slab during the cooling phase. While steel girders have good thermal conductivity, steel girders cool down quickly when the temperature drops. The range of T_2^- values reaches 9.43~17.12 °C, and the distribution characteristics are basically the same as that of T_1^- values. The negative thermal gradient is mainly affected by the daily temperature difference, with larger values in the western Tibetan Plateau region and smaller values in the eastern region.

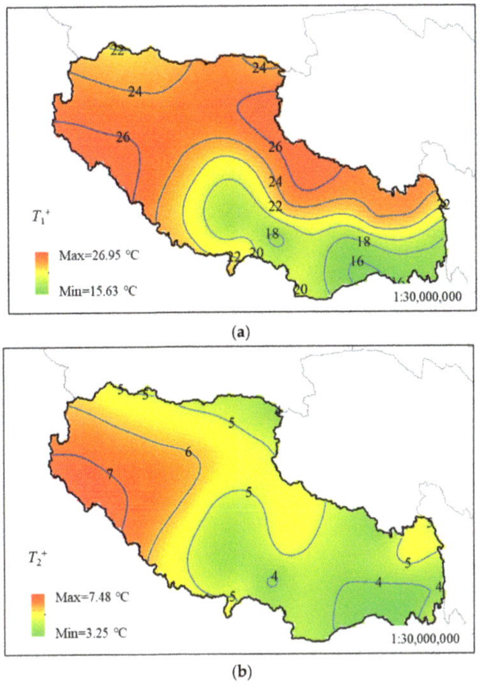

Figure 9. Positive thermal gradient extremum Isotherm maps: (**a**) $T_1{}^+$. (**b**) $T_2{}^+$.

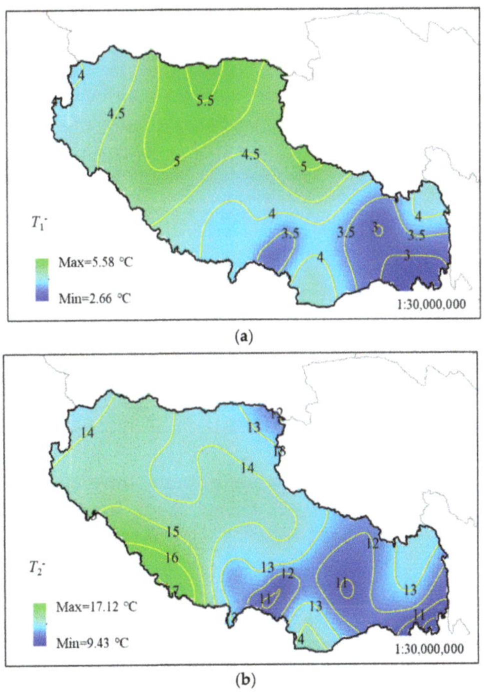

Figure 10. Positive thermal gradient extremum Isotherm maps: (**a**) $T_1{}^-$. (**b**) $T_2{}^-$.

The temperature difference at the top and bottom of the composite girder in Eurocode-1 is taken as −5 °C and −8 °C, respectively, the top temperature difference with the calculation of this paper basically coincides with the results, but this paper calculates the T_2^- much larger than the European norms, which may be caused by the more severe climatic environment of the Tibetan Plateau region.

5.3. Comparison of Temperature Effects

In order to further illustrate the differences in structural effects caused by the differences in temperature values in different regions. Taking 3 × 30 m span steel-mixed composite girder as a case study, the maximum stresses of concrete deck slabs and steel girders are extracted by calculating the temperature effects using the maximum and minimum values in the temperature maps proposed in this paper, respectively. The results are also compared with the stress calculations using the specification temperature values. The modulus of elasticity of concrete is taken as 3.55×10^4 MPa and the coefficient of linear expansion is taken as $1 \times 10^{-5}/°C$, while the modulus of elasticity of steel is taken as 2.06×10^5 MPa and the coefficient of linear expansion is taken as $1.2 \times 10^{-5}/°C$. The values of temperature actions are shown in Table 3.

Table 3. Comparison of temperature values.

Temperature Action (°C)	Uniform Temperature		Positive Thermal Gradient		Negative Thermal Gradient	
	$T_{e,max}$	$T_{e,min}$	T_1^+	T_2^+	T_1^-	T_2^-
Specification recommended value	39	−32	25	6.7	5	8
Maximum in Isotherm map	42.27	4.71	26.95	7.48	5.58	17.12
Minimum in Isotherm map	18.29	−41.07	15.63	3.25	2.66	9.43

The comparative results of concrete stresses are shown in Table 4. For static structures, the effects of uniform temperature are usually released through deformation and no thermal stresses are generated. For super-static structures, both axial expansion and temperature secondary stresses are generated. For the composite girder in this example, the maximum and minimum uniform temperatures generate tensile and compressive stresses in the deck slab, respectively. The maximum tensile stress in the concrete occurs at the upper edge of the deck slab at the pivot point. The maximum value calculated using the Isotherm map reaches 0.84 MPa, which is 12% larger compared to the calculated result of 0.75 MPa for the recommended value of the specification.

Table 4. Comparison of concrete stresses.

Stress (MPa)	Maximum Uniform Temperature	Minimum Uniform Temperature	Positive Thermal Gradient	Negative Thermal Gradient
Specification recommended value	0.75	−1.47	2.65	1.47
Maximum in Isotherm map	0.84	−0.13	2.91	1.55
Minimum in Isotherm map	0.39	−2.01	1.77	1.13

A positive thermal gradient generated a large tensile stress at the lower edge of the bridge deck slab. The calculated result using the suggested value of specification is 2.65 MPa, which is slightly smaller than the standard value of tensile strength of C50 concrete 2.74 MPa, while the calculated result using the maximum value of Isotherm map reaches 2.91 MPa, which has exceeded the standard value of tensile strength of concrete, and there is a significant risk of cracking. A negative thermal gradient in the fixed support at the upper edge of the bridge deck plate produces a smaller tensile stress, and the calculation results of this paper are closer to the specification.

The calculated results of the steel girder stresses are shown in Table 5. The uniform temperature change did not produce significant stresses in the steel girders, and neither tensile nor compressive stresses exceeded 2 MPa. Positive and negative thermal gradients produced tensile and compressive stresses in the steel girders, respectively. The compressive stresses in the steel girders calculated using the maximum value of the Isotherm map reached 19.35 MPa, which is an increase of 136% compared to the result of 8.20 MPa calculated from the recommended value of the specification. Although the temperature-generated stresses in the steel girders are not significant, they may still increase the risk of instability of the steel girders when superimposed with vehicle loads and other environmental loads.

Table 5. Comparison of steel girder stresses.

Stress (MPa)	Maximum Uniform Temperature	Minimum Uniform Temperature	Positive Thermal Gradient	Negative Thermal Gradient
Specification recommended value	0.81	−1.13	11.41	**−8.20**
Maximum in Isotherm map	0.88	−0.08	12.95	**−19.35**
Minimum in Isotherm map	0.43	−1.47	8.03	−9.89

6. Conclusions

Isotherm maps of uniform temperature and thermal gradient action extremes have been obtained based on the most unfavorable extreme meteorological conditions by investigating nearly 50 years of historical meteorological data from 26 meteorological observation stations in Tibet of China. Results show that:

(1) The temperature action of composite girder bridges in the cold region of the plateau is more unfavorable, and the suggested values in the specification are not safe enough for the design of composite girder bridges in the Tibetan region. The isotherm maps are more detailed and reasonable designs for composite girder bridges in Tibet.

(2) The uniform temperature of the composite girder is significantly affected by the climatic environment. The maximum and minimum uniform temperatures in Tibet are 18.28~42.27 °C and −41.07~4.71 °C, respectively.

(3) The positive thermal gradient in the Chinese code and the negative thermal gradient in the European code are used to describe the temperature difference characteristics of the composite girders. The values of T_1^+ and T_2^+ range from 15.63 °C to 26.95 °C and 3.25 °C to 7.48 °C, respectively. The temperature difference at the top is more significant.

(4) The temperature effects calculated using the most unfavorable values of isotherm maps are more unfavorable than the specification calculations, in which the tensile stress of the concrete reaches 2.91 MPa, with a significant risk of cracking.

(5) The isotherm maps of the temperature action of composite girder bridges in Tibet were obtained from meteorological data and the accuracy of isotherm maps should be verified and improved by placing sensors for long-term structural temperature monitoring.

Author Contributions: Conceptualization, Y.L. and Z.M.; methodology, Y.L. and Z.M.; software, Z.M.; validation, Z.M.; formal analysis, Y.L. and Z.M.; investigation, Y.L. and Z.M.; supervision, Y.L.; project administration, Y.L.; writing—original draft preparation, Y.L. and Z.M.; writing—review and editing, J.L. and Z.M. All authors have read and agreed to the published version of the manuscript.

Funding: This study is funded by the Special Fund for Basic Scientific Research of Central College of Chang'an University (Grant No. 300102213703, 300102214205), which is gratefully acknowledged.

Data Availability Statement: The original contributions presented in the study are included in the article, further inquiries can be directed to the corresponding author.

Conflicts of Interest: The authors declare no conflict of interest.

Appendix A

The total solar radiation heat flow density q_h in the horizontal plane consists of direct radiation q_{bh} in the horizontal plane and scattered radiation q_{dh} in the horizontal plane.

$$q_h = q_{bh} + q_{dh} \tag{A1}$$

The solar radiation heat flow density can be obtained from real measurements, and when only total daily radiation data are available, it can be calculated by decomposing it in the following equation [19]

$$q_h = I_r \left[\frac{\pi}{24} (a_1 + b_1 \cos\omega) \frac{\cos\omega - \cos\omega_s}{sin\omega_s - \omega_s \cos\omega_s} \right] \tag{A2}$$

$$q_{dh} = I_{dr} \left[\frac{\pi}{24} \frac{\cos\omega - \cos\omega_s}{sin\omega_s - \omega_s \cos\omega_s} \right] \tag{A3}$$

where: I_r and I_{dr} are the total daily solar radiation in the horizontal plane and the total daily scattered radiation in the horizontal plane, respectively, which can usually be obtained from the radiation meteorological station; ω and ω_s are the time angle and the corresponding time angle at sunset, respectively; a_1 and b_1 can be determined by the following formula [20].

$$a_1 = 0.409 + 0.5016 \sin(\omega_s - 1.047) \tag{A4}$$

$$b_1 = 0.6609 - 0.4767 \sin(\omega_s - 1.047) \tag{A5}$$

The solar radiation heat flow density q_θ absorbed by any surface of the structure is:

$$q_\theta = \alpha \left[q_{bh} R_b + q_{dh} R_d + \rho q_h R_\rho \right] \tag{A6}$$

where: α is the absorption rate; ρ is the ground reflectivity; R_b, R_d, R_ξ are the correction factors for direct, scattered and reflection on the surface at any inclination angle, which are calculated according to Equations (A5)–(A7), respectively.

$$R_b = \frac{\cos i}{\sin h} \tag{A7}$$

$$R_d = \frac{1 + \cos\theta}{2} \tag{A8}$$

$$R_\rho = \frac{1 - \cos\theta}{2} \tag{A9}$$

where: i is the solar incidence angle; h is the solar altitude angle; θ is the structural surface inclination. The solar radiation received by the web will be shaded by the cantilever flange, and the shading range changes continuously with time, the shading length l_s at any moment is:

$$l_s = l_c \frac{\tan(h)}{\sin(90 + \gamma_n - \gamma_s)\sin(\beta) - \cos(\beta)\tan(h)} \tag{A10}$$

where: l_c is the cantilever length; β is the web inclination; γ_n is the web face azimuth; γ_s is the solar azimuth.

Heat convection is calculated as follows:

$$q_c = h_c(T_a - T) \tag{A11}$$

where, T_a and T are the air temperature and structure temperature, respectively; h_c is the convective heat transfer coefficient, It can be calculated using the following empirical formula.

$$h_c = \begin{cases} 3.83v + 4.67 & \text{Top surface} \\ 3.83v + 2.16 & \text{Bottom surface} \\ 3.83v + 3.67 & \text{Lateral surface} \end{cases} \quad \text{(A12)}$$

References

1. Liu, J.; Liu, Y.J.; Zhang, C.Y.; Zhao, Q.H.; Lyu, Y. Temperature Action and Effect of Concrete-filled Steel Tubular Bridges: A Review. *J. Traffic Transp. Eng. (Engl. Ed.)* **2020**, *7*, 174–191. [CrossRef]
2. Liu, J.; Liu, Y.J.; Jiang, L.; Zhang, N. Long-term Field Test of Thermal gradients on the Composite Girder of a Long-span Cable-stayed Bridge. *Adv. Struct. Eng.* **2019**, *22*, 2785–2798. [CrossRef]
3. Kim, S.H.; Park, S.J.; Wu, J.; Won, J.H. Temperature variation in steel box girders of cable-stayed bridges during construction. *J. Constr. Steel. Res.* **2015**, *112*, 80–92. [CrossRef]
4. Elbadry, M.M.; Ghali, A. Temperature Variations in Concrete Bridges. *J. Struct. Eng.* **1983**, *109*, 2355–2374. [CrossRef]
5. Ding, Y.; Zhou, G.; Li, A.; Wang, G. Thermal Field Characteristic Analysis of Steel Box Girder Based on Long-term Measurement Data. *Int. J. Steel Struct.* **2012**, *12*, 219–232. [CrossRef]
6. Zhang, N.; Liu, Y.J.; Liu, J. Temperature Effects of H-shaped Concrete Pylon in Arctic-alpine Plateau Region. *J. Traffic Transp. Eng.* **2017**, *17*, 66–77.
7. Badar, J.; Umar, T.; Akbar, M.; Abbas, N.; Shahzad, Q.; Chen, W.; Arshid, M.U. Innovative Methods to Improve the Seismic Performance of Precast Segmental and Hybrid Bridge Columns under Cyclic Loading. *Buildings* **2024**, *14*, 1594. [CrossRef]
8. The American Association of State Highway and Transportation Officials (AASHTO). *AASHTO LRFD Bridge Design Specification*; American Association of State Highway and Transportation Officials: Washington, WA, USA, 2012.
9. *EN 1991-1-5*; Eurocode 1: Actions on structures-Part 1-5: General Actions-Thermal Actions. European Committee for Standardization (CEN): Brussels, Belgium, 2003.
10. *TB 10002-2017*; National Railway Administration of the People's Republic of China. Code for Design on Railway Bridge and Culvert. China Railway Publishing House: Beijing, China, 2017.
11. Umar, T. Making future floating cities sustainable: A way forward. *Proc. Inst. Civ. Eng.-Urban Des. Plan.* **2020**, *173*, 214–237. [CrossRef]
12. Potgieter, I.C.; Gamble, W.L. Nonlinear Temperature Distributions in Bridges at Different Locations in the United States. *PCI J.* **1989**, *34*, 80–103. [CrossRef]
13. Mirambell, E.; Aguado, A. Temperature and stress distributions in concrete box girder bridges. *J. Struct. Eng.* **1990**, *116*, 2388–2409. [CrossRef]
14. Liu, J.; Liu, Y.J.; Bai, Y.X. Study on Regional Difference and Zoning of the Thermal gradient Pattern of Concrete Box Girder. *China J. Highw. Transp.* **2020**, *33*, 73–84.
15. Quan, C. *Effects of Thermal Loads on Texas Steel Bridges*; University of Texas Library: Austin, TX, USA, 2008. Available online: https://www.proquest.com/openview/591e28df629caca6ddb99e845e1a143c/1?pqorigsite=gscholar&cbl=18750 (accessed on 25 November 2024).
16. Liu, J.; Liu, Y.; Ma, Z.; Zhang, G.; Lyu, Y. Thermal gradient Action of Steel-concrete Composite Girder Bridge: Action Pattern and Extreme Value Analysis. *China J. Highw. Transp.* **2022**, *35*, 269–286. [CrossRef]
17. Liu, Y.; Liu, J. Review on temperature action and effect of steel-concrete composite girder bridge. *J. Traffic Transp. Eng.* **2020**, *20*, 42–59. [CrossRef]
18. Zhang, C.; Liu, Y.; Liu, J.; Yuan, Z.; Zhang, G.; Ma, Z. Validation of long-term temperature simulations in a steel-concrete composite girder. *Structures* **2020**, *27*, 1962–1976. [CrossRef]
19. Li, Y.; He, S.H.; Liu, P. Effect of Solar Temperature Field on a Sea-crossing Cable-stayed Bridge Tower. *Adv. Struct. Eng.* **2019**, *22*, 1867–1877. [CrossRef]
20. Wang, Z.; Liu, Y.J.; Tang, Z.W.; Zhang, G.J.; Liu, J. Three-dimensional Temperature Field Simulation Method of Truss Arch Rib Based on Sunshine Shadow Recognition. *China J. Highw. Transp.* **2022**, *35*, 91–105.
21. Chen, D.S.; Wang, H.J.; Qian, H.L.; Li, X.Y.; Fan, F.; Shen, S.Z. Experimental and Numerical Investigation of Temperature Effects on Steel Members Due to Solar Radiation. *Appl. Therm. Eng.* **2017**, *127*, 696–704. [CrossRef]
22. Liu, H.B.; Chen, Z.H.; Zhou, T. Theoretical and experimental study on the temperature distribution of H-shaped steel members under solar radiation. *Appl. Therm. Eng.* **2012**, *37*, 329–335. [CrossRef]
23. Ministry of Transport of the People's Republic of China. *JTG D60-2015*; General Specifications for Design of Highway Bridges and Culverts. China Communications Press: Beijing, China, 2015; (In Chinese). Available online: https://xxgk.mot.gov.cn/jigou/glj/202006/t20200623_3312312.html (accessed on 25 November 2024).
24. Pearson, R.K.; Neuvo, Y.; Astola, J.; Gabbouj, M. Generalized Hampel filters. *EURASIP J. Appl. Signal Process* **2016**, *87*, 2016. [CrossRef]
25. Kehlbeck, F. *Effect of Solar Radiation on Bridge Structure*; China Railway Publishing House: Beijing, China, 1981.

26. Hottel, H.C. A Simple Model for Estimating the Transmittance of Direct Solar Radiation Through Clear Atmospheres. *Sol. Energy* **1976**, *18*, 129–134. [CrossRef]
27. Jiang, H.F.; Wen, D.Y.; Li, N.; Ding, Y.; Xiao, J. A New Simulation Method for the Diurnal Variation of Temperature--Sub-sine Simulation. *Metro. Disaster Red. Res.* **2010**, *33*, 61–65.
28. Thapar, V.; Agnihotri, G.; Sethi, V.K. Estimation of Hourly Temperature at a Site and its Impact on Energy Yield of a PV Module. *Int. J. Green. Energy* **2012**, *9*, 553–572. [CrossRef]
29. Zhou, K.; Han, L.-H. Modelling the behaviour of concrete-encased concrete-filled steel tube (CFST) columns subjected to full-range fire. *Eng. Struct.* **2019**, *183*, 265–280. [CrossRef]

MDPI AG
Grosspeteranlage 5
4052 Basel
Switzerland
Tel.: +41 61 683 77 34

Buildings Editorial Office
E-mail: buildings@mdpi.com
www.mdpi.com/journal/buildings

www.ingramcontent.com/pod-product-compliance
Lightning Source LLC
Chambersburg PA
CBHW061800210926
51850CB00006B/602
9783725848218